HORTICULTURAL REVIEWS
VOLUME 3

HORTICULTURAL REVIEWS

VOLUME 3

edited by
Jules Janick
Purdue University

AVI PUBLISHING COMPANY, INC.
Westport, Connecticut

ISSN-0163-7851
ISBN-0-87055-352-6

Printed in the United States of America
by The Saybrook Press, Inc., Old Saybrook, Connecticut

Horticultural Reviews is co-sponsored by the American Society for Horticultural Science and The AVI Publishing Company

Thomas W. Whitaker

Dedication

To know Thomas W. Whitaker is to be frequently surprised. For example, one day in 1970, shortly before his retirement from the USDA, he casually mentioned a forthcoming trip to Bolivia to look for new *Amaryllis* species. This seemed to be a bit bizarre, until I found out that Tom Whitaker is one of the world's acknowledged authorities on *Amaryllis*. Until that time, I had thought of him as one of the world's authorities on lettuce and the cucurbits, clearly his territory, because the breeding and genetics of lettuce and melons have occupied most of his professional life.

The *Amaryllis* revelation made me begin to wonder who this man—a colleague and supervisor for over 10 years, but whom I really did not seem to know very well—really was. I began paying more attention and asking questions.

Now, there is another side to Tom Whitaker's personality that is important. He is very modest, and therefore responses to personal questions are not always readily forthcoming. As an example, in 1978 he gave a talk about the history of lettuce breeding. In it he dutifully described the important milestones, crediting each one to the appropriate researcher. However, in the speech and in the article later published, he neglected to mention the name of the breeder responsible for the development of 'Great Lakes' lettuce. This was a major accomplishment in lettuce breeding, but it seemed to have just happened by itself. The name of the breeder, T.W. Whitaker, was simply not mentioned. This is typical.

Many of his accomplishments are well known. He was president of the American Society for Horticultural Science and Editor of the *Journal* and *HortScience*, President of the Society of Economic Botany, and Executive Secretary of the American Plant Society, and held a host of other offices, fellowships, research associateships, memberships, and consultant positions. Perhaps it is not so well known that he was a postdoctoral fellow among that most illustrious group of geneticists at the Bussey Institution at Harvard, a group that also included W. E. Castle and E. M. East, and many other geneticists of the classic period.

A measure of the tremendous breadth of interest and ability harbored in this man is found among his reviews of books. I have in my files 23 of these. The books reviewed range from the expected ones in the field of vegetables to also include such diverse titles as "Studies in Genetics," "Haldane and Modern Biology," "The Bering Land Bridge," and "The Sunflower."

This breadth of interest is carried over into his popular and semi-popular writings. The titles include "Collecting Amaryllis in the Bolivian Yungas," "Lettuce: Evolution of a Weedy Cinderella," "Gourds and Gardeners," "Gourds and People," "The Torrey Pines Association, Its Purpose and Program," "Agriculture Behind the Iron Curtain," and "J. T. Rosa, Jr. (1895–1928), Pioneer in Vegetable Crops Research."

This breadth of interest is not to be taken lightly. Tom Whitaker is no dilettante, no jack-of-all-trades. He is master of two important horticultural crops. No one knows more about the cucurbits and no one knows more about lettuce than he does. No one has mastered the sciences of these crops better than he has, as can be seen by perusing the many professional publications he has written on his own or in widespread collaboration. And few have contributed more to the industries of these crops.

It seems to me that the breadth of his interests comes from a spirit of adventure and a fearlessness often found among talented scientists. The depth of his accomplishments with lettuce and melons comes from a spirit of perseverance found in our most dedicated scientists. Breadth and depth together are found only in our best scientists. T. W. Whitaker belongs in that rare group.

This description may give the picture of a formidable man. Not so. He is kind, generous, and friendly. In conversation, these characteristics are wrapped around the ever-present spirit of inquiry, which brings forth many questions nearly always prefaced with "Say, Ed . . . " or "Say, Jim . . . ."

Life with Tom Whitaker is a delight. To him, Volume 3 of Horticultural Reviews is dedicated, with respect, affection, and awe.

Edward J. Ryder
U.S. Agricultural Research Station
U.S. Department of Agriculture
Science and Education Administration,
Agricultural Research
Salinas, California

Contributors

CARTER, JOHN V., Department of Horticultural Science and Landscape Architecture, University of Minnesota, St. Paul, Minnesota

COHEN, MORTIMER, Agricultural Research Center, Institute of Food and Agricultural Sciences, The University of Florida, Fort Pierce, Florida

COYNE, D.P., Department of Horticulture, University of Nebraska, Lincoln, Nebraska

CRANE, JULIAN C., Department of Pomology, University of California, Davis, California

EL-GOORANI, M.A., Faculty of Agriculture, Department of Plant Pathology, University of Alexandria, Alexandria, Egypt

EVANS, D.A., Campbell Institute for Research and Technology, Cinnaminson, New Jersey

FLICK, C.E., State University of New York, Center for Somatic-Cell Genetics and Biochemistry, Binghamton, New York

GRAY, D., National Vegetable Research Station, Wellesbourne, Warwick CV35 9EF, United Kingdom

HALEVY, ABRAHAM H., Department of Ornamental Horticulture, The Hebrew University of Jerusalem, Rehovot, Israel

HENDRIX, JAMES W., Department of Plant Pathology, University of Kentucky, Lexington, Kentucky

IWAKIRI, BEN T., Department of Pomology, University of California, Davis, California

KIERNAN, JENNIFER, Department of Plant Pathology, University of Kentucky, Lexington, Kentucky

LOCKARD, R.G., Department of Horticulture, University of Kentucky, Lexington, Kentucky

MARONEK, DALE M., Department of Horticulture and Landscape Architecture, University of Kentucky, Lexington, Kentucky

MAYAK, SHIMON, Department of Ornamental Horticulture, The Hebrew University of Jerusalem, Rehovot, Israel

PELLETT, HAROLD M., Department of Horticultural Science and Landscape Architecture, University of Minnesota, St. Paul, Minnesota

SCHNEIDER, G.W., Department of Horticulture, University of Kentucky, Lexington, Kentucky

SCHUSTER, M.L., Department of Horticulture, University of Nebraska, Lincoln, Nebraska

SHARP, W.R., Campbell Institute for Research and Technology, Cinnaminson, New Jersey

SOMMER, N.F., Department of Pomology, University of California, Davis, California

Contents

1

Fluid Drilling of Vegetable Seeds

D. Gray
National Vegetable Research Station, Wellesbourne, Warwick
CV35 9EF, U.K.

I. INTRODUCTION

Fluid drilling or sowing is a term that has been used loosely to refer to a number of ways in which seeds are sown using a fluid, including (1) spraying a suspension of dry seeds in a gel over the surface of the soil (Anon. 1974), (2) "hydro-seeding," (3) using the "flow sow" technique, or (4) "fluid drilling" pre-germinated seeds suspended in a gel. Hydro-seeders (Hydro-spacer, Agritek, Inc.) sow imbibed but not germinated

Note: Trade names of products and machines are used to simplify the presentation of information. No endorsement of named products and machines is intended nor is criticism implied of similar products and machines which are not mentioned.

seeds with a small quantity of water, whereas the "flow sow" technique (Flow-Sow Ltd.) was designed to enable dry seeds to be sown in a gel which, it is claimed, provides water for germination. In this review, the term "fluid drilling" is taken to mean the sowing of pre-germinated seeds—either suspended in a protective gel or delivered to the soil with a small amount of water. The potential advantage of a fluid drilling system over conventional drilling systems using dry seeds arises (1) because the seeds are germinated in ideal conditions before sowing, so eliminating the variable effects of the uncontrolled seedbed environment on germination, and (2) because the gel which suspends the seeds can be used also as a carrier for nutrients, plant growth regulators, and pesticides, so protecting the seedling in the early stages of growth. As a result of the earlier seedling emergence, fluid drilling also minimizes the risk of prolonged exposure to attack by pathogens and of restricted seedling growth caused by physical damage to the soil crumb structure from heavy rainfall after sowing.

II. HISTORICAL BACKGROUND

The sowing of pre-germinated seeds was used in the Netherlands, the United Kingdom, and the United States in the early part of this century to establish small areas of celery and carrots. The seeds were mixed with a moist compost in boxes and allowed to germinate in a warm place. After germination had occurred, the seed and compost mixture was sown by hand. It has proved to be difficult to adapt this technique for sowing on a large scale, although machines have been developed for sowing dry seeds mixed in a compost. For a discussion of the problems, see Ure and Loughton (1978) and Gray *et al.* (1979). This method of plant establishment, known as "plug-mix" planting (Hayslip 1974), is used commercially in North America to establish tomatoes and peppers.

The first reported attempt to sow pre-germinated seeds suspended in a protective fluid was by Elliott (1966, 1967), who used a sodium alginate gel that both suspended the seeds and carried them into the furrow. In these early studies, several prototype drills were made, and the gel and seed suspension was pumped from the reservoir through flexible hoses to the soil by either a peristaltic pump (Elliott 1966) (Fig. 1.1) or by positive volumetric displacement from cylinders (Elliott 1967) (Fig. 1.2). The improvements obtained in the establishment of grasses, fodder crops, and sugar beet indicated that the technique could offer considerable potential for improving the reliability of plant establishment of direct-drilled vegetables. In 1972, work was started at the National Vegetable Research Station, where most of the experimental work on fluid drilling has been carried out since, using Elliott's positive displacement drill but with modified coulters to enable vegetable seeds to be sown on a bed

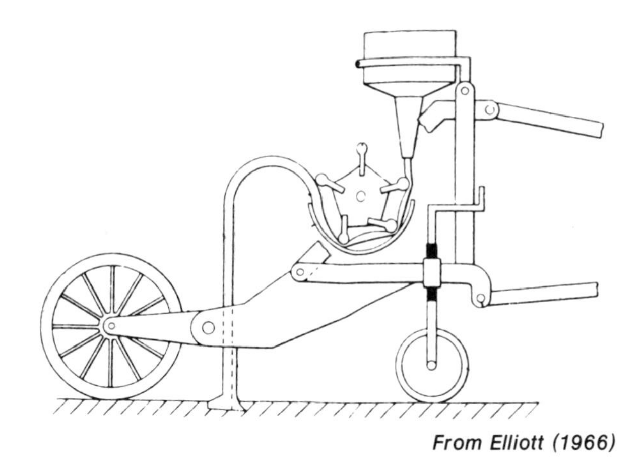

From Elliott (1966)

FIG. 1.1. AN EARLY FLUID DRILL FROM ELLIOTT'S PATENT

From Currah et al. (1974)

FIG. 1.2. AN EXPERIMENTAL FLUID DRILL BASED ON THE PISTON PRINCIPLE

system (Currah *et al.* 1974). These experiments were followed by the successful completion of larger-scale trials on onions in 1975 (Currah *et al.* 1976) and on celery in 1976 and 1977 (Salter and Darby 1977, 1978). A company, Fluid Drilling Ltd., was formed to develop a commercially-viable fluid drilling system using joint-venture finance provided by the National Research and Development Corporation. Further details of the background to, and the commercial development of, fluid drilling can be found in Gray (1975), Salter (1977, 1978a,b), Currah (1978), and Currah and Hurdley (1978).

III. THE FLUID DRILLING SYSTEM

The fluid drilling system is not solely a specialized field drilling operation; it is an integrated system involving (1) the treatment and germination of seeds prior to sowing, (2) the separation of germinated from ungerminated seeds, (3) the storage of germinated seeds, (4) the preparation of the gels for suspending the seeds, and (5) the drilling of the germinated seeds.

A. Seed Treatment and Germination Techniques

The precise effects of sowing pre-germinated seeds on seedling emergence is determined not only by the number of germinated seeds in the sample at the time of sowing but also by the uniformity of germination of the population of seeds (Salter 1978a,b). There is also an optimum radicle length providing the maximum advantage from fluid sowing. Gray (1978a) has shown that lettuce seeds sown when radicles were 1 to 2 mm long gave higher percentage seedling emergence than those sown with longer or shorter radicles than this. In the onion, sowing seeds with the radicles longer than 12 mm gave earlier emergence than those having radicles less than 2 mm in length (Table 1.1). Longden *et al.* (1979)

TABLE 1.1. EFFECT OF RADICLE LENGTH AT SOWING ON EMERGENCE OF PRE-GERMINATED 'WHITE LISBON' SALAD ONION SEEDS

Treatment	Days to 50% Emergence	Emergence (%)
Dry seeds	20.2	55
Pre-germinated seeds		
Seeds selected to have radicles <2mm long	15.4	70
Seeds selected to have radicles >12mm long	12.3	70
SED	± 0.6	± 4

Source: Gray (unpublished).

reported progressively earlier and higher emergence in sugar beet as the mean radicle length at sowing increased from 3 to 8 mm. A radicle length of at least 2 mm is necessary at the time of sowing of tomato to give higher and earlier emergence than can be obtained by sowing dry seeds (Bussell and Gray 1976) (p. 18). In other species, the effects of differences in radicle length on emergence have not been studied in detail.

Several techniques have been used to enhance the uniformity of germination under controlled conditions. For experimental plot work, small quantities of seeds have been germinated satisfactorily in trays lined with an absorbent non-medicated paper, covered with a layer of wet-strength paper to prevent the radicles from growing into the absorbent material (Currah *et al.* 1974; Anon. 1979). Some seeds, such as celery, need to be spread thinly (a density equivalent to 0.01 g cm^{-2}) over the surface of the papers to enable inhibitors to be leached out of the imbibing seeds.

An alternative technique is to germinate the seeds in aerated water held in perspex columns or glass jars (Darby and Salter 1976; Taylor 1977) (Fig. 1.3). The temperature of the water can be controlled easily and, by using an overflow pipe to pass fresh water continuously through the system, leachates inhibitory to germination can be removed from the germination medium. Asparagus, cabbage, carrot, celery, lettuce, onion,

From Darby and Salter (1976)

FIG. 1.3. BUBBLE COLUMNS FOR GERMINATING SEEDS

parsley, parsnip, pepper, and tomato have been germinated successfully by this technique (Darby and Salter 1976; Taylor 1977; Brocklehurst *et al.* 1980; Finch-Savage 1980).

For quantities of seeds of 2 kg or more, sand-bench techniques have been used successfully for germinating carrots and onions (Currah *et al.* 1976), although the depth of seeds should not exceed 5 mm. Satisfactory germination of up to 50 kg of onion, leek, lettuce, red beet, and sugar beet seeds also has been obtained by repeatedly soaking seeds held in nylon mesh bags in water for short periods, following each soaking by "spin-drying" to remove excess moisture and to promote adequate aeration (Currah *et al.* 1976; Salter 1978a). This process is repeated at intervals until the seeds have germinated. There are no published results comparing seed germination or seedling emergence from these systems of preparing "pre-germinated" seeds.

The techniques now adopted in commerce for germinating large quantities of seeds are based on germination in aerated water. Apparatus now has been devised for germinating up to 50 kg of seeds at one time and, while the technique has proved to be suitable for germinating vegetable seeds for commercial-scale trials (Anon. 1978), no critical comparative studies of the oxygen requirements of imbibing seeds of a wide range of species have been made yet.

Satisfactory germination at the time of sowing has been obtained for most vegetable seeds using temperatures between 18°C and 25°C (Currah *et al.* 1974; Biddington *et al.* 1975; Gray and Steckel 1977; Taylor 1977; Gray 1978a,b; Hardaker and Hardwick 1978). Lower temperatures have been used for sugar beet (Longden *et al.* 1979) and higher temperatures for onions (Lipe and Skinner 1979). Further information on the optimal temperatures for germination can be found in Edwards (1932) and Hegarty (1973), and additional relevant information has been published by Thompson (1974a,b,c).

Even at optimal temperatures for germination seeds begin to germinate at different times. In lettuce this "spread" of germination is short but in carrots, other Umbelliferae, onions, and leeks it can amount to several days. Because of this variability it is difficult to obtain a high proportion of the seeds in the population having radicles of the optimum length for sowing at any one time. If the full benefits of fluid sowing pre-germinated seeds are to be achieved, a greater synchrony of germination times will be required. Pre-treatment of seeds in dilute salt solutions or in solutions of polyethylene glycol (PEG) before germination ("priming") reduces the spread of germination in a large number of flower and vegetable seeds. This subject has been reviewed by Heydecker and Coolbear (1977). At present, no suitable apparatus has been devised to "prime" large quantities of seeds in PEG. However, Salter and Darby

(1976) have "primed" celery seeds in aerated dilute salt solutions in perspex columns. They showed that more than 50% of the "primed" seed had germinated within 2 days compared with less than 10% of untreated seeds. Twenty days were required to give 50% germination of untreated seeds, by which time many of those germinating early had radicles too long to be drilled without damaging them.

B. Seed Separation

For sowing seeds into soil blocks or for precision sowing, it is desirable to have a very high proportion of germinated seeds in the sample at the time of sowing, although this is not absolutely essential for crops sown with "extrusion" drills (see p. 10). Pre-treatment of seeds with growth regulators (Thomas *et al.* 1978; and see Khan *et al.* 1979 for a review), salts, or PEG (Heydecker and Coolbear 1977) can increase the percentage of germination and reduce the spread of germination times of the viable seeds, but a proportion of the pretreated seeds will be dead or of very low viability at the time of sowing. The ungerminated seeds can be identified during pre-germination, and attempts have been made to eliminate them from the sample before sowing takes place. Taylor *et al.* (1978) used sucrose density techniques to separate germinated from ungerminated seeds of celery and pepper. In celery, 71% of the seeds had radicles in the "germinated, unseparated" sample, but after separation this rose to 74%. The corresponding figures for pepper were 52% and 63%. Even though complete separation has not yet been achieved, "separated, germinated" celery seeds gave 95% seedling emergence compared with 83% for "unseparated, germinated" seeds. For pepper seeds, the corresponding figures were 98% and 76%. Separation of germinated from ungerminated seeds using mannitol and PEG also has been successful, but has given reduced seedling viability.

Currah (1977) reported that germinated seeds can be separated from ungerminated seeds hydraulically by making use of the differential resistance of germinated and non-germinated seeds to passage through water. With pre-germinated carrot seeds having radicles with a mean length of 4 mm it has been possible to produce samples of seeds having at least 95% of the seeds germinated. Separation of germinated from ungerminated onion seeds having radicles averaging 7 mm long was less efficient, and samples after separation contained only 84% of germinated seeds. As with the sucrose density method, the efficiency of separation depends on the proportion of germinated seeds in the sample at the time of separation. With a high proportion of germinated seeds in the sample, good separation can be achieved readily without discarding large quantities of germinated and ungerminated seeds. As the percentage of germinated seeds in the sample at the time of separation falls, the proportion of them which have to be discarded rises.

C. Storage of Pre-germinated Seeds

A suitable method of storing germinated seeds is essential in order to preserve their viability in the event of adverse weather delaying sowing. Furthermore, it is necessary to prevent excessive growth of the radicles so that they are not damaged during handling.

Two systems of storage have been examined: in humid air on tygan mesh trays inside an ice-bank cooled cabinet (Lindsay *et al.* 1974) and in aerated, cooled water (Salter 1978a). Pre-germinated seeds of cabbage, onion, lettuce, and carrot have been stored successfully under both conditions for up to 15 days, although seedling viability was reduced slightly in onion (Brocklehurst *et al.* 1980). Seedling viability was reduced by storage in unaerated water. Air-flow rates of 12 to 40 ml $min^{-1}g^{-1}$ seed are considered to be necessary to maintain the oxygen levels in the water to that found in the bulk of seeds prior to storage. Brocklehurst *et al.* (1980) have shown that storage in cold water does not result in materials leaching from the seeds, with the possible exception of cabbage at rates of aeration below 4 ml $min^{-1}g^{-1}$ seed.

Storage of pre-germinated carrot and parsnip seed for 15 days had no effect on percentage of seedling emergence compared with unstored seeds, but emergence of onion was reduced by storage for 12 days or longer and of cabbage for 3 days or longer (Finch-Savage 1980). Finch-Savage (1980) and Wurr *et al.* (1980) found that percentage of emergence of lettuce was unaffected by storage for up to 5 or 6 days. Storage of parsnip and cabbage for as short a period as 3 days significantly reduced the time from sowing to emergence compared with unstored seeds. Onions, lettuce, and carrot emerged more rapidly when stored for 15 days than for shorter periods. However, Wurr *et al.* (1980) found that storage of lettuce even for 5 days delayed emergence and increased the spread of emergence times. Finch-Savage (1980) showed that storage had no effect on the yield of carrot or parsnip but the interpretation of the effects of storage on the yields of onions, lettuce, and cabbage were complicated by variation in percentage of seedling emergence.

Comparisons between storage in aerated water and in moist air have given similar results with carrot, but comparisons of the two storage systems have not been made for other species.

Little work has been done to define the conditions of storage for pre-germinated vegetable seeds known to suffer damage from exposure to low but non-freezing temperatures. With tomato there is evidence that storage at 0°C for even one day is detrimental, although storage at

5°C for prolonged periods had no effect on seedling emergence (Gray *et al.* 1978).

D. Carrier Gels and Additives to Gels

Several materials have been used to make gels for suspending and carrying the pre-germinated seeds to the soil. The materials that have been most widely tested are (1) a formulation of sodium alginate (Agrigel-Alginate Industries Ltd.), (2) a hydrolyzed starch-polyacrylonitrile (H-SPAN, General Mills Chemicals, Inc.), (3) guar gums (K4492 and K59.5, Hercules Powder Company), (4) a synthetic clay (Laponite 508, Laporte Industries Ltd.), (5) a modified potato starch (Perfactamyl CMA 2K, Tunnel Avebe Starches), (6) a form of polyacrylamide (Magnafloc 511, Allied Colloids Ltd.), and (7) a polyacrylate (Viscalex, Allied Colloids Ltd.).

Agrigel requires a high shear force to mix it into water to form the gel, but the other materials are more readily mixed using simple equipment (Darby 1980; Anon. 1978).

Both H-Span and Agrigel are sensitive to the calcium hardness of water which causes variable suspension characteristics. Agrigel has suitable pumping characteristics, but it desiccates rapidly in dry soil, thereby trapping the seeds and reducing emergence (Gray 1978a; Darby 1980). Agrigel also is rapidly degraded by bacteria (Taylor and Dudley 1978a) on standing for 24 hours or more. The gelation of Laponite, guar, polyacrylamides, and polyacrylates is not as sensitive as Agrigel to water hardness, and these materials also have satisfactory seed suspension and pumping characteristics. Of these five gels, Darby (1980) has shown that only Laponite has given consistently higher and earlier emergence than Agrigel under a range of soil and weather conditions.

There are conflicting reports of the effects of gel extrusion rate during drilling on seedling emergence. Gray (1978a) reported reduced emergence of lettuce in warm (20°C), moist soils at gel extrusion rates in excess of 20 ml Agrigel per meter, but no significant depressions in emergence were reported by Darby (1980) for a number of vegetable species, including lettuce, over a range of gel extrusion rates from 6.7 to 26.7 ml per meter of row. Little work has been done on the oxygen requirements of seeds in gels, although Brocklehurst (1979) has shown that polyacrylate, guar gum, and Laponite reduce the rate of oxygen uptake by the seeds and marginally reduce the growth of the seedlings.

It is claimed that gels provide water for germination (Elliott 1967). There is no consistent evidence that this is true for Agrigel (Gray 1978a), although Darby (1980) has shown that Laponite gives higher emergence than other gels in dry conditions. These results raise the possibility that germinated or dry seeds could be sown on the surface of the soil (Darby 1980) (Table 1.2) and then covered with a gel which would not dry out rapidly.

The use of gels as a means of introducing small amounts of hormones or starter fertilizers, fungicides, and insecticides close to the seeds and developing seedlings has not been exploited fully yet, though there have been several preliminary reports that indicate the potential value of such a technique. Gray and Bryan (1978) reported that the growth of celery seedlings was increased when Cytex (a cytokinin-containing material) was incorporated into the gel prior to sowing (Fig. 1.4) at rates (13 ml Cytex per liter of gel) considerably lower than would be necessary to give similar increases in growth if applied to seedlings by traditional spraying techniques. Hardaker and Hardwick (1978) and Taylor and Dudley (1978b) have shown that nodulation in *Phaseolus vulgaris* by *Rhizobium phaseoli* can be increased significantly by incorporating the inoculum into the gel with the pre-germinated seeds compared with applications in a peat/lime mix to dry seeds. Entwistle (1979) has reported that the white rot disease in onions caused by *Sclerotium cepivorum* Berk. can be controlled effectively by incorporating iprodione into the gel at one-quarter of the rate used in normal dry seed dressings. A similar improvement in the efficiency of the control of seedling losses in Brassicas by *Phoma* spp. using metalaxyl (25 mg a.i. per liter of gel) has been reported by White (1979).

E. Drills and Sowing

For the original experiments on fluid drilling conducted by Elliott (1967) a tractor-mounted drill designed by the National Institute of Agricultural Engineering employing the "piston" principle was used. Pre-germinated seeds mixed with a gel were sown into killed grass swards, but the cylinder capacity allowed only 50 to 60 m of row to be sown be-

TABLE 1.2. EFFECT OF DIFFERENT GELS ON LETTUCE SEEDLING EMERGENCE WHEN SOWN ONTO THE SURFACE OF DRY SOIL

Treatment (pre-germinated seeds)	Seedling Emergence (%)
No gel	9
Guar[1]	43
Polyacrylamide	44
Starch	47
Alginate	48
HSPAN	72
Laponite	75
Seeds sown into moist soil	82

Source: Darby (1980).
[1] For all gels the rate of extrusion was 20 ml/m of row.

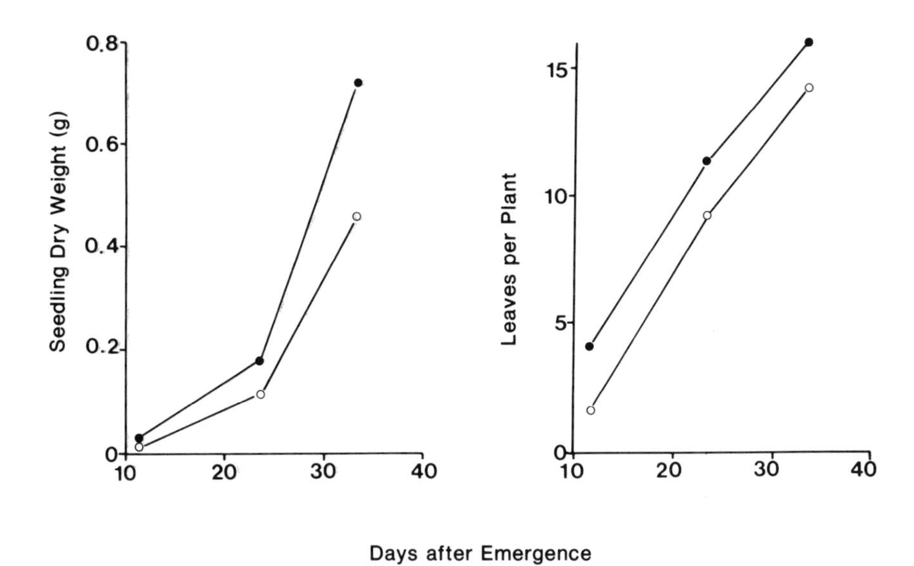

FIG. 1.4. EFFECT ON GROWTH OF CELERY OF ADDING CYTEX TO THE ALGINATE GEL (●-●, WITH CYTEX (A MATERIAL HAVING CYTOKININ-LIKE ACTIVITY); O-O, WITHOUT CYTEX).

tween refills. A similar device was used in the initial experiments at the National Vegetable Research Station (Currah *et al.* 1974) (Fig. 1.2), except that the coulter system was modified to correspond with that on dry seed drills with which it was being compared. For small-scale plot work, two hand-pushed fluid drills of similar design using the piston principle have been described. One (Lickorish and Darby 1976) is adapted to the chassis of a single-headed cone-seeder (Craftsman Machine Co., Winnepeg), and the other (Spinks *et al.* 1979) to the chassis of a Planet Junior single-row unit. Both drills have a cone-seeder attachment for sowing dry seeds, and the same coulter system is used for both the pre-germinated and dry seeds. In the fluid drilling mode both drills can sow approximately 15 m of row between refills of the cylinder with the gel/seed mixture. Small single-row, hand-pushed drills for sowing seeds into plant-raising beds and other small areas using peristaltic pumps (Flow Sow Ltd., Fluid Drilling Ltd.) to deliver the gel/seed mixture are now available.

For sowing areas of up to 0.4 ha between refills, several designs of drill have been manufactured. A drill using a centrifugal pump to deliver the mixture of gel and seeds to the ground from the reservoir via a manifold to each coulter (Currah *et al.* 1975; Salter and Darby 1977, 1978) has been used satisfactorily at low pump speeds. At high pump speeds, pre-germinated seeds are damaged (Darby and Gray 1976). No damage to pre-germinated seeds has been reported for drills using peristaltic pumps (Fluid Drilling Ltd.) or air pressure (Hiron and Balls 1978) to deliver the mixture of gel and seeds to the soil. Because it has been difficult to equalize the seed rates between each coulter using a single peristaltic pump and manifold, each coulter is supplied with the mixture of gel and seeds by its own peristaltic pump on commercially available drills. The spacing of the seeds along the row with these continuous extrusion drills is at random (Richardson and O'Dogherty 1972) and similar to the spacing distribution produced by "forced feed" or "brush feed" metering devices on dry seed drills.

Drills that deliver a small quantity of the mixture of seeds and gel at regular intervals along the row (Kevin Skipper Farm Supply Ltd.) have been made by modifying the back pressure plate of the peristaltic pump. Although the number of seeds delivered per hill cannot be controlled accurately, such drills are in use in the United States for establishing crops of bush tomatoes for the fresh market and for processing.

F. Seed Singulation and Precision Sowing

Currah (1975a) invented a prototype seed selection device capable of selecting an individual pre-germinated seed from a mass of such seeds and delivering it with a small quantity of water onto the soil surface. Seed selection heads (Fluid Drilling Ltd.) based on a similar principle are now in commercial use for sowing pre-germinated seeds into soil blocks. These devices will deliver seeds at the rate of 1 seed per second, and, on average, they achieve an efficiency of selection of 97% as single seeds, 2 to 3% as doubles, and less than 1% misses. It is unlikely that these particular devices could be incorporated into field drills for "precision" sowing for general usage, because the rate of seed selection is too slow and because constant "heads" of water must be maintained to ensure efficient seed selection. The hydro-spacer drill (Agritek, Inc.) is capable of precision sowing imbibed seeds or seeds with very small radicles (<1 mm in length), but it will not sow pre-germinated seeds having radicles longer than 1 mm without damaging them.

IV. EFFECTS OF FLUID DRILLING PRE-GERMINATED SEEDS ON SEEDLING ESTABLISHMENT, GROWTH AND YIELD

There are a number of benefits of fluid sowing pre-germinated seeds. Some are associated with the use of a carrier gel which, in addition to suspending and carrying the seeds to the soil, can be used as a means of introducing nutrients, growth regulators, and pesticides close to the developing seedling. These benefits are described in detail on p. 9–10. Direct benefits associated with the use of pre-germinated seeds include earlier, higher, and more uniform and predictable emergence as compared with dry seeds. The earlier growth and more predictable emergence have, in turn, given earlier and higher yields and more uniform crop maturity.

Although fluid drilling of pre-germinated seeds has improved seedling emergence characteristics in one or more ways, there is considerable variation in response within and between crops. This variation is partly due to differences in mean radicle length at the time of sowing, to differences in seed bed conditions, and to the type of carrier gel (Darby 1980) and drill used (Darby and Gray 1976; Wright 1978). These factors influence not only the time to emergence but also the proportion of the "germinated" seeds in the sample which actually emerge.

To date, seeds of over 20 crops, including a range of types of vegetables, flowers, cereals, grasses, fodders, and trees have been pre-germinated and fluid drilled. The detailed responses for a number of crops are reviewed below.

A. Carrot, Parsnip and Parsley

Sowing pre-germinated carrot seeds has given about 5 to 9 days earlier emergence compared with dry seeds (Currah *et al.* 1974; Hiron 1976; Darby 1980) for sowings made when soil temperatures were between 9°C and 20°C. This earlier emergence resulted in earlier growth and 18% higher yields of roots 64 days after sowing (Currah *et al.* 1974), although there was little yield advantage from fluid sowing after an additional 2 weeks of growth. These results suggest that fluid sowing could be useful for producing early crops of bunching carrots. However, there are no reports of the effects of fluid sowing pre-germinated compared with dry seeds early in the season on early yields. Hiron (1976) has reported that while fluid sowing reduced the incidence of separate "flushes" of emerging seedlings, it did not give significantly improved percentage of seedling emergence compared with sowing dry seeds.

With parsnip, between 4 days and 12 days earlier emergence and increases in percentage seedling emergence of 10 to 30% were obtained (Gray and Steckel 1977; Darby 1980) from fluid sowing pre-germinated compared with dry seeds, although no improvement in percentage of emergence was reported by Finch-Savage (1980). After approximately 90 days of growth, plants from fluid-sown crops were 20% larger than from those crops established from dry seeds (Gray and Steckel 1977), but at harvest 140 days later there was no effect on root yield.

Compared with sowing dry seeds, fluid sowing pre-germinated parsley seeds reduced the time to emergence by 7 days for an early season sowing and by 6 days for sowings made in mid-summer (Darby 1980). At both sowings, percentage of seedling emergence was the same or slightly higher than from dry seeds.

B. Celery

Fluid sowing pre-germinated seeds has given consistently earlier and higher seedling emergence than sowing dry seeds for both green and self-blanching celery (Currah *et al.* 1974; Biddington *et al.* 1975; Taylor 1977; Darby *et al.* 1980). In general, pre-germinated seeds emerged 10 days earlier than dry seeds and, contrary to what might be expected, this difference between the 2 methods diminished the earlier the sowing (Darby *et al.* 1980). With 'Florida 683', up to 60% of the seeds sown using pre-germinated seeds have emerged compared with 6% from dry seeds (Biddington *et al.* 1975; Taylor 1977). With 'Lathom Blanching', seedling emergence from pre-germinated seed was about 17% compared with 7% for dry seeds (Table 1.3). In the latter cultivar, it was evident that a considerable proportion of the germinated seeds did not emerge. This poor emergence was confirmed by experiments in which samples of seeds, consisting only of germinated seeds at the time of sowing (produced by hand selection or "priming"), were compared with a normal pre-germinated sample and with dry seeds; only 14% of the seeds in the pre-germinated sample emerged (Darby *et al.* 1980). The reasons for the failure of these pre-germinated seeds to emerge is not known. Certainly, poor emergence at the early sowings suggests cold damage, but Hegarty (1979) has shown in other species that disease and soil impedance are important causes of seedling losses. Nevertheless, the early emergence and growth from sowing pre-germinated seeds of 'Lathom Blanching' gave between 25% and 30% higher yields of marketable sticks compared with crops established from dry seeds.

Because of the difficulties of obtaining rapid and good establishment of celery from direct drilling, almost all of the crop in the United Kingdom is established from transplants and only 4% is established directly from seeds. The improved establishment and yield of crops obtained by fluid sowing compared with sowing dry seeds suggests that it could provide a viable and cheaper alternative to the use of transplants, particularly for late-season supplies.

C. Lettuce

Fluid sowing of pre-germinated seeds gave consistently higher per-

TABLE 1.3. EFFECT OF FLUID SOWING PRE-GERMINATED SEEDS OF 'LATHOM BLANCHING' CELERY ON EMERGENCE AND YIELD

			Days to 50% Emergence		Emergence (%)		Yield of Sticks (MT/ha)	
	Date	Soil Type	Dry Seeds	Pre-germinated Seeds	Dry Seeds	Pre-germinated Seeds	Dry Seeds	Pre-germinated Seeds
1975	Feb. 14	Peat	37	40	7	7	36	48
	Mar. 21	Peat	45	37	5	10	44	63
	Apr. 10	Peat	32	25	5	17	48	60
	Apr. 24	Mineral	34	20	10	24	—[2]	—
1976	Mar. 26	Peat	36	21	5	7	—	—
	May 13	Mineral	[1]	18	[1]	17	—	—
1977	Mar. 24	Peat	49	45	24	34	31	37
Mean			39	29	7	17	40	52

Source: Darby *et al.* (1980).
[1]No emergence.
[2]—=No harvest data.

centage of seedling emergence than dry seeds (Gray 1978a,b), particularly in mid-summer when high soil temperatures induce thermodormancy (Gray 1977; Darby 1980). Averaged over 16 sowings made at different times in the season, pre-germinated seeds gave 60% emergence and dry seeds 38%. These average figures, however, conceal considerable variation in the level of establishment from sowing to sowing which has been ascribed to soil crusting (Fig. 1.5) and also to dry seed beds (Hiron 1976). In all of the experiments, the percentage of germinated seeds in the sample at the time of sowing was greater than 95%.

Pre-germinated seeds also gave between 5 days and 8 days earlier emergence than from dry seeds, particularly early in the season (Gray 1978a; Finch-Savage 1980) when soil temperatures were below 10°C; for sowings made later in the season the differences between pre-germinated and dry seeds in time to seedling emergence were small (Gray 1978b; Darby 1980).

Variation in the time of head maturity within a crop of Butterhead lettuce is largely associated with the variation in the time to emergence of the seedlings (Gray 1976). Sowing pre-germinated seeds reduced this spread of emergence (Gray 1978a) (Fig. 1.6) and increased the proportion of the crop cut at successive harvests compared with crops established from dry seeds (Gray 1978b) (Table 1.4).

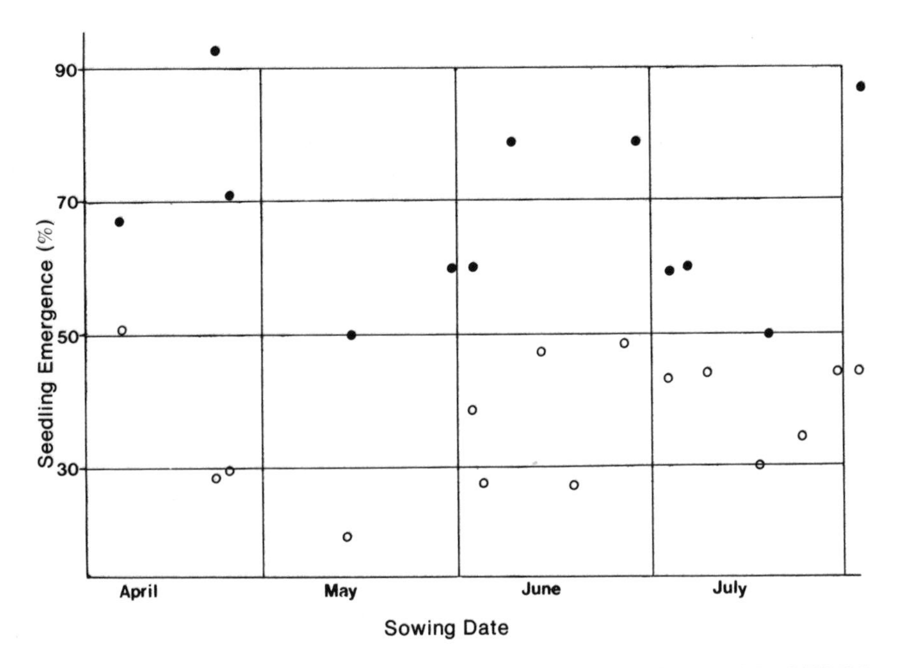

FIG. 1.5. EFFECT OF FLUID SOWING ON SEEDLING EMERGENCE OF LETTUCE SEEDS (●-●) COMPARED WITH SOWING DRY SEEDS CONVENTIONALLY (o-o)

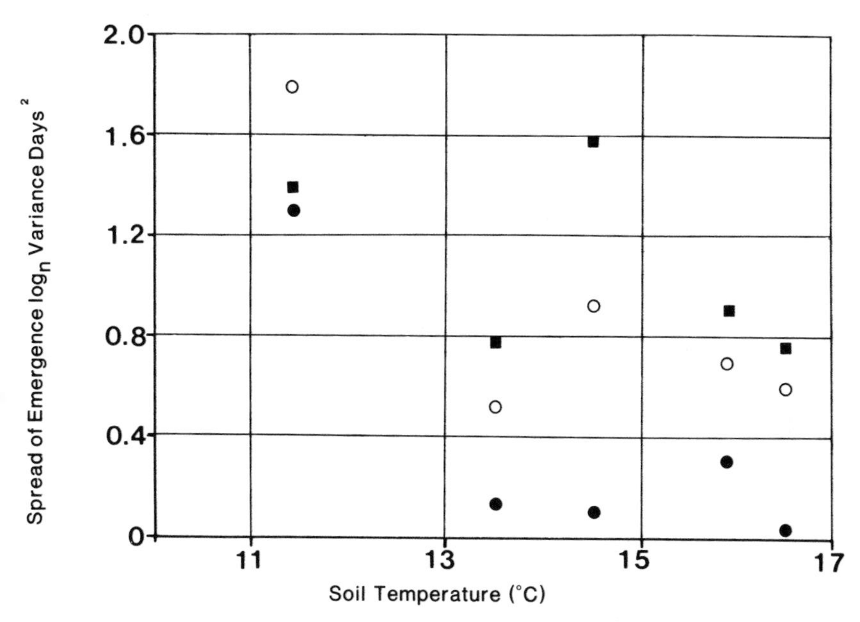

FIG. 1.6. EFFECT OF FLUID SOWING PRE-GERMINATED SEEDS (●-●) AND CONVENTIONALLY SOWING DRY (O-O) AND PELLETED SEEDS (■-■) ON THE SPREAD OF EMERGENCE OF LETTUCE

TABLE 1.4. EFFECT OF SOWING PRE-GERMINATED 'COBHAMGREEN' LETTUCE SEEDS ON THE PERCENTAGE OF THE CROP HARVESTED ON TWO OCCASIONS

	Crop Harvested (%)	
Sowing Date	Conventionally Sown Dry Seeds	Fluid Sown Pre-germinated Seeds
April 22	61[A]	95[A]
25	44	54
May 14	26	82
21	71	94
June 2	17	12
6	29	60
July 16	85	75
17	82	75
23	55	86
30	45	39
August 13	82	75
Mean	54	68[B]
SED A. ± 13.4		
B. ± 7.8		

Source: Gray (1978b).

D. Onion

In a crop such as bulb onions, where seedlings emerge poorly and the early seedling growth is slow, fluid sowing of pre-germinated seeds might be expected to give earlier growth and thus larger plants by the time bulbing started than from sowing dry seeds, thus producing higher yields.

Lipe and Skinner's (1979) work with New Mexico White Grano onions confirms this expectation. They showed that pre-germinated seeds sown using a "plug-mix" gave 2 weeks earlier emergence than dry seeds, even though only 20 to 30% of the seeds were germinated at the time of sowing. This early emergence resulted in 2 weeks earlier maturity and 30% higher yields compared with sowing dry seeds. Results with 'Rijnsburger' and other long-day types have been variable (Currah 1976; Taylor 1977). Sowing pre-germinated seeds of these types has given between 5 days and 35 days earlier emergence depending on soil temperature (Hiron 1976), but on occasions percentage of seedling emergence has been lower than from dry seeds, and particularly from very early sowings. There is some evidence that exposure of pre-germinated onion seeds to freezing temperatures for a short period (−1°C for 2 days) reduces subsequent seedling emergence, but imbibed seeds are not affected (Steckel and Gray 1980). The effects of fluid sowing on the yields of mature bulbs have been small and inconsistent (Currah 1976).

Fluid sowing salad onions ('White Lisbon' types) has given between 5 days and 6 days earlier emergence when sown in late spring and early summer compared with dry seeds (Darby 1980; Finch-Savage 1980), but 17 days earlier emergence when sown in early spring (Currah 1975b). Fluid sowing improved percentage of seedling emergence compared with dry seed by between 6% and 11% (Darby 1980; Finch-Savage 1980).

E. Sugar Beet and Red Beet

As with most small-seeded vegetables, reliable establishment of sugar beet is difficult to achieve and, in addition, emergence of the crop is slow in cool springs, so delaying leaf development and restricting the yield potential. To improve establishment and to advance growth, efforts have been made to establish the crop from transplants in the United Kingdom (Scott and Bremner 1966), but this technique is expensive. More recently, attention has been given to the possibility of fluid sowing pre-germinated seeds (Elliott 1967). Even under ideal conditions seeds (more correctly "fruits") of sugar beet germinate over a long period of time, and it is difficult to obtain a high proportion of the seeds germinated at the time of sowing. Consequently, pre-treatment of the seeds prior to germination has been investigated as a possible method of improving the percentage of germinated seeds and reducing the variation in radicle length. In two series of experiments conducted on several different soil types, sugar beet seeds treated either in small amounts of water ("advanced") or in salts or PEG ("primed") before pre-germination gave up to 10 days earlier emergence than from dry seeds. While this earlier emergence was reflected by an increased plant size in the early stages of growth, there was no significant improvement in final sugar yield compared with crops established from dry seeds (Longden *et al.* 1979). Sowing pre-treated, pre-germinated seeds did not improve percentage of seedling emergence consistently. Even when only germinated seeds were selected for sowing, lower emergence was obtained than from dry seeds, particularly with very early sowing in February.

In red (table) beet, seedling emergence from pre-germinated seeds was 10 to 40% higher than from dry seeds for sowings made in late May and early June (Darby 1980). Fluid sowing of red beet also gave about 5 days earlier emergence than from dry seeds, but effects on yield and root uniformity have not been reported.

F. Tomato and Other Chilling-Sensitive Species

Fluid sowing of pre-germinated tomato seeds has given considerably earlier emergence than from dry seeds, but the effects are dependent on soil temperatures after sowing (Bussell and Gray 1976; Gray *et al.* 1979; Bussell 1980). At mean soil temperatures of 9° to 11°C emergence was between 15 days and 17 days earlier from pre-germinated than dry seeds, but this advantage fell progressively to 6 days at soil temperatures of 12° to 15°C and to 2 days at 20°C. The precise effect also depended upon the mean radicle length at the time of sowing; seedlings developing from seeds with radicles shorter than 2 mm emerged only slightly earlier than those from dry seeds, even at low temperatures. This is because the stage of germination most sensitive to cold is when the radicle is just starting to elongate through the seed coat (Bussell and Gray 1976). In general, sowing pre-germinated seeds has given a higher percentage of seedling emergence than from sowing dry seeds, even at soil temperatures as low as 10°C (Bussell and Gray 1976; Taylor 1977; Gray *et al.* 1979). Bussell (1980) reported reduced emergence from pre-germinated seeds for sowings in very early spring. Although this reduction was considered to be due to soil crusting, the root tip of germinated tomato seeds can be damaged by prolonged exposure to low soil temperatures, resulting in poor emergence (Gray *et al.* 1978). At early sowings, when soil temperatures were between 9°C and 11°C, growth rates of the seedlings from fluid-sown crops were lower than those from dry seeds which emerged later into slightly warmer conditions. Nevertheless, the plants from the fluid-sown crops were always larger at flowering time than those from dry seeds in both early- and late-maturing cultivars. These plants also flowered earlier and fruits ripened up to 7 days earlier giving, in early-maturing cultivars, 13 to 110% higher yields from fluid-sown compared with dry-sown crops (Gray *et al.* 1979) (Table 1.5). Data on the response of late-maturing cultivars such as 'VF145-B7879' are conflicting. Improved earliness and yield from fluid sowing have been reported by Taylor (1977), while Bussell (1980) has shown that the advan-

With this system the pre-germinated seeds in the gel are sown into a "plug-mix," or the gel and seeds are first sown and then covered with a small amount of plug-mix.

The exploitation of tissue culture in commercial plant propagation is largely confined to high-value ornamental crops because the procedure is both labor- and capital-intensive. Although Galston (1972) has advocated its use for low-value crops, bulk-handling systems for the plantlets would be required. Fluid sowing of the plantlets in a protective gel incorporating a suitable growth medium could provide a cheap bulk-handling system, although initial experiments to test this have been disappointing (Drew 1979).

The use of the carrier gel as a means of applying growth regulators, nutrients, and pesticides (Salter 1978b) offers considerable potential for improving seedling growth and for controlling pests and diseases. Experiments suggest that disease control can be achieved by using smaller quantities of pesticides than are necessary for traditional methods of application (Entwistle 1979; White 1979).

One of the limitations of traditional sowing techniques is that sowing must be delayed until the surface of the soil is sufficiently dry so that the coulters do not become blocked. Soils in this condition are not ideal for good seed germination and seedling establishment. This conflict of differing requirements for efficient operation of machinery and seedling establishment need not arise with fluid sowing, as the gel and seed mixture is extruded under pressure. Because vehicles applying low ground pressures now have been developed to allow passage over wet soils without damaging them (Elliott 1978), this advantage for fluid drilling of sowing into "wet" soils could be exploited rapidly.

The development and use of gels (such as Laponite), which do not desiccate rapidly and which can reabsorb water from showers of rain, would enable a fluid drilling system for sowing seeds on the surface of the soil to be developed. Such a development could have a major impact on vegetable growing by eliminating soil crusting and by reducing the variation in depth of seed placement, a major problem with conventional drilling methods. It also could lead to the development of sowing techniques for small-grained cereals. Even if there were no biological advantage over traditional methods using dry seeds, it would permit higher drilling speeds without the loss of depth control normally found at high forward speeds with traditional systems.

The development of the singulation device now in commercial use for selecting single pre-germinated seeds (p. 12) could have a major impact on the economics of the raising of plants in blocks or containers and on the development of techniques for precision sowing of field crops. However, to take full advantage of this development so that a crop is established without gaps and with a single plant at each position, it will be necessary to have all of the seeds germinated uniformly at the time of sowing. Some progress in pre-treating seeds to improve the proportion of germinated seeds in the sample has been made but, with most samples of seeds, physical separation of germinated from ungerminated seeds will be necessary. As yet, however, no rapid and reliable separation techniques have been developed for large-scale use.

The experimental work on vegetable crops has shown that fluid sowing of pre-germinated seeds in comparison with traditional sowing methods using dry seeds gives earlier and, in most species, higher and more predictable seedling emergence. This advantage has been reflected in higher yields and, in lettuce, in greater uniformity of crop maturity. Fluid sowing offers greater scope for improving the growers' control over seedling establishment than do traditional systems of sowing, and this will become increasingly evident as the carrier gel is more fully exploited as a means of providing a "packaged environment" (Salter 1978b) for the seed and seedling.

VI. LITERATURE CITED

ANON. 1974. Anemones. Rosewarne Expt. Hort. Sta. Annu. Rpt., 1974. Agricultural Development and Advisory Services, U.K.

ANON. 1978. Grower trials 1978. Fluid Drilling Ltd., Queensway Industrial Estate, Leamington Spa, U.K.

ANON. 1979. Fluid sowing of pre-germinated seeds. Practical guide 5 (5th ed.). Natl. Veg. Res. Sta., Wellesbourne.

BIDDINGTON, N.L., T.H. THOMAS, and A.J. WHITLOCK. 1975. Celery yield increased by sowing germinated seeds. *HortScience* 10:620–621.

BROCKLEHURST, P.A. 1979. Fluid drilling—oxygen supply to the developing seedling. Rpt. Natl. Veg. Res. Sta., Wellesbourne, 1978. p. 100–101.

BROCKLEHURST, P.A., J. MILLS, and W. FINCH-SAVAGE. 1980. Effects of aeration during cold storage of germinated vegetable seeds prior to fluid drilling on seedling viability. *Ann. Appl. Biol.* 95:261–266.

BUSSELL, W.T. 1980. Emergence and growth of tomatoes after sowing chitted and untreated seeds. *N.Z. J. Expt. Agr.* 8:159–162.

BUSSELL, W.T. and D. GRAY. 1976. Effects of pre-sowing seed treatments and temperatures on tomato seed germination and seedling emergence. *Scientia Hort.* 5:101–109.

CURRAH, I.E. 1975a. Dispensing apparatus. U.K. patent application 41220/75.

CURRAH, I.E. 1975b. Establishment of crops by fluid drilling—onions. Rpt. Natl. Veg. Res. Sta., Wellesbourne, 1974. p. 81–82.

CURRAH, I.E. 1976. Establishment of crops by fluid drilling—onions. Rpt. Natl. Veg. Res. Sta., Wellesbourne, 1975. p. 66–67.

CURRAH, I.E. 1977. Hydraulic fractionation of seeds. Rpt. Natl. Veg. Res. Sta., Wellesbourne, 1976. p. 67.

CURRAH, I.E. 1978. Fluid drilling. *World Crops & Livestock* January/February, p. 22–24.

CURRAH, I.E., R.J. DARBY, A.S. DEARMAN, D. GRAY, W.E.F. RANKIN, and P.J. SALTER. 1976. Establishment of crops by fluid drilling—development of techniques. Rpt. Natl. Veg. Res. Sta., Wellesbourne, 1975. p. 64–65.

CURRAH, I.E., R.J. DARBY, D. GRAY, W.E.F. RANKIN, and P.J. SALTER. 1975. Establishment of crops by fluid drilling—development of techniques. Rpt. Natl. Veg. Res. Sta., Wellesbourne, 1974. p. 80–81.

CURRAH, I.E., D. GRAY, and T.H. THOMAS. 1974. The sowing of germinating vegetable seeds using a fluid drill. *Ann. Appl. Biol.* 76:311–318.

CURRAH, I.E. and G.M. HURDLEY. 1978. Fluid drilling. Bul. Natl. Res. Dev. Corp., U.K. Winter 1977/78 (47). p. 18–23.

DARBY, R.J. 1980. Effects of seed carriers on seedling establishment after fluid drilling. *Expt. Agr.* 16, 153–160.

DARBY, R.J. and D. GRAY. 1976. Establishment of crops by fluid drilling—damage assessments on fluid-drilled seeds. Rpt. Natl. Veg. Res. Sta., Wellesbourne, 1975. p. 68–69.

DARBY, R.J. and P.J. SALTER. 1976. A technique for osmotically pre-treating and germinating quantities of small seeds. *Ann. Appl. Biol.* 83:313–315.

DARBY, R.J., P.J. SALTER, and A.J. WHITLOCK. 1980. Effects of osmotic treatment and pre-germination of celery seeds on seedling emergence. *Expt. Hort.* 31:10–20.

DREW, R.L.K. 1979. The development of carrot (*Daucus carota* L.) embryoids (derived from cell suspension culture) into plantlets on a sugar-free basal medium. *Hort. Res.* 19:79–84.

EDWARDS, T.I. 1932. Temperature relations of seed germination. *Quart. Rev. Biol.* 7:428–443.

ELLIOTT, J.G. 1966. Improvements relating to sowing of seeds. U.K. Patent Specification 17886/62.

ELLIOTT, J.G. 1967. The sowing of seeds in aqueous fluid. Second Annu. Rpt. Weed Research Organisation, Begbroke Hill, Oxford. p. 31–32.

ELLIOTT, J.G. 1978. The implications of direct-drilling for vegetable production. *Acta Hort.* 72:93–100.

ENTWISTLE, A.R. 1979. Fluid drilled salad onions. Rpt. Natl. Veg. Res. Sta., Wellesbourne, 1978. p. 72–73.

FINCH-SAVAGE, W. 1980. Cold storage of germinated vegetable seeds prior to fluid drilling—Emergence and yield of field crops. *Ann. Appl. Biol.* (in press)

GALSTON, A.W. 1972. The immortal carrot. *Nat. Hist.* (N.Y.) 81(4):14, 16, 89, 90.

GRAY, D. 1975. Some developments in the establishment of drilled vegetable crops. XIXth Intern. Hort. Congr., Warsaw. Sept. 11–18, 1974. p. 407–418. Intern. Soc. Hort. Sci.

GRAY, D. 1976. The effect of time to emergence on head weight and variation in head weight at maturity in lettuce *(Lactuca sativa)*. *Ann. Appl. Biol.* 82: 77–86.

GRAY, D. 1977. Temperature sensitive phases during the germination of lettuce seeds. *Ann. Appl. Biol.* 86, 77–86.

GRAY, D. 1978a. The effect of sowing pre-germinated seeds of lettuce *(Lactuca sativa)* on seedling emergence. *Ann. Appl. Biol.* 88:185–192.

GRAY, D. 1978b. Comparison of fluid drilling and conventional establishment techniques on seedling emergence and crop uniformity in lettuce. *J. Hort. Sci.* 53:23–30.

GRAY, D. and H.H. BRYAN. 1978. Establishment of crops by fluid drilling. Incorporation of growth-promoting substances into the gel carrier. Rpt. Natl. Veg. Res. Sta., Wellesbourne, 1977. p. 66.

GRAY, D. and H.C. PRICE. 1978. Fluid drilling and plug-mix planting of tomatoes. Rpt. Natl. Veg. Res. Sta., Wellesbourne, 1977. p. 63–64.

GRAY, D., H.C. PRICE, D. BARTLETT, and F. TOGNONI. 1978. Chilling injury in pre-germinated tomato seeds. Rpt. Natl. Veg. Res. Sta., Wellesbourne, 1977. p. 63.

GRAY, D. and J.R.A. STECKEL. 1977. Effects of pre-sowing treatments of seeds on the germination and establishment of parsnips. *J. Hort. Sci.* 52: 525–534.

GRAY, D., J.R.A. STECKEL, and J.A. WARD. 1979. The effects of fluid sowing pre-germinated seeds and transplanting on emergence, growth and yield of outdoor bush tomatoes. *J. Agr. Sci. (Camb.)* 93:223–233.

GRAY, D., J.R.A. STECKEL, and J.A. WARD. 1980. A comparison of methods of growing tomato transplants. *Scientia Hort.* 12, 125–133.

HARDAKER, J.M. and R.C. HARDWICK. 1978. A note on Rhizobium inoculation of beans *(Phaseolus vulgaris)* using the fluid drilling technique. *Expt. Agr.* 14:17–21.

HAYSLIP, N.C. 1974. A 'plug-mix' seeding method for field planting of tomatoes and other small seeded hill crops. Fort Pierce ARC Res. Rpt. RL 1974-3.

HEGARTY, T.W. 1973. Temperature relations of germination in the field. p. 411–432. *In* W. Heydecker (ed.) Seed ecology. Proc. 19th Easter School in Agricultural Science, Univ. Nottingham 1972. Butterworths, London.

HEGARTY, T.W. 1979. Factors influencing the emergence of calabrese and carrot seedlings in the field. *J. Hort. Sci.* 54:199–207.

HEYDECKER, W. and P. COOLBEAR. 1977. Seed treatments for improved performance—survey and attempted prognosis. *Seed Sci. & Technol.* 5, 353–425.

HIRON, R.W.P. 1976. Experiments with pre-germinated seed. Kirton Expt. Hort. Sta. Annu. Rpt., 1976. Agricultural Development and Advisory Service, U.K.

HIRON, R.W.P. and R.C. BALLS. 1978. The development and evaluation of an air pressurised fluid drill. *Acta Hort.* 72:109–120.

KHAN, A.A., C.M. KARRSEN, E.F. LEUE, and C.H. ROE. 1979. Pre-conditioning of seeds to improve performance. p. 395–413. *In* T.K. Scott (ed.) Plant regulation and world agriculture. Plenum Press, New York.

LICKORISH, G.R. and R.J. DARBY. 1976. A hand-operated fluid drill for small plot experiments. *Expt. Agr.* 12:299–303.

LINDSAY, R.T., M.A. NEALE, and H.J.M. MESSER. 1974. The performance of an ice-bank cooling system for use in vegetable storage. *Acta Hort.* 38:421–442.

LIPE, W.N. and J.A. SKINNER. 1979. Effect of sowing pre-germinated onion seeds in cold soil on time of emergence, maturity and yield. *HortScience* 14: 238–239.

LONGDEN, P.C., M.G. JOHNSON, R.J. DARBY, and P.J. SALTER. 1979. Establishment and growth of sugar beet as affected by seed treatment and fluid drilling. *J. Agr. Sci. (Camb.)* 93, 541–552.

RICHARDSON, P. and M.J. O'DOGHERTY. 1972. Theoretical analysis of the seed spacing distribution produced by a fluid drill. National Institute of Agricultural Engineering, Rpt. 4, Silsoe.

SALTER, P.J. 1977. Fluid drilling—a new approach to crop establishment. British Assoc. for the Advancement of Science, Section M, London. September 1977.

SALTER, P.J. 1978a. Techniques and prospects for "fluid drilling" of vegetable crops. *Acta Hort.* 72:101–108.

SALTER, P.J. 1978b. Fluid drilling of pre-germinated seeds; progress and possibilities. *Acta Hort.* 83:245–249.

SALTER, P.J. and R.J. DARBY. 1976. Synchronisation of germination of celery seeds. *Ann. Appl. Biol.* 84:415–424.

SALTER, P.J. and R.J. DARBY. 1977. Fluid drilling of celery. Rpt. Natl. Veg. Res. Sta., Wellesbourne, 1976. p. 67–68.

SALTER, P.J. and R.J. DARBY. 1978. Fluid drilling of celery. Rpt. Natl. Veg. Res. Sta., Wellesbourne, 1977. p. 65–66.

SCOTT, R.K. and P.M. BREMNER. 1966. The effects on growth, development and yield of sugar beet of extension of the growth period by transplantation. *J. Agr. Sci. (Camb.)* 66:379–388.

SPINKS, J.L., L.O. ROTH, and J.E. MOTES. 1979. A combination fluid drill and dry seed planter for vegetable plot research. *HortScience* 14:170–171.

STECKEL, J.R.A. and D. GRAY. 1980. Fluid drilling—onions. Rpt. Natl. Veg. Res. Sta., Wellesbourne, 1979. p. 100.

TAYLOR, A.G. 1977. Comparative performance of pre-germinated, high moisture content and dry vegetable seed in greenhouse and field studies. *J. Seed Technol.* 2:52–61.

TAYLOR, A.G., J.E. MOTES, and H.C. PRICE. 1978. Separating germinated from ungerminated seed by specific gravity. *HortScience* 13:481–482.

TAYLOR, J.D. and C.L. DUDLEY. 1978a. Alginate degrading bacteria. Rpt. Natl. Veg. Res. Sta., Wellesbourne, 1977. p. 106.

TAYLOR, J.D. and C.L. DUDLEY. 1978b. *Rhizobium* inoculation of dwarf beans. Rpt. Natl. Veg. Res. Sta., Wellesbourne, 1977. p. 105.

THOMAS, T.H., N.L. BIDDINGTON, and D. PALEVITCH. 1978. Improving the performance of pelleted celery seeds with growth regulator treatments. *Acta Hort.* 83:235–243.

THOMPSON, P.A. 1974a. Germination of celery (*Apium graveolens* L.) in response to fluctuating temperatures. *J. Expt. Bot.* 25:156–163.

THOMPSON, P.A. 1974b. Effects of fluctuating temperatures on germination. *J. Expt. Bot.* 25:164–175.

THOMPSON, P.A. 1974c. Characterisation of the germination responses to temperature of vegetable seeds. I. Tomatoes. *Scientia Hort.* 2:35–54.

URE, G.B. and A. LOUGHTON. 1978. Plug-mix planting of pre-germinated seed in Ontario. *Acta Hort.* 72:125–126.

WHITE, J.G. 1979. Brassica seedling establishment. Rpt. Natl. Veg. Res. Sta., Wellesbourne, 1978. p. 76.

WILLIAMS, J.B. 1971. Runner beans (pinched): early production. Luddington Expt. Hort. Sta. Annu. Rpt., 1971, part II. Agricultural Development and Advisory Service.

WILLIAMS, J.B. 1972. Runner beans (pinched): early production. Luddington Expt. Hort. Sta. Annu. Rpt., 1972, part II. Agricultural Development and Advisory Service.

WILLIAMS, J.B. 1974. Beans, runner: effects of seed chitting and herbicides on early production of runner beans raised under low polythene tunnels. Luddington Expt. Hort. Sta. Annu. Rpt., 1974, part II. Agricultural Development and Advisory Service.

WRIGHT, W.J. 1978. Vegetables—fluid drilling of chitted seeds. ADAS Rpt. of Eastern Region Investigations, 1978. p. 165–167. Ministry of Agriculture, Fisheries and Food.

WURR, D.C.E., R.J. DARBY, and J.M. AKEHURST. 1980. Cold storage of pre-germinated seeds. Rpt. Natl. Veg. Res. Sta., Wellesbourne, 1980. p. 99–100.

2

Biology, Epidemiology, Genetics and Breeding for Resistance to Bacterial Pathogens of *Phaseolus vulgaris* L.[1]

M.L. Schuster and D.P. Coyne
Department of Horticulture, University of Nebraska,
Lincoln, Nebraska 68583

[1] Published as Paper *5878*, Journal Series, University of Nebraska Agricultural Experiment Station; work conducted under Project *20-042*. The authors wish to acknowledge Dr. T.P. Riordan, Dr. S.S. Salac, and Dr. R.C. Shearman for reviewing the manuscript, and Blythe Stickney, Jean Rice, and Waynetta Morningstar for their excellent secretarial services; financial assistance from Green Giant Corporation is appreciated.

I. INTRODUCTION

Bacterial pathogens cause important diseases of beans throughout the world (Zaumeyer and Thomas 1957). Satisfactory chemical control of these pathogens has not been achieved. Recommended control measures are use of disease-free seed, suitable rotations, deep plowing of bean debris, and use of tolerant/resistant cultivars, if available. This review covers various aspects of the biology of the bacterial pathogens and genetics and breeding for resistance.

II. VARIABILITY OF BACTERIAL PATHOGENS

Currently, about 15 species of bacteria that incite bean diseases have been reported. Of these, five are economically important: *Xanthomonas phaseoli* (E.F.S.) Dows., *Xanthomonas phaseoli* var. *fuscans* (Burkh.) Starr and Burkh., *Pseudomonas phaseolicola* (Burkh.) Dows., *Corynebacterium flaccumfaciens* (Hedges) Dows., and *Pseudomonas syringae* van Hall. Other lesser important bacterial pathogens are *Agrobacterium tumefaciens* (Smith & Townsend) Conn, *Pseudomonas tabaci* (Wolf & Foster) Stevens, *Rhizobium japonicum* (Kirchner) Buchanan, *Xanthomonas vignicola* Burkh, *Pseudomonas solanacearum* (E.F.S.) Smith, *Xanthomonas alfalfae* (Riker *et al.*) Dows., *Pseudomonas viridiflava* (Burkh.) Clara, *Xanthomonas phaseoli* var. *sojense* (Hedges) Starr & Burkh., *Erwinia lathyri* Manns and Taub., and *Agrobacterium tumefaciens* (E.F. Sm. & Towns.) Conn.

Positive identification of the principal bacterial pathogens is difficult on the basis of field symptoms (Schuster, Coyne, Hulluka, Brezina, and Kerr 1978). Thus, host range and etiological studies are employed to differentiate them more precisely.

Initially, isolations are made on appropriate media to identify and detect certain bacterial characteristics. A method of differentiating the isolates is schematically illustrated in Fig. 2.1.

Figure 2.1 illustrates differences among strains as well as species. Each bacterial species is a confederation of strains with characteristics that may show specific differences in some traits such as pathogenicity.

A. *Xanthomonas*

A case in point with regard to variability is the discovery of new strains of *X. phaseoli (Xp)* from South America and Uganda which were more virulent than certain Nebraska isolates (Schuster and Coyne 1975a, 1977a). Some of the strains of *Xp* and *Xp* var. *fuscans (Xpf)* are capable of infecting cowpeas (*Vigna unguiculata* [L.] Walp). The *Xanthomonas*

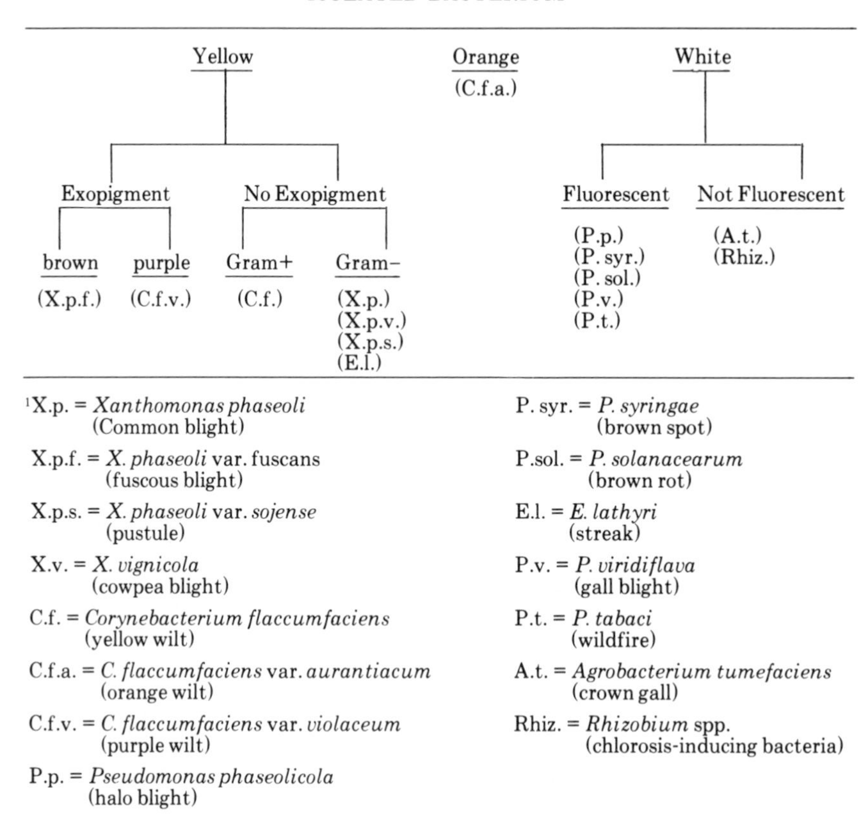

FIG. 2.1. METHOD OF DIFFERENTIATING BEAN BACTERIAL PATHOGENS

taxonomy is unclear (Buchanan and Gibbons 1974; Dye 1962) because many species are relegated into five species with pathotypes based on host reaction.

The current trend is to combine species of *Xanthomonas* because they are inseparable by the usual determinative biochemical tests (Dye 1962). Work at Centro Internacional de Agricultura Tropical (CIAT) indicated that there was no significant interaction between *Xp* isolates and bean (Anon. 1978). Valladares, Coyne, and Schuster (1977, 1979) reported that a strain from Brazil (*Xp*-Br) was virulent to all *P. vulgaris* germplasm which previously showed tolerance against other isolates tested in Nebraska. Yoshii *et al.* (1976) reported that a culture of *Xp* from Colombia was virulent in leaves of 4000 *P. vulgaris* lines. Some previously reported tolerant bean cultivars were susceptible to this culture. Sabet (1959) made a comparative study of several strains of *Xanthomonas* species of *Xp, Xpf, X.p.* var *sojense (Xps), X. vignicola (Xv), X. alfalfae, X. cajani* Kulkarni *et al.*, and *X. cassaie* Kulkarni *et al.* and found them indistinguishable on the basis of biochemical characters. Cross inoculations indicated that these pathogens, with the exception of *X. cajani*, overlapped and thus could be relegated to special forms of *X. phaseoli*. The dualism of pigment- and non-pigment-producing strains is common to many non-bean pathogens and does not warrant designation as a variety, e.g., *Xpf* (Dye 1962; Leakey 1973).

Results of pathogenicity and biochemical tests revealed that the *Xp* - *Xpf* - *Xv* group could be considered as strains of one species, *Xp*, on the basis of priority (Schuster and Coyne 1977a). It is also possible that *X.p.* var. *sojense*, which infects beans, cowpeas, soybeans, and dolichos, could be included in this group as indicated by the results of Sabet (1959). Vakili *et al.* (1975) maintain that strains virulent on cowpeas should be classified as *Xv*. Kaiser and Ramos (1979) reported that cowpea bacterial blight and cowpea bacterial pustule were indistinguishable, and they feel that both should be in *Xv*. Strains of the *Xp* - *Xpf* - *Xv* group infect several legumes in addition to beans and cowpeas (Schuster and Coyne 1977a; Zaumeyer and Thomas 1957). There is a tendency for strains of this group to be mutually exclusive in attacking specific hosts which may explain species differentiation on the basis of susceptible host species. If a sufficient number of strains are employed, combining into one species is preferred, regardless of differential pathogenicity (Schuster and Coyne 1977a).

Attempts in our laboratory to differentiate virulence of *Xp* on 2,3,5 triphenyltetrazolium chloride did not confirm the contention of Small and Worley (1956) that this could be done. The fallacy in using artificial media to demonstrate differences in virulence is variation of strains in accordance with bean cultivars employed. Toxin induction is not a prerequisite for growth *in vivo*, but may be a separate requirement for pathogenicity.

The variability in virulence among bean xanthomonads has been confirmed in several reports (Schuster and Coyne 1971, 1975b,c, 1977a; Schuster *et al.* 1973; Epko and Saettler 1976; Vakili *et al.* 1975). There is perhaps an infinite number of strains within the *Xp* - *Xpf* - *Xv* group. The strains need not be relegated to any one of the three species, but do possess different degrees of virulence. Schuster and Coyne (1975b, 1977a) and Epko and Saettler (1976) demonstrated that there are no clear-cut differences in host ranges of *Xp* and *Xv* because some strains of each "species" can infect certain cultivars of both beans and cowpeas.

Generally, fewer strains of *Xp* - *Xpf* infected cowpea cultivars and were less virulent than *Xv* strains. Burkholder (1944) reported that *Xp* and *Xv* could be differentiated on basis of energy sources, particularly in litmus milk. *Xv* produced a solid curd that lasted several weeks before peptonization, while *Xp* began to clear the medium on the second day without curd formation. Schuster and Coyne (1975b, 1977a) found similar degrees of differences in litmus milk utilization among *Xp* and *Xpf* isolates. Differences in utilization of mannitol also were cited specifically by Burkholder (1944). Schuster and Coyne (1977a) found that certain strains of *Xp* and *Xpf* utilize mannitol. Burkholder (1944) also reported differences in the utilization of other compounds such as glycerol, succinic acid, and xylose, but Schuster and Coyne (1977a) did not confirm this. It would appear that differentiation of *Xp* and *Xv* is tenuous on the basis of biochemical tests and use of too few isolates. These data confirm Dye's extensive work (1962) on biochemical tests of the xanthomonads in general. Dye and Lelliott (see Buchanan and Gibbons 1974) concluded that practical diagnosis of xanthomonad species will be centered around plant host specificity or pathogenicity.

B. *Pseudomonas*

Pseudomonas phaseolicola (Pp), the incitant of halo blight of beans, also attacks lima beans, soybeans, and other legumes, and has been studied extensively (Coyne and Schuster 1979). Halo blight occurs in the United States, Europe, Africa, Australia, New Zealand, Canada, and South America. It is not a predominant bacterial disease problem in Latin America compared to the *Xanthomonas*-incited diseases. Zaumeyer (1973) observed it once in Guatemala and near Bogota, Colombia. Schuster and Coyne (unpublished) isolated *Pp* from bean samples from Colombia and Guatemala. Many bacterial problems on beans in South America need to be resolved (Graham 1976, 1978; Echandi 1975), including the effect of climate as influenced by altitude on disease development (Schuster and Coyne 1977a,b; Wellman 1972). Halo blight is favored by lower temperatures and common blight by warm temperatures. This accounts for general occurrence of common blight on green beans in southern United States.

Variability in *Pp* was demonstrated first in Nebraska (United States) by Jensen and Livingston (1944). Differences in colony types (rough and smooth) were found in 1934 (Adam and Pugsley). These rough and smooth forms differed serologically and in reaction to bacteriophage. No mention was made of virulence, although rough forms of a bacterial pathogen lack virulence or have weaker virulence than those found in smooth colonies. Patel and Walker (1965) differentiated "race 1" from "race 2" on the basis of reaction of 'Red Mexican UI3'. These two races have been reported in several different countries. Jensen and Goss (1942) reported strains of *Pp* which induced small haloless spots on 'Red Mexican' and lesions with typical haloes on 'Red Kidney'. Schroth *et al.* (1970) concluded that there is an infinite number of strains within the *P. phaseolicola-P. glycinea-P. mori* group. They stated that neither *Pp* race 1 nor 2 is homogeneous with respect to virulence when tested on a number of cultivars, and that strains in Patel and Walker's "race 1" are weaker in virulence than those in "race 2." In central Nebraska, where halo blight devastated a 600-acre pilot snap bean acreage in 1964, strains were found to be more virulent than those reported for race 2 (Schuster *et al.* 1966) and thus could be considered as race 3. A similar situation was reported for *Pp* strains isolated in 1977, 1978, and 1979 from dry beans in western Nebraska (Schuster *et al.* 1979a; Coyne *et al.* 1979). These findings and the report of pathogenic and halo-producing strains of *P. glycinea* (Sinclair and Dhingra 1975) confirm the existence of an infinite number of strains ranging from mild to highly virulent (Schroth *et al.* 1971). Omer and Wood (1969) found a much higher multiplication of *Pp* in a susceptible cultivar than in a resistant cultivar. They found that pre- and simultaneous inoculation with a weak strain reduced the growth of a more virulent *Pp* strain. Therefore, strain interactions affect disease expression and incidence.

Pseudomonas syringae (P syr), which incites bean brown spot, has a wide host range. Brown spot is a serious problem on snap beans in Wisconsin (United States), according to Hagedorn and Patel (1965). The strains of *P syr* can be differentiated in ten cowpea cultivars according to a range of virulence (Lai and Hass 1973). 'California Blackeye' is susceptible to all *P syr* isolates tested.

Pseudomonas solanacearum (P sol), which is common in the tropics, consists of strains virulent for beans and cowpeas. *P sol* consists of strains virulent for beans and cowpeas. This bacterium, which may be seed-borne, causes brown rot and has been isolated from snap beans in Florida. It is usually not a serious problem in the United States. *P sol* consists of three heterogeneous groups or "races" which are separated along host lines (Kelman 1953). Ribeiro *et al.* (1979) reported a strain of *P. tabaci* that incites bean wildfire (BW) in Brazil. Cowpea is very susceptible to this BW organism. The authors are not aware of variability studies of *P. viridiflava* (gall blight bacterium).

C. *Corynebacterium*

The corynebacteriads pathogenic for plants are principally vascular and thus typically induce wilting. Three strains *(Cf, Cfa, Cfv)* of the bean wilt bacterium *(Corynebacterium flaccumfaciens)* have been reported and differ in pigment production (Schuster and Christiansen 1957;

Schuster and Sayre 1967). *Corynebacterium f.* var *aurantiacum* appears more virulent than *Cf* or *Cfv*. These strains can infect many legumes, including cowpeas and soybeans.

D. Miscellaneous Genera

Certain strains of *Rhizobium* spp. are capable of inducing chlorosis in beans and soybeans (Coyne and Schuster 1979). No reports on variability are known for *Erwinia lathyri* (streak bacterium) or *Agrobacterium tumefaciens* (crown gall) with regard to beans and cowpeas.

III. INOCULATION METHODS AND EVALUATION OF DISEASE REACTIONS

Several techniques have been employed for inoculation of beans with bacterial pathogens. Standardization of inoculation methods is needed for the individual diseases. Some of the techniques duplicate natural infections whereby bacteria gain entry into plants through natural openings and/or wounds.

Inoculation of leaves has been accomplished by spraying (general atomizing), and watersoaking by means of pressure (close atomizing) (Schuster 1955a) and/or in combination with wounding (multineedle, stabbing, rubbing, cutting, and sandblasting). Inoculation of stems, roots, pedicels, and pods is usually associated with wounding. Seeds are inoculated by dipping in a bacterial suspension under normal atmospheric conditions or negative pressures.

Spraying involves application of the bacterial suspension to leaf or other plant surfaces, and subsequently under high humidity the bacteria recede into the intercellular spaces. Watersoaking is a more ideal method in the field and greenhouse as it duplicates natural infections occurring during driving rainstorms. The bacteria are forced into the intercellular spaces or infection sites and maintenance of high humidity subsequent to inoculation is not required for infection.

Natural infection can result under field conditions by use of infested bean debris as a primary inoculum source. Spreader rows of a highly susceptible cultivar inoculated by the watersoaking method to initiate infection favors secondary spread.

The reaction of bean plants to inoculations may be influenced by a variety of factors including (1) *environmental factors* such as temperature, moisture, light, nutrition (Goss 1940; Patel and Walker 1965; Schuster and Christiansen 1957; Schuster and Sayre 1967; Daub and Hagedorn 1979); (2) *plant factors*, such as plant type and cultivar, age, number and size of stomatal openings, treatment prior or subsequent to inoculation, and physiological conditions (Coyne and Schuster 1973, 1974c,d; Patel and Walker 1963); and (3) *parasite factors*, such as concentration and age of inoculum and strains of the bacterium (reviewed by Schuster and Coyne 1975b; Coyne *et al.* 1973). Among workers standardization of inoculum concentration and other factors would be desirable for making comparisons more meaningful.

Disease rating schemes vary, depending on plant part inoculated as well as the individual making the assessments (Saettler 1977; Schuster and Coyne 1977b; Anon. 1977). For individual leaf inoculations, readings for *Xanthomonas* blights can be classified into five categories: 0 (no visible infection) to 4 (severe infection). In the field, ratings of whole plants can range from 0 (no visible symptoms) to 5 (very severe, plants chlorotic, necrotic, and largely defoliated). With regard to halo blight, the degree of systemic chlorosis must be classified with either the watersoaked or hypersensitive reaction. For pods, reactions are determined by measuring the extent of the watersoaked lesions (in millimeters) around the inoculation punctures. For cowpea stem inoculations, ratings are made on the basis of length (in millimeters) of the stem canker. For bacterial wilt, ratings range from no visible wilting to complete wilting and death. Appropriate susceptible and resistant genotypes are included as standard checks as a guide in assessing reactions.

IV. BACTERIAL POPULATIONS

Host-parasite interactions with respect to bacterial diseases are of concern to bean breeders. Following introduction into plant tissues, bacteria are confined to intercellular spaces or vascular parts and multiply initially in these areas. Symptoms are associated with bacterial multiplication in plant parts. Allington and Chamberlain (1949) initiated such a study on population trends of *X. phaseoli* and *P. glycinea* in bean and soybean leaves. The populations in the noncongenial host decreased markedly, whereas within the congenial host the populations continued to increase until death of tissues occurred. Since then other workers have found that bacterial populations of *P. phaseolicola* and *X. phaseoli* in leaves of resistant hosts decline at a time when rapid multiplication continues in congenial cultivars (Skoog 1952; Scharen 1959). The tolerant beans *(P. vulgaris)* possess factors that keep bacterial populations down. Daub and Hagedorn (1979) found large differences of *in vivo* multiplication of *P syr* in a resistant bean *(P. vulgaris)* cultivar compared to a susceptible cultivar. Different population trends of the three strains of *C. flaccumfaciens* were noted in bean *(P. vulgaris)* stems (Schuster and Coyne 1977a). Coyne *et al.* (1977) showed that bacterial populations of *Xp* multiplied at a higher rate in bean plants *(P. vulgaris)* during the period of pod development than in plants in the vegetative stages. Near-isogenic lines, differing in genes which determined different times of flowering under a different photoperiod regime, were used in this study. Bean plants are more susceptible in the reproductive stage than in the vegetative stage of growth. Differences in populations of three strains of *C. flaccumfaciens* were found in leaves of *Phaseolus* species (Hulluka *et al.* 1978; Schuster *et al.* 1978). *Cfa* retarded leaf expansion at the higher inoculum concentration. Population trends were correlated with host susceptibility but not leaf thickness. Epiphytic populations of *X. phaseoli* were not significantly different on resistant and susceptible bean genotypes (Coyne *et al.* 1977). High populations of *Xp* and *Xpf* were found on leaf surfaces of Navy bean *(P. vulgaris)* leaves (Weller and Saettler 1977).

V. SURVIVAL MECHANISMS OF BACTERIA

Plant parasitic bacteria must survive during dormancy of the host plant or unfavorable periods for infection and development (dry periods in the tropics and winter in the temperate zones) or the diseases they incite would disappear. Plant parasitic bacteria do not form resting spores but remain dormant during the quiescent period in association with animate or inanimate agencies: (1) seeds, (2) plant residues, (3) epiphytes, and (4) soil and other non-host materials. Recent reviews by Schuster and Coyne (1974, 1975c, 1977c,d) adequately cover survival factors of pathogenic bacteria affecting beans, and consequently all the citations were not included in this section; references are cited in Schuster and Coyne (1975c).

A. Association with Seed

Seed is an ideal site for survival of bacterial pathogens. For seed transmission, survival of pathogens must correspond to that of host seed viability. Bacterial pathogens that infect *Phaseolus vulgaris* serve as excellent examples of bacteria that survive in seeds beyond the period of seed viability: 5 to 24 years for *Cf, Cfa,* and *Cfv*, and 2 to 15 years for *Xp* and *Xpf*. Christow (1934) and Rapp (1919) found *Xp* nonviable in 2- to 3-year old seed. Longevity of bean bacteria is dependent on storage conditions and bacterial species and strains.

Control of bacterial diseases through production of near pathogen-free seed has provided an opportunity for the growth of a vigorous bean seed industry in semi-arid Idaho (United States), and is based on that area's relative freedom from several bacterial pathogens *(Pp, Xp, Xpf, Cf, Cfa, Cfv,* and *Psyr)*. Recent outbreaks of bacterial diseases of snap and dry edible bean may be attributed in part to favorable weather in seed production areas and to changes in strains of the pathogen.

Certain levels of seed infection are required to initiate an epidemic. A few infested seed per 100 lb of seed may result in considerable losses in "humid" bean production areas; this was illustrated under Wisconsin conditions where a dozen seeds infected with *Pp* per acre (0.02%) and distributed at random could promote a general epidemic. In Canada, 0.5% seed infection level of *Xpf* can be disastrous, although 2% of seed infection was not necessarily serious under other conditions. Conversely, Taylor reported that a high proportion of bean seeds infected with *Pp* failed to produce infected plants.

Mere detection of bacterial cells per seed is of little value if high numbers are required to initiate seedling infection. Saettler and Cafati (1979) determined minimal levels on seed required to initiate infection. Removal of infected seed based on "discolored" symptoms did not eliminate all infected seed. Saettler (1974) reported a bioassay method for internally borne blight bacteria in bean seed with the aid of a seedling injection test indicating that one infected seed in 500 to 1000 could be detected. Schuster and Coyne (1975a) found that by using the water-soaking method detection of pathogenic *Xp* could be demonstrated in seed at even lower population levels.

Internal seed infection among tolerant bean cultivars is of major concern to bean seed producers. Thomas and Graham (1952) found bacterial pathogens in symptomless plants grown from certified seed. Seed of tolerant/resistant cultivars can be internally infected with *Xp* and *Xpf* (Saettler and Cafati 1979; Schuster *et al.* 1979b). Tests to detect bacteria in seed should be an integral part of certified disease-free programs.

B. Association with Plant Residues

Survival of pathogenic bacteria in dead plant tissues may be important in the temperate climate, while in the tropics plants may not be decomposed between successive crops because of short time intervals between successive crops.

Because crop rotation is not practiced common blight still exists, although disease-free seed of *P. vulgaris* is commonly used for planting. It has been established that pathogenic *Xp* survives at least one winter in bean straw (Schuster and Harris 1957). *Pp* can overwinter in plant parts on the soil surface (Guthrie and Fenwick 1967; Natti 1967; Schuster *et al.* 1965; Schuster 1967, 1968, 1970). *Xp, Xpf, Cf, Cfa,* and *Psyr* were shown to overwinter in Nebraska in bean straw kept on soil surface (Schuster 1967, 1968, 1970; Schuster and Coyne 1975c). However, placement of the straw 8 in. below the soil surface eliminated most of these bacterial species. Greenhouse tests confirmed field experiments showing that infested bean debris kept in dry conditions favors survival compared to infested bean debris in moist conditions.

Many reports on survival of common blight in the field are based on circumstantial evidence. Survival in debris *per se* without exit and transmission does not provide for disease induction or perpetuation of the pathogen. Proof is necessary that under natural conditions infection can result from infested residues. Transmission to a susceptible host is essential to implicate the survival source. A realistic field demonstration in western Nebraska confirmed the findings that placement of infested straw on the soil surface of new ground favored survival of *X. phaseoli* (Schuster and Harris 1957).

C. Epiphytes as Inoculum Sources

Plant organs sustain a characteristic epiphytic bacterial flora which may be sources of primary inocula. Ercolani *et al.* (1974) and Hagedorn *et al.* (1972) recovered *P. syringae* throughout the year from leaf surfaces of healthy hairy vetch *(Vicia villosa)* and associated natural outbreaks of bean brown spot disease with epiphytes on hairy vetch. *Xpf* survived on the phylloplane of primary leaves of *P. vulgaris* but disappeared quickly from unifoliate leaves. *Xpf* colonized bean roots but disappeared after two weeks. Apparently, an alteration in plant metabolism occurred during the development of the bean seedling. Epiphytic populations of *Xp* were not significantly different on resistant and susceptible bean genotypes. High populations of *Xp* and *Xpf* also were found on surfaces of non-host plants (Schuster and Coyne 1975c; Cafati and Saettler 1980). *Pp* was unable to overwinter on wheat roots.

D. Association with Soil and Other Non-host Materials

Pathogenic bacteria may reside in non-host materials, such as soil or plant debris. *Xp, Xpf, Cf, Psyr,* and *Pp* overwinter in soil—probably in debris of diseased plants. Whether the bacteria survive in host parts buried within the soil or are free-living in the soil has not been clearly established. A case in point is the assumption that *X. phaseoli* survives in the "soil" with the exact inoculum source not ascertained. Populations of phytopathogenic bacteria decrease rapidly in the soil phase, and commonly do not contribute to the perpetuation of the pathogens.

A common explanation for poor survival in the soil is that phytopathogenic bacteria are inhibited by antagonistic microflora. Although common in soil associated with diseased plants, bacteriophages are not important in reducing bacterial pathogens in the soil. For example, Crosse

(1968) found that under most favorable conditions *P. syringae*, for example, rarely exceeded 10^2 phage particles per milliliter of soil suspension. This yield is equivalent to the lysis of only two or three bacterial cells. Therefore, it is abundantly clear that the initial concentration of phage is very low and that the chance for absorption onto a sensitive cell is very remote. A similar situation was found for phages of *X. phaseoli*. Bacteriocins may fall in the same category. Other organisms (nematodes, protozoans, bdellovibrios) might affect bacterial survival.

Planting and harvesting equipment can be survival sites for *Pp*. The non-host agencies should be considered seriously, especially with contagious organisms, such as bacteria that incite bean halo or common blight. Non-susceptible plants could be important in the epidemiology of diseases; for example, *Cfa* and *Xp* overwinter inside weeds. Reciprocal secondary spread of *Xp* between susceptible bean and non-host weeds occurred after inoculation (Cafati and Saettler 1980).

Some phases of survival need elaboration. The effects of host and non-susceptible crops on survival would be desirable. For example, *Xp* and *Cfa* overwintered in weeds. What are the relative degrees of competitive saprophytic abilities of pathogens and the substrates utilized by them? Soil is heterogeneous; therefore, different niches must exist.

VI. NATURE OF RESISTANCE

Before discussing mechanisms of resistance to bacterial diseases, we present some information on phytotoxin production. Patil (1974) has reviewed the efforts to elucidate the mode of action of bacterial toxins and bases for resistance, so individual citations in the review are not included in this section. Certain of these toxins can reproduce bacterial disease symptoms in beans in the absence of the pathogen. However, the bacterial pathogen and the toxin do not exhibit similar host specificity and thus the toxins are only one of the several determinants required in host specificity. Despite the fact that certain toxins have been isolated from bean pathogens, the exact mechanism (inherent or induced) that plants possess to counteract them has not been ascertained. The induced types constitute either the hypersensitive (HR) or protective reactions.

Protective reactions are not as readily detected as macroscopic HR symptoms. The HR type of induced resistance can be effected by employing various combinations of living bacterial cells, but the chemical basis cannot be determined by this procedure. Omer and Wood (1969) found that pre-inoculation with a weak strain reduced the growth of a more virulent strain of *P. phaseolicola*. Such interactions could affect disease expression under natural conditions. Pre- and simultaneous inoculations of beans with avirulent and virulent strains of *X. phaseoli* strains in the greenhouse and field did not provide control or impart resistance (Schuster and Coyne, unpublished). Thus, inoculating plants with mixtures of weak or avirulent strains of phytopathogenic bacteria does not always impart resistance. In some instances synergism or a "consortium" can occur, depending on the combination of bacteria used.

Most recent studies on induced resistance have indicated another approach to a possible mechanism. Sing and Schroth (1977) have found saprophytic but not pathogenic bacteria encapsulated in fibrillar structures in leaf intercellular spaces of *P. vulgaris*. They hypothesized that lectins, which are present in large concentrations in legume seeds, may be involved in the attachment process. The tobacco system differs from the bean in that avirulent but not virulent bacteria are encapsulated in tobacco cell walls (Sequeira *et al.* 1977).

A few toxins of bacterial pathogens of beans have been investigated (Patil 1974). An extracellular toxin produced by *P. phaseolicola*, phaseotoxin, induces chlorotic haloes in bean leaves. It is presumed that this peptide inhibits ornithine carbamoyl transferase (OCT) reaction. The toxin produced by *P. glycinea*, which is considered by Schroth *et al.* (1970) to be in the same taxospecies group as *P. phaseolicola*, is similar to phaseotoxin in biological activity. Despite large increases of *Pp* populations, phaseotoxin was not detected in resistant bean cultivars, and the inhibition of OCT coincided with the appearance of HR. In susceptible cultivars, OCT activity averaged 64% of that found in noninoculated leaves. The suppression of toxins in resistant tissues implicated phaseotoxin in pathogenicity of *Pp*. The manner by which OCT inhibition might induce chlorosis is debatable. Patil (unpublished) also isolated from *Pp* a high molecular weight endotoxin which induces HR in plants. Resistance to these pseudomonads appears to be a constitutive type, but the exact mechanism of action is not known.

Syringomycin (SR) is the endotoxin produced by *P. syringae* which is pathogenic for beans, cowpeas, and other crops. It is a low molecular weight polypeptide and plasmid phytotoxin which induces membrane disruption and involves its rapid "detergent-like" lysis. Effects of SR are counteracted by sterols and by Ca^{2+} and Mg^{2+}, and these effects appear to implicate the membranes. Toxin production by *P. syringae (Psyr)* isolates is not correlated with their relative pathogenicity. The application of the toxin reproduces characteristic disease symptoms in host tissues. The existence of nontoxigenic isolates of *Pp* and *Psyr* that are less virulent in the field than toxigenic isolates substantiates the hypothesis that toxigenicity is a prerequisite to pathogenicity. The nature of resistance to the above pseudomonads is not known, but correlations on differences in oxygen uptake, linoleic acid, and phenolics between resistant and susceptible cultivars have been presented.

Pseudomonas tabaci, incitant of wildfire of bean, cowpeas, etc., produces a mixture of tabtoxin (β lactam threonine) and 2-serine tabtoxin (β-lactamerine). Recent work using highly purified tabtoxin produced no inhibition of glutamine synthetase (GS). The thesis that tabtoxins affect plant tissues by inhibiting GS has been abandoned.

High molecular weight glycopeptide was isolated from certain coryne-bacteriads. Some glycopeptides have been considered to cause membrane disruption. It has been postulated that the mechanism of wilting is the restriction of water movement in the xylem by the phytotoxic polysaccharide. A toxin produced by bean wilt bacteria may be composed of glycopeptides.

Certain strains of *Rhizobium japonicum (Rj)*, a symbiont of soybean root nodules, are capable of inducing chlorosis-inciting substances. The toxin rhizobitoxine ($C_7H_{14}N_2O_4$), which is produced by *Rj*, inhibits β-cystathionase. To what extent inhibition of this enzyme *in vivo* is responsible for chlorosis is a moot question. Higher concentrations of rhizobitoxine are produced in nodules of susceptible cultivars than in resistant cultivars, even though populations of the bacterium are similar. Cultivars resistant to chlorosis induction required four times more rhizobitoxine to elicit chlorosis than did the susceptible cultivars. The resistant factors have not been assayed; unfortunately, no attempt has been made to define or to identify the mechanism involved in this resistance.

Induced types of resistance include HR and protective reactions. Attempts were made to associate "phytoalexin" production with bacterial induction of HR. Antibacterial isoflavonoids accumulated in hypersensitively reacting bean leaf tissues inoculated with *Pp* (Gnanamanickam and Patil 1977). Resistant bean cultivars inoculated with incompatible *Pp* and *Pg* strains yielded high levels of the phaseollin and hydroxyphaseollin, respectively. Pisatin, the first phytoalexin characterized, does not inhibit *X. phaseoli*; its effects on other bacteria are quite variable as is the ability of bacteria to stimulate pisatin formation. Results of studies associating substances in plants with resistance are not conclusive. Such correlations may be coincidental. It is necessary to prove that virulent and avirulent strains are differentially inhibited by the chemical substances under consideration. Such information is not available at present.

It could be presumed that differences in bean cultivars might affect penetration by bacteria; Daub and Hagedorn (1979) found no differences in penetration of *P. syringae* into susceptible and resistant bean cultivars. Leaves of tepary bean are readily watersoaked; tepary bean is susceptible to *P. phaseolicola* and resistant to *X. phaseoli*, both motile bacteria. Thus, not penetration, but rather some inherent factor(s) in the plant cells is involved in resistance.

Information on the effect of plant age and differences in susceptibility between plant parts, foliage vs. pods, for example, as well as difference between systemic chlorosis and watersoaked reactions are areas that need explanation (Hill *et al.* 1972; Coyne and Schuster 1974c,d). Another area that needs study is the nature of resistance to epiphytic bacteria on plant surfaces (Hulluka *et al.* 1978; Schuster *et al.* 1978; Weller and Saettler 1977). Microorganisms may stimulate production of phytoalexins on leaf surfaces as well as internally.

VII. SOURCES OF TOLERANT GERMPLASM

Several sources of tolerance to bacterial pathogens exist in different *Phaseolus* species (Table 2.1). Of the halo-blight *(Pp)* resistant types, the 'Red Mexican' bean was identified first by Jensen and Goss (1942) in the United States. It was employed as a resistant source in developing a number of dry bean cultivars, e.g., 'Red Mexican UI3' and 'UI34'. Breakdown of resistance in 'Red Mexican' was noted first by Ferguson *et al.* (1955). Walker and Patel (1964) described a new race of the pathogen and Patel and Walker (1965) found that 'Red Mexican UI3' and many other dry beans were resistant to race 1 and susceptible to race 2 isolates. Snap bean cultivars, except certain Blue Lake lines, were susceptible to both races. PI 150414 contained some plants with a high degree of tolerance (Patel and Walker 1965; Coyne *et al.* 1971). *Phaseolus vulgaris*, 'Great Northern (GN) Nebraska #1 sel. 27' had high tolerance to *Pp* races 1 and 2 (Coyne *et al.* 1971). Taylor *et al.* (1978) recognized four pathogenic groups in accessions of *P. vulgaris* and *P. coccineus* on the basis of reaction of primary leaves and pods to races 1 and 2 of *Pp*.

Leakey (1973) and Coyne and Schuster (1974a) have provided reviews of sources of tolerance to *X. phaseoli*. Coyne *et al.* (1963) found that 'GN Nebraska #1 sel. 27' was tolerant to all isolates of *X. phaseoli* tested in the United States. The 'GN Nebraska #1' cultivar, in which this selection was identified, was derived from the interspecific hybrid *P. vulgaris* × *P. acutifolius* (Honma 1956). *Phaseolus acutifolius* was first found resistant to *Xp* by Schuster (1955a). The tolerance of the selection #27 was confirmed by Pompeu and Crowder (1972) and by Saettler (personal correspondence). Schuster *et al.* (1973) later found that 'GN Nebraska #1 sel. 27' was susceptible to *X. phaseoli* isolates from Uganda and Colombia. However, PI 207262 (Colombia) had good foliage tolerance to isolates from Uganda, Colombia, and the United States. The authors since have observed that the pods are slightly susceptible to several U.S. isolates (Coyne and Schuster 1974d). Isolates from other areas have not been tested on the pods. Coyne and Schuster (1973) reported numerous *P. coccineus* PI lines which have high tolerance to isolates of *X. phaseoli* occurring in the United States.

TABLE 2.1. SOURCES OF RESISTANCE IN *PHASEOLUS* SPECIES TO BACTERIAL PATHOGENS

Legume	Pathogen	Source	Country
Phaseolus vulgaris	*Pseudomonas phaseolicola*	GN Nebraska #1 sel. 27	United States
		PI 150414, PI 181954	United States
		Red Mexican UI-3	United States
		OSU Bush Blue Lake lines	United States
		BO19	Germany
		Negro Vaine Blanca El Conge	Netherlands
	Pseudomonas syringae	PI 313537	United States
		Trugreen, Tempo	United States
	Corynebacterium flaccumfaciens	PI 165078	United States
	Xanthomonas phaseoli and *X. phaseoli* var. *fuscans*	GN Nebraska #1 Sel. 27, PI 207262	United States
		PI 313343, PI 282086	Colombia (CIAT)
		HW5, Lumarep	France
Phaseolus acutifolius		Tepary	United States
Phaseolus coccineus		PI Lines	United States
		Selected populations	Puerto Rico
		PI 165421, PI 181790	United States

Graham (1978) reported that *P. vulgaris* PI 313343 and PI 282086 had tolerance to *Xp*. Valladares *et al.* (1977) tested the following lines of *Phaseolus* species to a collection of isolates of *Xp* (E.F. Smith) Dowson (*Xps* and *Xp* 816 (Nebr.), *Xp* - 15 (Michigan), *Xp-Br* (Brazil), and *X.p. fuscans* (*Xpf*-UI, Uganda): *P. vulgaris* L.—PI 169727, PI 197687, PI 163117, PI 207262, PI 325677, PI 325684, PI 325691, 'Great Northern Nebr. #1 sel. 27'; *P. coccineus*—PI 165421; and *P. acutifolius* (Tepary Nebr. Acc. #10). The multiple needle method was used to inoculate leaves and a dissecting needle was used to inoculate pods of these plants. A differential reaction of isolate × lines was observed for each of the reactions on leaves and pods. In addition, a differential reaction occurred between pods and leaves within lines. All of the reportedly tolerant *P. vulgaris* lines were susceptible or moderately susceptible to the new virulent *Xp*-Br strain. Leaves and pods of *P. acutifolius* were highly tolerant to all isolates, while *P. coccineus* PI 165421 showed high tolerance for leaves and high susceptibility for pods. The internal reaction for pods was more severe than was the external reaction. *Phaseolus vulgaris* PI 207262 showed a uniform tolerance of leaf and pod (external and internal) to the U.S. isolates, while 'GN Nebr. #1 sel. 27' had a tolerant leaf and susceptible pod. These results suggested differential genetic control of pod and leaf reactions.

High tolerance to *P syr* was found by Hagedorn and Rand (1977) in *P. vulgaris* PI 313537, and Coyne and Schuster (1969a) observed moderate tolerance in two *P. vulgaris* cultivars.

High tolerance to *C. flaccumfaciens* and *Cfa* and Cfv was observed by Coyne, Schuster, and Estes (1966) in *P. vulgaris* PI 165078 at all stages of plant growth. This was the only tolerant line encountered in the 1500 lines tested.

VIII. GENETICS AND BREEDING

A. Halo Blight

Several patterns of qualitative genetic control were reported, depending on the host genotype and races of the pathogen used (Coyne and Schuster 1979). These studies involved a study of the reaction of the plant as a whole. No attempt was made to study the genetic relationships of the reactions of different plant parts or different expressions of the disease reaction. Patel and Walker (1965) showed the relation of temperature to systemic chlorosis development and considered that the systemic chlorosis reaction was not adequate in defining classes.

Results of a plant component genetic analysis of reaction to *Pseudomonas phaseolicola* were summarized by Coyne *et al.* (1971) and Hill *et al.* (1972). These workers found that it was necessary to grow the inoculated plants at 20°C in growth chambers to obtain a uniform development of systemic chlorosis in all susceptible plants. Coyne *et al.* (1971) found that the leaf systemic chlorosis and watersoaked leaf reactions to race 2 in the crosses 'Gallatin 50' (susceptible for both reactions) × 'GN Nebraska #1 sel. 27' (tolerant for both reactions) and 'Gallatin 50' × PI 150414 (tolerant for both reactions) were controlled primarily by separate major genes. Coupling linkage was detected between the genes controlling each of the reactions. Different genes controlled each of the tolerant reactions in each of the parents, the susceptible reaction being dominant in the first cross and recessive in the second cross. Hill *et al.* (1972) showed that the pod reaction to race 1 was controlled by another gene which is independent of the genes controlling the two leaf reactions mentioned previously. Coyne and Schuster (unpublished) recently found a fourth major gene, independent of the genes described previously, which controls a wilting reaction of the primary leaves. This component genetic analysis of disease reaction indicates the importance of selecting plants which have a tolerant pod, leaf, and non-systemic plant reaction using specific environmental conditions and inoculation procedures.

Taylor *et al.* (1978) presented a recent update on inheritance to halo blight. This review reconciles the apparent conflicting evidence on the inheritance of the resistant reaction obtained by earlier workers. Genetic background appeared to affect the expression of the genes controlling resistance or susceptibility.

B. Common Blight

Several workers found reaction to *Xanthomonas phaseoli* to be inherited quantitatively (Honma 1956; Coyne, Schuster, and Harris 1965, 1973; Pompeu and Crowder 1972; Valladares *et al.* 1977, 1979). Coyne, Schuster, and Harris (1965) reported a low narrow sense heritability estimate by regression of F_3 progeny means on individual F_2 plants. Genes controlling delayed flowering and tolerant reaction were linked in the crosses with 'GN Nebraska #1 sel. 27', but linkage was not apparent in a cross between that selection and PI 207262 (Coyne *et al.* 1974c). The delayed flowering of the parents was due to the interaction of long photoperiod and high temperature and was controlled also by different genes in these parents.

Coyne and Schuster (1974d) observed a differential reaction of pod and foliage to *Xp* isolates in some bean lines. PI 207262 had tolerant foliage but slightly susceptible pods, while 'GN 1140' had severely susceptible foliage and slightly susceptible pods. This indicated that the reaction of these lines may be due to a recombination of different genes controlling the response of different plant parts to bacterial infection and that it is important to obtain tolerance to the pathogen in both leaves and pods. Valladares (1979) made crosses between the bean lines/cultivars 'GN Nebraska #1 sel. 27' and PI 197687, PI 163117, PI 207262, and 'Guali' to determine if genes in these genotypes controlling the leaf reaction to the isolate *Xp*-Br were the same or different and to determine if transgressive segregation for increased tolerance could be obtained. Continuous distributions for disease ratings were found in field-grown F_2 plants derived from the crosses 'GN Nebraska #1 sel. 27' with PI 197687 and PI 297262. No transgressive segregation for increased tolerance was found in these crosses. Slight bimodal distributions of disease ratings and transgressive segregations for increased tolerance and susceptibility were observed in the field-grown F_2 generations derived from the crosses 'GN Nebraska #1 sel. 27' with PI 163117 and 'Guali'. Transgressive segregation for increased tolerance was confirmed in greenhouse-grown selected F_3 families. Leaf and pod tolerance were not associated in the F_3 families.

Valladares (1979) also studied a diallel cross of the lines/cultivars PI 163117, 'GN 1140', 'Guali', 'GN Nebraska #1 sel. 27', PI 207262, and 'Tacarigua' to determine the inheritance of leaf and external and internal pod reactions to the Nebraska isolate *Xp*-EK11. The Gardner and Eberhart (1966) model, Analysis II, was used to generate estimates of the genetic effects for each trait. Mainly additive effects were involved in the genetic control of the leaf and external and internal pod reactions to *Xp*. 'GN Nebraska #1 sel. 27' and PI 207262 contributed a considerable amount of additive genetic effects for leaf reaction. The external pod reaction of 'Tacarigua' and the internal pod reactions of PI 207262 and 'Tacarigua' were controlled mainly by additive gene action. The crosses PI 163117 × 'GN Nebraska #1 sel. 27', 'Guali' × 'GN Nebraska #1 sel. 27', and 'GN 1140' × 'Guali' showed significant specific heterosis effects for leaf reaction. A significant genotype × location interaction was detected in the field experiments.

C. Bacterial Wilt

The inheritance of the reaction to *Corynebacterium flaccumfaciens* var. *aurantiacum* in *P. vulgaris* was reported by Coyne, Schuster, and Young (1965). The susceptible reaction was found to be determined primarily by two complementary dominant genes between the late maturing wilt tolerant PI lines 165078 × late maturing common blight tolerant 'GN Nebraska #1 sel. 27'. Early maturing segregates were observed in F_2 populations, indicating that different genes controlled the late flowering (late maturity) response in the two parents. Some of these early segregates also combined high tolerance to wilt bacteria. The blight tolerant parent 'GN Nebraska #1 sel. 27' was used then as the recurrent parent in a backcross program, because the inheritance of the blight reaction was complex while earliness and wilt tolerance were primarily under qualitative genetic control. The genes controlling the latter traits were readily transferred to 'GN Nebraska #1 sel. 27' in a backcross breeding program.

Success also has been realized in developing lines and cultivars of dry beans with high tolerance to *Xp, Xpf* (Coyne and Schuster 1974a,b, 1976; Zaumeyer and Meiners 1975), and *Cf* (Schuster and Coyne 1975b) using pedigree selection.

IX. BREEDING METHODS

A. Approaches

The backcross breeding method has been used successfully to transfer major genes controlling disease reactions caused by several bacterial pathogens into acceptable *P. vulgaris* cultivars. Coyne and Schuster (1976) successfully combined resistance to *Xp*, which is quantitatively inherited, and *Cf*, which is simply inherited, into a cultivar 'GN Star'. They did so by utilizing the common blight-tolerant line 'GN Nebraska #1 sel. 27' as the recurrent parent in a backcross program transferring resistance to *Cf* and earliness.

Valladares *et al.* (1979) observed transgressive segregation for a high level of leaf tolerance to the virulent *Xp*-Br strain (watersoaking method of inoculation) in field-grown *P. vulgaris* F_2 plants derived from the crosses 'GN Nebraska #1 sel. 27' × PI 163117, and 'GN Nebr. #1 sel. 27' × 'Guali', but not from the crosses 'GN Nebraska #1 sel. 27' × PI 197687 or 'GN Nebraska #1 sel. 27' × PI 207262. Transgressive segregation was confirmed in greenhouse-grown selected F_3 families. High leaf tolerance was not associated with pod tolerance. Linkage was detected among the major genes controlling late maturity and plant habit, and the polygenes controlling common blight tolerance. No early-maturing, determinate or indeterminate, blight-tolerant recombinant plants were observed in the F_2. Different major genes controlled late maturity in the parents, and early-maturing segregates in the F_2 were found to be determined by duplicate recessive genes.

The results indicate that it is not always necessary to screen large collections of *P. vulgaris* to identify tolerance to new virulent strains of *Xp*. Tolerance can be obtained by transgressive segregation in progeny derived from crosses of some susceptible parents. The technique would permit recombination of the favorable genes present in the original susceptible or moderately susceptible germplasm in order to develop high levels of tolerance to new virulent strains of the pathogen in populations.

Honma (1956) and Mok *et al.* (1978) made the interspecific cross *P.*

vulgaris × *P. acutifolius* by using embryo culture; the former investigator obtained fertile F_1 hybrids. Coyne and Schuster (1974a) discovered a highly resistant *Xp* single plant selection, classified as #27, in the 'GN Nebraska #1' cultivar which was derived from the interspecific cross made by Honma (1956). This selection has been widely used since as source of resistance to *Xp, Xpf,* and *Pp* in North and South America.

Several breeding approaches are suggested in order to develop more effective bacterial disease-tolerant cultivars. Different genes controlling the tolerant reaction in available germplasm can be recombined to provide a broader genetic base of tolerance. This will reduce the genetic vulnerability of beans to these pathogens. Genes controlling bacterial tolerant reactions in different plant parts and different expressions of the disease in the same or different plant parts can be recombined through appropriate breeding procedures to increase the total plant tolerance. Increased attention to the component genetic approach is recommended. The opportunity to obtain increased levels of tolerance through transgressive segregation should be explored more frequently. Bacterial disease tolerance exists in *Phaseolus* species other than *P. acutifolius*, but relatively little use has been made of these sources. The barriers to crossability among *Phaseolus* species require further investigation in order to utilize effectively genes for tolerance in these species.

B. Cultivars

Several bacterial disease-resistant cultivars have been released in recent years (Table 2.2). The light red kidney cultivars 'Redkote' and 'Redkloud', resistant to halo blight, were developed by Wallace *et al.* at Cornell University (personal correspondence). 'Starland', a snap bean cultivar, resistant to halo blight, was developed by Schuster (1955b). The dark red kidney cultivars 'Montcalm' and 'Mecosta', resistant to halo blight, were released by the Michigan Agricultural Experiment Station (Copeland and Erdman 1977). The moderately late-maturing, vigorous, viny (indeterminate), high-yielding, common blight-tolerant cultivars 'GN Tara' and 'GN 'Jules' were derived by pedigree selection from the cross 'GN1140' × 'GN Nebraska #1 sel. 27' (Coyne and Schuster 1969b, 1970). Both cultivars have performed well on lighter soils when planted early. On heavy soils the cultivars are susceptible to white mold because of the vigorous plant canopy under these conditions. The moderately early cultivar 'GN Valley' was derived from a backcross program with 'GN Nebraska #1 sel. 27' as the recurrent parent (Coyne and Schuster 1974c). Coyne and Schuster (1971) released the larger, brighter, white-seeded indeterminate 'GN Emerson', possessing high tolerance to the three strains of the wilt bacterium and moderate tolerance to common

TABLE 2.2. CULTIVARS AND LINES DEVELOPED FOR RESISTANCE TO BACTERIAL PATHOGENS

Legume	Pathogen	Cultivar	Line
Phaseolus vulgaris	*Pseudomonas phaseolicola*	Redkote, Redkloud Montcalm, Mecosta Lumarep, Starland	Wis. HBR-40, Wis HBR-72 OSU 1040 Cornell Lines
	Corynebacterium flaccumfaciens	GN Star, GN Emerson	
	Xanthomonas phaseoli; X. phaseoli var. *fuscans*	GN Tara, GN Valley GN Star, GN Harris	Pea Bean MSU Lines CIAT Lines Wis. 71-3938
	Pseudomonas syringae		BBSR 130, WBR 133
Phaseolus coccineus	*Xanthomonas phaseoli; X. phaseoli* var. *fuscans*		Populations-Puerto Rico

blight. This cultivar was derived by pedigree selection from the cross 'GN 1140' × PI 165078. Coyne and Schuster (1976) also developed 'GN Star', resistant to common blight and bacterial wilt. Coyne *et al.* (1980) released 'GN Harris' which is tolerant to common blight.

X. GENETIC VULNERABILITY

Nearly all of the predominant dry bean and green bean cultivars used in the United States are genetically uniform and have a narrow genetic base (National Research Council 1972).

About 80% of the seed production of green beans and most of the certified seed of Pinto, Red Mexican, and Great Northern dry beans along with foundation seed of pea beans are produced in southern Idaho (United States). This entire bean seed crop could be destroyed if conditions were favorable for an epidemic. Serious bean seed losses due to the halo blight bacterial disease occurred in Idaho during the years 1963 to 1967. This indicated the need for a decentralization of seed production, more intensive surveillance of changes in virulence of bacterial pathogens, and broadening the genetic base of bean cultivars. In cases where major gene resistance to different bacterial pathogens has been utilized, most cultivars of dry beans, green beans, and cowpeas possess one or two genes from the same source that controls the resistant reaction. There is a good probability that new virulent strains of the pathogens will develop to overcome host resistance. Efforts need to be made to recombine different resistance genes in the same genetic background or possibly introduced as different lines into genetic blends to provide increased stability of resistance. A polygenic system may not always provide long term resistance to a bacterial pathogen. *Phaseolus vulgaris* 'GN Nebraska #1 sel. 27', although highly resistant to strains of *Xp* in the United States, was found to be susceptible to strains from Brazil, Colombia, and Uganda (Schuster and Coyne 1975b).

In the future increased efforts are needed to broaden the genetic base of most of the cultivated beans by utilizing exotic germplasm sources to improve bean populations. New research also should focus on improved breeding strategies to reduce the genetic vulnerability of beans and cowpeas.

XI. LITERATURE CITED

ADAM, D.B. and A.T. PUGSLEY. 1934. 'Smooth-rough' variation in *Phytomonas medicaginis phaseolicola* Burk. *Austral. J. Expt. Biol. Med. Sci.* 12: 193–202.

ALLINGTON, W.B. and D.W. CHAMBERLAIN. 1949. Trends in the population of pathogenic bacteria within leaf tissues of susceptible and immune plant species. *Phytopathology* 39:656–660.

ANON. 1977. Standardization of disease screening procedures for *Xanthomonas* bacterial blights of beans. *In* A.W. Saettler (ed.) Rpt. Bean Improvement Coop. and Natl. Dry Bean Counc. Mtg. Nov. 5–7, 1975. (Mimeo.) Mich. State Univ., East Lansing.

ANON. 1978. Bean production problems. Bean Program 1977 Rpt. 02E1R-77, Sept. CIAT, Cali, Colombia.

BAGGETT, J.R. and W.A. FRAZIER. 1967. Sources of resistance to halo-blight in *Phaseolus vulgaris*. *Plant Dis. Rptr.* 51:661–665.

BUCHANAN, R.E. and N.E. GIBBONS (eds.) 1974. Bergey's manual of determinative bacteriology, 8th ed. Williams & Wilkins Co., Baltimore, Md.

BURKHOLDER, W.H. 1944. *Xanthomonas vignicola* sp. nov. pathogenic on cowpeas and beans. *Phytopathology* 34:430–432.

BURKHOLDER, W.H. and K. ZALESKI. 1932. Varietal susceptibility of beans to an American and European strain of *Phytomonas medicaginis* var. *phaseolicola*, and a comparison of the strains in culture. *Phytopathology* 22: 85–94.

CAFATI, C.R. and A.W. SAETTLER. 1980. Role of non-host species as alternate inoculum sources of *Xanthomonas phaseoli*. *Plant Dis.* 64:194–196.

CHRISTOW, A. 1934. Einige Versuche uber Die Bakterienkrankheiten bei Bohnen. *Phytopathol. Ztcher.* 7:534–544.

COPELAND, L.O. and M.H. ERDMAN. 1977. Montcalm and Mecosta—Halo blight tolerant kidney bean varieties from Michigan. *Mich. Coop. Ext. Serv. Ext. Bul.* E.-957.

COYNE, D.P., D.S. NULAND, M.L. SCHUSTER, and F.N. ANDERSON. 1980. Great Northern Harris dry bean. *HortScience* 15:531.

COYNE, D.P. and M.L. SCHUSTER. 1969a. Moderate tolerance of bean varieties to brown spot bacterium *(Pseudomonas syringae)*. *Plant Dis. Rptr.* 53:677–680.

COYNE, D.P. and M.L. SCHUSTER. 1969b. 'Tara' a new Great Northern dry bean variety tolerant to common blight bacterial disease. *Univ. Nebr. Agr. Expt. Sta. Bul.* 506.

COYNE, D.P. and M.L. SCHUSTER. 1970. 'Jules' a Great Northern dry bean variety tolerant to common blight bacterium *(Xanthomonas phaseoli)*. *Plant Dis. Rptr.* 54:557–559.

COYNE, D.P. and M.L. SCHUSTER. 1971. 'Emerson'—A new large-seeded Great Northern dry bean variety tolerant to bacterial wilt disease. *Univ. Nebr. Agr. Ext. Sta. Bul.* 516.

COYNE, D.P. and M.L. SCHUSTER. 1972. 'Great Northern Star' dry bean tolerant to bacterial diseases. *HortScience* 11:621.

COYNE, D.P. and M.L. SCHUSTER. 1973. *Phaseolus* germ plasm tolerant to common blight bacterium *(Xanthomonas phaseoli)*. *Plant Dis. Rptr.* 57:111–114.

Pseudomonas syringae on hairy vetch in relation to epidemiology of bacterial brown spot of bean. *Phytopathology* 62:672. (Abstr.)

HILL, K., D.P. COYNE, and M.L. SCHUSTER. 1972. Leaf, pod and systemic chlorosis reactions in *Phaseolus vulgaris* to halo blight controlled by different genes. *J. Amer. Soc. Hort. Sci.* 97:494–498.

HONMA, S. 1956. A bean interspecific hybrid. *J. Hered.* 47:217–220.

HUBBELING, N. 1973. Field resistance in *Phaseolus* beans to *Pseudomonas phaseolicola. Meded. van de Fakul. Landbouw., Gent* 38:1351–1363.

HULLUKA, M., M.L. SCHUSTER, J.L. WEIHING, and D.P. COYNE. 1978. Population trends of *Corynebacterium flaccumfaciens* strains in leaves of *Phaseolus* species. *Fitopatol. Brasil.* 3:13–26.

JENSEN, J.H. and R.W. GOSS. 1942. Physiological resistance to halo blight in beans. *Phytopathology* 32:246–253.

JENSEN, J.H. and J.E. LIVINGSTON. 1944. Variation in symptoms produced by isolates of *Phytomonas medicaginis* var. *phaseolicola. Phytopathology* 34:471–480.

KAISER, W.J. and A.H. RAMOS. 1979. Two bacterial diseases of cowpea in East Africa. *Plant Dis. Rptr.* 63:304–308.

KELMAN, A. 1953. A bacterial wilt caused by *Pseudomonas solanacearum*: a literature review and bibliography. *N. C. Expt. Sta. Tech. Bul.* 99.

LAI, M. and B. HASS. 1973. Reaction of cowpea seedlings to phytopathogenic bacteria. *Phytopathology* 63:1099–1103.

LEAKEY, D.L.A. 1970. Need one grow pure lines in developing countries? *In* D.P. Coyne (ed.) Annu. Rpt. Bean Improvement Coop. 13:62–63. (Mimeo.) Univ. of Nebraska, Lincoln.

LEAKEY, C.L.A. 1973. A note on *Xanthomonas* blight of beans (*Phaseolus vulgaris* (L.) Davi) and prospectus for its control by breeding for its tolerance. *Euphytica* 22:132–140.

MOK, D., W.S. MACHTEED, C. MOK, and A. RABAKOARIHANTA. 1978. Interspecific hybridization of *Phaseolus vulgaris* with *P. lunatus* and *P. acutifolius. Theor. Appl. Genet.* 52:209–215.

NATIONAL RESEARCH COUNCIL. 1972. Soybeans and other edible legumes. p. 207–252. *In* J.G. Horstall *et al.* (eds.) Genetic vulnerability of major crops. Natl. Acad. Sci. Agr. Board, Washington, D.C.

NATTI, J.J. 1967. Overwinter survival of *Pseudomonas phaseolicola* in New York. *Phytopathology* 57:343. (Abstr.)

OMER, M.E.H. and R.K.S. WOOD. 1969. Growth of *Pseudomonas phaseolicola* in susceptible and in resistant bean plant. *Ann. Appl. Biol.* 63:103–116.

PATEL, P.N. and J.C. WALKER. 1963. Relation of air temperature and age and nutrition of the host to the development of halo and common bacterial blights of bean. *Phytopathology* 53:407–411.

PATEL, P.N. and J.C. WALKER. 1965. Resistance in *Phaseolus* to halo blight in bean. *Phytopathology* 56:681–682.

PATIL, S.S. 1974. Toxins produced by phytopathogenic bacteria. *Annu. Rev. Phytopathol.* 12:259–278.

POMPEU, A.S. and L.V. CROWDER. 1972. Inheritance of resistance of *Phaseolus vulgaris* L. (dry beans) to *Xanthomonas phaseoli* (common blight). *Cienciae Cultura* 24:1055–1063.

PRYKE, P.I. 1966. Halo blight-resistance studies. *In* W.A. Frazier (ed.) Annu. Rpt. Bean Improvement Coop. 9:31–32. (Mimeo.) Univ. of Nebraska, Lincoln.

RAPP, V.V. 1919. Aged bean seed, a control for bacterial blight of beans. *Science* 50:568.

RIBEIRO, R. DE L., D.J. HAGEDORN, R.D. DURBIN, and T.F. UCHYTIL. 1979. Characterization of the bacterium inciting bean wild fire in Brazil. *Phytopathology* 69:208–212.

RUSSEL, P.E. 1976. Potential sources of resistance to halo-blight of beans. *In* M.H. Dickson (ed.) Annu. Rpt. Bean Improvement Coop. 19:65–67. (Mimeo.) Univ. of Nebraska, Lincoln.

SABET, K.A. 1959. Studies in the bacterial diseases of Sudan crops. III. On the occurrence, host range and taxonomy of the bacteria causing leaf blight diseases of certain leguminous plants. *Ann. Appl. Biol.* 47:318–331.

SABET, K.A. and F. ISHAG. 1969. Studies on the bacterial diseases of Sudan crops. VIII. Survival and dissemination of *Xanthomonas phaseoli* (E.F. Smith) Dowson. *Ann. Appl. Biol.* 64:65–74.

SAETTLER, A.W. 1974. Testing bean seed for internal bacterial blight contamination. *In* D.P. Coyne (ed.) Annu. Rpt. Bean Improvement Coop. 17: 73–74. (Mimeo.) Univ. of Nebraska, Lincoln.

SAETTLER, A.W. 1977. Breeding dry edible beans (*Phaseolus vulgaris* L.) for tolerance to *Xanthomonas* bacterial blights. *Fitopatol. Brasil.* 2:179–186.

SAETTLER, A.W. and C.R. CAFATI. 1979. Transmission of bean common blight bacteria *(Xanthomonas phaseoli)* in seed of resistant and susceptible bean genotypes. IX Intern. Congr. Plant Pathol. Aug. 5–11, 1979. Washington, D.C. (Abstr. 224.) The American Phytopathological Society, St. Paul, Minn.

SCHAREN, A.L. 1959. Comparative population trends of *Xanthomonas phaseoli* in susceptible, field tolerant and resistant hosts. *Phytopathology* 49: 425–428.

SCHROTH, M.N., V.B. VITANZA, and D.C. HILDEBRAND. 1970. Pathogenic and nutritional variation in the halo blight group of fluorescent pseudomonads of beans. *Phytopathology* 61:852–857.

SCHUSTER, M.L. 1955a. A method of testing resistance of beans to bacterial blights. *Phytopathology* 45:519–520.

SCHUSTER, M.L. 1955b. Starland, a new hybrid wax bean. *Nebr. Farmer* 97 (May 7):44.

SCHUSTER, M.L. 1967. Survival of bean bacterial pathogens in the field and greenhouse under different environmental conditions. *Phytopathology* 57: 830. (Abstr.)

SCHUSTER, M.L. 1968. Survival of bean bacterial pathogens in the field and greenhouse under different environmental conditions. *In* D.P. Coyne (ed.) Ann. Rpt. Bean Improvement Coop. 11:43–44. (Mimeo.) Univ. of Nebraska, Lincoln.

SCHUSTER, M.L. 1970. Survival of bacterial pathogens of beans. *In* D.P. Coyne (ed.) Ann. Rpt. Bean Improvement Coop. 13:68–70. (Mimeo.) Univ. of Nebraska, Lincoln.

SCHUSTER, M.L. and D.W. CHRISTIANSEN. 1957. An orange-colored strain of *Corynebacterium flaccumfaciens* causing bean wilt. *Phytopathology* 47:51–53.

SCHUSTER, M.L. and D.P. COYNE. 1971. New virulent strains of *Xanthomonas phaseoli*. *Plant Dis. Rptr.* 55:505–506.

SCHUSTER, M.L. and D.P. COYNE. 1974. Survival mechanisms of phytopathogenic bacteria. *Annu. Rev. Phytopathol.* 12:199–221.

SCHUSTER, M.L. and D.P. COYNE. 1975a. Detection of bacteria in bean seed. *In* D.P. Coyne (ed.) Annu. Rpt. Bean Improvement Coop. 18:71. (Mimeo.) Univ. of Nebraska, Lincoln.

SCHUSTER, M.L. and D.P. COYNE. 1975b. Genetic variation of bean bacterial pathogens. *Euphytica* 24:143–147.

SCHUSTER, M.L. and D.P. COYNE. 1975c. Survival factors of plant pathogenic bacteria. *Univ. Neb. Agr. Expt. Sta. Res. Bul.* 268.

SCHUSTER, M.L. and D.P. COYNE. 1977a. Characterization and variation of *Xanthomonas* and *Corynebacterium* incited diseases of beans (*Phaseolus vulgaris* L.). *Fitopatol. Brasil.* 2:199–209.

SCHUSTER, M.L. and D.P. COYNE. 1977b. Standardization of screening beans to Xanthomonads. *In* M.H. Dickson (ed.) Annu. Rpt. Bean Improvement Coop. 20:73–74. (Mimeo.) N.Y. Agr. Expt. Sta. Geneva, N.Y.

SCHUSTER, M.L. and D.P. COYNE. 1977c. Supervivencia pathogenos bacteriales de plantas en el tropico con enfasis en frijol *(Phaseolus vulgaris). Fitopatol. Colomb.* 6:101–111.

SCHUSTER, M.L. and D.P. COYNE. 1977d. Survival of plant parasitic bacteria of plants grown in tropics with emphasis on beans *(Phaseolus vulgaris). Fitopatol. Brasil.* 2(2):177–130.

SCHUSTER, M.L., D.P. COYNE, and B. HOFF. 1973. Comparative virulence of *Xanthomonas phaseoli* strains from Uganda, Colombia, and Nebraska. *Plant Dis. Rptr.* 57:74–75.

SCHUSTER, M.L., D.P. COYNE, M. HULLUKA, L. BREZINA, and E.D. KERR. 1978. Characterization of bean bacterial diseases and implications in control by breeding for resistance. *Fitopatol. Brasil.* 3:149–161.

SCHUSTER, M.L., D.P. COYNE, and E.D. KERR. 1965. New virulent strain of halo blight bacterium overwinters in the field. *Phytopathology* 55:1075. (Abstr.)

SCHUSTER, M.L., D.P. COYNE, and E.D. KERR. 1966. Pathogenicity tests of Nebraska isolates of halo blight bacteria in legumes. *In* D.P. Coyne (ed.) Annu. Rpt. Bean Improvement Coop. 9:35–36. (Mimeo.) Univ. of Nebraska, Lincoln.

SCHUSTER, M.L., D.P. COYNE, D.S. NULAND, and C.C. SMITH. 1979b. Transmission of *Xanthomonas phaseoli* and other bacterial species or varieties in seeds of tolerant bean *(Phaseolus vulgaris)* cultivars. *Plant Dis. Rptr.* 63:955–959.

SCHUSTER, M.L., D.P. COYNE, and K. SINGH. 1964. Population trends and movement of *Corynebacterium flaccumfaciens* var. *aurantiacum* in tolerant and susceptible beans. *Plant Dis. Rptr.* 48:823–827.

SCHUSTER, M.L., D.P. COYNE, and C.C. SMITH. 1979a. New strains of halo blight bacterium in Nebraska. *In* M.H. Dickson (ed.) Annu. Rpt. Bean Improvement Coop. 22:19–20. (Mimeo.) N.Y. Agr. Expt. Sta. Geneva, N.Y.

SCHUSTER, M.L. and L. HARRIS. 1957. Find new ground for your 1958 bean crop. *Univ. Nebr. Agr. Expt. Sta. Quart.* (summer).

SCHUSTER, M.L. and R.M. SAYRE. 1967. A coryneform bacterium induces purple-colored seed and leaf hypertrophy of *Phaseolus vulgaris* and other Leguminosae. *Phytopathology* 57:1064–1066.

SEQUEIRA, L., G. GAARD, and G.A. DEZOETEN. 1977. Interactions of bacteria and host cell walls: its relation to mechanisms of induced resistance. *Physiol. Plant Pathol.* 10:43–50.

SINCLAIR, J.B. and O.D. DHINGRA. 1975. An annotated bibliography of soybean diseases. INTSOY Ser. 7. Univ. Illinois, Urbana-Champaign.

SING, V.O. and M.N. SCHROTH. 1977. Bacteria-plant cell surface interaction: active immobilization of saprophytic bacteria in plant leaves. *Science* 197(4305):759–761.

SKOOG, H.A. 1952. Studies on host-parasite relations of bean varieties resistant and susceptible to *Pseudomonas phaseolicola* and toxin production by the parasite. *Phytopathology* 42:475. (Abstr.)

SMALL, B.C. and J.F. WORLEY. 1956. Evaluation of 2,3,5-triphenyl-tetrazolium chloride for obtaining pathogenic types from stock cultures of halo blight and common blight organism. *Plant Dis. Rptr.* 40:628.

TAYLOR, J.D., N.L. INNES, C.L. DUDLEY, and W.A. GRIFFITHS. 1978. Sources and inheritance of resistance to halo blight of *Phaseolus* beans. *Ann. Appl. Biol.* 90:101–110.

THOMAS, W.D., JR. and R.W. GRAHAM. 1952. Bacteria in apparently healthy pinto beans. *Phytopathology* 42:214.

VAKILI, N.G., W.J. KAISER, J. ENRIQUE PEREZ, and A. CORTES-MONLLOR. 1975. Bacterial blight of beans caused by two *Xanthomonas* pathogenic types from Puerto Rico. *Phytopathology* 65:401–403.

VALLADARES, N.E. 1979. Transgressive segregation for *Xanthomonas phaseoli* tolerance from crosses of *Phaseolus vulgaris* germ plasm and inheritance of leaf, external and internal pod reactions in a diallel cross. PhD dissertation, Univ. of Nebraska, Lincoln.

VALLADARES, N.E., D.P. COYNE, and M.L. SCHUSTER. 1977. Reaction of *Phaseolus* germ plasm to different strains of *Xanthomonas phaseoli* and *X. phaseoli* var. *fuscans* and transgressive segregation for increased tolerance. *HortScience* 12:391.

VALLADARES, N.E., D.P. COYNE, and M.L. SCHUSTER. 1979. Differential reaction of leaves and pods of *Phaseolus* germ plasm to strains of *Xanthomonas phaseoli* and transgressive segregation for tolerance from crosses of susceptible germ plasm. *J. Amer. Soc. Hort. Sci.* 104:648–655.

VIEIRA, C. 1973. Potentials of field beans and other food legumes in Latin America. *Centro Intern. de Agr. Tropical Series Seminars* 2E. p. 239–252. CIAT, Cali, Colombia.

WALKER, J.C. and P.N. PATEL. 1964. Inheritance of resistance to halo blight of bean. *Phytopathology* 54:952–954.

WELLER, D.M. and A.W. SAETTLER. 1977. Population studies of *Xanthomonas phaseoli* var. *fuscans* in field grown navy pea beans. *In* M.H. Dickson (ed.) Annu. Rpt. Bean Improvement Coop. 20:76–79. (Mimeo.) N.Y. Agr. Expt. Sta., Geneva, N.Y.

WELLMAN, F.L. 1972. Tropical American plant disease. The Scarecrow Press, Metuchen, N.J.

WILLIAMS, R.J. 1975. The control of cowpea diseases in the IITA grain legume improvement program. p. 139–147. *In* J. Bird and K. Maramorosch (eds.) Tropical diseases of legumes. Academic Press, New York.

YOSHII, K., G.E. GALVEZ, and G. ALVAREZ. 1976. Highly virulent strains of *Xanthomonas phaseoli* from Colombia. *Proc. Amer. Phytopathol. Soc.* 3: 299. (Abstr.)

ZAUMEYER, W.J. 1973. Goals and means for protecting *Phaseolus vulgaris* in the tropics. *In* P.H. Graham (ed.) Potentials of field beans and other food legumes in Latin America. CIAT Ser. Seminars *2E*, Feb. 26–Mar. 1, 1973. CIAT, Cali, Colombia.

ZAUMEYER, W.J. and J.P. MEINERS. 1975. Disease resistance in beans. *Annu. Rev. Phytopathol.* 13:313–334.

ZAUMEYER, W.J. and H.R. THOMAS. 1957. A monographic study of bean diseases and methods for their control. *USDA Tech. Bul.* 868.

3

Senescence and Postharvest Physiology of Cut Flowers—Part 2[1]

Abraham H. Halevy and Shimon Mayak
Department of Ornamental Horticulture,
The Hebrew University of Jerusalem, Rehovot, Israel

[1]We are grateful to Dr. F.T. Addicott, Dr. A.M. Kofranek, and Dr. R. Nichols for reviewing sections of this paper.

I. INTRODUCTION

The first part of this review was published in *Horticultural Reviews, Volume 1* (Halevy and Mayak 1979). Subjects covered were:

Measurement of Keeping Quality
Variability, Selection, and Breeding
Effect of Preharvest Conditions
Stage of Development and Time of Harvest
Ultrastructural, Biochemical, and Biophysical Changes during Petal Senescence
Changes in Pigmentation
Carbohydrate and Nitrogen Metobolism and the Role of Applied Sugars

The search of literature for Part 1 was terminated at the end of 1977, and of Part 2 in September, 1980. In some relevant cases we also cite here recent literature of subjects discussed in detail in Part 1.

II. WATER RELATIONS

The termination of vase life of many cut flowers is characterized by wilting, even though they are constantly held in water. Many studies, therefore, have been aimed at evaluation of the events leading to this phenomenon. In a gross analysis the following components of water balance can be distinguished: water uptake and transport, water loss, and the capacity of the flower tissue to retain its water. These physiological processes are interrelated, but for discussion each process will be examined separately.

After flowers are cut and placed in water they exhibit changes in fresh weight. Typically, cut flowers initially increase and subsequently decrease in fresh weight (Rogers 1973). Water uptake and water loss may fluctuate cyclically with an overall declining trend (Mayak *et al.* 1974; Carpenter and Rasmussen 1973; De Stigter (1980, 1981b). However, the balance of the two processes affects the typical fresh weight change described previously. In roses it has been shown that the loss of petal turgidity and fresh weight was preceded by a decreased rate of water uptake (Durkin and Kuc 1966; Burdett 1970). Since the continued water loss and the reduced water potential that developed in the flower were not realized in an increased water uptake, a stem blockage was suggested. When flowers were cut and allowed to age in water, there was a gradual decrease in water conductivity; but when flowers were allowed to develop and senesce on the plant, the rate of conductivity remained constant (Durkin and Kuc 1966; Mayak *et al.* 1974). Although a decline in conductivity seems to be a general phenomenon in many aging cut flowers, conductivity was not the cause for differences in longevity between two rose cultivars (Mayak *et al.* 1974). No correlation was found between the effect of several chemicals on water conductance in carnation stems and the effect of the same chemicals on longevity of the flowers (Camprubi and Fontarnau 1977). Some cut flowers such as tulips and narcissus do not show a reduction in water conductivity with time.

The reduction in stem conductivity is apparently caused by several factors. Microbial growth paralleled the increase in stem resistance to water flow (Larsen and Frolich 1969). Sterile water and germicides controlled microbial growth and partially decreased the resistance to water flow (Aarts 1957a; Larsen and Cromarty 1967; Burdett 1970; Marousky 1969; Van Meeteren 1978a,b). Therefore, microorganisms were considered to be one of the main causes of reduced water uptake by cut flowers. It was demonstrated (Aarts 1957a; Mayak and Accati-Garibaldi 1979; Accati *et al.* 1981) that metabolites produced by certain bacteria reduce longevity and water conductivity in carnations. Metabolites of some bacteria were more toxic than others.

Decrease in water conductivity ("stem blockage") is not wholly dependent on microorganisms' population. Cut roses held in sterile water had decreased rates of water uptake (Marousky 1969). Durkin and Kuc (1966) suggested that vascular blockage was the result of oxidative processes induced from harvesting injury. This idea has been supported by others (Aarts 1957; Burdett 1970; Marousky 1971).

Other studies made attempts to identify and describe the substances involved in stem plugging. Some authors have demonstrated occlusions of microbial origin (Aarts 1957a; Burdett 1970). Others, however, described occlusions as a gummy substance (Aarts 1957), pectinaceous or carbohydrate in nature (Burdett 1970; Parups and Molnar 1972), suggested to be composed of breakdown products of cell walls (Rasmussen and Carpenter 1974). This was supported by the fact that an increase in cellulase activity paralleled the decline in conductivity of cut roses and application of cellulase decreased water uptake (Mayak *et al.* 1974).

These seemingly conflicting findings seem to have been resolved by Lineberger and Steponkus' (1976) findings demonstrating the existence of two types of vascular occlusions in rose stems. Microbial occlusion was located at the base of the cut stem, while a second type, gum deposition, was found always above the solution level.

Rasmussen and Carpenter (1974) questioned the relevance of the occlusions to the increased resistance to water flow. In their study, only 4% of rose stem vessels were "blocked" when flowers wilted. This small percentage of blockage cannot explain the much higher reduction in water conduction measured in cut rose flowers (Durkin and Kuc 1966; Gilman and Steponkus 1972; Mayak *et al.* 1974).

Water uptake by cut chrysanthemum flowers varies with the season and the lignification of stem (Marousky 1973). Marousky suggested that products associated with lignification impaired water uptake. Stems harvested close to soil level were more lignified and absorbed less water than those cut higher on the plants. Addition of extracts from stem bases to the holding water also reduced water uptake by chrysanthemums and gerberas (Marousky 1973; Buys 1979). Similar mechanisms were suggested in roses, lilacs, and mimosa (Buys 1969; Moe 1975). It was suggested that some chemicals, mainly polyphenols, leach out of the stem bases and submerged leaves into the holding water, and are then oxidized to quinones. These oxidative products poison the cells and plug the xylem vessels (Buys 1969, 1979). Buys also suggested that the improved uptake by boiled water was due to the reduced oxygen level and lower oxidized metabolites in the vase. However, Durkin (1979a) demonstrated that gassing the holding water with nitrogen was even more detrimental than gassing it with air or oxygen.

Another source of obstruction to water flow may be very low concentrations of some macromolecules, which are known to be involved in causing some wilt diseases. Van Alfen and Allard-Turner (1979) found that picomole quantities of macromolecules such as dextran induced stem plugging and wilting in alfalfa. Similar compounds may be leached from the cut flower stems and submerged leaves and also may be part of bacterial metabolites, which were shown to induce wilting of several flowers (Aarts 1957a; Mayak *et al.* 1977). The reduction in "bent neck" and the increase in water conductivity in cut roses by a small column of ion exchange resin (Sacalis 1974) also may be related to the removal of metabolites in the holding water. De Stigter (1980) reached a similar conclusion by comparing water uptake and water loss of intact and cut roses in light and darkness.

Disruption of water columns in the stem vessels by air embolism long has been considered to be one of the factors causing water deficit (Crafts 1968). Air entering the base of the cut stems during shipment or storage may be a factor disturbing the rehydration of flowers. Durkin (1979a,b) showed that removing air from water by vacuum increased water flow rates through rose stem sections. He suggested "that flow of water through a vascular system where elements and pits are dissimilar in size will create instabilities which result in the appearance of gas bubbles, thus decreasing flow rates." Although the work was conducted in reference to short term hydration, it offers a new approach to this problem.

The reduction in water uptake, coupled with continuous transportation, leads to water deficit and reduced turgidity in the cut flowers. This may cause the stem to bend under the weight of the flower. The bending occurs often just below the flower, a phenomenon known as "bent neck" (Burdett 1970). Bending resistance depends on the development of secondary thickening and lignification of the vascular elements in the peduncle area subtending the flower head (Kohl 1961; Parups and Voisey 1976; Zieslin *et al.* 1978). The thickening of the vascular elements takes place at a relatively late stage of flower development, and in some cultivars occurs after the normal harvesting stage (Kohl 1961; De Jong 1978; Wilberg 1973; Zieslin *et al.* 1978).

Zieslin *et al.* (1978) analyzed factors of water balance involved in the "bent neck" of several rose cultivars. They concluded that three factors influence the water deficit in the neck tissue: (1) the transpiration rate, which is related to the presence of leaves on the flower shoot and the ability of the stomates to close in reaction to water deficit (Halevy *et al.* 1974a; Mayak *et al.* 1974); (2) the rate of water uptake and transport; and (3) the ability of the different organs on the cut flower shoot to compete for the water which may be in limited supply (Halevy 1972). It was demonstrated (Zieslin *et al.* 1978) that part of the water loss from the peduncle and leaves on the cut flower shoot is due to water absorption by the flower. Various rose cultivars vary in the intensity of each of these factors.

Factors which improve water balance or promote the mechanical strength of the stem should reduce bent neck. In roses, the bent neck tendency increased strongly at low contents of foliar K, in plots which received low K fertilizer (Johanssen 1979). This may be related to the known phenomenon that K-deficient plants have thinner walls and smaller and poorer xylem vessels, and are more sensitive to lodging and water stress (Humbert 1969). In Hippeastrum, however, the bent neck was associated with low Ca, but rather high K content (Carow and Röber 1979).

Improved water uptake and hydration of the cut flowers thus can be achieved by using deaerated and microorganism-free water. The addition of a wetting agent and acidification of the water greatly improve water uptake (Durkin 1981). The mode of action of the acidification in promoting hydration is not clear.

Cut flowers lose water from all tissues depending on environmental and internal factors. Simultaneous measurement of water uptake and loss was determined in some cut flowers (Burdett 1970; Mayak and Halevy 1974; Mayak *et al.* 1974; Halevy *et al.* 1974a; De Stigter 1980, 1981b; Van Meeteren 1978a). These two water parameters certainly influenced each other and their interaction determined the water balance.

In roses and other cut flowers it was demonstrated that after cutting, water loss decreased sharply due to stomatal closure (Mayak *et al.* 1974); water loss then paralleled water uptake (Burdett 1970; Halevy *et al.* 1974a), and finally an increase in water loss occurred before wilting (Halevy *et al.* 1974a; Hsiang 1951; Van Meeteren 1978a).

Loss of water depends largely on the pressure deficit (VPD), which is a function of temperature and water content. We have found, e.g., (unpublished) that at a VPD of 16,000 mm Hg, roses and carnations transpire 9 to 12 g/24 hours and 6 to 7 g/24 hours per flower, respectively, while at 6700 mm Hg the loss is 4 to 6 g and 3 to 4 g/24 hours per flower, respectively.

Light also promotes water loss (Aarts 1957a; Kofranek and Halevy 1972; Marousky 1969), presumably by causing stomatal opening. Carpenter and Rasmussen (1973) reported that roses held under constant light or alternating 12 hours of light and 12 hours of darkness lost 5 times more water than those held in complete darkness. De Stigter (1980) found that water uptake of cut roses held in darkness did not decline with time, but remained constant and similar to that of intact flowers.

Leaf removal reduced water loss by 78% in roses and only 60% in carnations, apparently because carnation stems contain a relatively large number of functional stomates (Carpenter and Rasmussen 1974).

Inducing stomatal closure by ABA (Kohl and Rundle 1972; Halevy *et al.* 1974a) or Al ions (Schnabl 1976; Schnabl and Ziegler 1974a) improved water balance and delayed wilting of roses. Mayak *et al.* (1974)

explained the difference in longevity between two cut rose cultivars on the basis of the differences in the efficiency of their stomatal closure mechanism. Similar results were found for other rose cultivars (Durso 1979).

In several cases, the use of film forming an antitranspirant has been tried on cut flowers. Beneficial effect was found with cut green foliage and anthuriums (Sherwood and Hamner 1948; Watson and Shirakawa 1967). However, Besemer (1977), Davenport *et al.* (1971), and Nichols (1967b) obtained variable results with other cut flowers.

Whenever the amount of transpiration exceeds absorption, a water deficit and wilting will develop. This deficit will be reflected in corresponding reduction in water potential of plant tissues (Barrs 1968). In intact roses, water potential remained constant throughout anthesis and senescence (Mayak *et al.* 1974), but a slight decrease in water potential was found in gerbera flowers (Van Meeteren 1979b). However, in cut flowers held in water, a constant water potential or a gradual decline was found the first day followed subsequently by a sharp decline. Similar patterns were found in cut roses (Durso 1979; Mayak *et al.* 1974), gerbera (Van Meeteren 1979b), and carnations (Acock and Nichols 1979).

The main reason for the deficit is apparently the resistance to water flow which develops in the stem. Indeed, preventing the increase in this resistance resulted in unchanged water potential in cut gerberas (Van Meeteren 1978b). However, even when the water potential remained constant, the water content of petals and turgidity of cut gerbera flowers declined with time (Van Meeteren 1978, 1979a,b), indicating changes in cellular physiology associated with senescence.

The water holding capacity of cut flowers changes as they age (Van Meeteren 1978b, 1979b; Acock and Nichols 1979). Changes in membrane properties of senescing petals have been reviewed recently (Halevy and Mayak 1979). These changes are associated with loss of semipermeability leading to increased ion and water leakage, and terminating in desiccation of the tissue. Factors which maintain membrane integrity also maintain cellular integrity and delay ion and water leakage. Acock and Nichols (1979) suggested that sugar's ability to delay senescence of cut carnation flowers is related to its ability to support cell metabolism and to maintain membrane integrity.

In *Iris germanica* flowers, a temporary water stress caused a decline in protein synthesis and a decrease in protein content. Rehydration induced a short and partial increase in protein content (Paulin 1972, 1975c). This indicated that water stress caused an irreversible metabolic change leading to senescence of cut flowers similar to that found in some intact herbaceous plants (Gates 1968). Temporary water stress did not induce a decrease in hexoses and sucrose contents in cut carnations (Le Masson and Paulin 1979).

Another factor in water relations is the concentration of osmotic components in the cell. In cut carnations a gradual decline in osmotic concentration with senescence was described (Acock and Nichols 1979; Mayak *et al.* 1978). In cut gerberas, however, the osmotic potential initially increased (i.e., the osmotic concentration decreased) and subsequently decreased. With intact flowers the osmotic potential was constant in the first days and then it decreased (Van Meeteren 1979b).

The effect of sugars on improving the water balance and delaying wilting of cut flowers was attributed in part to its contribution to the osmotic pool (Halevy 1976). This view is supported by the observation that some mineral ions also increased the cellular osmotic concentration and delayed flower senescence (Halevy 1976; Mayak *et al.* 1978; Van Meeteren 1980).

Van Meeteren (1980, 1981) emphasized the importance of pressure potential to flower longevity. He concluded that changes in pressure potential of gerbera plant cells can induce increased membrane permeability and ion leakage. He attributed the increased longevity in cut carnations grown under water stress (see Halevy and Mayak 1979, p. 210) and the promotion of longevity by osmotic components to their ability to increase pressure potential, and lower the water content.

The combined effects of the changes in membrane permeability and osmotic potential are a decrease in the ability of the cells to retain water. At this stage the decline in turgor pressure results in an irreversible wilting. In ephemeral flowers, where the changes are very rapid, infiltration of the intracellular spaces of the petals was observed (Horie 1962). In most other flowers, where a more gradual process occurs, tissue infiltration is not observable.

Senescence is the final phase of flower development which can be influenced or controlled by growth regulators. Water relation changes associated with senescence are therefore affected by growth regulators. This aspect of water relations is discussed in the following section of this review.

III. ETHYLENE AND FLOWER LONGEVITY

Members of each of the five groups of plant hormones have been implicated in the regulation of senescence. However, ethylene seems to have been dealt with most extensively, apparently because of its potential hazard as a pollutant. Most of the early reports are concerned with the flower damage caused by ethylene (Crocker and Knight 1908; Zimmerman *et al.* 1931; Hitchcock *et al.* 1932; Fischer 1950; Uota 1969;

Hasek *et al.* 1969; Barden and Hanan 1972), i.e., the premature senescence and wilting of the corolla. A wide range of flowers is, to various degrees, affected by ethylene, with some responding with typical symptoms.

Some striking examples include: (1) "sleepiness," the in-rolling of carnation and *Kalanchoe* petals (Nichols 1968a; Marousky and Harbaugh 1979); (2) fading and in-rolling of the corolla of *Ipomoea* (Kende and Baumgartner 1974); (3) fading and wilting of sepal tips in orchid (Akamine 1963) and induction of anthocyanin formation in both gynostemia and labella (Arditti 1979; Arditti *et al.* 1973); and (4) abscission of flowers and petals (see "Flower Bud and Petal Abscission"). These symptoms are well defined and can be used to assay the effects of ethylene. In fact, the in-rolling of the midrib of *Ipomoea's* corolla is used to quantitatively measure the response to ethylene (Kende and Baumgartner 1974).

Ethylene applied to plant tissue can catalyze the synthesis of ethylene. This phenomenon is termed autocatalysis and is common to both flower senescence (Burg and Dijkman 1967; Kende and Baumgartner 1974; Nichols 1968a) and ripening of climacteric fruits (Sacher 1973).

In many cases the visual symptoms induced by ethylene and the autocatalytic ethylene production seem to be coupled. However, in some situations the two events are not coupled. Exposure of mature carnation flowers to ethylene resulted in a dual response, i.e., in-rolling of petals and increased synthesis of ethylene. In young carnation buds, exposure to ethylene did not stimulate ethylene production, but a promotion of petal elongation was observed (Camprubi and Nichols 1978, 1979; Nowak and Rudnicki 1979). Similarly, exposure of mature open *Ipomoea* flowers to ethylene resulted in the dual response. However, younger buds (a day before opening) responded only with the in-rolling of the corolla without a concomitant increase in ethylene synthesis (Kende and Hanson 1976, 1977).

Carbon dioxide at 4%, a competitive inhibitor of ethylene, completely prevented the development of in-rolling of carnation petal tissue associated with naturally occurring changes in ethylene production (Smith and Parker 1966; Nichols 1968a; Uota 1969; Mayak and Dilley 1976b). By contrast, 4% CO_2 delayed but did not prevent the in-rolling of the corolla in *Ipomoea* flowers (Kende and Hanson 1976). In *Tradescantia* petals, CO_2 delayed but did not prevent the rise in ethylene evolution and pigment leakage (Suttle and Kende 1978). This difference in the responses indicates that the responses regulated by ethylene are complex. In *Ipomoea* flowers, the seemingly well-correlated events, ethylene-induced ethylene synthesis and the in-rolling of the corolla, are not conditionally coupled. A somewhat different situation exists in carnation: for in-rolling to occur, ethylene is a prerequisite.

A wide assortment of flowers produce ethylene (Fischer 1950; Hall and Forsyth 1967) and have the capacity to metabolize substrates into ethylene (Phan 1970). In many studies, ethylene emanation by vegetative or flower tissues was thought to reflect changes in synthesis. However, this is not necessarily so, because the emanation also might be influenced by changes either in the capacity of ethylene binding sites (Jerie *et al.* 1978) or of ethylene metabolism rates (Beyer 1978). Although this interpretation of ethylene measurements helps to explain its possible involvement in the regulation of senescence processes, it should be regarded as tentative.

The time course of ethylene production follows a typical profile composed of three distinct phases: (1) a low steady rate, (2) an accelerated rise to maximum emanation, and (3) a last phase in which production is declining. Various events associated with senescence can be studied with reference to these phases. Associated with the second phase, a rise in CO_2 evolution resembling the climacteric respiration of fruits has been demonstrated (Maxie *et al.* 1973).

Usually, the visual symptoms of the effects of ethylene can be distinguished at the end of the second phase. Thus, the onset of the second phase is important, because it signals the terminal stage of senescence. The transition to the second phase is well correlated with longevity of rose flowers. The second phase occurred earlier in short-lived roses than in long-lived flowers (Mayak and Halevy 1972).

The onset of the second phase may be induced by various means, such as exposure to ethylene (Nichols 1968a), pollination (Arditti 1979; Hall and Forsyth 1967; Burg and Dijkman 1967; Nichols 1977), and abscisic acid (Mayak and Dilley 1976b). By contrast, it can be delayed by CO_2 (Smith and Parker 1966; Nichols 1968b; Uota 1969; Mayak and Dilley 1976b), low concentration of oxygen (Nichols 1968a), a hypobaric pressure environment (Mayak and Dilley 1976b), or inhibitors of ethylene synthesis or action.

Senescence includes processes leading to cell disorganization. A possibility that can be considered is that change in the degree of organization could occur either spontaneously during senescence or be initiated by a plant hormone, such as ethylene (McGlasson 1970). Cell organization is generally maintained by spatial separation of reactants by partially permeable protoplasmic membranes. Matile and colleagues (Matile and Winkenbach 1971; Winkenbach and Matile 1970) demonstrated cell membrane disintegration, especially the tonoplast. These events lead to uncontrolled mixing of vacuole contents with the cytoplasm (for review see Halevy and Mayak 1979, Matile 1978).

Kende and Baumgartner (1974) have proposed a model based on

changes in compartmentation to account for autocatalytic ethylene synthesis: by increasing the permeability of the tonoplast, ethylene could enhance the flow of substrates from the vacuole to the cytoplasm, where an ethylene-generating system is located. This model is supported by experiments in which flower tissue was preloaded with radioactive tracers, representing either the vacuole or cytoplasm content, then exposed to ethylene and subjected to compartmental analysis (Macrobbie 1971). The results indicate that one of the first changes to take place in response to ethylene was an enhanced efflux across the tonoplast (Hanson and Kende 1975; Mayak *et al.* 1977c).

There is some experimental evidence, however, that does not support this model. A rise in ethylene production was measured in *Tradescantia* petals without corresponding increase in pigment leakage, a natural indicator of changes in permeability (Suttle and Kende 1978). Thus, permeability changes are not a precondition for the rise in ethylene production. Careful examination of the results of Mayak *et al.* in carnations (1977c, Fig. 4) also reveals that the rise in ethylene production started earlier than the onset of changes in the permeability of the tonoplast. Recently, Suttle and Kende (1980) stated that there is a lag of about 2 hours between ethylene application and increased permeability in *Tradescantia* petals. The increase in permeability is accompanied by a massive loss of phospholipids as was found earlier is senescing rose petals (Borochov *et al.* 1978). They concluded that ethylene does not affect membrane integrity directly, but rather indirectly through its effect on cellular metabolism and phospholipids loss.

Numerous studies on the biosynthesis of ethylene have been conducted, at which time several ethylene-forming systems were evaluated. Methionine is generally accepted as a common precursor of ethylene in the tissue of higher plants (Owens *et al.* 1971; Yang 1974; Hanson and Kende 1976; Varner and Ho 1976; Lieberman 1979). The sequence of the pathway for ethylene biosynthesis recently was demonstrated (Adams and Yang 1979) to be methionine → S′-adenosylmethionine → (SAM) → 1-aminocyclopropane-1-carboxylic acid (ACC) → ethylene.

An important question is, "Why is the model of ethylene-induced ethylene synthesis not operative during most of the young and early maturation phases of flower petal tissue, while there is a constant low emanation of ethylene, yet the transition into the second phase of ethylene emanation does not occur?" It seems that an additional factor is needed for the rise in ethylene evolution to occur. Another possibility is that some components like SAM or ACC or attendant enzymes may be deficient at this stage.

The presence of separate storage and metabolic amino acid pools has been suggested in higher plants (Oaks and Bidwell 1970). Although demonstrated clearly in yeast (Wiemken and Nurse 1973), the vacuoles of higher plants probably serve as a storage place for amino acids, including methionine.

Van der Westhuizen and de Swardt (1978) conducted a study on the changes in free amino acids in petals of cut carnations. A rise in the level of methionine was found at about the physiological stage associated with commencement of the rise in ethylene production. These results suggest that the substrate for ethylene becomes available only at a specific physiological stage.

When young flower tissue of *Ipomoea* (Kende and Hanson 1976) is cut or bruised, ethylene production begins to rise soon afterward. Later, after relatively short periods, ethylene production declines without the development of typical senescent symptoms. In carnation flowers (Smith *et al.* 1964) damaging fungal infection advances the typical ethylene burst at senescence.

Exposure of mature carnation flower tissue to ethylene generally hastens the onset of senescence and the development of typical aging symptoms. These symptoms are irreversible and last even when the ethylene was ventilated away (Mayak and Kofranek 1976). Young flowers develop the same symptoms upon exposure to ethylene but quickly recover and resume their natural course of development when ethylene is vented away. Hence, the tissues vary in their response to ethylene. Young flowers are not responsive, but mature flowers respond and transition into the second phase occurs. Mor and Reid (1981) found that petals detached from mature flowers senesced and showed the climacteric rise in ethylene simultaneously with the parent flowers. However, petals detached from young flowers at the day of flower opening senesced much later and produced much less ethylene than did detached petals of older flowers.

It seems that while ethylene content in flower tissue is not markedly augmented until the senescence phase, sufficient gas is present during the presenescence phase. The autostimulation of ethylene production occurring during senescence will be initiated once the sensitivity to ethylene has changed to respond to the existing ethylene. The sensitivity to ethylene can be the result of an intricate, complex interaction among internal factors such as plant hormones (Sacher 1973) like abscisic acid (Mayak and Dilley 1976a), carbohydrate reserve (Mayak and Dilley 1976a), and osmotic concentration of the petal tissue (Mayek *et al.* 1978). The flower is composed of various organs—petals, sepal, style, ovary, etc.—each at a different stage of its development. The compounded result of processes and internal interactions, occurring at each phase of development, may affect the sensitivity to ethylene. The responding system in each organ gradually becomes operative at different rates and times as the flower organs mature.

A plausible possibility, though not yet demonstrated in flowers, is that the compounded result leads to a higher oxidation state of the tissue, which may be in the form of accumulation of peroxides (Brennan and Frenkel 1977) or in lipoxygenase activity, resulting in lipid hydroperoxides. A substantial rise in lipoxygenase activity was demonstrated in tomatoes prior to an upsurge in hydroperoxide, which in turn precedes ethylene biosynthesis (Frenkel *et al.* 1976). Ethylene synthesis could be initiated from methional in the presence of linolenate and lipoxygenase (Mapson and Wardale 1968). These results suggest that lipid hydroperoxides may be required for the methionine-dependent synthesis of ethylene (Beauchamp and Fridovich 1970; Eskin and Grossman 1977). However, further research is required to establish these enzymes' role in the production of ethylene by aging flowers. The oxidation of ethylene, demonstrated also in carnations, was suggested as an important aspect of ethylene action (Beyer 1977, 1978, 1979).

The entire process of ethylene-induced senescence may be under programmed regulation or control. However, the responding system may be affected by environmental factors. High temperature (Maxie *et al.* 1973; Marousky and Harbaugh 1979), water stress to cut flowers, and cold storage used in the practice of handling flowers (Mayak and Kofranek 1976) resulted in higher sensitivity to ethylene. However, water stress, imposed by low irrigation of carnation plants, rendered the cut flowers less sensitive to ethylene (Mayak and Kofranek 1976). We assume that during growth and development of the flower osmotic adaption probably occurred, rendering the flower less ethylene-sensitive. This assumption is supported by experiments with cut flowers in which treatments with sugars and some mineral ions induced osmotic adjustment of cut carnation and chrysanthemum flowers, and also decreased their sensitivity to ethylene (Halevy 1976; Mayak *et al.* 1978).

Generally, the responding system gradually becomes more sensitive as the flower organs mature. The petals may respond to ethylene generated in the style. The latter was demonstrated to have high capacity for ethylene synthesis (Burg and Dijkman 1967; Hall and Forsyth 1967; Lipe and Morgan 1973; Nichols 1977). Burg and Dijkman (1967), using *Vanda* orchids in their studies, proposed that following pollination a transfer of auxin from the pollen to the stigma takes place and is spread through the column. The auxin induces ethylene synthesis in the cells of the column. Thereafter, the increased synthesis of ethylene, triggered by ethylene, is spread sequentially to other flower parts with the net result of petal fading. A detailed review of the post-pollination phenomena in orchids has been published recently (Arditti 1979). A similar mechanism was proposed to explain the wilting of carnation occurring after pollination (Nichols 1977). Here the style, through its endogenous ethylene production, is thought to trigger petal wilting. The production of ethylene by the style accounts for 40 to 50% of the flower's total ethylene production. The petals produce a similar quantity, with most of the production in the elongated base of the petal (Nichols 1977). Gilissen (1977), using petunia flowers, reported that the penetrating pollen tubes activate the stylar tissue which then leads to the wilting of the flower. The rate of wilting depends on the number of viable pollen grains penetrating the stigma. Similar results have been found in *Digitalis* (Stead and Moore 1979), demonstrating also that a stimulus was transmitted from the pollinated stigma through the style and the ovary stimulating petal abscission. The stigmas of blueberry have been reported to produce ethylene following pollination (Hall and Forsyth 1967). In staminate annona flowers, however, most of the ethylene is produced by the anthers (Blumenfeld 1975).

In conjunction with petal wilting, rapid growth of the ovary occurs (Nichols 1968a, 1975). Nichols (1976b) also found that ethylene promotes an accumulation of sugars and inorganic materials in the ovary accompanied by loss of fresh and dry weight of the petals. The swelling of the ovary appears to result from the enlargement of existing cells. He suggested that by induction of petal wilting ethylene promotes mobilization of carbohydrates from the petals and stem, providing for the development of the ovary. Mor *et al.* (1980a) showed, however, that although the start of ovary growth precedes wilting, the ovary does not control carnation petal senescence, since the removal of the ovary did not alter the time of petal senescence. The ovary seems to compete with the petals for metabolites, and removal of the petals promotes ovary growth. Silver thiosulfate (STS) delayed petal senescence and ovary growth apparently by strengthening the sink of the petals.

The hormonal regulation of a physiological process may be accomplished by a proportionate mechanism (Varner and Ho 1976). Thus, the process is enhanced as concentration increases and diminishes as the concentration drops. Accordingly, the changes in plant hormone concentrations may be interpreted as being signals controlling the processes. Indeed, the tissue response is proportional to the log of the ethylene concentration present. The level of ethylene in the tissue, as changed with time, as well as being part of a controlling signal, is determined by the rate of biosynthesis, metabolism, and its diffusion along a concentration gradient into the surrounding tissue and into the atmosphere.

Since ethylene is usually the major chemical environmental factor affecting flower senescence, it is important to locate and monitor the sources of ethylene in the environment of the cut flowers (for a review see Hasek *et al.* 1969).

In an industrial urban area automobiles account for 90% of the ethylene released to the atmosphere (Abeles *et al.* 1971). Highest level of ethylene corresponded to intensive motor activity (Gordon *et al.* 1968). Ethylene also varies with the season. Using "dry-sepal" orchid damage as a criterion, highest levels of ethylene were found during the fall and winter when photochemical degradation is reduced (James 1963). The winds also influence the level of ethylene, which emphasizes the importance of natural or artificial ventilation of closed spaces in which the flower is located, such as greenhouses and packing and storage rooms (Staby and Thompson 1978). Polyethylene manufacturing plants are a rich source of ethylene. A concentration of 3 μl/liter was found in the immediate vicinity of such a plant, with corresponding damage to cotton plants grown there (Gordon *et al.* 1968). These levels of ethylene are much above the concentration known to affect the quality of flowers. Levels of up to 1 to 1.5 μl/liter were found in the atmosphere of an air terminal in Israel, with a parallel high incidence of damage to carnation flowers. The source of the ethylene, gas-operated forklifts, was located and removed, resulting in a lower level of ethylene (Mayak *et al.* 1977b).

In addition to internal combustion engines, plant material also contributes to elevated ethylene levels (Rogers 1973). The major sources are ripe fruits (Wintz 1954), mechanically damaged leaf tissue (Denny and Miller 1935), and diseased plants including flowers, such as those infected by Botrytis (Dimock and Baker 1950). The source of ethylene can be the tissue infected by the pathogens as well as the pathogens themselves (Primrose 1979; Swanson *et al.* 1979).

Reduction of ethylene level may be accomplished by use of ethylene scrubbers, which act either by absorbing the gas or by a combination of absorbing and reacting. Brominated activated charcoal which combines a reactant with adsorber has been used widely in controlled atmosphere storage of fruits (Southwick and Smock 1943; Forsyth *et al.* 1967). It is corrosive and too expensive to be practical (Smock 1979). The use of potassium permanganate to oxidize ethylene has been reported also. The chemical is adsorbed on a carrier like silica gel, celite, or vermiculite (Scott *et al.* 1970). The commercial preparation Purafil is potassium permanganate coated on a granular form of aluminum silicate. Liu (1970) reported high removal efficiency of ethylene by Purafil. Another absorbent found to be effective on cut flowers is mercury perchlorate (Parups 1969).

The absorbents are marketed as fillers for traps equipped with pumps, in large blanket sheets for hanging in storage rooms, and as small bags for inserting in flower containers. We have evaluated the efficiency of the blanket and have found that the area required to lower the ethylene level in cold storage, from 1 ppm to a safe level, is so large that it renders this method impractical (Peles *et al.* 1977). This seems to be due mainly to the relatively small air circulation through the blanket as compared to the storage air volume. We also tested the efficiency of the scrubber bags on ethylene levels in flower containers (unpublished), and found the effects to be negligible at a distance of 6 cm away from the bag, presumably because there is no air movement through the scrubber bag when it is in the box. Adding purafil to the flower cartons during shipment had no effect on longevity of narcissus and iris (Anon. 1979).

The use of an ultraviolet scubber which oxidizes ethylene by producing ozone was suggested by Scott and Wills (1973). A high temperature catalyst, which oxidizes light hydrocarbons including ethylene, was evaluated by Eastwell *et al.* (1978). These methods are not yet used commercially.

Another approach to reduce ethylene damage is the use of chemicals which block ethylene synthesis or action (see "Ethylene Inhibitors," p. 91).

Rhizobitoxine and some of its analogues inhibited the synthesis of ethylene from methionine (Lieberman 1979). They also reduced ethylene production and extended longevity of carnations (Baker *et al.* 1977) and snapdragons (Wang *et al.* 1977), which are sensitive to ethylene. However, rhizobitoxine analogues also extended longevity of daffodil, iris, and chrysanthemums (Wang and Baker 1979), which are considered not to be sensitive to ethylene. In contrast, a rhizobitoxine analogue reduced ethylene production by roses, which respond to ethylene at higher concentration (Burg 1962; Mayak and Halevy 1972); but longevity was unaffected (Wang and Baker 1979). Recently Yu *et al.* (1979) and Amrheim and Wenkler (1979) reported that aminooxyacetic acid (AOA) is an inhibitor of ACC synthase, a key enzyme in ethylene biosynthesis (Yang 1980). Addition of AOA to the vase solution of carnations greatly extended longevity and suppressed the climacteric rise in respiration and ACC and ethylene production (Brown and Mayak 1980; Bufler *et al.* 1980; Fujino *et al.* 1981; Wang and Baker 1980).

Several polyamines, e.g., putrescine, cadaverine, spermidine, and spermine, which were reported to retard the senescence of leaf protoplasts (Altman *et al.* 1977), also inhibited ethylene production by senescing petals of *Tradescantia* (Suttle 1980) and mildly extended longevity of cut carnations (Wang and Baker 1980).

Beyer (1976a,b) reported that silver ions specifically and effectively inhibited ethylene action in plants. Ag affected several known ethylene actions including the senescence of cattleya orchids. Following this, it was shown (Halevy and Kofranek 1977) that $AgNO_3$ applied directly to the flowers by spray or dip delayed senescence of carnation flowers and prevented the damage of ethylene generated by ethephon. Treated flowers failed to show the in-rolling symptom typical of ethylene effect. Only direct treatment of the flowers with $AgNO_3$ was effective, since $AgNO_3$

applied through the stem base moved only a short distance in the stem and did not reach the flowers (Kofranek and Paul 1972, 1974).

Unlike other silver salts, silver thiosulfate (STS) is very mobile, and travels from the base of the cut stem to the flower, blocking ethylene action and extending longevity of carnations (Veen and Van de Geijn 1978). Beyer (1976a,b) claimed that Ag is an inhibitor of ethylene action but not synthesis. This is supported by findings that senescence and in-rolling of *Ipomoea* (Beutelmann and Kende 1977) and carnation (Ronen 1979) flowers were inhibited by $AgNO_3$ spray but ethylene evolution was not. Veen (1979a,b), however, reported that STS applied through the cut stem inhibited both ethylene action and synthesis. These conflicting results perhaps may be explained on the basis of the better penetrating ability of STS to the inner cells' compartments and the sites of ethylene biosynthesis.

Ethylene oxide was reported (Lieberman *et al.* 1964) to be an inhibitor of ethylene production, and to delay the senescence of cut roses and carnations (Asen and Lieberman 1963). However, detailed evaluation employing seven rose cultivars showed that in no case was longevity of the flowers extended, and in most cases opening and growth of the flowers were inhibited (Ben-Yehoshua *et al.* 1966). In fact, Jerie and Hall (1978) reported ethylene oxide as a major metabolite of ethylene, and ethylene action was shown to be related to its metabolism and oxidation (Beyer 1977, 1978, 1979).

Some other chemicals acting as ethylene inhibitors are listed under "Ethylene Inhibitors" (p. 91). Nichols (1966) reported that ethanol extended carnation longevity. It also inhibited ethylene synthesis, if applied before the onset of rise in ethylene production (Heins 1980). Heins (1980) did not detect any ethanol in untreated flowers. Recent results from our laboratory have demonstrated ethanol accumulation in carnation flowers exposed to a mixture of ethylene and CO_2, which may indicate the involvement of ethanol in the antagonistic effect of CO_2 to ethylene.

The inhibition of ethylene production under anaerobic or low-oxygen conditions has been observed by many researchers. The reduction of oxygen level is apparently one of the main effects of low pressure storage (see "Storage," p. 96). It was found recently (Adams and Yang 1979; Yang 1980) that oxygen participates directly in the conversion of ACC to ethylene.

CO_2 has long been known to antagonize ethylene action (Burg 1962). It also delays senescence of cut carnations and the onset of the rise in ethylene production (Mayak and Dilley 1976b). The antagonistic effect of high CO_2 levels on ethylene may account for some of the beneficial effects of modified atmosphere in storage and transit (see "Storage" and "Transport").

IV. INVOLVEMENT OF OTHER GROWTH REGULATORS IN FLOWER SENESCENCE

Other plant hormones are also involved in the regulation of petal senescence and they do seem to abide to the principle of proportionate mechanism (Varner and Ho 1976). The effect of growth regulators in solutions on cut flowers is described below.

The level of cytokinins in rose petals decreased as the flower aged and the level was lower in a short-lived cultivar than in a long-lived cultivar (Mayak and Halevy 1970). In gerbera, however, Van Meeteren and Van Gelder (1980) found no correlation between cytokinin activity of petal and longevity of three cultivars differing in their potential vase life. Results with external applications of cytokinins which delay senescence of various flowers (Heide and Oydvin 1969; Mayak and Halevy 1970; Mayak and Dilley 1976a) support the possibility that the diminishing of internal levels of the growth regulator is associated with senescence processes in plants. Further support for this view may be drawn from experiments (Mayak and Halevy 1974) in which applied kinetin slowed down the reduction in dry weight of aging rose petals. In addition, cytokinins reduced water stress damage in carnation flowers (Paulin and Muloway 1979) and enhanced water uptake in cut roses (Mayak and Halevy 1974). It was postulated that this was achieved by maintenance of cell integrity. This is supported by the finding that benzyl amino purine (BA) delayed the decrease in water content and the increase in ion leakage associated with senescence of gerbera flowers (Van Meeteren 1979a). Later Van Meeteren and Van Gelder (1980) found, however, that the kinetics of endogenous cytokinin activity with time was not correlated with the changes in membrane permeability during aging.

In *Cymbidium* flowers, pollination as well as the removal of the pollinia and anther caps induces anthocyanin accumulation in the lips and columns, followed by senescence of petals and sepals (Arditti 1979). These changes were correlated with a decrease in endogenous cytokinins in all parts of the flower after emasculation. The cytokinin level in the anther was much higher than in the rest of the flower, which may point to the regulating effect of this organ (Van Staden 1979).

A rise in cytokinin levels was found in the ovary of senescing carnation flowers (Van Staden and Dimalla 1980). The rise was prevented in flowers treated with silver thiosulfate, which retarded senescence. This rise in cytokinin was suggested to be related to the mechanism which mobilizes metabolites from the senescing flower parts to the developing ovary.

Cytokinins' mode of action in delaying senescence is not known. Recently, cytokinins were shown to lower the lipoxygenase activity in senescing tissue, thus decreasing the peroxidation of membrane lipids.

There are also indications that cytokinins may act also as free radical scavengers in cells (Leshem *et al.* 1979).

The involvement of auxin and gibberellin in the control of flower senescence is not clear. Gibberellins (GA) have been extracted from carnation flowers and have been shown to play a role in the control of carnation petal growth (Jeffcoat and Harris 1972). Application of GA to isolated carnation petals delayed their senescence (Garrod and Harris 1978). However, treatment to whole flowers had little or no effect on longevity but caused some petal enlargement (Nichols 1968b). GA also prolonged the life of intact flowers of Easter lilies (Kelley and Schlamp 1964), but had little or no effect on orchids (Arditti 1979) and most other flowers tested (see "Growth Regulators," p. 92).

Gilbart and Sink (1970, 1971) assigned a central role to auxin in the control of senescence in poinsettia flowers. Application of IAA delayed senescence and substituted for the blade in inhibiting abscission. The auxin level decreased with age in two poinsettia cultivars, but declined faster in the short-lived one. Also, the activity in IAA-oxidase and the level of hydrogen peroxide increased with aging. Gilbart and Sink hypothesized that auxin delays senescence by inducing the synthesis of peroxidase, which then prevents the accumulation of free peroxide associated with aging. NAA delayed petal senescence of *Nicotiana offinis* flowers cultured *in vitro*, while cytokinins alone had little or no effect (Deaton *et al.* 1980). Auxin induces post-pollination phenomena in orchids, e.g., folding, wilting, and anthocyanin synthesis (Arditti 1979). Some of these effects may be mediated by ethylene as discussed previously.

The role attributed to abscisic acid (ABA) in petals is that of regulating the process of senescence. Endogenous level of ABA increased in roses as petals senesced, and the level was higher in a short-lived cultivar than in a long-lived cultivar (Mayak and Halevy 1972). Endogenous ABA was correlated with senescence and abscission of *Hibiscus* flowers (Swanson *et al.* 1975). Exogenous application accelerated senescence of carnations (Mayak and Dilley 1976b) as well as of roses (Mayak and Halevy 1972). The development of senescence was reflected in changes such as red petal bluing, decrease in protein content, and increase in ribonuclease activity (Borochov *et al.* 1976b; Mayak and Halevy 1971; Halevy *et al.* 1974a). *De novo* synthesis of this enzyme is probably involved in its activity increase during the final stages of petal senescence of *Ipomoea* (Baumgartner *et al.* 1975, 1977). ABA induced some, but not all, of the post-pollination symptoms in orchids (Arditti 1979). However, when leaves were attached to the flowering stem of a rose, opposite effects were observed. By causing stomatal closure on the leaves, ABA reduced water loss from the flowering shoot, thereby delaying senescence in the flower (Halevy *et al.* 1974a).

Although not always studied in the same flower, each plant hormone was demonstrated to have a potential role in the hormonal controlling system. It is generally accepted that control is exerted through a balance among plant hormones interacting with each other and with other internal factors. Combined application of BA and auxin to daffodils was more effective in delaying senescence and climacteric-like respiration (Ballantyne 1965, 1966) than was the additive effect of the two growth regulators.

Kende and Hanson (1976) reported that BA delayed ethylene production by isolated rib segments of *Ipomoea* flowers. Similar results were reported with carnation flowers, and their responsiveness to ethylene also was reduced (Eisinger 1977). A plant hormone thus may affect the action or the level of another plant hormone, thereby having a more intensive effect on the hormonal balance. Evidence for this kind of interaction comes from studies with carnations. ABA induced an earlier increase in ethylene production (Mayak and Dilley 1976b). We hypothesized (Mayak and Halevy 1980) that whenever a rise in ABA is induced ABA will accelerate developmental processes associated with aging. It increased the sensitivity to ethylene, enabling the tissue to respond to the existing ethylene level. Transition into the phase of ethylene autostimulation may occur at an earlier stage. ABA may also be involved in later stages of senescence; the permeability of the tonoplast (Hanson and Kende 1975; Mayak *et al.* 1977c) increased, leading to cell disorganization accompanied by a reduction in water uptake (Mayak *et al.* 1977c) and plausible development of water or ion stress. The well-established stress response, i.e., increase in the level of ABA (Livne and Vaadia 1972), might follow soon. In this respect the rise in the level of ABA, in response to ethylene as observed with cut rose flowers (Mayak and Halevy 1972), can be interpreted as a secondary result via an effect on membrane permeability and water stress imposed by ethylene.

Membrane modification was suggested to be controlled by the cytokinin/ABA balance (Itai and Benzioni 1976). Events reflecting the effect of ethylene on the level of ABA are mediated by membrane modifications. It will be of interest to study whether or not the reciprocal effects of cytokinin and ABA on each other's activity, suggested by Itai and Benzioni (1976) to be a widely occurring phenomenon in plants, are also mediated by membrane modifications. The observation that kinetin delays wilting of cut rose flowers by protecting cell integrity (Mayak and Halevy 1974) supports this possibility. It is conceivable that if cell integrity is maintained, the development of water stress will be delayed and abscisic acid will be kept at a low level.

A tentative model for the various endogenous factors interacting in the control of flower senescence has been presented recently (Mayak and Halevy 1980).

V. CHEMICAL SOLUTIONS FOR PROMOTING KEEPING QUALITY

The use of preservative solution to promote the quality and prolong the life of cut flowers has been known for many years. Flower preservatives are composed mainly of sugar and germicides and sometimes include other ingredients. Our review will describe studies with known chemicals and not with those commercial preparations in which the ingredients are not revealed.

Early work emphasized the importance of flower preservative use by all those handling flowers, i.e., the grower, wholesaler, retailer, and consumer. Little or no benefit was found by only the grower treating flowers (Clausen and Kristensen 1974; Parvin and Krone 1963). However, in these studies the same solution was used at all stages of flower handling, while different solutions may have been required for optimal results for different purposes, flower species, and handling levels (Halevy and Mayak 1974a,b, 1979).

Four main uses of water and chemical solutions are common: conditioning, pulsing, bud-opening, and holding.

A. Conditioning or Hardening

The main purpose of this treatment is to restore the turgidity of the cut flowers by saturating them with water after they have suffered from water stress during handling in the field, greenhouse, or grading room or during storage and transport. Conditioning normally was done initially with warm water at room temperature and then overnight in the cooler (Rogers 1963). It was found, however, that during the first 4 to 8 hours, water uptake by rose and carnation flowers was not significantly different at 3°C and 20°C (Sytsema 1975). It is therefore advisable to condition flowers in the cooler throughout. When flowers are wilted, turgidity can be regained by immersing the entire flower in water for 1 hour before conditioning as above.

Conditioning is preferably done with deionized water containing a germicide, but no sugar (Lancaster 1975a). Hydration is considerably promoted when water is deaerated or acidified or when a wetting agent (at 0.1 to 0.01%) is added (Durkin 1979b, 1981).

B. Pulsing or Loading

This is a preshipment short-term treatment by growers or shippers, the effect of which should last for the entire shelf-life of the flower, even when the flowers are held in water (Halevy 1976; Halevy and Mayak 1974a,b). Specific pulsing formulations have been developed for different flowers and sometimes even for different cultivars (Halevy *et al.* 1978a; Halevy and Mayak 1974a; Kofranek and Halevy 1972). The main ingredient of the various pulsing solutions is sucrose, which is often used in concentrations that are several times higher than those used in preservative formulations. However, the optimum concentration varies from 20% sucrose and higher for gladiolus (Mayak *et al.* 1973), gerbera (Van Meeteren 1981), and *Eremurus* (Krebs and Zimmer 1977), 10% for carnations (Halevy and Mayak 1974a; Nichols 1974), *Strelitzia* (Halevy *et al.* 1978b), and *Gypsophila* (Farnham *et al.* 1978), to 2 to 5% for roses (Halevy *et al.* 1978a; Halevy and Mayak 1974b), chrysanthemum (Posner *et al.* 1979), and *Leucospermum* (Criley *et al.* 1978).

The duration of treatment as well as temperature and lighting conditions during pulsing is also important for optimum effects. These have been specified for a number of different flowers (Halevy and Mayak 1974a,b; Kofranek and Halevy 1972; Mayak *et al.* 1973; Monnet and Paulin 1975). Pulsing time is generally between 12 hours and 24 hours, with light of ca. 1000 lux and temperature of 20° to 27°C. Relative humidities between 35% and 100% had no effect on the efficiency of carnation pulsing (Neumannova and Paulin 1974). Some interactions of pulsing duration, temperature, and sucrose concentration were found. With shorter pulsing time and higher temperatures, the optimal sucrose concentration was higher (Borochov *et al.* 1975d; Bravdo *et al.* 1974; Peles *et al.* 1974).

In roses pulsing at high temperatures will cause excessive opening of the flower buds during the treatment. The recommended procedure is, therefore, pulsing for 3 to 4 hours at 20°C followed by 12 to 16 hours in the cooler (Halevy *et al.* 1978a).

Pulsing was found to be of great value in prolonging life, promoting opening, and improving the color and size of petals in gladiolus (Kofranek and Halevy 1976; Mayak *et al.* 1974), miniature carnations (Borochov *et al.* 1975d; Halevy and Mayak 1974a), standard carnations (Farnham *et al.* 1971; Halevy *et al.* 1978a), chrysanthemums (Posner *et al.* 1980), and roses (Halevy and Mayak 1974a,b; Halevy *et al.* 1978a).

If the optimal procedure (time, concentration, temperature, light) for pulsing is not used, little or no effect is found and sometimes damage even occurs. For example, Farnham *et al.* (1978b, 1979) found no effect of pulsing roses prior to truck shipment, but they pulsed their flowers for only 4 hours. Pulsing at too high a temperature or with too high sucrose concentration may damage the flowers or leaves.

Impregnation of the cut bases of flowers with high concentration (ca. 1000 ppm) of $AgNO_3$ or other silver salts for 5 to 10 minutes greatly promoted longevity of several flowers (Kofranek and Paul 1974). The silver travels only a short distance in the stem; therefore, the bases of treated flowers should not be re-cut after treatment. Silver impregnation can be followed by sugar pulsing soon after treatment or even after a few days of transport or storage (Halevy *et al.* 1978a). Since the Ag^+ ion stays at the base, it does not act as an ethylene antagonist (Halevy and Kofranek 1977) but probably as a bactericide. Best results were obtained with carnations, but beneficial effects were found also with chrysanthemum, gladiolus, gerbera (Farnham 1973; Kofranek and Paul 1974), statice (Shillo and Halevy 1980), phalaenopsis orchid (Aharoni and Halevy 1975), China aster (Kofranek *et al.* 1978), gerbera (Van Meeteren 1978a), Bougainvillea (Borochov *et al.* 1975b), and marguerite daisy (Byrne *et al.* 1979). In phalaenopsis impregnation with $NiCl_2$ gave better results than with $AgNO_3$ (Aharoni and Halevy 1977). Base treatment with silver thiosulfate acts as an ethylene antagonist since the Ag^+ travels upwards (Veen 1979a). Details are presented on p. 86.

C. Bud-opening

This is a procedure for harvesting flowers at a stage earlier than that normally considered as "cutting stage," and then opening them off the plant. The advantages and possible uses of this procedure were described in the first part of this review (Halevy and Mayak 1979, p. 212).

The chemical solutions and environmental conditions used for bud opening are in most cases similar to or identical with those of pulsing. However, since the time required for opening is much longer than that for pulsing (several days), the sugar concentration used for opening is often lower than the concentration for pulsing, and the optimal temperature may be somewhat lower. Too high a sugar concentration may cause desiccation of tender foliage, like that of roses and chrysanthemums, by sugar accumulating in the free space at the ends of the leaf vessels and creating osmotic gradient between the cells and the free space (Halevy 1976). Foliage desiccation can be prevented by covering the foliage of the bunched flowers with a plastic sleeve during pulsing and opening, thus reducing transpiration and sugar uptake by the leaves (Borochov *et al.* 1975c; Gay and Nichols 1977; Nichols 1976a).

The stage of harvesting for bud-opening is critical for each flower and sometimes even for cultivars. Buds cut at a too early stage will not develop properly or will require too much time for opening, rendering the method impractical (see, e.g., Kofranek and Halevy 1972, Kofranek *et al.* 1975b).

For a long time bud opening was used to force flowering branches of certain woody plants like mimosa (Laumonnier 1959; Ravel d'Esclapon 1962), *Forsythia* (Broertjes 1955), and lilac (Sytsema 1962, 1964), after the dormancy of the flower buds had been broken naturally or by cold storage. A method for bud opening of a major cut flower was described first for carnation by Kohl and Smith (1960). So far methods have been developed for the following flowers: roses (Halevy and Mayak 1974a,b; Mayak *et al.* 1969b, 1970c; Paulin *et al.* 1979), standard and miniature carnations (Halevy and Mayak 1974a,b; Kofranek 1976; Kofranek and Kubota 1972; Mayak *et al.* 1970c; Monnet and Paulin 1975; Nichols 1974), standard and spray chrysanthemums (Borochov *et al.* 1975c; Kofranek 1976; Kofranek and Halevy 1972; Marousky 1973; Nichols 1976a), gladiolus (Borochov *et al.* 1975d; Halevy and Mayak 1974a,b; Kofranek and Halevy 1976; Mayak *et al.* 1970c), *Gypsophila* (Apelbaum and Katchansky 1977; Farnham 1975; Farnham *et al.* 1978a), gerbera (Borochov *et al.* 1975c), statice (Shillo and Halevy 1980), and *Strelitzia* (Halevy *et al.* 1978b).

D. Holding (Vase) Solutions

Many commercial preparations are available for the wholesaler or the retailer to hold flowers until they are sold or for the consumer to use continuously in the vase. Unlike pulsing and bud-opening formulations, consumer preparations are made for a wide variety of flowers. They normally contain low concentrations (0.5 to 2%) of sugar.

Some flowers excrete substances from the cut stem base or the submerged foliage that may damage themselves or other flowers placed in the same vase. Daffodils, therefore, should stay alone in water for 1 hour, and the water should be changed before placing them with other flowers, especially tulips (Hekstra 1967). Polyphenols were reported to be leached from rose, lilac, and mimosa foliage submerged in the vase water, enhancing stem blockage (Buys 1969, 1973). Therefore, they should be removed from the lower part of the stem before placing the flowers in the vase.

VI. INGREDIENTS USED IN CHEMICAL SOLUTIONS

The most important and universal ingredient is, of course, water. Sugar is included in almost all preparations, but the other ingredients vary greatly in the various formulations.

A. Water

The composition of "tap water" varies greatly in various locations. This may influence the longevity of the flowers kept in tap water, as well as the efficiency of chemical solutions used for holding, pulsing, or bud-opening (Rogers 1973; Farnham *et al.* 1972; Mayak *et al.* 1972a; Staby and Erwin 1978; Waters 1966b). Deionized or distilled water increased longevity and enhanced the effect of the preservative used (Farnham *et al.* 1971; Lancaster 1975b; Mayak *et al.* 1972a; Staby and Erwin 1978). Millipore filtration of water was much more effective than deionizing on cut roses. Millipore filtration enhanced flow rate of water through rose stems, decreased blockage, and increased the hydration of the peduncle tissue, thus reducing the rate of "bent neck" (Durkin 1979a,b). The major effect of millipore filtration was not of filtration *per se*, but the removal of air bubbles under reduced pressure, thus decreasing air blockage of the vessels. When water was filtered through the same millipore using pressure instead of vacuum, no beneficial effect was found. Reducing air content in water by boiling also reduced blockage (Buys 1969). This was attributed by Buys (1969) to the removal of O_2, oxidizing the polyphenols extruded from the submerged foliage. However, Durkin (1979a) has shown that gassing with N_2 was as detrimental as gassing with O_2.

The sensitivity to water quality varied with various flowers. Carnations (Farnham *et al.* 1971), roses (Mayak *et al.* 1972a; Waters 1968b), and chrysanthemums (Staby and Erwin 1978; Waters 1968a) are very sensitive, but tulips (Staby *et al.* 1978) are not.

The detrimental effect of tap water depends on several factors: acidity (pH), total dissolved solids (TDS), and the presence of specific toxic ions. Because of the complexity of the various factors and their interactions, it is difficult to predict the effect of certain tap waters on longevity just from the mineral analysis of the water. In a few cases tap water was found to be better than deionized water for increasing keeping quality of some flowers (Coorts and Gartner 1963; Staden and Molenaar 1975).

1. Acidity (pH).—The benefits of low pH water (3 to 4) have long been recognized (Aarts 1957a). Most preservative formulations contain an acid to reduce the pH. The types of acid added seem unimportant (Aarts 1957a). Acidification of alkaline water with sulfuric acid increased longevity of roses, carnations, and stocks (Pokorny and Kamp 1953). The effect of low pH was attributed mainly to the reduction of microbial population (Aarts 1957a). However, Marousky (1971) demonstrated that low pH retarded stem blockage of roses even in bacteria-free water, and Durkin (1979a) found an increase in flow rate of water through rose stem segments, with decrease in pH from 6 to 3.

2. Wetting Agents.—Wetting agents at 0.1 to 0.01% greatly improved the flow of water and hydration in roses and chrysanthemums (Durkin 1981).

3. Total Dissolved Solutes (TDS).—The severity of foliage deterioration of chrysanthemums increased in direct relation to TDS in the water (Waters 1966b, 1968a). A similar but less severe effect was found also with roses (Waters 1968b). Sensitivity of flowers varies; longevity of carnations (Farnham *et al.* 1971), roses, and chrysanthemums (Waters 1968a,b) was reduced by low TDS of 200 ppm, while gladiolus lasting quality decreased only when water salinity was higher than 700 ppm (Waters 1968a). In some other cases no direct relationship between total salinity and longevity was found (Marousky 1974; Staby and Erwin 1978).

4. Specific Ions.—Hitchcock and Zimmerman (1929) found that most of the inorganic chemicals commonly found in tap water were toxic to cut flowers and reduced water uptake. This was confirmed by Waters (1968a). However, some ions are more toxic than others. Soft water in which Na was substituted for Ca and Mg is worse than hard water for both carnations (Holley and Matthews 1962) and roses (Mayak *et al.* 1972a; Staden and Molenaar 1975). $NaHCO_3$ was more toxic to roses than was NaCl, but was not toxic to carnations (Lancaster 1975b). Fe^{++} was toxic to chrysanthemums (but not to gladiolus) at 12 ppm, and B (8 to 14 ppm) toxic to both chrysanthemums and gladiolus (Kulwiec 1967; Waters 1968a). The most toxic ion to certain flowers is fluoride (F). Gladiolus and freesia are most sensitive to fluoride, and were injured even when held in 1 ppm F, as in fluoridated drinking tap water (Spierings 1969; Sytsema 1972; Waters 1968a). Gerbera are also injured by 1 ppm F (Marousky and Woltz 1975). Chrysanthemum, roses, poinsettia, and snapdragons were injured at levels higher than 5 ppm (Marousky and Woltz 1975; Waters 1968a,b; Woltz and Marousky 1972). No damage was found with lilac, cymbidium, and some cultivars of roses and narcissi (Sytsema 1972). Preservative solutions containing HQC and sucrose were unable to overcome the toxic effects of F (Marousky and Woltz 1975; Woltz and Marousky 1972).

B. Sugars

Sucrose is included in most preservative formulations, but other metabolic sugars like glucose and fructose are similarly effective. Lactose and maltose were active only in low concentrations, while the nonmetabolic sugars mannitol and mannose were inactive or harmful (Aarts 1957a; Halevy and Mayak 1974b; Kofranek and Halevy 1972).

The optimal concentration of sugar varies with the treatment and the flower. Generally for a given flower the longer the exposure to the chemical solution, the lower the concentration required. Therefore, high concentrations are used for pulsing, intermediate for bud-opening, and low for holding solutions.

In some flowers sugar has little or no benefit and is sometimes even damaging. This is the case with convallaria (Aarts 1957a), stock, pyrethrum (Aarts 1964), daffodil (Freeland 1974; Nichols 1975), lupine (Mohan Ram and Rao 1977), and *Oncidium* orchids (Yong and Ong 1979). Controversial or inconsistent results were obtained with tulips (Aarts 1957a; Sytsema 1971) and cyclamen (Aarts 1957a; Kohl 1975). In narcissus, the applied sugar was utilized for ovary growth (Nichols 1975) and in tulips, for stem elongation.

Excessively high sugar concentrations can damage foliage and petals. One of the main reasons for the variability in optimal sugar concentration for different flowers is the sensitivity of the foliage of some plants. Green leaves are more sensitive to high sugar concentration than are the petals, probably because their ability for osmotic adjustment is less than that of the petals (Halevy 1976). The fact that externally applied sugars are first accumulated in rose leaves and only then are translocated from the leaves to the flowers (Halevy and Mayak 1979; Paulin 1979) may contribute to the sensitivity of rose foliage to excess sugar. Once externally applied sugars are absorbed by cut rose shoots, they follow the same translocation pattern as naturally forming carbohydrates; that is, the sugars move from the leaves to the petals (Nichols and Ho 1979).

The role of applied sugars in delaying senescence has been reviewed (Halevy and Mayak 1979). Acock and Nichols (1979) recently confirmed that sugar improved the water balance and osmotic potential of carnation flowers. They confirmed earlier results (Nichols 1973) that absorbed sucrose is rapidly converted in petals to reducing sugars which accumulate in the corolla. Paulin (1980) demonstrated similar processes in rose flowers. Wagner (1979), using direct analysis of isolated vacuoles from petals of hippeastrum and tulip, demonstrated that glucose and fructose were located primarily in the vacuole.

C. Mineral Solutes

The detrimental effect of saline water has been described. However, certain nontoxic mineral salts can increase the osmotic concentration and the pressure potential of the petals' cells, thus improving their water balance and promoting longevity (Halevy 1976; Mayak *et al.* 1978; Van Meeteren 1980, 1981). These were mainly potassium salts (KCl, KNO_3, K_2SO_4), but also $Ca(NO_3)_2$ and NH_4NO_3. Potassium ions and their associated anions were also the main natural inorganic osmotic constituents of carnation flowers (Acock and Nichols 1979). When nitrogenous compounds are used with sugar in a holding solution they promote the growth of bacteria, making it more difficult to control the bacteria than in solutions containing sugars alone (Staden 1974).

1. Calcium.—$Ca(NO_3)_2$ (0.1%) was reported to prolong the life of cut bulb flowers (Widmer and Struck 1973). Ca also was used combined with

$AgNO_3$ to extend the longevity of some flowers (Aarts 1957a; Nichols and Kulwiec 1967; Pastac and Driguet 1947). Ca was used with several potassium salts in carnations to prevent stem softening and bending which were prevalent without Ca (Mayak *et al.* 1978).

$CaCO_3$ (10 ppm) was used in a tulip preservative with sugar and a bactericide (Staden 1976).

2. Aluminum.—$Al_2(SO_4)_3$ (50 to 100 ppm of Al) has been used in many preservative formulations for roses (Halevy *et al.* 1978a; Weinstein and Laurencot 1963), gladiolus (Mayak *et al.* 1973), and other flowers (Aarts 1957a; Mohan Ram and Rao 1977). Weinstein and Laurencot (1963) attributed the effect of Al to lowering the rose petal pH and stabilizing the anthocyanins (for review see Halevy and Mayak 1979, p. 216–219). Al sulfate also acidifies the holding water, thus reducing bacterial growth and improving water uptake. Mayak and Bar-Yosef (1972) showed that roses exposed to Al for only 12 hours had reduced bent neck and wilting. Schnabl and Ziegler (1974, 1975) studied the role of Al and saw no effect in the dark. Al in the holding water reduced transpiration and improved water balance of cut roses by inducing stomatal closure. In Al-treated epidermis strips, no accumulation of K in the guard cells and starch mobilization were observed. They ascribed the effect of Al to the inhibition of starch hydrolysis and the promotion of starch synthesis by precipitation of Pi as $Al_2(PO_4)_3$ (Schnabl and Ziegler 1975). In roses and carnations Al was effective in reducing transpiration and increasing longevity, also when applied as a foliar spray at 0.1%. Spray applications were ineffective on tulips, iris, and gladiolus (Schnabl 1976). In chrysanthemum, however, Al in the pulsing and bud opening solutions promoted foliage wilting (Kofranek and Halevy 1972).

3. Boron.—Boric acid or borax (100 to 1000 ppm) was found to be useful to carnations and other *Dianthus* species (Aarts 1957a, 1964; Nichols and Kulwiec 1967).

B acted mainly together with sugar by directing the translocation of sugar to the corolla and away from the ovary. However, B also delayed senescence of isolated petals (Aarts 1957a). B was also beneficial to *Convallaria*, sweet pea, lilac (Aarts 1957a), and lupine (Aarts 1958), but was toxic to snapdragon, chrysanthemum, *Cosmos bipinnatus, Lilium henryi, Scabiosa atropurpurea* (Aarts 1957a), and gladiolus (Waters 1968a).

Camprubi and Fontarnau (1977) extended the longevity and reduced ethylene production of carnations with $NaBH_4$ (500 ppm). They attributed the increase in longevity to the antioxidant activity of the sodium bromohydride (Camprubi *et al.* 1981).

4. Silver.—$AgNO_3$ and Ag-acetate (10 to 50 ppm) are two of the most effective bactericides used in preservative formulations (Aarts 1957a).

The main disadvantage of these Ag salts is that they are photooxidized to form black insoluble compounds which precipitate. Ag also reacts with the chlorine in many tap waters to form insoluble AgCl. A few minutes of stem base impregnation with high concentration (1000 to 1500 ppm) was effective in extending longevity in several flowers (see "Pulsing"). $AgNO_3$ is relatively immobile in the stems (Kofranek and Paul 1974), but Ag thiosulphate (STS) moves readily in the stem to the corolla (Veen and Van de Geijn 1978). STS acted as an ethylene antagonist, reduced ethylene production (Veen 1979a) and respiration (Veen 1979b) of standard carnations, and extended flower longevity (Reid *et al.* 1980b; Staby *et al.* 1979; Veen and Van de Geijn 1978). Five minutes to 24 hours base treatment of STS (1 to 4 mM) was effective on carnations, lilies, and some other flowers, but not on tulips (Swart and Kamerbeek 1979). Preplanting dip of 'Enchantment' lily bulbs in STS promoted the quality and longevity of the flowers as postharvest treatment to the cut flower (Swart 1981). STS had no effect on roses when applied alone, but prevented the damage induced by application of ethephon (De Stigter 1981a). STS prevented normal and ethylene-induced florets shattering in snapdragon (Farnham *et al.* 1981), *Delphinium*, and sweet peas (Mor *et al.* 1980b; Shillo *et al.* 1980). Reid *et al.* (1980a) reported good response of spray carnations to STS, but Sytsema (1981) found that some spray carnation cultivars were less responsive to STS than were standard carnations. Excessive concentration or time of STS treatment may damage the petals and foliage.

An Ag-ammonium complex was used successfully for pulsing and bud opening of carnations (Nowak and Rudnicki 1979). A complex of Ag and EDTA extended longevity and prevented ethylene damage in gerbera flowers, while STS was toxic to gerbera (Nowak 1979).

5. **Nickel.**—A stem base treatment with $NiCl_2$ (1500 ppm for 10 minutes) was more effective than $AgNO_3$ in increasing longevity (Aharoni and Halevy 1977) and promoting stem water conductance of phalaenopsis (Aharoni and Mayak 1977). Ni may act as a germicide and as an inhibitor of ethylene production (Lau and Yang 1976). Ni had no effect when added to holding solutions for tulips (Staden 1974) or when used as pretreatment for carnations (Kofranek and Paul 1972).

6. **Zinc.**—Zinc ions were effective as a germicide for some flowers (Rogers 1963; Ryan 1957).

7. **Copper.**—It was reported (Ratsek 1935) that flowers lasted longer in copper containers due to the small quantity of copper ions released to the water which served as a germicide. The effect of Cu salts on longevity varied in different flowers; some benefited; in some there was no effect; in carnation, reduction of longevity was observed (Laurie 1936). It was shown, however, that 10 to 80 ppm Cu^{+} did not provide any bacterial

control and injured chrysanthemum stem and leaves (Marousky 1976).

D. Germicides

All preservative formulations include at least one compound with germicidal activity; bactericides are almost always included and sometimes also fungicides. The use of mineral ions which act as germicides, the most common and active of which is Ag^+, has been described. Some of the more common organic germicides will be described in the following sections.

1. 8-Hydroxyquinoline (HQ).—HQ base or its esters—mainly sulphate (HQS) and citrate (HQC) at 200 to 600 ppm—were the most commonly used germicides in the last decade (Rogers 1973; Marousky 1972, 1973). Apart from being a broad spectrum bactericide and fungicide, HQC was shown also to reduce "physiological" stem blockage in sterile tissues. It was suggested that this effect is related to the chelating properties of the quinoline esters, which may chelate metal ions of enzymes active in creating the stem blockage (Marousky 1972). The chelating complex of HQ with divalent metals (mainly Fe and Cu) may be the basis for its antibacterial activity (Albert *et al.* 1953). HQS and HQC also may affect flower longevity by acidifying the water.

Part of HQS' beneficial effect on water balance of flowers was attributed to its effect on stomatal closure (Stoddard and Miller 1962). However, the opposite was found with cut chrysanthemums using concentrations (up to 200 ppm) relevant to cut flowers (Gay and Nichols 1977).

Chua (1970) suggested that HQS has cytokinin-like activity in retarding senescence. This claim was not confirmed by further studies (Kraft and Ellis 1971; Rameshwar and Steponkus 1970; Ringe 1972).

Parups and Peterson (1973) found that HQ inhibits ethylene evolution in rose stamens and apple fruit slices, and they attributed the senescence-retarding effect of HQ to inhibition of ethylene production. HQ delayed ethylene emanation also in carnation flowers (Wilkins and Swanson 1975).

Another effect of HQ is as an inhibitor of the alternative cyanide-resistant final oxidase in isolated mitochondria (Bendall and Bonner 1971). If this respiratory pathway is indicative of the final senescence stage (see Halevy and Mayak 1979, p. 219), then the inhibition of this process may delay senescence.

In some flowers deleterious effects of HQ were observed, which reduced or prevented the use of HQ in these flowers. In chrysanthemums HQ caused leaf damage and stem browning (Gladon and Staby 1976; Kofranek and Halevy 1972) and similar damage was observed in *Gypsophila* (Katchansky 1979) and marguerite daisy (Byrne *et al.* 1979). In white flowers HQC accumulated in the petals and caused an undesirable yellowish tint.

2. Slow-release Chlorine Compounds.—Several organic, stabilized, slow-release halogen compounds (principally chlorine) are available as swimming pool disinfectants. These compounds are very effective bactericides and have been included in some recent preservative formulations at 50 to 400 ppm Cl. Compounds used were sodium dichloro-5-triazine-trione (synonym, sodium dichloroisocyanurate, Gurdex, ACL-60) (Kofranek *et al.* 1974; Halevy *et al.* 1978a), and 1,3-dichloro-5,5-dimethylhydantoin (Marousky 1976, 1977). They were tested and found to be effective on many flowers, but high concentrations induced foliar chlorosis and stem bleaching on roses, snapdragons, and chrysanthemums (Kofranek *et al.* 1974; Marousky 1976). Another disadvantage of these compounds is that they disintegrate after a few days in the solution. In gerbera, 50 ppm chlorine in the holding solution was effective in controlling bacteria, with no damage to flowers (Barendse 1978).

3. Quaternary Ammonium Compounds.—Several preparations containing quaternary ammonium compound have been used recently at 5 to 300 ppm in an attempt to overcome some of the disadvantages of HQ. These compounds are relatively non-toxic and are more stable and durable than HQ, especially in tap or hard water (Farnham *et al.* 1978c; Levy and Hanan 1978).

The preparations used on flowers are (1) n-alkyl dimethylbenzyl ammonium-chloride, when the alkyl group is made of C_8 to C_{18} components (called benzalkone or benzalkonium chloride); (2) a mixture of compound 'a' and n-alkyl (68% C_{12} and 32% C_{14}) dimethyl ethylbenzyl ammonium chloride (Physan-20); (3) lauryldimethylbenzyl ammonium chloride (Vantoc CL); (4) n-alkyl trimethylbenzyl ammonium bromide (Vantoc AL).

Physan-20 at 100 to 200 ppm with sucrose was optimal for pulsing and bud opening of carnations (Farnham *et al.* 1978c; Levy and Hanan 1978), *Gypsophila* (Farnham *et al.* 1978a), chrysanthemum (Kofranek and Halevy 1981), and marguerite daisy (Byrne *et al.* 1979). Optimal concentrations in holding solution for carnations (Nichols 1978b) were 50 to 100 ppm (Vantoc AL) and 25 to 50 ppm (Vantoc CL), and for tulip (Staden 1976) 5 to 10 ppm (benzalkone). Physan was ineffective or detrimental when used with sugar for pulsing China aster (Kofranek *et al.* 1978) and roses (Zieslin and Kofranek 1980).

4. Thiabendazole (TBZ).—TBZ is a broad-spectrum fungicide and is usually used at 300 ppm together with a bactericide, HQ (Apelbaum and Katchansky 1977; Halevy *et al.* 1978b), or benzalkone (Katchansky 1979). TBZ is insoluble in water, and this may be the reason for its ineffectiveness in the study of Levy and Hanan (1978). It is used in preservative formulations as glycolate which is soluble (Apelbaum and Katchansky 1977). Apart from its effect as a fungicide, TBZ showed some cytokinin-like activity (Thomas 1974), retarded ethylene evolution, and reduced the sensitivity of cut carnations to ethylene (Apelbaum and Katchansky 1978).

5. Dichlorophen (Panacide).—At 10 to 250 ppm, this was found to be a useful germicide for pulsing and bud opening of carnations (Nichols 1974) and chrysanthemums (Nichols 1976a), and for holding gerberas (Van Meeteren 1978a). It has the advantage that it is colorless and inexpensive.

6. Chlorhexidine.—Digluconate (Hibitane) and diacetate (Nolvasan) at 50 to 100 ppm were commonly used in England for carnations (Nichols and Kulwiec 1967; Smellie and Brincklow 1963). Some formulations of chlorhexidine were found to be comparable but not better than HQC (Staby and Erwin 1977).

E. Organic Acids and Salts, and Antioxidants

Many preservatives contain an acid to reduce the pH (see previous discussion). This may be one of the main roles of organic acids that are included in preservative formulations. However, some acids may have additional specific functions, since they were more effective than other acids at the same pH.

Citric acid (CA) is the most widely used acid at 50 to 800 ppm. It was found to be very effective on roses (Durkin 1979a,b; Weinstein and Laurencot 1963), chrysanthemums (Kofranek and Halevy 1972), carnations (Halevy *et al.* 1978a), lupine (Mohan Ram and Rao 1977), gladiolus (Reist 1977), *Strelitzia* (Halevy *et al.* 1978b), and marguerite daisy (Byrne *et al.* 1979). Van Meeteren (1978b) used a buffer of CA (150 ppm) and $Na_2HPO_4 \cdot 2H_2O$ (50 ppm) for gerbera. CA improved water balance and reduced stem plugging (Durkin 1979b; Mohan Ram and Rao 1977; Van Meeteren 1978b).

In some cases *tartaric acid* (Sims 1976; Staby and Erwin 1977) or *glycolic acid* (Apelbaum and Katchansky 1977) was used with sugar in preservative preparations.

Benzoic acid (500 ppm) was effective for extending longevity of anthurium (Akamine and Goo 1975). Na-benzoate (150 to 300 ppm) delayed senescence of carnations (Baker *et al.* 1977) and daffodils (Wang and Baker 1979), but had little or no effect on snapdragons (Wang *et al.* 1977), iris, chrysanthemums, or roses (Wang and Baker 1979). This chemical is included in the patented formulation of "Everbloom" (Biggs 1961). Benzoate's effects on longevity of carnations and daffodils were additive to the effects of sucrose, HQC, and the acidity of the solution (Baker *et al.* 1977; Wang and Baker 1979; Wang *et al.* 1977). The researchers attributed benzoate's activity to its action as an antioxidant and a free radical scavenger and on the reduction of ethylene production in treated flowers.

Iso-ascorbic acid or Na-ascorbate at 100 ppm was included in the "Ottawa" solution together with sucrose and HQS (Parups 1975a; Parups and Chan 1973), and was reported to be an effective holding solution for roses, carnations, and snapdragons. The beneficial effect of Na-ascorbate was attributed not only to its antioxidative activity (Leibovitz and Siegel 1980), but also to its effect as an electron carrier and growth-promoting activity (Parups and Chan 1973).

The antioxidant *hydrazine sulfate* (100 ppm) was included in holding solution formulations (Weinstein and Laurencot 1963).

The use of the antioxidant *sodium bromohydride* has been described under "Boron," p. 86.

Na-phytate (100 to 500 ppm), when added to sucrose and HQS, extended the longevity of roses (Carpenter and Dilley 1975).

F. Ethylene Inhibitors

Several of the chemicals listed previously under different headings were reported also to inhibit ethylene action (Ag) or production (Ni, Co, HQ, TBZ, benzoic acid). Many of these are not specific inhibitors of ethylene, since it seems that any treatment which delayed senescence also decreased or delayed ethylene evolution (Wilkins and Swanson 1975). In the following are other chemicals which inhibit ethylene production or action.

Aminoethoxyvinyl glycine (AVG) and *methoxyvinyl glycine* (MVG) are *rhizobitoxine* analogues. They were shown to be specific inhibitors of ethylene synthesis from methionine (Lieberman 1979), by blocking the conversion of S-adenosylmethionine to 1-aminocyclopropane-1-carboxylic acid.

AVG and MVG (0.07 to 0.13 mM) extended the longevity of carnations (Baker *et al.* 1977), snapdragons (Wang *et al.* 1977), iris, daffodils, and chrysanthemums, but not roses (Wang and Baker 1979).

Aminooxyacetic acid (AOA) acts in the same way as rhizobitoxine to inhibit ethylene biosynthesis (Yang 1980). AOA at 0.2 to 2mM in the holding solution or at 100 mM as a 10-minute pulse considerably extended longevity of carnations (Brown and Mayak 1980; Fujino *et al.* 1981; Wang and Baker 1980).

The *benzothiadiazole*, 5-methyl-7-chloro-4-ethoxy-carbonyl-methoxy-2,1,3-benzothiadiazole (TH-6241), extended carnations' vase life when included at 5 ppm in holding solution with HQS and sucrose (Carpenter and Dilley 1973). TH-6241 appears to antagonize several ethylene actions (Parups 1973).

Benzylisothiocyanate at 0.18 mM (Parups 1975b) and *ethanol* at 0.5 to 4% (Heins 1980) in the holding solution greatly decreased ethylene evolution and increased longevity of carnations. The effective concentrations of ethanol varied in different experiments and the response appeared to depend on the physiological stage of the flower. Earlier, Nichols (1966) found that ethanol (2 to 4%) increased vase life of carnations, but methanol had no effect on flower longevity. The effect of ethanol may be related to its action as an antioxidant (Leibovitz and Siegel 1980).

G. Growth Regulators

Although the involvement of several growth regulators in the control of flower senescence has been demonstrated, there is relatively limited use of growth substances in preservative solutions, apart from some use of cytokinins.

1. Cytokinins.—The cytokinins used on cut flowers were kinetin (KI), 6-benzylamino purine (BA), isopentenyl adenosine (IPA), and 6-(benzylamino)-9-(2-tetrahydropyranyl) 9-H purine (PBA). The effective concentration for carnations depended on the time of treatment and varied from 10 to 100 ppm in extended treatments in holding or bud opening solutions (Cywinska-Smoter *et al.* 1978; Eisinger 1977; Hasegawa *et al.* 1976; Jeffcoat 1977; Mayak and Dilley 1976a; Mayak and Kofranek 1976), to 100 ppm for pulsing (Halevy *et al.* 1978a), to 250 ppm for a 2-minute dip of whole stems (Heide and Oydvin 1969). Smith (1967) reported best results with a 5-minute dip in 20 ppm BA. Too high a concentration or too long a treatment may be detrimental (Heide and Oydvin 1969).

It seems that best results were obtained with carnations. Not only was longevity promoted but also the sensitivity to ethylene was reduced (Eisinger 1977; Mayak and Kofranek 1976), and the burst in ethylene production delayed (Eisinger 1977; Kende and Hanson 1976). Paulin and Muloway (1979), however, found little or no value of adding cytokinin to chemical solutions containing sugar for carnations. Other flowers in which delay in senescence was reported by using cytokinins in solutions are rose (Halevy *et al.* 1978a; Mayak and Halevy 1974) and iris (Wang and Baker 1979). Sytsema (1971) reported good results with 2 to 4 hours' treatment of BA (10 to 50 ppm) on tulip, while Staden (1973) obtained detrimental results with BA in holding solution.

A momentary dip of *Anthurium* flowers in BA (10 ppm) extended longevity and increased the tolerance of the flowers to chilling (Shirakawa *et al.* 1964). A 2-minute dip of gerbera flowers in BA (25 ppm) delayed the decrease in water content and the increase in ion leakage typical of senescent flowers (Van Meeteren 1979a). This is similar to the improvement of water balance found in roses held in KI (60 ppm), mainly under water stress conditions, which was manifested by increased water uptake and in maintaining petal turgidity for an extended period (Mayak and Halevy 1974).

In daffodils, a dip with a mixture of BA (100 ppm) and auxin (2,4-D at 22 ppm) retarded senescence, while BA or 2,4-D used alone was not effective (Ballantyne 1965). A mixture of BA (5 ppm) and NAA (20 ppm) accelerated carnations' bud opening after storage (Goszczynska and Nowak 1979).

2. Auxin.—Moderate concentrations (1 to 100 ppm) of auxin enhanced carnation flower senescence apparently by promoting ethylene production. However, 500 ppm 2,4-D damaged the vegetative tissue of carnation flower shoots, but retarded petal senescence and inhibited ethylene evolution (Sacalis and Nichols 1980).

3. Gibberellin (GA_3).—At 1 ppm this extended longevity of stocks with and without sucrose (Aarts 1957a). Garrod and Harris (1978) reported that application of GA_3 (200 ppm) to isolated carnation petals growing in agar or liquid medium delayed senescence. However, Nichols (1968b) found little or no effect on longevity by dipping carnations in GA_3 solutions (0.1 to 200 ppm). GA_3 (100 to 400 ppm) in opening solution of carnations promoted opening but decreased longevity and caused discoloration of flowers (Cywinska-Smoter *et al.* 1978). GA (20 to 35 ppm) accelerated bud opening of carnations (Goszczynska and Nowak 1979) and gladiolus (Ramanuja Rao and Mohan Ram 1979) after storage.

4. Abscisic Acid (ABA).—Kohl and Rundle (1972) showed that ABA at 1 ppm in continuous holding solution or 10 ppm for only 1 day delayed wilting and extended longevity of cut roses by inducing stomatal closure. On the other hand, ABA accelerates senescence of flowers (Borochov *et al.* 1976a; Halevy and Mayak 1975; Mayak and Dilley 1976b; Mayak and Halevy 1972; Mayak *et al.* 1972c). The opposing effects of ABA on longevity of roses have been demonstrated by Halevy *et al.* (1974a). Under conditions promoting water loss (light and high atmospheric VPD) the effect on stomatal closure prevailed and longevity was extended; but in darkness, when stomata are closed, ABA enhanced senescence.

H. Growth Retardants

Halevy and Wittwer (1965, 1966) first reported on the effect of the growth retardants butanedioic acid mono-(2,2-dimethylhydrazide) (daminozide, B-Nine) and (2-chloroethyl)trimethylammonium chloride (CCC, chloromequat) on delaying senescence of several plant tissues including snapdragon and carnation flowers. The optimal concentration for each growth regulator varied with the cultivar and the season. Larsen and Scholes (1965, 1966, 1967) and Larsen and Frolich (1969) have shown that the inclusion of daminozide in a holding solution with sugar and HQC improved longevity of several flowers. Optimal concentrations varied from 10 to 50 ppm for snapdragons, to 25 ppm for stocks, to 500 ppm for carnations and roses (Dilley and Carpenter 1975; Metzger 1972). Hasegawa *et al.* (1976) found 500 ppm optimal for pulsing carnations and 125 ppm for holding solution.

Inclusion of chloromequat at 25 to 50 ppm in the bud opening solution and 10 ppm in the holding solution improved the keeping qualities of gladiolus (Shillo 1969). A holding solution containing chloromequat (50 ppm) with HQS and sucrose was shown to increase the vase life of tulips, sweet peas, stocks, snapdragon, carnations (Nowak and Rudnicki 1975), and gerberas (Rudnicki and Nowak 1976).

The mode of action of growth retardants in delaying senescence is not known. It was suggested (Halevy 1967a) that the effect on extending longevity of flowers and vegetative tissues is related to their effect on increasing the tolerance to environmental stresses, by affecting sensitive metabolic processes and interacting with native phytohormones.

I. Inhibitors

Inhibitors were included in many early preservative formulations. Generally they reduce the respiration of the flowers and slow down their metabolism including some aspects of aging and development. However, inhibitors often prevent the full development and opening of the flowers and may hasten some phenomena associated with senescence such as bluing of red flowers. When beneficial effects were observed, the difference between effective concentrations and toxic levels was narrow. These are presumably the reasons why inhibitors are less common in recent formulations for promoting keeping quality of flowers.

Maleic hydrazide (MH)—mainly the diethanolamine of MH—was probably the most commonly recommended inhibitor for cut flowers. Weinstein and Laurencot (1963) reported that a pulsing treatment which included a 30-minute dip of 0.5 to 1% MH followed by 24 hours at 4°C with a solution of Al (100 ppm) and citric acid (800 ppm) gave best results with roses. A preservative containing 2500 MH (with Al, citric acid, hydrazine sulfate, and glucose) gave satisfactory results with roses, carnations, chrysanthemums, and snapdragons. Some beneficial effects of MH (250 to 500 ppm) in holding solutions were reported for snapdragon (Kelley and Hamner 1958) and lupine (Mohan Ram and Rao 1977), and 50 ppm was effective on dahlia (Bar-Sella 1959).

Morphactins (derivatives of fluorene-α-carboxylic acid, 100 ppm), applied as a pretreatment at 4°C for 20 hours, prevented undesirable opening of roses during uncontrolled transport, but also prevented the subsequent full opening of the flowers (Mayak *et al.* 1972b).

Unlike the effect of the former inhibitors, the growth inhibitor *ammonium ethyl carbamoylphosphonate* (Kermite) was reported to promote opening and delay bluing of roses when added (1000 ppm) to preservative solutions (Brantley 1975). However, later observations with this chemical gave variable results (Brantley, personal communication).

Cyclohexamide, the protein synthesis inhibitor, extended longevity of carnations when added (10 to 20 ppm) to a solution of sucrose and HQS, but was toxic to roses (Dilley and Carpenter 1975). A 24-hour pretreatment of narcissus with 1 ppm delayed senescence, and pretreatment of tulips with 0.2 ppm prevented undesirable stem elongation (Nichols 1978a). However, slightly higher concentrations or longer treatments were toxic. Pretreatment with cycloheximide inhibited growth and opening of *Iris germanica* (Paulin 1973, 1975b).

1. Enzyme Inhibitors.—*Sodium azide* (NaN_3) (10 ppm) and *2,4-dinitrophenol* (10 ppm) reduced stem plugging in some cut woody shoots and extended longevity of *Convallaria* and dahlia (Aarts 1957a). NaN_3 (9 ppm) also extended longevity and reduced ethylene emanation of carnations (Wilkins and Swanson 1975). NaN_3 (50 ppm) was used with sugar in a holding solution for gladiolus (Lukaszewska 1978).

VII. PRETREATMENTS OTHER THAN CHEMICAL SOLUTIONS

A. Treatment for Stems Excreting "Milky" Fluid

Flowers with stems producing "milky" latex (such as Euphorbias, dahlia, poppy) tend to have very short vase lives. This generally was attributed to the latex plugging the conducting tissues. To prevent this, pretreatments which coagulate the latex are recommended, e.g., dipping the cut base in alcohol, searing the base of the stem over a flame or steam, or holding the stem in boiling water (Rogers 1963; Widmer and Struck 1973). However, immersing the base of dahlia stems in boiling water was found to be detrimental, while warm water (50°C) was slightly beneficial (Bar-Sella 1959). In poinsettia (Sytsema 1966) a stem base dip in boiling water for 1 second and in Iceland poppy (Stirling 1950) a dip for 30 seconds were beneficial.

B. Gamma Irradiation

Studies to prolong the lives of flowers with gamma rays were in most cases negative, detrimental, or variable. Kohl (1967) irradiated roses in doses of 10 to 200 krad; lower doses had little or no effect, while higher doses prevented full opening and enhanced senescence. Results with carnations and roses were variable with different cultivars (Dupuy 1975) and no effect was found on tulips (Hekstra 1966). In several proteas irradiation with 10 to 50 krad decreased opening of the flowers and enhanced foliage blackening (Haasbroek and Rousseau 1971; Haasbroek *et al.* 1973).

VIII. STORAGE

Storage of flowers is an important procedure in supply and demand regulation. Methods evaluated for cut flowers include cold storage, controlled or modified atmosphere, and low pressure storage.

A. Cold Storage

This is the most common method used commercially today for cut flowers. "Wet" and "dry" methods have been developed (Lutz and Hardenburg 1968; Nichols 1967a, 1971b; Nowak and Rudnicki 1979; Paulin 1975a; Rogers 1963; Ulrich 1979). Normally, the "wet" method (flowers held during storage with their bases in water) is for short term (1 to 3 days) storage, and the "dry" storage method is for longer periods. Some flowers like freesia, dahlia (Aarts 1957b), and iris (De Hertogh and Springer 1977) keep better when stored wet in water or in preservative solution, even for long terms.

Storage temperature of temperate flowers (e.g., roses, carnations, chrysanthemums) should be as close as possible to the freezing point of the tissue or of the water (if stored "wet"). Since freezing temperature of most flowers is below 0.5°C (Whiteman 1957), they would be best stored "dry" at this temperature, if temperature control is good and air circulation is provided to prevent "cold spots."

In "dry" storage flowers are normally harvested early in the morning when fully turgid, handled dry, graded, and sealed in plastic bags or boxes. Since the cooling of the flowers in the closed boxes is very slow (Halevy and Mayak 1974a; Halevy *et al.* 1978a; Nichols 1971b), flowers should be pre-cooled before packing and storing (see "Transport"). When taken out of storage, the bases of the flower stems should be recut and the flowers pulsed in a proper solution.

Some flowers like carnations (Kofranek *et al.* 1975a; Souter *et al.* 1977) and narcissus (Nichols and Wallis 1972) are best stored dry in the tight-bud stage, and then opened in water or in a proper opening solution to obtain quality similar to that of fresh flowers. Flowers like gladiolus (Kofranek and Halevy 1976) or *Strelitzia* (Halevy *et al.* 1978a) respond better to pulsing prior to storage. Other flowers, like chrysanthemums (Kofranek *et al.* 1975b) and iris ('Prof. Blaauu') (unpublished results), however, failed to open properly and to reach the desired quality of those stored in the bud stage, although they opened when fresh.

It has been suggested that frost damage of plant tissues at normal freezing temperatures is triggered by ice-nucleating activity derived mainly from bacteria (Arny *et al.* 1976). The possibility of storing flowers at sub-zero temperatures and preventing freezing damage caused by bacteria therefore was explored (Mayak and Accati-Garibaldi 1979). Indeed, it was found that inoculation of flowers with microbial suspension rendered them more susceptible to frost damage. Streptomycin reduced the freezing damage of inoculated flowers, but did not affect the susceptibility of freshly cut untreated flowers.

Tropical flowers are damaged by chilling at fairly high temperatures of 10° to 15°C (e.g., *Anthurium*, tropical orchids like *Cattleya* and *Vanda*, *Heliconia, Alpinia, Eucharis*, and several *Euphorbias* including Poinsettia). They therefore should be transported and stored at temperatures of 12° to 18°C (Akamine 1976; Lutz and Hardenburg 1968). Cymbidium orchids, however, kept well at −0.5°C (Sheehan 1954).

Subtropical flowers such as gladiolus (Waters 1966a), *Strelitzia* (Halevy *et al.* 1978b), jasmine (Bose and Raghava 1975), proteas (Paull *et al.* 1981), anemones, and gloriosa (Lutz and Hardenburg 1968) should be stored at medium temperatures of 2° to 8°C.

Humidity control is of great importance in storage. Moisture loss in stored flowers is directly related to vapor pressure deficits (i.e., to both relative humidity (RH) and temperature). Therefore, temperature variations in storage should be kept to a minimum. RH should be maintained as high as possible without causing free water on the flowers. Practically, 90 to 95% RH is normally recommended (Lutz and Hardenburg 1968; Paulin 1975a). Gentle air movement is needed for temperature control, but excessive air movement may cause water stress to flowers even at relatively high RH (Franklin 1977). Studies have shown that quality and longevity of several perishable vegetables and apples were better and decay by pathogens was less when they were stored at high RH of 99 to 100% (Van den Berg and Lentz 1978). We are unaware of experiments evaluating this technique for flowers. In view of the success of very high humidity storage of leafy vegetables, the method should be evaluated for cut flowers.

Light in storage has little or no effect on most flowers (Hekstra 1967) except those in which foliage yellowing is a problem (see "Foliage Discoloration").

Storage rooms should be as free as possible from ethylene. This can be achieved by several means (see "Ethylene," p. 66). (1) Provide air exchange in storage. One complete air exchange every hour was suggested (Staby and Thompson 1978). (2) Avoid storage with fruits, vegetables, and flowers which evolve much ethylene (see "Transport," p. 100). (3) Avoid internal combustion vehicles in storage and the vicinity. (4) Remove all diseased and decayed tissues promptly. (5) Use ethylene scrubbers (like brominated charcoal or permanganate). Although the use of ethylene scrubbers has been suggested, their efficiency is doubtful because excessive amounts of the materials are required to control the ethylene level (Peles *et al.* 1977).

B. Controlled and Modified Atmosphere Storage (CA, MA)

CA and MA storage of fruits (Smock 1979) and vegetables (Isenberg 1979) have been reviewed recently. Longevity of many vegetables and fruits is increased and quality is maintained by CA, providing reduced oxygen and increased CO_2 levels, which decrease respiration and destructive oxidation processes, and also reduce ethylene effects.

CA for cut flower storage is not commonly used. The most extensive work on CA in flowers was carried out with carnations. Some authors reported beneficial effects from CA (Flanzy 1975; Longley 1933; Pratella *et al.* 1967; Tonini and Tesi 1969; Uota and Garazsi 1967), while others have shown that none of the gaseous combinations tried was better than air circulation (Hanan 1967; Nichols 1971b; Smith 1967, 1968). Results with other flowers, including roses (Flanzy 1975; Longley 1933; Thornton 1930), were also inconclusive. Daffodils are an exception: atmosphere of 100% nitrogen was reported to be the best (Parsons *et al.* 1967). André *et al.* (1980) obtained good results with CA storage of gladiolus, iris, carnations and two rose cultivars ('Visa' and 'Baccara'), but not with 'Sonia' roses, chrysanthemum, and gerbera.

High CO_2 levels were found to be toxic to some flowers (Bunemann and Dewey 1956; Hanan 1967). Injury from CO_2 was greater at low temperatures than at high temperatures (Rogers 1962).

The "dry" storage method of sealed flowers is, in effect, a MA storage, with reduced O_2 and elevated CO_2 created by the flower's respiration. It is therefore preferable to store flowers in partially or selectively permeable material to prevent accumulation of excessive CO_2 levels. However, the factors involved are, at present, not accurately defined. Thus, close control is impossible, and results occasionally may be erratic.

C. Low Pressure Storage (LPS)

The principles and application of low pressure (hypobaric) storage for cut flowers have been reviewed (Burg 1973; Dilley 1977; Dilley *et al.* 1975; Lougheed *et al.* 1978). In this process "perishable commodities are maintained under refrigeration while being continuously ventilated with water saturated air at a controlled subatmospheric pressure". (Dilley 1977).

The desirable effect of LPS was attributed to the reduction in internal oxygen, ethylene, and possibly other volatiles' levels. Results have indicated that LPS at 40 to 60 mm Hg greatly extended longevity of several cut flowers, ornamental cuttings, and pot plants (Burg 1973; Dilley *et al.* 1975).

The extended storage of rooted and unrooted cuttings by LPS has been confirmed (Eisenberg *et al.* 1978) and seems to have found initial commercial application. However, although the process was first introduced 14 years ago (Burg and Burg 1966), it has not yet been used commercially for cut flowers (Lougheed *et al.* 1978).

Best results of LPS were obtained with carnations—mainly with those stored in the bud stage (Dilley *et al.* 1975). However, conventional cold storage of carnation buds followed by opening in a proper solution gives comparable results, with probably less expense (Kofranek *et al.* 1975a). In roses, for which long term conventional cold storage is unreliable (Halevy *et al.* 1968; Jensen and Hansen 1972; Mayak *et al.* 1972b; Paulin 1975a), LPS also gives variable and erratic results (Bangerth 1973; Dilley 1977; Dilley *et al.* 1975). Good results with 'Belinda' roses were obtained recently with LPS at 24 mm Hg. Other rose cultivars responded less favorably to LPS (Bredmose 1979, 1981).

Several problems are associated with storage of cut flowers:

(a) *Loss of longevity.* Even under the best known storage conditions, senescence processes are slowed down but not stopped (Halevy and Mayak 1979). Flowers may look "fresh" when they come out of storage, but they do not last as long as fresh flowers. Evaluation of storage therefore should include longevity measurements of fresh flowers as controls; unfortunately, this was not done in many studies.

(b) *Failure of buds to open after storage.* This is a major problem in storage of roses (Halevy *et al.* 1968), and is also a problem in other flowers like narcissus (Nichols and Wallis 1972), chrysanthemums (Kofranek *et al.* 1975a), and some iris cultivars (e.g., 'Prof. Blaauu' synonym 'Blue Ribbon'). Some improvement was obtained by storing iris without the bulbs, recutting the stem bases after storage, and "conditioning" them in warm water for a few hours (Mayak and Halevy 1971). However, full opening of stored 'Prof. Blaauu' irises was obtained only when the flowers were placed after storage in an opening-holding solution (Mayak and Halevy, unpublished results). Pulsing of gladiolus (Kofranek and Halevy 1976) and *Strelitzia* (Halevy *et al.* 1978b) flowers prior to storage ensured full opening and good longevity.

(c) *Opening of flowers to an unacceptable commercial stage.* This is mainly associated with too high temperatures in storage. The use of growth inhibitors like MH (Weinstein and Laurencot 1963) or morphactin (Mayak *et al.* 1972b) was suggested to retard precocious opening of roses. However, the inhibitors often prevent the full opening of the flowers for the consumer.

(d) *Petal discoloration.* This is a common problem with red roses and carnations which often "blue" or turn dark in storage (Halevy and Mayak 1979). Petal darkening in 'Ilona' roses was reduced considerably when flowers were not placed immediately in cold storage after picking, but held in water in the shed for a few hours before cooling (Marsbergen and Barendse 1977). On the other hand, a desirable increase in orange pigmentation of 'Soleil D'Or' narcissus is promoted by cold storage (Smith and Wallis 1967).

(e) *Foliage yellowing.* This is a severe problem in some leafy flowers (see "Foliage Discoloration"). Yellowing normally starts from the lower older leaves. Therefore it was suggested that in addition to treatment with BA, chrysanthemums (and probably also other leafy flowers) should be stored with all their leaves on the stems, and the lower leaves removed after storage (Posner *et al.* 1979). Using light in storage and packing flowers with transparent lids were found to be beneficial in delaying yellowing in chrysanthemums (Woltz and Waters 1967, 1976).

(f) *Spread of diseases, mainly botrytis during cold storage.* Only disease-free flowers should be chosen for storage. As a preventive measure, it is recommended to treat the flowers with fungicides before and soon after harvest (Halevy *et al.* 1978b; Magie 1963; Vigodsky-Haas 1979). The use of ozone gas also has been suggested (Magie 1961), but it is not used commercially.

IX. TRANSPORT

Flowers are tender and short-lived commodities which traditionally have been grown close to the centers of consumption, where they were delivered within a few hours after harvest. Long-distance transport of flowers became prevalent because of advances in airplane transport and the development of advanced methods of handling, packing, and cooling.

Flowers transported short distances are normally harvested at a more advanced stage and are often shipped in water. This is the case mainly with gladiolus, gerbera, and roses. Using this procedure for harvesting and shipping enables those growers also to use cultivars that are not suitable for long term dry shipments.

Some tropical flowers (like orchids and anthuriums) that can not be refrigerated before or during transport are often supplied with water also in long-term transport. The base of the flower stem is placed in a vial or rubber balloon filled with water or wrapped in wet tissue (Akamine 1976). Most other flowers are, however, shipped dry.

Long distance transport has been mainly by air, and most studies on transportation of cut flowers have involved air-shipped flowers (Halevy and Mayak 1974a,b; Hardenburg *et al.* 1970; Harvey *et al.* 1962; Maxie *et al.* 1973). The recent increases in cost of air freight promoted research on alternative methods of transport by land and sea (Borochov *et al.* 1975a; Danieli *et al.* 1972; Farnham *et al.* 1979; Halevy *et al.* 1978a; Mayak *et al.* 1969a, 1970b, 1980; Rij *et al.* 1979; Yvernel and Paulin 1976; Yvernel and Souter 1975).

Problems associated with long term transit of flowers are similar to those associated with storage: opening of flower buds during transit, petal and foliage drying and discoloration, spread of diseases, failure of buds to open after transit, and reduction in longevity. A common symptom of long term transit of dry flowers is low turgidity of flower petals, also termed "tired" flowers when they reach the wholesaler or retailer. They often may recover turgidity after conditioning and may show good keeping quality. By contrast, long term transit of flowers in water will maintain flower turgidity, but these flowers will senesce earlier than flowers transported dry and hydrated properly later. Methods evaluated for long term transit are also related to those evaluated for storage.

Proper handling of flowers for transit starts at harvest. Flowers should be cut and delivered to coolers as quickly as possible. If there are conditions favoring botrytis infection, flowers should be treated with fungicides before and soon after harvest and before cooling (Halevy *et al.* 1978b; Magie 1963; Vigodsky-Haas 1979). Dipping the cut stem bases with pesticides to control aphids and mites also has been suggested (Poe and Marousky 1971). For crops in which foliage yellowing is a problem, spray or dip application with cytokinin is useful (see "Foliage Discoloration"). After spraying, flowers should be well dried before further handling.

Some flowers are best handled "dry" after harvest (if they are not "pulsed" with solutions). This is the case with standard carnations (Lancaster 1975a). Most cut flowers, including roses, chrysanthemums, gerberas, and miniature carnations (Mayek *et al.* 1977b), are best held in water under refrigeration before packing.

Harvesting flowers at the tight bud stage and opening them after transit in proper solutions (see "Bud Opening") was suggested (Borochov and Tirosh 1975; Halevy *et al.* 1978a; Halevy and Mayak 1974a; Hardenburg *et al.* 1970; Lundquist and Mongelli 1971). This may save space during transport and will reduce the flowers' sensitivity to ethylene and to temperature and drought stresses (Barden and Hanan 1972; Borochov and Tirosh 1975; Camprubi and Nichols 1978; Maxie *et al.* 1973; Yvernel and Paulin 1976).

Short term pulsing of flowers with specific solutions containing optimal levels of sucrose and other chemicals before packing has been demonstrated to be of great value to most flowers (see "Pulsing"). Pulsing is especially important before long term truck and sea shipments (Borochov *et al.* 1975a; Danieli *et al.* 1972; Halevy *et al.* 1978a,b; Halevy and Mayak 1974a,b; Yvernel and Paulin 1976).

Pulsing also reduces the sensitivity of the flowers to ethylene (Mayak and Kofranek 1976). Recently it was reported (Reid *et al.* 1980; Staby *et al.* 1979; Veen 1979a,b; Veen and Van de Geijn 1978) that a few minutes of treating carnations' and other flowers' bases with silver thiosulfate greatly reduced ethylene damage.

Other pretreatments are similar to those described in "Storage." Of special importance is pretreatment with BA to delay foliage yellowing in sensitive flowers (see "Foliage Discoloration"). Dipping *Anthurium andreanum* flowers with BA also decreased the chilling injury to the flowers during transport (Shirakawa *et al.* 1964).

Ethylene is one of the main sources of trouble during transport. Especially high levels of ethylene (more than 1000 ppm) were found in air terminals (Mayak *et al.* 1977b; Peles *et al.* 1977). Ethylene scrubbers in shipping boxes have been suggested, and some commercial preparations containing brominated activated charcoal and/or $KMnO_4$ are available. It was shown, however, that a scrubber is effective only when used in large quantities, which renders the use of ethylene-scrubbers impractical (Akamine 1976; Peles *et al.* 1977).

Nichols (1979) found that, in some instances, one or two pollinated carnation flowers in the box caused reduced longevity in the unpollinated flowers, apparently because of the increased ethylene production by pollinated flowers. Akamine (1976) reported a similar phenomenon in orchid marketing.

The question of mixed shipment of flowers with fruits and vegetables has been studied very little in spite of its great practical importance. It is normally expected that mixed shipment with fruits producing high levels of ethylene should be avoided. The highest levels of ethylene were recorded for melon (2400 μl/kg/hour at 25°C), followed by apple, pear, avocado, banana, and tomato (ca. 1000 μl/kg/hour), broccoli and celery (750 μl/kg/hour), pepper, eggplant, and citrus (ca. 500 μl/kg/hour) to lettuce and cauliflower (0.050 μl/kg/hour) (Mayak *et al.* 1977b). Wintz (1954) found that cabbage, cucumber, carrot, sweet potato, lettuce, potato, grapes, and cauliflower did not damage carnation flowers.

The heat generated by the respiration of cut rose, forsythia, and iris flowers is about twice that recorded for produce such as peppers, chicory and celery (Anon. 1980). The heat generated by cut flowers greatly increased with the rise in temperature (Anon. 1980; Lutz and Hardenburg 1968; Mayak *et al.* 1980).

Other than tropical flowers, the best handling of most cut flowers is under refrigeration from the grower to final consumer, apart from the periods when they undergo special pretreatments with chemical solutions (pulsing, opening, or tinting), which are normally done at high temperatures. Controlled temperatures are difficult to maintain in air shipments, but easier in refrigerated land and sea transport.

It was suggested that ice be included in shipping boxes, placed not in the middle of the box as is commonly done but at the two ends of the box close to the flowers (Halevy and Mayak 1974a; Harvey *et al.* 1962). A special box was designed for this purpose (Harvey *et al.* 1962, 1963). However, ice in the boxes had only very slow cooling effect on the flowers, even when transported in refrigerated trucks (Halevy and Mayak 1974a; Halevy *et al.* 1978a). It may take 2 days or more under these conditions to remove the "field heat" of the flowers and overcome the heat of respiration. For the ice and the refrigeration systems to be effective during transit, flowers should be *precooled* prior to loading (Farnham *et al.* 1978; Halevy and Mayak 1974a; Halevy *et al.* 1978a; Rij *et al.* 1979). The simplest way of precooling is by holding the unpacked bunched flowers or the packed *open* boxes for several hours in the cooler, until the flowers reach the desired temperature, since cooling of flowers in closed boxes is a very slow process (Halevy and Mayak 1974a; Halevy *et al.* 1978a). Ice should be put in the boxes after precooling. Flower packing should be done at low temperature, since rewarming of unpacked flowers occurs quickly (Halevy and Mayak 1974a; Halevy *et al.* 1978a; Maxie *et al.* 1973).

Several methods of fast precooling have been suggested recently and some are already in operation. Most involve forced air for precooling with air temperature of 0° to 1°C and 95 to 98% RH. Variations include: (1) "Hydrair" cooling—forced air with fine mist of water; (2) blowing cold air over open boxes moving on conveyor belts through a cooler (Borochov and Mayak 1973), with box closing after precooling; (3) forcing cold air through perforated plastic tubes placed in the boxes during packing; the tubes are taken out of the box after precooling and reused; this is the most extensively evaluated method in Israel (Peles *et al.* 1978); (4) forcing cold air through vented boxes, done either by pulling or by pushing the cold air through the boxes; several types of these forced air coolers have been installed recently in flower packing houses in California (Farnham *et al.* 1978; Rij *et al.* 1979); (5) vacuum cooling. This method, commonly used for some leafy vegetables, was suggested for precooling of flowers more than 20 years ago (Perry *et al.* 1958). It was tested experimentally in several places (André *et al.* 1980; Wiersma and Boer 1970; Danieli *et al.* 1972; Duvekot 1973; Flanzy 1975; Greidanus 1971) with good results, but as far as we know, is not being used commercially for precooling of flowers. The main disadvantage of this method is that it cools by evaporation, thus causing water loss from the flowers.

Cooling time by forced-air or vacuum varies from 8 to 60 minutes, with the type of flowers, size of boxes, and the method used.

Flowers should be kept constantly under refrigeration, after precooling and during transit, to maintain the cold temperature of the flowers. When this is not feasible it is advisable to put ice in the boxes close to the flower heads and use insulated boxes (Borochov and Mayak 1973; Halevy *et al.* 1978a).

In several simulated and actual long term (5 to 28 days) experimental shipments of flowers by land or sea, flower quality was comparable and in some cases better than that of uncontrolled air-shipped flowers for a much shorter period by combining pulsing, precooling, and transport under refrigeration (Borochov *et al.* 1975a; Danieli *et al.* 1972; Halevy *et al.* 1978a,b; Halevy and Mayak 1974a,b; Mayak *et al.* 1969a, 1970a,b, 1980).

Controlled or modified atmosphere shipment of flowers has been tested and is done commercially on a limited scale by prepacking flowers for consumer use in inflated plastic bags. This package is said also to form a protective air cushion around delicate flowers like gloriosa or lilies, and also prevent the flowers from drying out (Boer and Hatkema 1977). Controlled atmosphere is created by flushing the package before sealing with air or with the desired atmosphere. Modified atmosphere is created in the sealed package by the flower respiration (Lutz and Hardenburg 1968). No advantage was found from filling the pack with several mixtures of CO_2 and air as compared with air alone (Smith 1967).

X. SPECIAL PROBLEMS

A. Flower Bud and Petal Abscission

Abscission of flower buds is one of the main troubles of cut flowers, mainly of those having racemous inflorescences, bearing flowers of various stages of development, like snapdragon, sweet pea, *Delphinium* and *Lupinus* species, and some orchids. Some beautiful garden flowers are not commonly used commercially because they drop their flower buds and petals during handling and shipping.

The final stage of flower senescence is, in many instances, abscission. Shedding may involve the entire inflorescence, florets, or parts of flowers. Most studies on abscission were carried out with leaves and fruits (Addicott 1970; Addicott and Lyon 1973; Morrison Baird and Webster 1979), but some work also has been done with flower bud abscission (Simons 1973). The physiological, biochemical, anatomical, and ultrastructural processes accompanying abscission seem to be similar to those of leaves and fruits (Addicott and Wiatr 1977; Gilliland *et al.* 1976; Simons 1973; Valdovinos *et al.* 1974; Webster and Chiu 1975). The physiology of petal shedding has been greatly ignored by researchers. An important work was published by Fitting in 1911, but the next major work on petal abscission is probably that of Addicott in 1977. A question is thus raised as to what extent petal abscission is similar to or different from that of leaf abscission. Some facts seem to indicate that the process in petals may not be identical to that of leaves.

When whole flower buds are shed, an abscission layer is normally formed as with leaves (Hänisch Ten Cate *et al.* 1973; Simons 1973). However, cell division usually does not precede petal shedding; no clear abscission layer is apparent and shedding of petals is caused by softening of the middle lamella (Esau 1965; Wiatr 1978). There are indications, however, that petal shedding involves a rising activity of cell wall hydrolytic enzymes as in leaves (Addicott 1977; Wiatr 1978).

Fitting (1911) already had observed that external factors such as shaking, wounding, high temperature, and some gases induced very rapid petal abscission (in sensitive species). The response with petals is much faster than with leaves. Promotion of petal shattering by high temperatures was observed in roses (Halevy and Kofranek 1976) and in geraniums (Armitage *et al.* 1980). Ethylene promotes abscission of flower buds and petals of many flowers. This has been known since the work of Zimmerman *et al.* (1931) on roses. However, CO_2, which is known to antagonize ethylene and retard leaf abscission (Addicott and Lyon 1973), did not delay petal shattering of geranium (Armitage *et al.* 1980) and sometimes even promoted petal abscission (Fitting 1911). A rise in ethylene production was associated with flower or bud abscission in several plants (Blanpied 1972; Israeli and Blumenfeld 1980; Morgan *et al.* 1974; Swanson *et al.* 1975). Flower buds of *Ecballium* respond to ethylene by abscising only after they have reached a critical stage in their development. This was shown to be associated with the differentiation of specific "target cells" for ethylene in the separations region (Wong and Osborne 1978).

In many species, pollination and especially fertilization promote the abscission of flower parts, especially the petals (Fitting 1911). Breeding for "shatterproof" cultivars of some flowers such as snapdragon and geranium is based on inducing pollen sterility which prevents fertilization (Marks 1969), producing "double" flowers which lack sexual organs (Kofranek 1958; Wallner *et al.* 1979), and selecting for resistance to moderate concentrations of ethylene (Haney 1958).

Stead and Moore (1979) found in *Digitalis* flowers a correlation between longevity of the corolla and the amount of pollen loaded on the stigma. IAA application to the stigma or style did not reproduce the pollination effect. Pollination of the receptive stigma leads to a rapid and progressive weakening of the abscission zone between the corolla and the receptacle and the eventual abscission of the turgid corolla. Weakening of the abscission zone was detectable within 8 hours after pollination, while the pollen tubes were still growing through the stigma. Weakening of the abscission zone is therefore independent of fertilization. This and the fact that removal of the pollinated stigma was sufficient to prevent accelerated corolla abscission suggest movement of the stimulus for corolla abscission through the style and ovaries.

The promotion of abscission by pollination is apparently associated with a rise in ethylene evolution at this stage (Akamine 1963; Burg and Dijkman 1967; Hall and Forsyth 1967; Wallner *et al.* 1979). Aminoethoxyvinyl glycine (AVG), an inhibitor of ethylene synthesis, reduced ethylene production and flower abscission of snapdragon (Wang *et al.* 1977), and treatment with an ethylene antagonist, a benzothiodiazole (TH6241), prevented abscission in flowers exposed to 0.5 ppm of ethylene (Parups 1973). Pretreatment of the base of cut snapdragon (Farnham *et al.* 1981), sweet peas, and delphinium (Mor *et al.* 1980b; Shillo *et al.* 1980) with silver thiosulfate prevented the natural or ethylene-induced florets' abscission. Other inhibitors of ethylene biosynthesis ($CoCl_2$ and AVG) retarded petal abscission in *Linum lewisii* (Wiatr 1978). Lining containers with filter paper moistened with 0.01% mercuric perchlorate (an ethylene absorbent) prevented flower drop in snapdragon (Parups 1969).

Gilbart and Sink (1970, 1971) showed that abscission of poinsettia bracts is related to the endogenous levels of auxin. Abscission was correlated with increased activity of IAA-oxidase and decreased levels of auxin. Application of IAA to debladed petioles inhibited their abscission, while other growth regulators were not effective.

The most common auxin used to reduce flower abscission in cut flowers and potted plants is NAA in spray or dip at 30 to 50 ppm. The treatment was found to be useful in *Dendrobium* orchid (Boyd 1965), sweet pea (Halevy 1967b), *Bougainvillea* (Borochov *et al.* 1975b; Hackett *et al.* 1972), peony (Whiteman 1949), and Geraldton wax flower *(Chamaelaucium uncinatum)* (Halevy *et al.* 1974b). Much higher concentrations of auxins (250 to 800 ppm) were used for *Lupinus polyphyllus* (Warne 1947; Aarts 1958) and *Bougainvillea* (Pearse 1976). NAA, however, did not delay petal abscission in geranium (Armitage *et al.* 1980). 2,4-D at 10 to 300 ppm delayed flower drop of cut snapdragon, clarkia, delphinium, and *Eschscholtzia* (Ruge 1957). In sweet pea 2,4-D (0.2 to 2 ppm) prevented bud and petal drop but caused spike distortion (Kulwiec 1968). In roses cytokinin (50 to 100 ppm), rather than auxin, was a better inhibitor of flower bud and petal abscission (Halevy and Kofranek 1976).

Application of chemical solutions through the cut stem base also was found to reduce abscission. Maleic hydrazide (200 to 400 ppm) and citric acid (500 ppm) were active in *Lupinus hartwegii* Lindl (Mohan Ram and Rao 1977), and a solution of 2% sucrose and 0.1% boric acid or higher concentration (6 to 8%) of sucrose alone in *L. polyphyllus* (Aarts 1958). This suggests that carbohydrate levels may affect the efficiency of the auxin treatment, since the effect of auxin was enhanced when flowers were treated with sucrose (Aarts 1958). Also, NAA inhibited petal drop of cut peonies in a normal year but not in a very rainy one, in which the endogenous carbohydrate level was probably low (Whiteman 1949). The antiauxin p-chlorophenoxyisobutyric acid (1 ppm) inhibited petal drop in roses (Carpenter and Dilley 1975).

Abscission of begonia flowers (Hänisch Ten Cate and Bruinsma 1973) and *Linum* petals (Addicott 1977) was promoted by external application of ABA, as with leaves (Addicott 1970). A rise in endogenous-free abscisic acid of *Hibiscus rosa-sinensis* was observed at petal drop stage, increasing sharply to a peak at calyx abscission (Swanson *et al.* 1975). However, the endogenous ABA contents of pistachio (Takeda and Crane 1980) and begonia (Hänisch Ten Cate *et al.* 1975) flower buds were not correlated with abscission. In *Lupinus luteus* flowers endogenous ABA was correlated with abscission (Porter 1977). However, direct application of ABA to flowers and injection into the stem failed to promote abscission. Only foliar application increased flower abscission. The role of ABA as a primary regulator of flower abscission is still unclear.

Oviposition by the cotton boll weevil caused inevitable abscission of the flower buds. The active agent was found (King and Lane 1969) to be a protein, which accelerated abscission when injected into flower buds. It was tentatively identified later as endo-polymethylgalacturonase (King 1973).

The summary of flower and petal abscission reveals similarities and dissimilarities with analogous processes in leaves and fruits. Much more work is needed, especially on petal shedding, to clarify the situation.

B. Foliage Discoloration: Yellowing and Darkening

Normally green leaves on flower shoots as well as bracts and the green calyx last much longer than the non-chlorophyllous petals. Flowers bearing green petals such as green dahlia *(D. viridiflora)* or the green rose *(R. Chinensis viridiflora)* last longer than the normally colored flowers. In some flowers, such as in tulips, it sometimes happens that one of the petals is green and it remains attached to the stem longer than the normal colored petals (Tames 1978). It is obvious that in most flowers bearing leaves, the longevity of the cut flowering shoot is determined by petal longevity. In a few instances, the foliage may deteriorate before the flower, and in these cases the longevity of the cut unit is determined mainly by the foliage. Two foliage discoloration processes are known: yellowing and darkening.

Yellowing is characterized by breakdown of chlorophylls, proteins, and nucleic acids in the detached leaves. Yellowing is increased when cut flowers are held in darkness and at high temperatures (see Thimann 1978, Thomas and Stoddart 1980). Foliar yellowing is a major problem in certain spray chrysanthemum cultivars (e.g., 'Polaris', 'Fandango', 'Marbles'), statice, miniature gladiolus ('Charm'), lilies, alstromeria, marguerite daisy, and some other minor crops.

High temperatures promote senescence processes, including leaf yellowing, in almost all flowers. Light, which has little or no effect on the longevity of most flowers, significantly retarded senescence and leaf yellowing of chrysanthemum (Woltz and Waters 1967, 1976), dahlia, stock, and pyrethrum (Aarts 1957a, 1964). In chrysanthemum high nutrients (especially nitrogen) during the growing period enhanced postharvest yellowing (Posner and Halevy 1980; Waters 1967).

It is well known that cytokinins can delay the whole syndrome of senescence in leaves. In chrysanthemum marguerite daisy, alstromeria, and statice *(Limonium sinuatum)*, a foliar spray with the cytokinins PBA or BA at 20 to 100 ppm, prior to shipment or storage, greatly delayed yellowing and maintained the turgidity of the leaves (Borochov *et al.* 1975a; Nichols 1968b; Shillo and Halevy 1980; Posner *et al.* 1980; Uota and Harris 1964). Dip treatment of the foliage was found to be more effective than spray and a lower concentration could be used. Higher concentrations of cytokinins (250 ppm) were needed to control yellowing in miniature gladiolus ('Charm') and *Ixia* (Gvilli and Mayak 1970). A 15-minute dip in 50 ppm BA was also very effective. In treating cut flower shoots with cytokinin care should be taken to cover only the foliage and not the flower buds, since the cytokinin may retard the development and opening of the flowers.

In some plants cytokinins had little or no effect, and gibberellins were more effective in delaying senescence (Thimann 1978). This is also the case with some lily cultivars (Zieslin, unpublished). In some other plants growth retardants were more effective in delaying yellowing than were cytokinins and gibberellins (Halevy 1967a). In *Euphorbia fulgens* foliage yellowing was retarded by immersing the bases of the stems in solution of citric acid for one hour (Staden 1978).

Leaf darkening (browning or blackening) is a major problem mainly in some of the Protaceae. The rate of discoloration varies widely among different species—from a few days after harvest, to full blackening in some species (e.g., *Protea repens*), to over a month in others (e.g., *P. longifolia*) (Akamine *et al.* 1979).

The darkening is caused by oxidation of native phenols, mainly leucoanthocyanins, which react with other cell constituents to produce many dark condensation products (De Swardt 1977). Oxidation is catalyzed by polyphenol oxidase which is a copper-containing enzyme (Meyer and Harel 1979). The process is promoted by wounding and by ionizing radiation (Haasbroek and Rousseau 1971; Haasbroek *et al.* 1973). The rate of leaf darkening was found to be significantly negatively correlated with the content of Mn, Zn, and N. No correlation was found to other minerals, nor to sugars, dry weight, or moisture content (Akamine *et al.* 1979).

Blackening was promoted by high temperatures (Jacobs and Minnaar 1977a) and darkness (Jacobs and Minnaar 1977b). Since blackening is retarded even by low light intensities (400 lux), its effect is not directly related to photosynthesis. Blackening was promoted also by water droplets on the leaf surface. Darkening occurred much more rapidly in flowers picked and packed in the morning, compared to those cut in the afternoon (Jacobs and Minnaar 1980). This may be related to the moisture on stems cut in the morning. Water stress also promoted the blackening reaction in the leaves (De Swardt 1977; Paull *et al.* 1981).

Preshipping, pulsing, or holding flower stem bases in several commercial preservative solutions delayed darkening. Best results were obtained with a solution of 2 to 3% sucrose and 200 ppm HQC (Akamine *et al.* 1979; Paull *et al.* 1981). Another effective solution contained 3.5% sucrose, 2500 ppm MH, 100 ppm $Al_2(SO_4)_3$, 1000 ppm citric acid, and 100 ppm hydrazine sulphate (Haasbroek *et al.* 1973).

C. Geotropic Bending

Flower stem and inflorescence bending as a response to gravity is a major problem of some cut flowers, mainly during transport, when flowers are held horizontally.

Flowers with long spike inflorescences, such as gladiolus, snapdragon, larkspur, wallflower, *Kniphofia, Molucella* (Bells of Ireland), and lupine, are sensitive to geotropic bending, but the problem is known also in other flowers such as gerbera, tulips, anemones, and some rose cultivars (e.g., 'Garnetts').

Flowers normally have a negative geotropic response, as do vegetative stems. Some flowers, such as *Fritillaria* (Kaldeway 1962) and poppy (Kohji *et al.* 1979), change their response to gravity as they age; first they have a negative response, then positive, and finally negative again just before anthesis.

The geotropic response of plants has been reviewed several times in recent years (Firn and Digby 1980; Juniper 1976; Wilkins 1978). It is caused by asymmetric growth due (especially of monocotyledonous shoots) to downward lateral movement of auxin. Some other hormones like gibberellins and ABA also may be involved in the control of this response.

The geotropic response is retarded by low temperatures and low humidity. However, even at low temperatures the reaction is perceived by the tissue, and stem curvature may occur later when gladiolus flowers are placed vertically in water at higher temperatures (Whiteman and McClellan 1946).

Since geotropic bending is a growth process, it occurs only in flower shoot regions capable of linear growth after harvest. Young inflorescences of lupine *(Lupinus polyphyllus)* were very sensitive to geotropic bending, but after flowers were fertilized, the stems lost their geotropic reactivity (Brauner and Diemer 1968). In roses, flowers cut at an early stage were more sensitive to bending than those cut at later stages (Borochov *et al.* 1974). In snapdragon (Teas and Sheehan 1957) and gerbera (Shoub and Mayak 1969), however, little or no difference in bending was found among flowers cut at various stages. In the poppy *(Papaver rhoeas)*, the bending zone of the flower stalk moved acropetally from the base, as the flower aged, until finally the whole stalk was negatively geotropic (Kohji *et al.* 1979). Unlike the responses of coleoptiles, young seedlings, and some flowers (Kaldeway 1962; Kohji *et al.* 1979), removing the inflorescence tip of snapdragon did not diminish its geotropic response (Teas and Sheehan 1957). The geotropic response was very quick. Snapdragon (Teas and Sheehan 1957) and lupine tips (Brauner and Diemer 1968) started bending up 10 to 15 minutes after they were placed in a horizontal position, and were bent 80° to 90° from the horizontal axis in 8 hours.

Several solutions have been suggested or tried to prevent geotropic bending:

(a) Handle and ship flowers vertically in specially designed containers. This practice is common in shipping gladioli.

(b) Snip out the top two to three buds of the spike. This practice is done sometimes by florists.

(c) Hold flowers dry in a vertical position in cold storage for one day before shipping. This method reduced the geotropic reactivity of roses, gladioli (Borochov *et al.* 1974), anemones (Kanterovitz *et al.* 1973), and gerberas (Shoub and Mayak 1969).

(d) Reduce the turgidity of the flowers before shipment to decrease their ability to elongate and also to respond to gravity (Borochov *et al.* 1974; Shoub and Mayak 1969; Kanterovitz *et al.* 1973; Whiteman and McClellan 1946).

(e) Irradiate with gamma rays at high doses (20 krad) to inhibit geotropic bending of snapdragons (Teas *et al.* 1959).

(f) Treat flowers with growth inhibitors, antiauxins, and chemicals blocking auxin transport.

Of several chemicals tried on snapdragons, the most active was N-1-naphthylphtalmate (NP) at 250 ppm, applied as stem base treatment for 4 hours. The treatment reduced curvature to half that of control flowers. However, in commercial shipments treated flowers were only

40% marketable because of bent stems (Teas and Sheehan 1957; Joiner *et al.* 1977).

Treatment of gladiolus flowers with NP and other chemicals including 2,3,5-triiodobenzoic acid (TIBA) (Teas *et al.* 1959) and morphactins (Borochov *et al.* 1974) had little or no effect. NP (Morgan and Söding 1958) as well as TIBA (Christie and Leopold 1956) and morphactins (Bopp 1972) are known to be inhibitors of polar transport of auxin.

In gerberas, a base treatment of flowers with 0.5% chloromequat for 16 hours prior to shipment completely controlled geotropic bending (Shoub, unpublished results). Stem base treatment of anemone flowers with the growth retardants chloromequat and daminozide controlled bending, but effective concentrations also caused foliage injury (Kanterovitz *et al.* 1973).

Unfortunately, the chemicals tried so far have not given completely safe control of geotropic bending.

D. Other Stem Problems

The phenomenon of *bent neck*, a severe problem mainly in roses and gerberas, already has been described (see "Water Relations").

Stem tip breakage in the florets area of the stem is a problem in snapdragon, mainly during periods of low light. The break point was found to be related to the end of the lignified xylem (Adams and Urdahl 1972). Breakage appeared to be related also to anthocyanin content which is produced from the same precursors as lignin. Red cultivars producing anthocyanins throughout the entire plant showed less breakage, and are apparently capable of producing both the pigment and lignins. In other cultivars, there is an apparent competition for the precursors and an increase in stem breakage.

Excessive stalk elongation associated with bending is one of the main problems of many tulip cultivars (Benschop and De Hertogh 1969, 1971). Postharvest elongation is predominantly of the last internode, and occurs mainly in the first 24 hours in the vase. Removal of the flower bud inhibited stem elongation. Auxin applied to the debudded stalk also inhibited stem elongation, and cytokinins and gibberellins were ineffective (Op den Kelder *et al.* 1971). Similar results were obtained with intact flowers (Hanks and Rees 1977).

A-cyclopropyl-α-(4-methoxyphenyl)-5-pyrimidinemethanol (Ancymidol) as a preharvest spray (50 to 200 ppm) (Einert 1975) or at 25 ppm in the holding solution was effective in reducing elongation, with no detrimental effects (Einert 1971). Ethephon (1 to 100 ppm) in the holding solution or as a 24-hour pulse (Nichols 1971a) and BA (10 to 50 ppm) for 2 to 4 hours (Sytsema 1971) were also effective.

Stem base splitting and upward curling is a problem in *Hippeastrum*. It can be prevented with 1% chloromequat (Carow and Röber 1979).

XI. CONCLUDING REMARKS

The study of petal senescence and postharvest physiology of cut flowers lagged for many years behind similar studies on other plant organs and horticultural crops. This is in some way reflected by the fact that the two parts of the present review are the first comprehensive review published on the subject. The situation has changed somewhat in the last two decades, so that these subjects have attracted increasing numbers of researchers, exploring both the basic and the applied aspects of flower senescence.

We now have systems that greatly improve the longevity and quality of the major flower crops, and have good understanding of some aspects of flower senescence. However, many of the practical problems have not yet been solved and a host of basic questions remain unanswered. While reviewing the present knowledge of flower senescence and postharvest physiology of cut flowers, we attempted also to point out the many unresolved questions that should be the subjects of future research.

XII. LITERATURE CITED

AARTS, J.F.TH. 1957a. Over de houdbaarheid van snijbloemen. *Meded van de Landbouwhogeschool te Wageningen* 57:1–62.

AARTS, J.F.TH. 1957b. De ontwikkeling en houdbaarheid van afgesneden bloemen. *Meded. Dir. Tuinb.* 20:690–701.

AARTS, J.F.TH. 1958. Influence of alpha naphthyl acetic acid, sucrose, and boric acid on the flower drop of *Lupinus polyphyllus* Ldf. *Koninkl. Nederl. Wetensch. (Amsterdam)* 61:325–333.

AARTS, J.F.TH. 1964. The keepability of cut flowers. p. 46–53. *In* A. Lecrenier and P. Goeseels (eds.) Proc. 16th Intern. Hort. Congr. Aug. 31–Sept. 8, 1962, Brussels. J. Duculot, Gembloux, Belgium.

ABELES, F.B., L.E. FORRENCE, and G.R. LEATHER. 1971. Ethylene air pollution. *Plant Physiol.* 48:504–505.

ACCATI, E., S. MAYAK, and I. ABBATISTAGENTILE. 1981. The role of bacterial metabolites in affecting water uptake by carnations flowers. *Acta Hort.* 113 (in press).

ACOCK, B. and R. NICHOLS. 1979. Effects of sucrose on water relations of cut, senescing, carnations—flowers. *Ann. Bot.* 44:221–230.

ADAMS, D.G. and W.A. URDAHL. 1972. Snapdragon stem tip breakage as related to stem lignification and flower color. *J. Amer. Soc. Hort. Sci.* 97:474–477.

ADAMS, D.O. and S.F. YANG. 1977. Methionine metabolism in apple tissue. *Plant Physiol.* 60:892–896.

ADAMS, D.O. and S.F. YANG. 1979. Ethylene biosynthesis: identification of 1-aminocyclopropane-1-carboxylic acid as an intermediate in the conversion of methionine to ethylene. *Proc. Nat. Acad. Sci. (USA)* 76:170–174.

ADDICOTT, F.T. 1970. Plant hormones in the control of abscission. *Biol. Rev.* 45:485–524.

ADDICOTT, F.T. 1977. Flower behavior in *Linum lewissii*: some ecological and physiological factors in opening and abscission of petals. *Amer. Midland Naturalist* 97:321–332.

ADDICOTT, F.T. and J.L. LYON. 1973. Physiological ecology of abscission. p. 85–124. *In* T.T. Kozlowski (ed.) Shedding of plant parts. Academic Press, New York.

ADDICOTT, F.T. and S. WIATR. 1977. Hormonal control of abscission: biochemical and ultrastructural aspects. p. 249–257. *In* P.E. Pilet (ed.) Plant growth regulation. Springer Verlag, Berlin.

AHARONI, M. and A.H. HALEVY. 1975. Postharvest handling of *Phalaenopsis* orchids. Annu. Rpt. Dept. Orn. Hort. Hebrew Univ. for 1974–75 (Hebrew). p. 58–59. Rehovot.

AHARONI, M. and A.H. HALEVY. 1977. Experiments on postharvest handling of *Phalaenopsis* flowers. Annu. Rpt. Dept. Orn. Hort. Hebrew Univ. for 1975–77 (Hebrew). p. 74–75. Rehovot.

AHARONI, M. and S. MAYAK. 1977. Water conductance in stems of *Phalaenopsis* cut flowers. Annu. Rpt. Dept. Orn. Hort. Hebrew Univ. for 1975–77 (Hebrew). p. 76–77. Rehovot.

AKAMINE, E.K. 1963. Ethylene production in fading vanda orchid blossoms. *Science* 140:1217–1218.

AKAMINE, E.K. 1976. Postharvest handling of tropical ornamental cut crops in Hawaii. *HortScience* 11:125–126.

AKAMINE, E.K. and T. GOO. 1975. Vase life extension of anthurium flowers with commercial floral preservatives, chemical compounds and other materials. *Flor. Rev.* 155(4027):14–15, 56–60.

AKAMINE, E.K., T. GOO, and R. SUEHISA. 1979. Relationship between leaf darkening and chemical composition of leaves of species of protea. *Flor. Rev.* 163(4236):62–63, 107–108.

ALBERT, A., M.I. GIBSON, and S.D. RUBBO. 1953. The influence of chemical constitution on antibacterial activity. The bactericidal action of 8-hydroxyquinoline (oxine). *British J. Expt. Pathol.* 34:119–130.

ALTMAN, A., R. KAUR-SAWHNEY, and A.W. GALSTON. 1977. Stabilization of oat leaf protoplasts through polyamide-mediated inhibition of senescence. *Plant Physiol.* 60:570–574.

AMRHEIM, N. and D. WENKLER. 1979. Novel inhibitors of ethylene production in higher plants. *Plant Cell Physiol.* 20:1635–1642.

ANDRÉ, P., C. FLANZY, M. BURET, Y. CHAMBROY, P. DAUPLÉ, C. PÉLISSE, and R. BLANC. 1980. Étude de la conservation après récolte de fleurs, fruits et légumes au moyen de la préréfrigération par le vide associée à des atmosphères modifées. *P.H.M. Revue Hort.* 204:23–32.

ANON. 1979. The use of proprietary material 'Purafil E.S.' to reduce ethylene from flower pack. Annu. Rpt. Rosewarne Expt. Hort. Sta. 1978. p. 39–41, 46–47.

ANON. 1980. Cool answer to longer life. *Grower* 93(16):73–75.

APELBAUM, A. and M. KATCHANSKY. 1977. Improving quality and prolonging vase life of bud cut flowers by pretreatment with thiabendazole. *J. Amer. Soc. Hort. Sci.* 102:623–625.

APELBAUM, A. and M. KATCHANSKY. 1978. Effects of thiabendazole on ethylene production and sensitivity to ethylene of bud cut flowers. *HortScience* 13:593–594.

ARDITTI, J. 1979. Aspects of the physiology of orchids. p. 422–656. *In* H.W. Woolhouse (ed.) Advances in botanical research, Vol. 7. Academic Press, London.

ARDITTI, J., N.M. HOGAN, and A.V. CHADWICK. 1973. Post-pollination phenomena in orchid flowers. IV. Effect of ethylene. *Amer. J. Bot.* 60:883–888.

ARMITAGE, A.M., R. HEINS, S. DEAN, and W. CARLSON. 1980. Factors influencing flower petal abscission in the seed-propagated geranium. *J. Amer. Soc. Hort. Sci.* 105:562–564.

ARNY, D.C., S.E. LINDOW, and C.D. UPPER. 1976. Frost sensitivity of *Zea mays* increase by application of *Pseudomonas syringae*. *Nature* 262:282–284.

ASEN, S. and M. LIEBERMAN. 1963. Ethylene oxide experimentation aimed at cut flower longevity. *Flor. Rev.* 131(3398):27–28.

BAKER, J.E., C.Y. WANG, M. LIEBERMAN, and R. HARDENBURG. 1977. Delay of senescence in carnations by a rhizobitoxine analog and sodium benzoate. *HortScience* 12:38–39.

BALLANTYNE, D.J. 1965. Senescence of daffodil (*Narcissus pseudonarcissus* L.) cut flowers treated with benzyladenine and auxin. *Nature* 205:819.

BALLANTYNE, D.J. 1966. Respiration of floral tissue of the daffodil (*Narcissus pseudonarcissus* L.) treated with benzyladenine and auxin. *Can. J. Bot.* 44:117–119.

BANGERTH, F. 1973. Zur Wirkung eines reduzierten Drucks auf Physiologie, Qualitat und Lagerfähigheit von Obst, Gemüse und Schnittblumen. *Gartenbauwissenshaften* 38:479–508.

BARDEN, L.E. and J.J. HANAN. 1972. Effect of ethylene on carnation keeping life. *J. Amer. Soc. Hort. Sci.* 97:785–788.

BARENDSE, L.V.J. 1978. The addition of chlorine to gerbera is no longer experimental (Dutch). *Vakblad v.d. Bloemsterij* 33(15):29. (*Hort. Abstr.* 5: 1321, 1979.)

BAR-SELLA, P. 1959. Causes of rapid wilting of cut dahlias and means to improve their keeping qualities. *Bul. Res. Counc. Israel.* 7D:35–38.

BARRS, H.D. 1968. Determination of water deficits on plant tissues. p. 235–368. *In* T.T. Kozlowski (ed.) Water deficits and plant growth, Vol. 1. Academic Press, New York.

BAUMGARTNER, B., H. KENDE, and P. MATILE. 1975. Ribonuclease in senescing morning glory. *Plant Physiol.* 55:734–737.

BAUMGARTNER, B. and P. MATILE. 1977. Isoenzymes of RNAse in senescing morning glory petals. *Z. Pflanzenphysiol.* 82:371–374.

BEAUCHAMP, C. and I. FRIDOVICH. 1970. A mechanism for the production of ethylene from methional. The generation of the hydroxyl radical by xanthine oxidase. *J. Biol. Chem.* 245:4641–4646.

BENDALL, D.S. and W.D. BONNER. 1971. Cyanide-insensitive respiration in plant mitochondria. *Plant Physiol.* 47:236–245.

BEN-YEHOSHUA, S., B. JUVEN, M. FRUCHTER, and A.H. HALEVY. 1966. Effect of ethylene oxide on opening and longevity of cut rose flowers. *Proc. Amer. Soc. Hort. Sci.* 89:677–682.

BENSCHOP, M. and A.A. DE HERTOGH. 1969. An analysis of the postharvest characteristics of cut tulips. *Flor. Rev.* 145:24–25, 62–65.

BENSCHOP, M. and A.A. DE HERTOGH. 1971. Postharvest development of cut tulip flowers. *Acta Hort.* 231:121–126.

BESEMER, S.T. 1977. Handling of cut roses—conditioning after harvest, use of antitranspirant and various consumer solutions. *Flor. Rev.* 160(4136):80–81, 128–132.

BEUTELMANN, P. and H. KENDE. 1977. Membrane lipids in senescing flower tissue of *Ipomoea tricolor*. *Plant Physiol.* 59:888–893.

BEYER, E.M. 1976a. A potent inhibitor of ethylene action in plants. *Plant Physiol.* 58:268–271.

BEYER, E.M. 1976b. Silver ion: a potent antiethylene agent in cucumber and tomato. *HortScience* 11:195–196.

BEYER, E.M. 1977. $^{14}C_2H_4$: Its incorporation and oxidation to $^{14}CO_2$ by cut carnations. *Plant Physiol.* 60:203–206.

BEYER, E.M. 1978. $^{14}C_2H_4$ metabolism in morning glory flowers. *Plant Physiol.* 61:896–899.

BEYER, E.M. 1979. Effect of silver ion, carbon dioxide and oxygen on ethylene action and metabolism. *Plant Physiol.* 63:169–173.

BIGGS, P.R. 1961. Composition and method for preserving cut flowers. U.S. Patent No. 3,122,432.

BLANPIED, G.D. 1972. Study of ethylene in apple, red raspberry and cherry. *Plant Physiol.* 49:627–630.

BLUMENFELD, A. 1975. Ethylene and the annona flower. *Plant Physiol.* 55:265–269.

BOER, W.C. and H. HATKEMA. 1977. Cut flowers in inflated packages (Dutch). *Verkblad v. d. Bloemisterij* 32:52–53.

BOPP, M. 1972. On the effect of morphactin. p. 333–348. *In* H. Kaldeway and Y. Vardar (eds.) Hormonal regulation in plant growth and development. Verlag Chemie, Weinheim.

BOROCHOV, A., G. GRUPER, A. PELES, T. TIROSH, R. SHILLO, M. HERMAN, A. GVILLI, and A.H. HALEVY. 1975a. Evaluation of methods for sea-transport of carnation, gladiolus and statice cut flowers. Annu. Rpt. Dept. Orn. Hort. Hebrew Univ. for 1974–75A (Hebrew). p. 46–48. Rehovot.

BOROCHOV, A., A.H. HALEVY, H. BOROCHOV, and M. SHINITZKY. 1978. Microviscosity of rose petal's plasmalemma as affected by age and environmental factors. *Plant Physiol.* 61:812–815.

BOROCHOV, A., A.H. HALEVY, and T. TIROSH. 1975b. Post harvest treatments of Bougainvillea cut branches. Annu. Rpt. Dept. Orn. Hort. Hebrew Univ. for 1974–75 (Hebrew). p. 56–57. Rehovot.

BOROCHOV, A., A.H. HALEVY, T. TIROSH, J. SHOUB, and A. GVILLI. 1975c. Evaluation of methods for bud opening of various cut flowers. Annu. Rpt. Dept. Orn. Hort. Hebrew Univ. for 1974–75A (Hebrew). p. 43–45. Rehovot.

BOROCHOV, A., S. MAYAK, and A.H. HALEVY. 1976a. Combined effects of abscisic acid and sucrose on the growth and senescence of rose flowers. *Physiol. Plant.* 36:221–224.

BOROCHOV, A. and T. TIROSH. 1975. Effect of harvesting stage and temperature during air-transit on quality of spray carnations. Annu. Rpt. Dept. Orn. Hort. Hebrew Univ. for 1974–75A (Hebrew). p. 49. Rehovot.

BOROCHOV, A., T. TIROSH, G. GRUPER, and A.H. HALEVY. 1974. Treatments to control stem bending of gladiolus and rose flowers during shipment. Annu. Rpt. Dept. Orn. Hort. Hebrew Univ. for 1973–74 (Hebrew). p. 97–101. Rehovot.

BOROCHOV, A., T. TIROSH, and A.H. HALEVY. 1976b. Abscisic acid content of senescing petals on cut rose flowers as affected by sucrose and water stress. *Plant Physiol.* 58:175–178.

BOROCHOV, A., T. TIROSH, A. PELES, G. GRUPER, and A.H. HALEVY. 1975d. Commercial application of pulsing treatments to spray carnations and gladiolus flowers. Annu. Rpt. Dept. Orn. Hort. Hebrew Univ. for 1974–75A (Hebrew). p. 50–52. Rehovot.

BOROCHOV, M. and S. MAYAK. 1973. Effect of temperature during air-transport of rose flowers and suggestions for improvement of flower handling before and during transit. Annu. Rpt. Dept. Orn. Hort. Hebrew Univ. for 1972–73 (Hebrew). p. 10–37. Rehovot.

BOSE, T.K. and S.P.S. RAGHAVA. 1975. Low temperature prolongs the storage life of jasmine flower. *Sci. & Cult.* 41:115–116.

BOYD, H. 1965. Bud drop of orchids. *Gard. Chron.*, Nov., p. 510.

BRANTLEY, R.K. 1975. A new postharvest chemical treatment for roses. *HortScience* 10:178–179.

BRAUNER, L. and R. DIEMER. 1968. Ueber das geoelektrische Reaktionsvermoegen der Infloresczenzachsen von *Lupinus polyphyllus*. *Planta* 81: 113–131.

BRAVDO, B., S. MAYAK, and Y. GRAVRIELI. 1974. Sucrose and water uptake from concentrated sucrose solutions by gladiolus shoots and the effect of these treatments on floret life. *Can. J. Bot.* 52:1271–1281.

BREDMOSE, N. 1979. The influence of subatmospheric pressure on storage life and keeping quality of cut flowers of 'Belinda' roses. *Acta Agr. Scandinovica* 29:287–290.

BREDMOSE, N. 1981. Effects of low pressure on storage life and subsequent keeping quality of cut roses. *Acta Hort.* 113 (in press).

BRENNAN, T. and C. FRENKEL. 1977. Involvement of hydrogen peroxide in the regulation of senescence in pear. *Plant Physiol.* 59:411–416.

BROERTJES, C. 1955. The forcing of *Forsythia intermedia spectabilis* Khne. p. 1065–1071. *In* J.P. Nieuwstraten (ed.) Proc. Intern. Hort. Congr., Aug. 29–Sept. 9, 1953, The Hague. Veenman and Zonen, Wageningen.

BROWN, R. and S. MAYAK. 1980. Aminooxyacetic acid as an inhibitor of ethylene synthesis and senescence in carnation flowers. *Scientia Hort.* (in press).

BUFLER, G., Y. MOR, M.S. REID, and S.F. YANG. 1980. Changes in 1-aminocyclopropane-1-carboxylic acid content of cut carnations. *Planta* (in press).

BUNEMANN, G. and D.H. DEWEY. 1956. Cold storage of cut tulips with and without water. *Mich. Agr. Expt. Sta. Quart. Bul.* 38:580–587.

BURDETT, A.N. 1970. The cause of bent neck in cut roses. *J. Amer. Soc. Hort. Sci.* 95:427–431.

BURG, S.P. 1962. The physiology of ethylene formation. *Annu. Rev. Plant Physiol.* 13:265–302.

BURG, S.P. 1973. Hypobaric storage of cut flowers. *HortScience* 8:202–205.

BURG, S. and E.A. BURG. 1966. Fruit storage at subatmospheric pressure. *Science* 153:314–315.

BURG, S.P. and M.J. DIJKMAN. 1967. Ethylene and auxin participation in pollen induced fading of vanda orchid blossoms. *Plant Physiol.* 42:1648–1650.

BUYS, C. 1969. Blaetter beeinflussen die Haltbarkert der Schnittblumen. *Zierpflanzenbau* 15:639–640.

BUYS, C. 1973. Frischhaltenmittel oder Post Harvest Physiologie. *Gartenwelt* 11:243–244.

BUYS, C. 1979. Quality problems in gerbera. p. 255–261. *In* L. Quagliotti and A. Baldi (eds.) Proc. Eucarpia Meeting on Carnation and Gerbera, Apr. 24–28, 1978, Allassio. Minerva Press, Turino.

BYRNE, T.G., D.S. FARNHAM, and L.S. PYETT. 1979. Postharvest studies with marguerite daisies. *Flor. Nurs. Rpt. (Univ. of Calif.)* Winter, p. 1–4.

CAMPRUBI, P., A. ALDRUFEU, M. PAGÉS, R. BARGALLO, and J. LOPEZ. 1981. Effect of sodium borohydride on the ethylene production and carbohydrates status in petals of carnation (*Dianthus caryophyllus* L.) cut flower. *Acta Hort.* 113 (in press).

CAMPRUBI, P. and R. FONTARNAU. 1977. Relationship between the vase life of the cut flower and the plugging of the xylem vessels of carnations. *Acta Hort.* 71:233–240.

CAMPRUBI, P. and R. NICHOLS. 1978. Effects of ethylene on carnation flowers *(Dianthus caryophyllus)* cut at different stages of development. *J. Hort. Sci.* 53:17–22.

CAMPRUBI, P. and R. NICHOLS. 1979. Ethylene-induced growth of petals and styles in the immature carnation inflorescence. *J. Hort. Sci.* 54:225–258.

CAROW, B. and R. RÖBER. 1979. Schaftknicken bei Hippeastrum. *Gartenbauwissenshaften* 44:67–70.

CARPENTER, W.J. and D.R. DILLEY. 1973. A benzothiadiazole extends cut carnation flower life. *HortScience* 8:334–335.

CARPENTER, W.J. and D.R. DILLEY. 1975. Investigations to extend cut flower longevity. Mich. State Univ. Res. Rpt. 263, East Lansing. p. 1–10.

CARPENTER, W.J. and H.P. RASMUSSEN. 1973. Water uptake rates by cut roses *(Rosa hybrida)* in light and dark. *J. Amer. Soc. Hort. Sci.* 98:309–313.

CARPENTER, W.J. and H.P. RASMUSSEN. 1974. The role of flower and leaves in cut flower water uptake. *Scientia Hort.* 2:293–298.

CHATTERJEE, S. and S.K. CHATTERJEE. 1971. Analysis of abscission process in cotton (*Gossypium barbadense* L.). Leaves of different maturity. *Indian J. Expt. Biol.* 4:485–488.

CHRISTIE, A.E. and A.C. LEOPOLD. 1956. On the manner of triiodobenzoic acid inhibition of auxin transport. *Plant Cell Physiol.* 6:337–345.

CHUA, S.E. 1970. Phytokinin-like activity of 8-quinolinol sulphate. *Nature* 225:101.

CLAUSEN, G. and K. KRISTENSEN. 1974. Keeping quality of roses, III–V. *Tidsskrift. Plant.* 78:666–681.

COORTS, G.D. and J.B. GARTNER. 1963. The effect of various solutions on keeping quality of Better Times rose with and without "hooks." *Proc. Amer. Soc. Hort. Sci.* 83:833–838.

CRAFTS, A.S. 1968. Water deficits and physiological processes. p. 85–190. *In* T.T. Kozlowski (ed.) Water deficits and plant growth, Vol. 2. Academic Press, New York.

CRILEY, R.A., C. WILLIAMSON, and R. MASUTANI. 1978. Vase life determinations for *Leucospermum cordifolium. Flor. Rev.* 163(4228):37–38, 80–82.

CROCKER, W. and L.I. KNIGHT. 1908. Effect of illuminating gas and ethylene upon flowering carnations. *Bot. Gaz.* 46:259–276.

CYWINSKA-SMOTER, K., R.M. RUDNICKI, and D. GOZZEZYNSKA. 1978. The effect of exogenous growth regulators in opening tight carnation buds. *Scientia Hort.* 9:155–165.

DANIELI, S., G. GRUPER, S. MAYAK, and A.H. HALEVY. 1972. Evaluation of transport of gladiolus flowers by sea. Annu. Rpt. Dept. Orn. Hort. Hebrew Univ. for 1971–72 (Hebrew). p. 92–95. Rehovot.

DAVENPORT, D.C., K. URIN, M.A. FISHER, and R.M. HAGEN. 1971. Anti-transpirant—effects and uses in horticulture. *Amer. Hort.* 50:110–113.

DEATON, M., J.W. BUXTON, and T.R. KEMP. 1980. Senescence studies in *Nicotiana offinis* flowers cultured *in vitro. HortScience* 15:433–434. (Abstr.)

DE HERTOGH, A.A. and G. SPRINGER. 1977. Care and handling of spring bulb flowers and plants and suggestions on the use and marketing of bulb flowers and plants. Holland Flower-Bulb Tech. Serv. Bul. 4. Netherland Flower-Bulb Inst. Hillegen.

DE JONG, J. 1978. Dry storage and subsequent recovery of cut gerbera flowers as an aid in selection for longevity. *Scientia Hort.* 9:389–397.

DE STIGTER, H.C.M. 1980. Water balance of cut and intact 'Sonia' rose plants. *Z. Pflanzenphysiol.* 99:131–140.

DE STIGTER, H.C.M. 1981a. Ethephon effects in cut 'Sonia' roses after pretreatment with silver thiosulfate. *Acta Hort.* 113 (in press).

DE STIGTER, H.C.M. 1981b. Water-balance aspects of cut and intact 'Sonia' rose plants, and effects of glucose, 8-hydroxyquinoline sulfate and aluminum sulfate. *Acta Hort.* 113 (in press).

DE SWARDT, G.E. 1977. Metodes om verbruining van proteablare te keer. *Landbauweekblad (Capetown)* 30:30–33.

DENNY, F.E. and L.P. MILLER. 1935. Production of ethylene by plant tissue as indicated by epinastic response of leaves. *Contrib. Boyce Thomp. Inst. Plant Res.* 7:97–102.

DILLEY, D.R. 1977. Hypobaric storage of perishable commodities—fruits, vegetables, flowers and seedlings. *Acta Hort.* 62:61–70.

DILLEY, D.R. and W.J. CARPENTER. 1975. The role of chemical adjuvants and ethylene synthesis on cut flower longevity. *Acta Hort.* 41:117–132.

DILLEY, D.R., W.J. CARPENTER, and S. BURG. 1975. Principles and application of hypobaric storage of cut flowers. *Acta Hort.* 41:149–258.

DIMOCK, A.W. and K.F. BAKER. 1950. Ethylene produced by diseased tissue injures cut flowers. *Flor. Rev.* 106(2754):27–28.

DUPUY, P. 1975. Conservation des fleurs par irradiation. *In* La conservation des fleurs coupees. *C.N.I.H. Flor. Orn. Pepin. Mon. Tech.* 75(2):83–91.

DURKIN, D. 1979a. Some characteristics of water flow through isolated rose stem segments. *J. Amer. Soc. Hort. Sci.* 104:777–783.

DURKIN, D. 1979b. Effect of millipore filtration, citric acid, and sucrose on peduncle water potential of cut rose flowers. *J. Amer. Soc. Hort. Sci.* 104:860–863.

DURKIN, D. 1981. Factors affecting hydration of cut flowers. *Acta Hort.* 113 (in press).

DURKIN, D. and R.H. KUC. 1966. Vascular blockage and senescence of the cut rose flower. *Proc. Amer. Soc. Hort. Sci.* 89:683–688.

DURSO, M. 1979. The relation of water stress to bent neck of cut roses. PhD Thesis, Cornell University, Ithaca, N.Y.

DUVEKOT, W.S. 1973. Transport of cut flowers. Annu. Rpt. Sprenger Inst. 1972. p. 66–67. Wageningen.

EASTWELL, K.C., P.K. BASSI, and M.E. SPENCER. 1978. Comparison and evaluation of methods for the removal of ethylene and other hydrocarbons from air for biological studies. *Plant Physiol.* 62:723–726.

EINERT, A.E. 1971. Reduction in last internode elongation of cut tulips by growth retardants. *HortScience* 6:459.

EINERT, A.E. 1975. Effects of ancymidol on vase behaviour of cut tulips. *Acta Hort.* 41:97–181.

EISENBERG, B.A., G.L. STABY, and T.A. FRETZ. 1978. Low pressure and refrigerated storage of rooted and unrooted ornamental cuttings. *J. Amer. Soc. Hort. Sci.* 103:732–737.

EISINGER, W. 1977. Role of cytokinins in carnation flower senescence. *Plant Physiol.* 59:707–709.

ESAU, K. 1965. Plant anatomy. John Wiley & Sons, New York.

ESKIN, M.N.A. and S. GROSSMAN. 1977. Biochemistry of lipoxygenase in relation to food quality. *Food Sci. & Nutr.* 9:1–40.

FARNHAM, D.S. 1973. Silver-impregnated stems aid chrysanthemum flower longevity. *Flower & Nurs. Rpt.* July, p. 1–5.

FARNHAM, D.S. 1975. Bud opening and overnight conditioning of gypsophila 'Bristol Fairy'. *Flower & Nurs. Rpt.* Nov.–Dec., p. 2–6.

FARNHAM, D.S., C. BARR, and A.H. HALEVY. 1971. The value of using chemical solutions for conditioning and bud opening carnations. *Flor. Rev.* 148(3846):27–28, 63–65.

FARNHAM, D.S., T.G. BYRNE, M.J. MAROUSKY, D. DURKIN, R. RIJ, J.F. THOMPSON, and A.M. KOFRANEK. 1979. Comparison of conditioning, precooling, transit method, and use of a floral preservative on cut flower quality. *J. Amer. Soc. Hort. Sci.* 104:483–490.

FARNHAM, D.S., A.M. KOFRANEK, and J. KUBOTA. 1978a. Bud opening of *Gypsophila paniculata* L. cv. Perfecta with Physan-20. *J. Amer. Soc. Hort. Sci.* 103:382–384.

FARNHAM, D.S., M.S. REID, and D.W. FUJINO. 1981. Shattering of snapdragon—effects of silver thiosulfate and ethephon. *Acta Hort.* 113 (in press).

FARNHAM, D.S., J.F. THOMPSON, R.F. HASEK, and A.M. KOFRANEK. 1977. Forced-air cooling for California flower crops. *Flor. Rev.* 161(4162):36–38.

FARNHAM, D.S., J.F. THOMPSON, and A.M. KOFRANEK. 1978b. Temperature management of cut roses during simulated transit: the effect of summer stress temperatures on flower quality delivered to consumers. *Flor. Rev.* 162(4195):26–28, 65–68.

FARNHAM, D.S., T. UEDA, A.M. KOFRANEK, and A.H. HALEVY. 1978c. Physan-20, an effective biocide for conditioning and bud opening of carnations. *Flor Rev.* 162(4190):24–26, 58–60.

FIRN, R.D. and J. DIGBY. 1980. The establishment of tropic curvatures in plants. *Annu. Rev. Plant Physiol.* 31:131–148.

FISCHER, C.W. 1950. Ethylene gas, a problem in cut flower storage. *N.Y. State Flower Growers Bul.* 61:1,4.

FITTING, H. 1911. Untersuchungen uber die varzeitige Entbatterung von Blueten. *Jahrb. Wiss. Bot.* 49:187–263.

FLANZY, C. 1975. Prerefrigeration par le vide et atmosphere controlle. *In* La conservation des fleurs coupees. *C.N.I.H. Mon. Tech.* 75(2):49–64.

FORSYTH, F.R., C.A. EAVES, and C.L. LOCKHART. 1967. Controlling ethylene levels in the atmosphere of small containers of apples. *Can. J. Plant Sci.* 47:717–718.

FRANKLIN, E.W. 1977. Flower storage. *Roses Inc. Bul.* Nov. p. 13–14.

FREELAND, P.W. 1974. An attempt to increase the longevity of cut daffodil *(Narcissus pseudonarcissus)* by chemical treatment. *J. Biol. Educ.* 8:161–166.

FRENKEL, C., C. CHEE-KOK, and N.A.M. ESKIN. 1976. Relation of ethylene metabolism to peroxide formation in fruits and potato tubers. *HortScience* 11:300. (Abstr.)

FUJINO, D.W., M.S. REID, and S.F. YANG. 1981. Effects of aminooxyacetic acid on postharvest characteristics of carnations. *Acta Hort.* 113 (in press).

GARROD, J.F. and G.P. HARRIS. 1978. Effect of gibberellic acid on senescence of isolated petals of carnation. *Ann. Appl. Biol.* 88:309–311.

GATES, C.T. 1968. Water deficits and growth of herbaceous plants. p. 135–190. *In* T.T. Kozlowski (ed.) Water deficits and plant growth, Vol. 2. Academic Press, New York.

GAY, A.P. and R. NICHOLS. 1977. The effects of some chemical treatments on leaf water conductance of cut flowering stems of *Chrysanthemum morifolium. Scientia Hort.* 6:167–177.

GILBART, D.A. and K.C. SINK. 1970. The effect of exogenous growth regulators on keeping quality in poinsettia. *J. Amer. Soc. Hort. Sci.* 95:784–787.

GILBART, D.A. and K.C. SINK. 1971. Regulation of endogenous indoleacetic acid and keeping quality of poinsettia. *J. Amer. Soc. Hort. Sci.* 96:3–7.

GILISSEN, L.J.W. 1977. Style controlled wilting of the flower. *Planta* 133: 275–280.

GILLILAND, M.G., C.H. BORNMAN, and F.T. ADDICOTT. 1976. Ultrastructure and acid phosphatase in pedicel abscission of *Hibiscus. Amer. J. Bot.* 63:925–935.

GILMAN, F.K. and P.L. STEPONKUS. 1972. Vascular blockage in cut roses. *J. Amer. Soc. Hort. Sci.* 97:662–667.

GLADON, R.J. and G.L. STABY. 1976. Opening of immature chrysanthemums with sucrose and 8-hydroxyquinoline citrate. *HortScience* 11:206–208.

GORDON, R.J., H. MAYRSOHN, and R.M. INGELS. 1968. C_2-C_5 hydrocarbons in the Los Angeles atmosphere. *Environ. Sci. Tech.* 2:1117–1120.

GOSZCZYNSKA, D. and J. NOWAK. 1979. The effect of growth regulators on quality and vase-life of dry-stored carnation buds. *Acta Hort.* 91:143–146.

GREIDANUS, P. 1971. Economic aspects of vacuum cooling. Annu. Rpt. Sprenger Inst., 1970. p. 76–77. Wageningen.

GVILLI, A. and S. MAYAK. 1970. Treatment with benzyladenine for control of yellowing in leaves and bracts of miniature gladiolus and *Ixia*. Annu. Rpt. Dept. Orn. Hort. Hebrew Univ. for 1969–70 (Hebrew). p. 128. Rehovot.

HAASBROEK, F.J. and G.G. ROUSSEAU. 1971. Effect of gamma-rays on cut *Protea compacta* R. Br. blooms. *Agroplantae* 3:5–6.

HAASBROEK, F.J., G.G. ROUSSEAU, and J.F. DE VILLIERS. 1973. Effects of gamma-rays on cut blooms of *Protea compacta* R. Br., *P. longiflora* Lamarck and *Leucospermum cordifolium* Salisb ex Knight. *Agroplantae* 5:33–49.

HACKETT, W.P., R.M. SACHS, and J. DE BIE. 1972. Growing bougainvillea as a flowering pot plant. *Flor. Rev.* 150(3886):21, 56–57.

HALEVY, A.H. 1967a. Effects of growth retardants on drought resistance and longevity of various plants. *Proc. 17th Intern. Hort. Congr.* 3:277–283.

HALEVY, A.H. 1967b. Reducing flower drop of cut sweet peas. Annu. Rpt. Dept. Orn. Hort. Hebrew Univ. for 1966–67 (Hebrew). p. 26. Rehovot.

HALEVY, A.H. 1972. Water stress and the timing of irrigation. *HortScience* 7:113–116.

HALEVY, A.H. 1976. Treatments to improve water balance of cut flowers. *Acta Hort.* 64:223–230.

HALEVY, A.H., T.G. BYRNE, A.M. KOFRANEK, D.S. FARNHAM, J.F. THOMPSON, and R.E. HARDENBURG. 1978a. Evaluation of postharvest handling methods for transcontinental truck shipments of cut carnations, chrysanthemums, and roses. *J. Amer. Soc. Hort. Sci.* 103:151–155.

HALEVY, A.H., M. FRUCHTER, S. MAYAK, and S. BEN-YEHOSHUA. 1968. Experiment in storage of rose flowers (Hebrew). *Hassadeh* 48:160–163.

HALEVY, A.H. and A.M. KOFRANEK. 1976. The prevention of flower bud and leaf abscission in pot roses during simulated transport. *J. Amer. Soc. Hort. Sci.* 101:658–660.

HALEVY, A.H. and A.M. KOFRANEK. 1977. Silver treatment of carnation flowers for reducing ethylene damage and extending longevity. *J. Amer. Soc. Hort. Sci.* 102:76–77.

HALEVY, A.H., A.M. KOFRANEK, and S.T. BESEMER. 1978b. Postharvest handling methods for bird-of-paradise flowers (*Strelitzia reginae* Ait.). *J. Amer. Soc. Hort. Sci.* 103:165–169.

HALEVY, A.H. and S. MAYAK. 1974a. Transport and conditioning of cut flowers. *Acta Hort.* 43:291–306.

HALEVY, A.H. and S. MAYAK. 1974b. Improvement of cut flower quality opening and longevity by pre-shipment treatments. *Acta Hort.* 43:335–347.

HALEVY, A.H. and S. MAYAK. 1975. Interrelationship of several phytohormones in the regulation of rose petal senescence. *Acta Hort.* 41:103–116.

HALEVY, A.H. and S. MAYAK. 1979. Senescence and postharvest physiology of cut flowers—part 1. p. 204–236. *In* J. Janick (ed.) Horticultural reviews, Vol. 1. AVI Publishing, Westport, Conn.

HALEVY, A.H., S. MAYAK, T. TIROSH, H. SPIEGELSTEIN, and A.M. KOFRANEK. 1974a. Opposing effects of abscisic acid on senescence of rose flowers. *Plant & Cell Physiol.* 15:813–821.

HALEVY, A.H., R. SHILLO, A. BOROCHOV, Y. KIRSCHOLTZ, and T. TIROSH. 1974b. Geraldton wax flower: growth and postharvest treatments.

Annu. Rpt. Dept. Orn. Hort. Hebrew Univ. for 1973–74 (Hebrew). p. 89–92. Rehovot.

HALEVY, A.H. and S.H. WITTWER. 1965. Prolonging cut flower life by treatment with growth retardants B-Nine and CCC. *Flor. Rev.* 136(3516):39–40.

HALEVY, A.H. and S.A. WITTWER. 1966. Effect of growth retardants on longevity of vegetables, mushrooms and cut flowers. *Proc. Amer. Soc. Hort. Sci.* 88:582–590.

HALL, I.V. and F.R. FORSYTH. 1967. Production of ethylene by flowers following pollination and treatment with water and auxin. *Can. J. Bot.* 45:1163–1166.

HANAN, J.J. 1967. Experiments in controlled atmosphere storage of carnations. *Proc. Amer. Soc. Hort. Sci.* 90:370–376.

HANEY, W.J. 1958. Snapdragons that are shatter resistant. *Flor. Exchange* 130:17.

HÄNISCH TEN CATE, CH. H. and J. BRUINSMA. 1973. Abscission of flower bud pedicels in begonia. 1. Effects of plant growth regulating substances on the abscission with intact plants and with explants. *Acta Bot. Neerl.* 22:666–674.

HÄNISCH TEN CATE, CH. H., J.J.L. VAN DER PLOEG-BOOGD, and J. BRUINSMA. 1973. Abscission of flower bud pedicels in begonia. III. Anatomical pattern of abscission. *Acta Bot. Neerl.* 22:681–685.

HÄNISCH TEN CATE, CH. H., J. BERGHOF, A.M.H. VAN DER HOORN, and J. BRUINSMA. 1975. Hormonal regulation of pedicel abscission in begonia flower buds. *Physiol. Plant.* 33:280–284.

HANKS, G.R. and A.R. REES. 1977. Stem elongation in tulip and narcissus: the influence of floral organs and growth regulators. *New Phytol.* 78: 579–591.

HANSON, A.D. and H. KENDE. 1975. Ethylene-enhanced ion and sucrose efflux in morning glory flower tissue. *Plant Physiol.* 55:663–669.

HANSON, A.D. and H. KENDE. 1976. Methionine metabolism and ethylene biosynthesis in senescent flower tissue of morning glory. *Plant Physiol.* 57: 528–537.

HARDENBURG, R.E., H.C. VAUGHT, and G.A. BROWN. 1970. Development and vase life of bud-cut Colorado and California carnations in preservative solutions following air shipment to Maryland. *J. Amer. Soc. Hort. Sci.* 95:18–22.

HARVEY, J.M., M. UOTA, R.H. SEGALL, and M.J. CEPONIS. 1963. Transit times and temperature of transcontinental cut-flower shipments. *USDA-ARS Mktg. Res. Rpt.* 592:1–16.

HARVEY, J.M., M. UOTA, R.H. SEGALL, J. M. LUTZ, M.J. CEPONIS, and H.B. JOHNSON. 1962. Transit temperatures in cut flowers shipped from California. *USDA-ARS-AMS* 459:1–11.

HASEGAWA, A., M. MANABE, M. GOI, and Y. IHARA. 1976. Studies on the keeping quality of cut flowers. II. On the floral preservative 'Kagawa Solution' and its practical use for cut carnation. *Tech. Bul. Fac. Agr. Kagawa Univ.* 27:85–94.

HASEK, R.F., J.A. HOWARD, and R.H. SCIARONI. 1969. Ethylene—its effect on flower crops. *Flor. Rev.* 144(3722):16–17, 53–56.

HEIDE, O.M. and J. OYDVIN. 1969. Effects of 6-benzylamino-purine on the keeping quality and respiration of glasshouse carnations. *Hort. Res.* 9:26–36.

HEINS, R.D. 1980. Inhibition of ethylene synthesis and senescence in carnations by ethanol. *J. Amer. Soc. Hort. Sci.* 105:141–144.

HEKSTRA, G. 1966. Houdbaarheid van tulpbloemen op water. *Weekbl. Bloemboll. Cult.* 77:305–306.

HEKSTRA, G. 1967. Die Haltbarkeit geschnittener Zwiebelblumen. *Dtsche. Gaertnerb.* 67:143–145.

HITCHCOCK, A.E., W. CROCKER, and P.W. ZIMMERMAN. 1932. Effect of illuminating gas on the lily, narcissus, tulip, and hyacinth. *Contrib. Boyce Thomp. Inst.* 4:155–176.

HITCHCOCK, A.E. and P.W. ZIMMERMAN. 1929. Effect of chemicals, temperature, and humidity on the lasting qualities of cut-flowers. *Amer. J. Bot.* 4:155–176.

HOLLEY, W.D. and B. MATTHEWS. 1962. Miscellaneous tests on cut flower life of carnations. *Colorado Flower Growers Bul.* 142:3–4.

HORIE, K. 1962. Studies on the flowering of *Tradescantia reflexa* with special reference to petal behavior. *Mem. Hyogo Univ. Agr.* 14(5):1–54.

HSIANG, T.H.T. 1951. Physiological and biochemical changes accompanying pollination in orchid flowers. I. General observations and water relations. *Plant Physiol.* 26:441–555.

HUMBERT, R.P. 1969. Potassium in relation to food production. *HortScience* 4:35–36.

ISENBERG, F.M.R. 1979. Controlled atmosphere storage of vegetables. p. 337–394. *In* J. Janick (ed.) Horticultural reviews, Vol. 1. AVI Publishing, Westport, Conn.

ISRAELI, Y. and A. BLUMENFELD. 1980. Ethylene production by banana flowers. *HortScience* 15:187–189.

ITAI, C. and A. BENZIONI. 1976. Water stress and hormonal response. p. 225–242. *In* O.L. Lange, L. Kappen and E.D. Scholzo (eds.) Ecological studies, analysis and synthesis, Vol. 19. Springer-Verlag, New York.

JACOBS, G. and H.R. MINNAAR. 1977a. Effect of temperature on the blackening of protea leaves. *SAPPEX Newsl. (Capetown)* 17:20–24.

JACOBS, G. and H.R. MINNAAR. 1977b. Effect of light on the blackening of protea leaves. *SAPPEX Newsl. (Capetown)* 18:18–20.

JACOBS, G. and H.R. MINNAAR. 1980. Effect of free water on leaves on blackening of protea. *SAPPEX Newsl. (Capetown)* (in press).

JAMES, H.A. 1963. Flower damage. A case study. Bay Area Air Poll. Contr. Dist. Inform. Bul. 8–63. San Francisco, Calif.

JEFFCOAT, B. 1977. Influence of the cytokinin, 6-benzylamino-9-(tetrahydropyran-2-yl)-9H-purine, on the growth and development of some ornamental crops. *J. Hort. Sci.* 52:143–153.

JEFFCOAT, B. and G.P. HARRIS. 1972. Hormonal regulation of the distribution of ^{14}C-labelled assimilates in the flowering shoot of carnation. *Ann. Bot.* 36:356–361.

JENSEN, H.E.K. and W. HANSEN. 1972. Keeping quality of roses. II. The influence of cold storage on the keeping quality and the opening of the flower. *Tidsskrift Planteavl.* 76:117–120.

JERIE, P.H. and M.A. HALL. 1978. The identification of ethylene oxide as a major metabolite of ethylene in *Vicia faba* L. *Proc. Roy. Soc. London,* Ser. B. 200:87–94.

JERIE, P.H., M.A. HALL, and M. ZERONI. 1978. Aspects of the role of ethylene in fruit ripening. *Acta Hort.* 80:325–332.

JOHANSSON, J. 1979. Main effects and interactions of N, P, and K applied to greenhouse roses. *Acta Agr. Scandinavica* 29:191–208.

JOINER, J.N., T.J. SHEEHAN, and K.F. MITCHELL. 1977. Control of ageotropic response in snapdragon flower spikes. *Fla. Flower Growers* 14(6):1–4.

JUNIPER, B.E. 1976. Geotropism. *Annu. Rev. Plant Physiol.* 27:385–406.

KALDEWEY, H. 1962. Plagio- und Diageotropismus der sprosse und Blaetter einschleisslich Epinastie, Hyponastie, Entfaltungsbewegungen. p. 200–321. *In* W. Ruhland (ed.) Encyclopedia of plant physiology, Vol. 17. Springer-Verlag, Berlin.

KANTEROVITZ, R., I. FINKILSTEIN, and A.H. HALEVY. 1973. Control of geotropic bending in anemones. Annu. Rpt. Dept. Orn. Hort. Hebrew Univ. for 1972–73 (Hebrew). p. 54–55. Rehovot.

KATCHANSKY, M. 1979. Bud opening and pulsing of gypsophila flowers with TOG II (Hebrew). *Hassadeh* 59:2312–2314.

KELLEY, J.D. and C.L. HAMNER. 1958. The effect of chelating agents and maleic hydrazide on the keeping qualities of snapdragon *(Antirrhinum majus). Quart. Bul. Mich. Agr. Expt. Sta.* 41:332–343.

KELLEY, J.D. and A.L. SCHLAMP. 1964. Keeping quality, flower size and flowering response of three varieties of Easter lilies to gibberellic acid. *Proc. Amer. Soc. Hort. Sci.* 85:631–634.

KENDE, H. and B. BAUMGARTNER. 1974. Regulation of aging in flowers of *Ipomoea tricolor* by ethylene. *Planta* 116:279–289.

KENDE, H. and A.D. HANSON. 1976. Relationship between ethylene evolution and senescence in morning glory flower tissue. *Plant Physiol.* 57:523–527.

KENDE, H. and A.D. HANSON. 1977. On the role of ethylene in aging. p. 172–180. *In* P.E. Pilet (ed.) Plant growth regulation. Springer-Verlag, Berlin.

KING, E.E. 1973. Endo-polymethygalacturonase of boll weevil larvae, *Anthonomus grandis*: an initiator of cotton flower bud abscission. *J. Insect Physiol.* 19:2433–2437.

KING, E.E. and H.C. LANE. 1969. Abscission of cotton flower buds and petioles caused by protein from boll weevil larvae. *Plant Physiol.* 44:903–906.

KINGHAM, H.G. (ed.). 1967. A manual of carnation production. Ministry Agr. Fish & Food Bul. 151, London.

KOFRANEK, A.M. 1958. Double snapdragons for shipping to Eastern markets. *Calf. State Flor. Assoc. Mag.* 8:4–5.

KOFRANEK, A.M. 1976. Opening flower buds after storage. *Acta Hort.* 64: 231–237.

KOFRANEK, A.M., E. EVANS, J. KUBOTA, and D.S. FARNHAM. 1978. Chemical pretreatments for China asters to increase flower longevity. *Flor. Rev.* 162(4206)26:70–72.

KOFRANEK, A.M., D.S. FARNHAM, E.C. MAXIE, and J. KUBOTA. 1975a. Long term storage of carnation buds. *Flor. Rev.* 151(3906):29–30, 72–73.

KOFRANEK, A.M. and A.H. HALEVY. 1972. Conditions for opening cut chrysanthemum flower buds. *J. Amer. Soc. Hort. Sci.* 97:578–584.

KOFRANEK, A.M. and A.H. HALEVY. 1976. Sucrose pulsing of gladiolus stems before storage to increase spike quality. *HortScience* 11:572–573.

KOFRANEK, A.M. and A.H. HALEVY. 1981. Chemical pretreatment of chrysanthemum before shipment. *Acta Hort.* 113 (in press).

KOFRANEK, A.M., A.H. HALEVY, and J. KUBOTA. 1975b. Bud opening of chrysanthemums after long term storage. *HortScience* 10:378–380.

KOFRANEK, A.M., H.C. KOHL, and J. KUBOTA. 1974. A slow-release chlorine compound as a vase water additive. *Flor. Rev.* 154(4000):21, 63–65.

KOFRANEK, A.M. and J. KUBOTA. 1972. Bud opening solutions for carnations. *Flor. Rev.* 151(3905):30–31.

KOFRANEK, A.M. and J.L. PAUL. 1972. Silver impregnated stems and carnation flower longevity. *Flor. Rev.* 151(3913):24–25.

KOFRANEK, A.M. and J.L. PAUL. 1974. The value of impregnating cut stems with high concentrations of silver nitrate. *Acta Hort.* 41:199–206.

KOHJI, J., H. HAGIMOTO, and Y. MASUDA. 1979. Georeaction and elongation of the flower stalk in a poppy, *Papaver rhoeas* L. *Plant & Cell Physiol.* 20:375–386.

KOHL, H.C. 1961. Rose neck droop. *Calif. State Flower Assoc.* 10:4–5.

KOHL, H.C. 1967. Irradiated roses. *Flower Nurs. Rpt. Calif.* Sept. p. 5–6.

KOHL, H.C. 1975. Cyclamen as cut flowers. *Flower Nurs. Rpt. Calif.* March. p. 6.

KOHL, H.C. and D.L. RUNDLE. 1972. Decreasing water loss of cut roses with abscisic acid. *HortScience* 7:249.

KOHL, H.C. and D.E. SMITH. 1960. Development of carnation flowers cut before fully open. *Carnation Craft* 53:7–8.

KRAFT, A.S. and R.J. ELLIS. 1971. Lack of cytokinin activity of quinolinol sulfate. *Plant Physiol.* 48:645–647.

KREBS, O. and K. ZIMMER. 1977. Eremurus im Schnittblumen-Sortiment. *Deutscher Gartenbau* 31:1534–1535.

KULWIEC, Z.J. 1967. Keeping quality of cut flowers. Annu. Rpt. GCRI, Littlehampton 1966. p. 121–122. Littlehampton, U.K.

KULWIEC, Z.J. 1968. Keeping quality of cut flowers. Annu. Rpt. GCRI, Littlehampton 1967. p. 119–120. Littlehampton, U.K.

LANCASTER, D.M. 1975a. Postharvest handling of cut carnations from harvest through consumer use. *Colorado Flower Growers Assoc. Bul.* 295:1–5.

LANCASTER, D.M. 1975b. Effects of saline water on keeping life of cut roses and carnations—a preliminary report. *Colorado Flower Growers Assoc. Bul.* 296:3–4.

LARSEN, F.E. and R.S. CROMARTY. 1967. Micro-organism inhibition by 8-hydroxyquinoline citrate as related to cut flower senescence. *Proc. Amer. Soc. Hort. Sci.* 90:546–549.

LARSEN, F.E. and M. FROLICH. 1969. The influence of 8-hydroxyquinoline citrate, N-dimethylamino succinamic acid, and sucrose on respiration and water flow in 'Red Sim' carnations in relation to flower senescence. *J. Amer. Soc. Hort. Sci.* 94:289–291.

LARSEN, F.E. and J.F. SCHOLES. 1965. Effects of sucrose, 8-hydroxyquinoline citrate, and N-dimethylamino succinamic acid on vase-life and quality of cut carnations. *Proc. Amer. Soc. Hort. Sci.* 87:458–463.

LARSEN, F.E. and J.F. SCHOLES. 1966. Effects of 8-hydroxyquinoline citrate, N-dimethylamino succinamic acid, and sucrose on vase life and spike characteristics of cut snapdragons. *Proc. Amer. Soc. Hort. Sci.* 89:694–701.

LARSEN, F.E. and J.F. SCHOLES. 1967. Effect of 8-hydroxyquinoline citrate, sucrose and Alar on vase life and quality of cut stocks. *Flor. Rev.* 139(3608): 46–47, 117.

LAU, O.L. and S.F. YANG. 1976. Inhibition of ethylene production by cobaltus ion. *Plant Physiol.* 58:114–117.

LAUMONNIER, R. 1959. Cultures florales Mediterraneennes. J. B. Bailliere et Fils, Paris.

LAURIE, A. 1936. Studies on the keeping qualities of cut flowers. *Proc. Amer. Soc. Hort. Sci.* 34:595–597.

LE MASSON, B. and A. PAULIN. 1979. Influence d'un déficit temporaire en eau sur le métabolism glucidique de l'Oeillet coupé (*Dianthus caryophyllus* L. var Scania). *C.R. Acad. Sci. Paris* Ser. D. 289:1097–1100.

LEIBOVITZ, B.E. and B.V. SIEGEL. 1980. Aspects of free radical reactions in biological systems. *J. Gerontol.* 35:45–56.

LESHEM, Y., S. GROSSMAN, A. FRIMER, and J. ZIV. 1979. Endogenous lipoxygenase control and lipid associated free radical scavenging as model of cytokinin action in plant senescence retardation. p. 193–198. *In* L. Apelquist and C.L. Liljenberg (eds.) Advances in the biochemistry and physiology of plant lipids, Vol. 3. Elsevier, Amsterdam.

LEVY, M. and J.J. HANAN. 1978. Some effects of floral preservatives on carnation keeping life. *Colorado Flower Growers Assoc. Bul.* 341:2–4.

LIEBERMAN, M. 1979. Biosynthesis and action of ethylene. *Annu. Rev. Plant Physiol.* 30:533–541.

LIEBERMAN, M., S. ASEN, and L. MAPSON. 1964. Ethylene oxide an antagonist of ethylene in metabolism. *Nature* 204:756–758.

LINEBERGER, R.D. and P.L. STEPONKUS. 1976. Identification and localization of vascular occlusions in cut roses. *J. Amer. Soc. Hort. Sci.* 101:246–250.

LIPE, J.A. and P.W. MORGAN. 1973. Location of ethylene production in cotton flowers and dehiscing fruits. *Planta* 115:93–96.

LIU, F.W. 1970. Storage of bananas in polyethylene bags with an ethylene absorbent. *HortScience* 5:25–27.

LIVNE, A. and Y. VAADIA. 1972. Water deficits and hormone relations. p. 255–275. *In* T.T. Kozlowski (ed.) Water deficits and plant growth, Vol. 3. Academic Press, New York.

LONGLEY, L.E. 1933. Some effects of storage of flowers in various gases at low temperatures on their keeping quality. *Proc. Amer. Soc. Hort. Sci.* 30: 607–609.

LOUGHEED, E.C., D.P. MURR, and L. BERARD. 1978. Low pressure storage for horticultural crops. *HortScience* 13:21–27.

LUKASZEWSKA, A. 1978. The effect of exogenously applied chemicals on keeping qualities of cut gladioli (Polish, English). *Prace Inst. Sadownictwa* Ser. B, 3:69–79.

LUNDQUIST, A.L. and R.C. MONGELLI. 1971. Saving possible by marketing standard chrysanthemums in the bud stage. *USDA-ARS Bul.* 52–67:1–11.

LUTZ, J.M. and R.E. HARDENBURG. 1968. The commercial storage of fruits, vegetables and florist and nursery stocks. *USDA Agr. Handb.* 66.

MACROBBIE, E.A. 1971. Fluxes and compartmentation in plant cells. *Annu. Rev. Plant Physiol.* 22:75–90.

MAGIE, R.O. 1961. Controlling gladiolus Botrytis bud rot with ozone gas. *Proc. Fla. State Hort. Soc.*:373–375.

MAGIE, R.O. 1963. Botrytis disease control of gladiolus, carnations and chrysanthemums. *Proc. Fla. State Hort. Soc.* 76:458–461.

MAPSON, L.W. and D.A. WARDALE. 1968. Biosynthesis of ethylene; enzymes involved in its formation from methional. *Biochem. J.* 107:433–442.

MARKS, G.E. 1969. Production of non-shattering forms of *Antirrhinums*. Annu. Rpt. John Innes Inst. 60:16. Norwich, U.K.

MAROUSKY, F.J. 1969. Vascular blockage, water absorption, stomatal opening, and respiration of cut 'Better Times' roses treated with 8-hydroxyquinoline citrate and sucrose. *J. Amer. Soc. Hort. Sci.* 94:223–266.

MAROUSKY, F.J. 1971. Inhibition of vascular blockage and increased moisture retention in cut roses induced by pH, 8-hydroxyquinoline citrate and sucrose. *J. Amer. Soc. Hort. Sci.* 96:38–41.

MAROUSKY, F.J. 1972. Water relations, effects of floral preservatives on bud opening and keeping quality of cut flowers. *HortScience* 7:114–116.

MAROUSKY, F.J. 1973. Recent advances in opening bud-cut chrysanthemum flowers. *HortScience* 8:199–202.

MAROUSKY, F.J. 1974. Influence of soluble salts and floral preservatives on open and bud-cut chrysanthemum and snapdragon flowers. *Proc. Trop. Reg. Amer. Soc. Hort. Sci.* 18:247–256.

MAROUSKY, F.J. 1976. Control of bacteria in vase water and quality of cut flowers as influenced by sodium dichloroisocyanurate, 1,3-dichloro-5.5-dimethylhydantoin, and sucrose. USDA-ARS-S-115.

MAROUSKY, F.J. 1977. Control of bacteria in gypsophila vase water. *Proc. Fla. State Hort. Soc.* 90:297–299.

MAROUSKY, F.J. and B.K. HARBAUGH. 1979. Ethylene-induced floret sleepiness in *Kalanchoe blossfeldiana* Poelln. *HortScience* 14:505–507.

MAROUSKY, F.J. and S.S. WOLTZ. 1975. Relationship of floral preservatives to water movement, fluoride distribution and injury in gladiolus and other cut flowers. *Acta Hort.* 41:171–182.

MARSBERGEN, W.V. and L. BARENDSE. 1977. Oorzaken zwarte bloemknoppen in Ilona opegspoord. *Vakblad v.d. Bloemisterij.* 32(46):34–35.

MATILE, P. 1978. Entwiklung einer Buete. Naturforschenden Gessellschaft, Zürich.

MATILE, P. and F. WINKENBACH. 1971. Function of lysosomes and lysosomal enzymes in the senescing corolla of the morning glory *(Ipomoea purpurea). J. Expt. Bot.* 22:759–771.

MAXIE, E.C., D.S. FARNHAM, F.G. MITCHELL, N.F. SOMMER, R.A. PARSONS, R.G. SNYDER, and H.L. RAE. 1973. Temperature and ethylene effects on cut flowers of carnation (*Dianthus caryophyllus* L.). *J. Amer. Soc. Hort. Sci.* 98:568–572.

MAYAK, S. and E. ACCATI-GARIBALDI. 1979. The effect of microorganisms on susceptibility to freezing damage in petals of cut rose flowers. *Scientia Hort.* 11:75–81.

MAYAK, S., E. ACCATI-GARIBALDI, and A.M. KOFRANEK. 1977a. Carnation flower longevity: microbial population as related to silver nitrate stem impregnation. *J. Amer. Soc. Hort. Sci.* 102:637–639.

MAYAK, S. and A. BAR-YOSEF. 1972. Effect of conditioning on bent-neck of Bridal Pink roses. Annu. Rpt. Dept. Orn. Hort. Hebrew Univ. for 1971–72 (Hebrew). p. 100–101. Rehovot.

MAYAK, S., A. BAR-YOSEF, and A.H. HALEVY. 1972a. Effect of water quality on cut roses. Annu. Rpt. Dept. Orn. Hort. Hebrew Univ. for 1971–72 (Hebrew). p. 102–103. Rehovot.

MAYAK, S., G. BITAN, and T. BEERI. 1980. Shipment of flowers in controlled temperature containers. Annu. Rpt. Dept. Orn. Hort. Hebrew Univ. for 1978–80. (Hebrew). p. 88–96. Rehovot.

MAYAK, S., A. BOROCHOV, and A. PELES. 1977b. Postharvest factors affecting the decrease in quality of carnation flowers. Annu. Rpt. Dept. Orn. Hort. Hebrew Univ. for 1975–77 (Hebrew). p. 63–65. Rehovot.

MAYAK, S., B. BRAVDO, A. GVILLI, and A.H. HALEVY. 1973. Improvement of opening of cut gladioli flowers by pretreatment with high sugar concentrations. *Scientia Hort.* 1:357–365.

MAYAK, S. and D.R. DILLEY. 1976a. Effect of sucrose on response of cut carnation to kinetin, ethylene and abscisic acid. *J. Amer. Soc. Hort. Sci.* 101:583–585.

MAYAK, S. and D.R. DILLEY. 1976b. Regulation of senescence in carnation *(Dianthus caryophyllus).* Effect of abscisic acid and carbon dioxide on ethylene production. *Plant Physiol.* 58:663–665.

MAYAK, S., A. GVILLI, and A. BAR-YOSEF. 1972b. Experiments on storage of rose flowers. Annu. Rpt. Dept. Orn. Hort. Hebrew Univ. for 1971–72 (Hebrew). p. 95–98. Rehovot.

MAYAK, S., A. GVILLI, and G. GRUPER. 1970a. Effect of pretreatment and transit conditions on rose flowers. Annu. Rpt. Dept. Orn. Hort. Hebrew Univ. for 1969–70 (Hebrew). p. 96–98. Rehovot.

MAYAK, S., A. GVILLI, A.H. HALEVY, and J. ZAFRIR. 1969a. Simulation experiment on sea-transport of cut flowers. Annu. Rpt. Dept. Orn. Hort. Hebrew Univ. for 1968–69 (Hebrew). p. 65–72. Rehovot.

MAYAK, S., A. GVILLI, A.H. HALEVY, J. ZAFRIR, and H. KLAUSNER. 1970b. Experiments on transport of flowers. Annu. Rpt. Dept. Orn. Hort. Hebrew Univ. for 1969–70 (Hebrew). p. 104–110. Rehovot.

MAYAK, S., A. GVILLI, R. SHILLO, A.H. HALEVY, S. SIMCHON, and D. EISENSTEIN. 1970c. Bud opening of cut flowers. Annu. Rpt. Dept. Orn. Hort. Hebrew Univ. for 1969–70 (Hebrew). p. 111–120. Rehovot.

MAYAK, S. and A.H. HALEVY. 1970. Cytokinin activity in rose petals and its relation to senescence. *Plant Physiol.* 46:497–499.

MAYAK, S. and A.H. HALEVY. 1971. Water stress as the cause for failure of flower bud opening in iris. *J. Amer. Soc. Hort. Sci.* 96:482–483.

MAYAK, S. and A.H. HALEVY. 1972. Interrelationships of ethylene and abscisic acid in the control of rose petal senescence. *Plant Physiol.* 50:341–346.

MAYAK, S. and A.H. HALEVY. 1974. The action of kinetin in improving the water balance and delaying senescence processes of cut rose flowers. *Physiol. Plant.* 32:330–336.

MAYAK, S. and A.H. HALEVY. 1980. Flower senescence. p. 131–156. *In* K.V. Thimann (ed.) Plant senescence. CRC Press, Boca Raton, Fla.

MAYAK, S., A.H. HALEVY, and M. KATZ. 1972c. Correlative changes in phytohormones in relation to senescence in rose petals. *Physiol. Plant.* 27:1–4.

MAYAK, S., A.H. HALEVY, S. SAGIE, A. BAR-YOSEF, and B. BRAVDO. 1974. The water balance of cut rose flowers. *Physiol. Plant.* 32:15–22.

MAYAK, S., A. HOROVITZ, and A.H. HALEVY. 1969b. Opening of rose flowers cut at a tight bud stage. Annu. Rpt. Dept. Orn. Hort. Hebrew Univ. for 1968–69 (Hebrew). p. 79–81. Rehovot.

MAYAK, S. and A. KOFRANEK. 1976. Altering the sensitivity of carnation flowers (*Dianthus caryophyllus* L.) to ethylene. *J. Amer. Soc. Hort. Sci.* 101:503–506.

MAYAK, S., A.M. KOFRANEK, and T. TIROSH. 1978. The effect of inorganic salts on the senescence of *Dianthus caryophyllus* flowers. *Physiol. Plant.* 43:282–286.

MAYAK, S., Y. VAADIA, and D.R. DILLEY. 1977c. Regulation of senescence in carnation *(Dianthus caryophyllus)* by ethylene mode of action. *Plant Physiol.* 54:591–593.

MCGLASSON, W.B. 1970. The ethylene factor. p. 475–519. *In* A.C. Hulme (ed.) The biochemistry of fruits and their products, Vol. 1. Academic Press, New York.

METZGER, B. 1972. Effects of growth regulators, HQC and sugar on cut rose vase-life. *Colorado Flower Growers Assoc. Bul.* 226:1–4.

MEYER, A.M. and E. HAREL. 1979. Polyphenol oxidases in plants. *Phytochemistry* 18:193–215.

MOE, R. 1975. The effect of growing temperature on keeping quality of cut roses. *Acta Hort.* 41:77–88.

MOHAN RAM, H.Y. and I.V. RAO. 1977. Prolongation of vase-life of *Lupinus hartwegii* Lindl. by chemical treatments. *Scientia Hort.* 7:377–382.

MONNET, Y. and A. PAULIN. 1975. Conditions d'eclairement a réaliser dens des chambres de floraison d'oeillets. *Pepin. Hort. Marach.* 159:17–25.

MOR, Y. and M.S. REID. 1981. Isolated petals—a useful system for studying flower senescence. *Acta Hort.* 113 (in press).

MOR, Y., M.S. REID, and A.M. KOFRANEK. 1980a. Role of the ovary in carnation senescence. *Scientia Hort.* 13:377–383.

MOR, Y., M.S. REID, and A.M. KOFRANEK. 1980b. Pulse treatment with silver thiosulfate and sucrose improve the vase life of sweet peas. *HortScience* (in press).

MORGAN, D.G. and H. SÖDING. 1958. Ueber die Wirkungsweise von Phthals-aeuremono-a-naphthylamid auf das Wachstum der Haferkoleoptile. *Planta* 52:235–249.

MORGAN, P.W., J.I. DURHAM, and J.A. LIPE. 1974. Ethylene production by cotton flower, role and regulation. p. 1062–1068. *In* Y. Sumiki (ed.) Plant growth substances. Hirokawa Publ., Tokyo.

MORRISON BAIRD, L.A. and B.D. WEBSTER. 1979. The anatomy and histochemistry of fruit abscission. p. 172–203. *In* J. Janick (ed.) Horticultural reviews, Vol. 1. AVI Publishing, Westport, Conn.

MURR, D.P., T. VENKATARAYAPPA, and M.J. TSUJITA. 1979. Counteraction of bent-neck of cut roses with cobalt nitrate. *Can. J. Plant Sci.* 59:1169–1171.

NEUMANNOVA, B. and A. PAULIN. 1974. Conditions de temperature et d'humidite relative a appliquer dans les chambres de floraison d'oeillets. *Ann. Tech. Agr.* 28:467–480.

NICHOLS, R. 1966. The storage and physiology of cut flowers. Factors affecting water uptake. Ditton Lab Annu. Rpt. 1965–66, p. 16–17. East Malling, England.

NICHOLS, R. 1967a. Investigations on the cold storage of cut flowers. Proc. 12th Intern. Congr. Refrig., Madrid. 3:329–337.

NICHOLS, R. 1967b. The storage and physiology of cut flowers. Factors affecting water uptake and senescence. Ditton Lab. Annu. Rpt. 1966–67. p. 24–28. East Malling, England.

NICHOLS, R. 1968a. The response of carnations *(Dianthus caryophyllus)* to ethylene. *J. Hort. Sci.* 43:335–349.

NICHOLS, R. 1968b. The storage and physiology of cut flowers. Ditton Lab. Annu. Rpt. 1967–68. p. 37–41. East Malling, England.

NICHOLS, R. 1971a. Control of tulip stem extension. Annu. Rpt. GCRI Littlehampton 1970. p. 74. Littlehampton, U.K.

NICHOLS, R. 1971b. Refrigeration and storage of cut flowers. *Refrig. & Air Cond.* 74:36–39.

NICHOLS, R. 1973. Senescence of the cut carnation flower: respiration and sugar status. *J. Hort. Sci.* 48:111–121.

NICHOLS, R. 1974. Developments in bud-opening and flower-preservative solutions for carnations (*Dianthus caryophyllus* L.). Annu. Rpt. GCRI Littlehampton 1973. p. 136–142. Littlehampton, U.K.

NICHOLS, R. 1975. Senescence and sugar status of the cut flower. *Acta Hort.* 41:21–29.

NICHOLS, R. 1976a. Observations of the effects of bud-opening solutions on spray chrysanthemum. Annu. Rpt. GCRI Littlehampton 1975. p. 129–133. Littlehampton, U.K.

NICHOLS, R. 1976b. Cell enlargement and sugar accumulation in the gynoecium of the glasshouse carnation *(Dianthus caryophyllus)* induced by ethylene. *Planta* 130:47–52.

NICHOLS, R. 1977. A descriptive model of the senescence of the carnation *(Dianthus caryophyllus)* inflorescence. *Acta Hort.* 71:227–232.

NICHOLS, R. 1978a. Quarternary ammonium compounds and flower preservatives. Annu. Rpt. GCRI Littlehampton 1977. p. 62. Littlehampton, U.K.

NICHOLS, R. 1978b. Cycloheximide and senescence of bulb flowers. Annu. Rpt. GCRI Littlehampton 1977. p. 60–61. Littlehampton, U.K.

NICHOLS, R. 1979. Ethylene, pollination and senescence. *Acta Hort.* 91:93–98.

NICHOLS, R. and L.C. HO. 1979. Respiration, carbon balance and translocation of dry matter in the corolla of rose flowers. *Ann. Bot.* 44:19–25.

NICHOLS, R. and L.J. KULWIEC. 1967. Flower preservatives and flower dyes. p. 71–73. *In* H. Kingham (ed.) A manual of carnation production. Ministry Agr. Fish & Food Bul. 151. London.

NICHOLS, R. and L.W. WALLIS. 1972. Cool storage of cut narcissus. *Expt. Hort.* 24:64–76.

NOWAK, J. 1979. Transport and distribution of silver ions in cut gerbera inflorescences. *Acta Hort.* 91:105–110.

NOWAK, J. and R.M. RUDNICKI. 1975. The effect of "Proflovit-72" on the extension of vase life of cut flowers. *Pr. Inst. Sadow.* Ser. B., Vol. 1, p. 173–179.

NOWAK, J. and R.M. RUDNICKI. 1979. Long term storage of cut flowers. *Acta Hort.* 91:123–134.

OAKS, A. and R.G.S. BIDWELL. 1970. Compartmentation of intermediary metabolites. *Annu. Rev. Plant Physiol.* 21:43–66.

OP DEN KELDER, P., M. BENSCHOP, and A.A. DE HERTOGH. 1971. Factors affecting floral stalk elongation of flowering tulips. *J. Amer. Soc. Hort. Sci.* 96:603–605.

OWENS, L.D., M. LIEBERMAN, and A. KUNISHI. 1971. Inhibition of ethylene production by rhizobitoxine. *Plant Physiol.* 48:1–4.

PARSONS, C.S., S. ASEN, and N.W. STUART. 1967. Controlled-atmosphere storage of daffodil flowers. *Proc. Amer. Soc. Hort. Sci.* 90:506–514.

PARUPS, E.V. 1969. Prolonging the life of cut flowers. Mercuric perchlorate absorbents found to prevent injury to flowers. *Can. Agr.* 14(4):18–19.

PARUPS, E.V. 1973. Control of ethylene-induced responses in plants by a substituted benzothiadiozole. *Physiol. Plant.* 29:365–370.

PARUPS, E.V. 1975a. Effects of flower care floral preservative on the vase life and bloom size of *Rosa hybrida* cv. 'Forever Yours' roses. *Can. J. Plant Sci.* 55:775–781.

PARUPS, E.V. 1975b. Inhibition of ethylene synthesis by benzylisothiocyanate and its use to delay the senescence of carnations. *HortScience* 10: 221–222.

PARUPS, E.V. and A.P. CHAN. 1973. Extension of vase life of cut flowers by use of isoascorbate-containing preservative solution. *J. Amer. Soc. Hort. Sci.* 98:22–26.

PARUPS, E.V. and J.M. MOLNAR. 1972. Histochemical study of xylem blockage in cut roses. *J. Amer. Soc. Hort. Sci.* 97:532–534.

PARUPS, E.V. and E.A. PETERSON. 1973. Inhibition of ethylene production in plant tissues by 8-hydroxyquinoline. *Can. J. Plant Sci.* 53:351–353.

PARUPS, E.V. and P.W. VOISEY. 1976. Lignin content and resistance to bending of the pedicel in greenhouse grown roses. *J. Hort. Sci.* 51:253–259.

PARVIN, P.E. and P.R. KRONE. 1963. Double vase life of roses with proper care. p. 74–88. *In* M.N. Rogers (ed.) Living flowers that last. Univ. of Missouri Press, Columbia.

PASTAC, I. and V. DRIGUET. 1947. Un probléme simplifie de macrobiotique: la prolongation de la vie des fleurs coupées. *C. R. Acad. Agr. France* 33:625–626.

PAULIN, A. 1972. Influence d'un déficit temporaire en eau sur le metabolism azote des fleurs coupées d'*Iris germanica*. *C. R. Acad. Sci.* 272:209–212.

PAULIN, A. 1973. Action de la cycloheximide sur la croissance et le metabolism azote des fleurs coupées d'*Iris germanica* L. *C. R. Acad. Agr. France* 11:900–907.

PAULIN, A. 1975a. La conservation frigorifique des fleurs coupées. *Bul. Inf. Tech. Min. Agr.* 265:64–79.

PAULIN, A. 1975b. Action exercée sur le métabolisme azote de la fleur coupée d'*Iris germanica* L. par l'apport exogere de glucose ou le cycloheximide. *Physiol. Vég.* 13:501–515.

PAULIN, A. 1975c. Effects of watering following a drought period on nitrogen metabolism of cut *Iris germanica* flowers. *Acta Hort.* 41:13–20.

PAULIN, A. 1979. Evolution des glucides dans les divers organes de la Rose coupée (var Carina) alimentee temporairement avec une solution glucosee. *Physiol. Vég.* 17:129–143.

PAULIN, A. 1980. Influence d'une alimentation continue avec une solution glucosée sun l'evolution des glucides et des protéines dans les divers organes de la rose coupée (*Rosa hybrida* cv. Carina). *Physiol. Plant.* 49:55–61.

PAULIN, A., M.J. DROILLARD, and D. SOUTER. 1979. La suppression de cueillettes des week-ends (application aux roses). *P.H.M. Rev. Hort.* 196:1–7.

PAULIN, A. and K. MULOWAY. 1979. Perspective in the use of growth regulators to increase the cut flower vase life. *Acta Hort.* 91:135–141.

PAULL, R., R.A. CRILEY, P.E. PARVIN, and T. GOO. 1981. Leaf blackening in cut *Protea eximia*; importance of water relations. *Acta Hort.* 113 (in press).

PEARSE, H.L. 1976. Control of flower and bract drop in potted Bougainvillea plants. *Agroplantae* 8:61–62.

PELES, A., A. BOROCHOV, T. TIROSH, and A.H. HALEVY. 1974. Commercial pulsing of gladiolus flowers. Annu. Rpt. Dept. Orn. Hort. Hebrew Univ. for 1973–74 (Hebrew). p. 104–106. Rehovot.

PELES, A., G. VALIS, Y. KIRSCHOLTZ, B. COHEN, and S. MAYAK. 1977. Experiment on handling of cut flowers in packing houses and air terminals (Hebrew). Agrexco and Flower Mkt. Board Bul. Tel Aviv, Israel.

PELES, A., G. VALIS, Y. KIRSCHOLTZ, B. COHEN, and S. MAYAK. 1978. Studies on postharvest problems of flowers (Hebrew). *Chamamot Uprachim* 8(3):23–25.

PERRY, R.L., H.C. KOHL, and A.M. KOFRANEK. 1958. Vacuum cooling of flowers. *Southern Calif. Flower Growers Bul.* May 12.

PHAN, C.T. 1970. Conversion of various substrates to ethylene by flowers. *Physiol. Plant.* 23:981–984.

POE, S.L. and F.J. MAROUSKY. 1971. Control of aphids and mites on cut chrysanthemums by postharvest absorption of Azodrin and Demeton. *Proc. Fla. State Hort. Soc.* 84:432–435.

POKORNY, F.A. and J.R. KAMP. 1953. Keeping cut flowers. *Ill. State Flor. Assoc. Bul.* 164:4–10.

PORTER, N.G. 1977. The role of abscisic acid in flower abscission of *Lupinus luteus.* *Physiol. Plant.* 40:50–54.

POSNER, B., A. BOROCHOV, and A.H. HALEVY. 1980. Postharvest handling of chrysanthemum flowers. Annu. Rpt. Dept. Orn. Hort. Hebrew Univ. for 1978–80 (Hebrew). p. 105–108. Rehovot.

POSNER, B. and A.H. HALEVY. 1980. Effect of nutrient levels during the growth periods on cut flowers and foliage longevity of chrysanthemum cv. Yellow Polaris. Annu. Rpt. Dept. Orn. Hort. Hebrew Univ. for 1978–80 (Hebrew). p. 64–67. Rehovot.

POST, K. 1956. Florist crop production and marketing. Orange Judd Publ., New York.

POST, K. and C.W. FISCHER. 1952. Commercial storage of cut flowers. *Cornell Univ. Ext. Bul.* 853:1–14.

PRATELLA, G.C., G. TOMARI, and R. TESI. 1967. La conservazione dei garofani in atmosfera controllata. *Rev. Ortoflorofrut Ital.* 51:301–310.

PRIMROSE, S.B. 1979. Ethylene and agriculture: the role of the microbe. *J. Appl. Bacteriol.* 46:1–25.

RAMANUJA RAO, I.V. and H.Y. MOHAN RAM. 1979. Interaction of gibberellin and sucrose in flower bud opening of gladiolus. *Indian J. Expt. Biol.* 17: 447–448.

RAMESHWAR, A. and P.L. STEPONKUS. 1970. Cytokinin-like activity of 8-quinolinol sulphate. *Nature* 228:1224–1225.

RASMUSSEN, H.P. and W.J. CARPENTER. 1974. Changes in the vascular morphology of cut rose stems: a scanning electron microscope study. *J. Amer. Soc. Hort. Sci.* 99:454–459.

RATSEK, J.C. 1935. Test metal containers in attempt to increase life of cut flowers. *Flor. Rev.* 76(1954):9,11 (cited from Post 1956).

RAVEL D'ESCLAPON, M.G. 1962. Les mimosas sur le littoral Meditterranéen. *Rev. Hort.* 134:332–339.

REID, M.S., D.S. FARNHAM, and E.P. MCENROE. 1980a. Effect of silver thiosulfate and preservative solutions on the vase life of miniature carnations. *HortScience* 15:807–808.

REID, M.S., J.L. PAUL, M.B. FARHOOMAND, A.M. KOFRANEK, and G.L. STABY. 1980b. Pulse treatments with the silver thiosulfate complex extend the vase life of cut carnations. *J. Amer. Soc. Hort. Sci.* 105:25–27.

REIST, A. 1977. Influence du stade au moment la coupe, de la composition et du pH d'une solution nutrative sur la qualite et la duree de vie des fleurs coupées de glaïeul (*Gladiolus* L. cv. Oscar). *Rev. Suisse Vitic. Arboric. Hort.* 9:221–225.

RIJ, R.E., J.F. THOMPSON, and D.J. FARNHAM. 1979. Handling, precooling and temperature management of cut flower crops for truck transportation. USDA-SEA Adv. Agr. Tech. West. Ser. 5:1–26.

RINGE, F. 1972. A further contribution to the question of cytokinin-like activity of 8-quinolinol sulphate. *Experientia* 28:234–235.

ROGERS, M.N. 1962. Sell flowers that last (reprints of 9 papers). *Flor. Rev.* 130 & 131:3378–3385.

ROGERS, M.N. (ed.). 1963. Living flowers that last—a national symposium. Univ. of Missouri Press, Columbia.

ROGERS, M.N. 1973. A historical and critical review of post-harvest physiology research on cut flowers. *HortScience* 8:189–194.

RONEN, M. 1979. The effect of abscisic acid on cut carnation flowers, and the relation to ethylene. MS Thesis (Hebrew), The Hebrew University of Jerusalem, Rehovot, Israel.

RUDNICKI, R.M. and J. NOWAK. 1976. Vase-life of *Gerbera jamesonii* Bolus cut flowers depending upon media, mineral nutrition, their morphological attributes and treatment with flower preservatives. *Acta Agrabotanica* 29:289–296.

RUGE, U. 1957. Über dan Abfall von Blueten und Bluetenblaettern. *Angew. Bot.* 31:126–129.

RYAN, W.L. 1957. Flower preservatives using silver and zinc ions as disinfectants. *Flor. Rev.* 121(3129):59–60.

SACALIS, J.N. 1974. Inhibition of vascular blockage and extension of vase life in cut roses with an ion exchange column. *HortScience* 9:149–151.

SACALIS, J.N. and R. NICHOLS. 1980. Effects of 2,4-D uptake on petal senescence in cut carnation flowers. *HortScience* 15:499–500.

SACHER, J.A. 1973. Senescence and postharvest physiology. *Annu. Rev. Plant Physiol.* 24:197–310.

SCHNABL, H. 1976. Aluminumioen als Frischhaltenmittel für Schnittblumen. *Deutsch. Gartenbau* 30:859–860.

SCHNABL, H. and H. ZIEGLER. 1974. Der Einfluss des Aluminium auf den Gasaustausch und das Welken van Schnittpflanzen. *Ber. Deutsch. Bot. Ges.* 87:13–20.

SCHNABL, H. and H. ZIEGLER. 1975. Über die Wirkung van Aluminiumionen auf die Stomatabewegung von *Vicia faba*—Epidermen. *Z. Pflanzenphysiol.* 74:394–403.

SCOTT, K.J., W.B. MCGLASSON, and E.A. ROBERTS. 1970. Potassium permanganate as an ethylene absorbent in polyethylene bags to delay ripening of bananas during storage. *Austral. J. Expt. Agr. & Animal Husb.* 10:237–240.

SCOTT, K.J. and R.B.H. WILLS. 1973. Atmospheric pollutants destroyed in an ultraviolet scrubber. *Lab. Practice* 22:103–106.

SHEEHAN, T.J. 1954. Orchid flower storage. *Bul. Amer. Orchid Soc.* 23: 579–584.

SHERWOOD, C.H. and C.L. HAMNER. 1948. Lengthening the life of cut flowers and floral greens by the use of plastic coatings. *Mich. Agr. Expt. Sta. Quart. Bul.* 30:272–276.

SHILLO, R. 1969. Post-harvest treatment of cut gladiolus flowers. Annu. Rpt. Dept. Orn. Hort. Hebrew Univ. for 1968–69 (Hebrew). p. 85–90. Rehovot.

SHILLO, R. and A.H. HALEVY. 1980. Postharvest handling of statice flowers. Annu. Rpt. Dept. Orn. Hort. Hebrew Univ. for 1978–80 (Hebrew). p. 99–102. Rehovot.

SHILLO, R., Y. MOR, and A.H. HALEVY. 1980. Prevention of flower drop in cut sweet peas and delphiniums (Hebrew, English Summary). *Hassade* 61: 274–276.

SHIRAKAWA, T., R.R. DEDOLPH, and D.P. WATSON. 1964. N6-benzyladenine effects on chilling injury, respiration and keeping quality of *Anthurium andraeanum*. *Proc. Amer. Soc. Hort. Sci.* 85:642–646.

SHOUB, J. and S. MAYAK. 1969. Harvesting stage and postharvest handling of gerbera flowers. Annu. Rpt. Dept. Orn. Hort. Hebrew Univ. for 1968–69 (Hebrew). p. 91–96. Rehovot.

SIMONS, R.K. 1973. Anatomical changes in abscission of reproductive structures. p. 383–434. *In* T.T. Kozlowski (ed.) Shedding of plant parts. Academic Press, New York.

SIMS, T. 1976. Once-over harvesting is now reality with chrysanthemums. *The Grower* 85:679–680.

SMELLIE, H. and P. BRINCKLOW. 1963. The use of antiseptics for delaying decomposition of cut flowers in a hospital ward. *Lancet* 2:777–778.

SMITH, W.H. 1967. Flower storage and marketing. p. 154–155. *In* H.G. Kingham (ed.) A manual of carnation production. Ministry Agr. Fish & Food Bul. 151, London.

SMITH, W.H. 1968. Storage in reduced oxygen atmospheres. Ditton Lab. Annu. Rpt. 1966–67. p. 19–20. East Malling, England.

SMITH, W.H., D.F. MEIGH, and J.C. PARKER. 1964. Effect of damage and fungal infection on the production of ethylene by carnations. *Nature* 204:92–93.

SMITH, W.H. and J.C. PARKER. 1966. Prevention of ethylene injury to carnations by low concentrations of carbon dioxide. *Nature* 211:100–101.

SMITH, W.H. and L.W. WALLIS. 1967. Use of low temperature to intensify colour of cut blooms of narcissus 'Soleil d'Or'. *Expt. Hort.* 17:21–26.

SMOCK, R.M. 1979. Controlled atmosphere storage of fruits. p. 301–336. *In* J. Janick (ed.) Horticultural reviews, Vol. 1. AVI Publishing, Westport, Conn.

SOUTER, D., J.M. BUREAU, and A. PAULIN. 1977. Vues nouvelles sur la conservation et le transport des oeillets. *Acta Hort.* 71:265–272.

SOUTHWICK, F.W. and R.G. SMOCK. 1943. Lengthening of storage life of apples by removal of volatile materials from the storage atmosphere. *Plant Physiol.* 18:716–717.

SPIERINGS, F. 1969. Injury to cut flowers of gladiolus by fluoridated water. *Neth. J. Plant Pathol.* 75:281–286.

STABY, G.L., M.S. CUNNINGHAM, B.A. EISENBERG, M.P. BRIDGEN, and J.W. KELLY. 1979. Protecting harvested carnation flowers with silver salts of sodiumthiosulfate. *HortScience* 14:445.

STABY, G.L. and T.D. ERWIN. 1977. Floral preservatives—it's time to clean the files. *Flor. Rev.* 159(4120):35, 79–82.

STABY, G.L. and T.D. ERWIN. 1978. Water quality, preservative, grower source and chrysanthemum flower vase-life. *HortScience* 13:155–157.

STABY, G.L., T.D. ERWIN, and A.A. DE HERTOGH. 1978. Cut flower life of tulips as influenced by preservative and water quality. *Flor. Rev.* 162(4200): 22–23.

STABY, G.L. and J.F. THOMPSON. 1978. An alternative method to reduce ethylene levels in coolers. *Flor. Rev.* 163(4229):30–31, 71.

STADEN, O.L. 1973. Physiology of cut flowers. Annu. Rpt. Sprenger Inst. 1972. p. 54–56. Wageningen.

STADEN, O.L. 1974. Vase life of cut flowers. Annu. Rpt. Sprenger Inst. 1973. p. 50. Wageningen.

STADEN, O.L. 1976. Vase life of cut flowers. Annu. Rpt. Sprenger Inst. 1974. p. 54–55. Wageningen.

STADEN, O.L. 1978. Cause and control of leaf yellowing of some cut flowers. Annu. Rpt. Sprenger Inst. 1977. p. 108. Wageningen.

STADEN, O.L. and W.H. MOLENAAR. 1975. Invloed verschillend leidingwater op vaasleven van snijbloemen. *Vakblad voor de Bloemisterij* 30(42):21.

STEAD, A.D. and K.G. MOORE. 1979. Studies on flower longevity in *Digitalis*. Pollination induced corolla abscission in *Digitalis* flowers. *Planta* 146:406–414.

STIRLING, W.F. 1950. Boiling water keeps Iceland poppies fresh. *Grower* 34:1059. (*Hort. Abstr.* 21:1892.)

STODDARD, E.M. and P.M. MILLER. 1962. Chemical control of water loss in growing plants. *Science* 137:224–225.

SUTTLE, J.C. 1980. Effect of polyamines on ethylene production. *Plant Physiol.* (Suppl.) 65:34.

SUTTLE, J.C. and H. KENDE. 1978. Ethylene and senescence in petals of *Tradescantia. Plant Physiol.* 62:267–271.

SUTTLE, J.C. and H. KENDE. 1980. Ethylene action and loss of membrane integrity during petal senescence in *Tradescantia. Plant Physiol.* 65:1065–1072.

SWANSON, B.T., H.F. WILKINS, and B.W. KENNEDY. 1979. Factors affecting ethylene production by some plant pathogenic bacteria. *Plant & Soil* 51:19–26.

SWANSON, B.T., H.F. WILKINS, C.F. WEISER, and I. KLEIN. 1975. Endogenous ethylene and abscisic acid relative to phytogerontology. *Plant Physiol.* 55:370–376.

SWART, A. 1981. Improvement of longevity and quality in *Lilium* 'Enchantment' flowers. *Acta Hort.* 113 (in press).

SWART, A. and G.A. KAMERBEEK. 1979. Keeping quality of bulb flowers. Annu. Rpt. Bulb Res. Center. 1978. p. 25–26, 96. Lisse.

SYTSEMA, W. 1962. Bloei van sering aan afgesneden takken. *Meded. Landbouwhogesch. Wageningen* 62(2):1–57.

SYTSEMA, W. 1964. Forcing. p. 38–45. *In* A. Lecrenier and P. Goeseels (eds.) Proc. 16th Intern. Hort. Congr. Aug. 31–Sept. 8, 1962, Brussels. J. Duculot, Gembloux, Belgium.

SYTSEMA, W. 1966. De houdbaarheid van poinsettia. *Vakbl. Bloemist.* 20: 301.

SYTSEMA, W. 1971. De houbdaarheid van tulpen. *Vakbl. v.d. Bloemisterij.* 26(15):8–9.

SYTSEMA, W. 1972. Fluor en snijbloemen. *Vakblad. v.d. Bloemisterij.* 27(28, 29, 30):9, 13.

SYTSEMA, W. 1975. Conditions for measuring vase life of cut flowers. *Acta Hort.* 41:217–225.

SYTSEMA, W. 1981. Vase life and development of carnations as influenced by silver thiosulfate. *Acta Hort.* 113 (in press).

TAKEDA, F. and J.C. CRANE. 1980. Abscisic acid in pistachio as related to inflorescence bud abscission. *J. Amer. Soc. Hort. Sci.* 105:573–576.

TAMES, P.M.L. 1978. Phloem loading versus unloading, with reference to white leaves and flower leaves. *Agrabiol. Res. Cent. Wageningen. Bul.* 17:1–12.

TEAS, H.J. and T.J. SHEEHAN. 1957. Chemical modification of geotropic bending in the snapdragon. *Proc. Fla. State Hort. Soc.* 70:391–398.

TEAS, H.J., T.J. SHEEHAN, and T.W. HOLMSEN. 1959. Control of geotropic bending in snapdragon and gladiolus inflorescences. *Proc. Fla. State Hort. Soc.* 72:437–442.

THIMANN, K.V. 1978. Senescence. *Bot. Mag. Tokyo Special Issue* 1:19–43.

THOMAS, T.H. 1974. Investigation into the cytokinin-like properties of benzimidazole-derived fungicides. *Ann. Appl. Biol.* 76:237–241.

THOMAS, H. and J.L. STODDART. 1980. Leaf senescence. *Annu. Rev. Plant Physiol.* 31:83–111.

THORNTON, N.C. 1930. The use of carbon dioxide for prolonging the life of cut flowers with special reference to roses. *Contrib. Boyce Thomp. Inst.* 2:535–547.

TONINI, G. and R. TESI. 1969. Risultati di una prova semicommerciale di conservazione in atmosfera controllata dei garofani recisi. *Rev. Ortoflorfrut. Ital.* 53:520–526.

ULRICH, R. (ed.). 1979. Recommendations for chilled storage of perishable produce. Intern. Inst. of Refrigeration, Paris.

UOTA, M. 1969. Carbon dioxide suppression of ethylene-induced sleepiness of carnation blooms. *J. Amer. Soc. Hort. Sci.* 94:598–601.

UOTA, M. and M. GARAZSI. 1967. Quality and display life of carnation blooms after storage in controlled atmospheres. *USDA Mktg. Res. Rpt.* 796:1–9.

UOTA, M. and C.H. HARRIS. 1964. Quality and respiration rates in stock flowers treated with N 6-benzylaminopurine. *USDA Mktg. Qual. Res. Div. Publ.* AMS-357:1–8.

VALDORINOS, J.E., T.E. JENSEN, and L.M. SICKO. 1974. Abscission: cellular changes at the ultrastructural level. p. 1034–1041. *In* Y. Sumiki (ed.) Plant growth substances 1973. Hirokawa Publishing, Tokyo.

VAN ALFEN, N.K. and V. ALLARD-TURNER. 1979. Susceptibility of plants to vascular disruption by macromolecules. *Plant Physiol.* 63:1072–1075.

VAN DEN BERG, L. and C.P. LENTZ. 1978. High humidity storage of vegetables and fruits. *HortScience* 13:565–569.

VAN DER WESTHUIZEN, A.J. and G.H. DE SWARDT. 1978. Changes in the free amino acid concentrations in the petals of carnations during the vase life of the flowers. *Z. Pflanzenphysiol.* 86:125–134.

VAN MEETEREN, U. 1978a. Water relations and keeping quality of cut gerbera flowers. I. The cause of stem break. *Scientia Hort.* 8:65–74.

VAN MEETEREN, U. 1978b. Water relations and keeping quality of cut gerbera. II. Water balance of aging flowers. *Scientia Hort.* 9:189–197.

VAN MEETEREN, U. 1979a. Water relations and keeping quality of cut gerbera flowers. III. Water content, permeability and dry weight of aging petals. *Scientia Hort.* 10:262–269.

VAN MEETEREN, U. 1979b. Water relations and keeping quality of cut gerbera flowers. IV. Internal water relations of aging petal tissue. *Scientia Hort.* 11:83–93.

VAN MEETEREN, U. 1980. Water relations and keeping quality of cut gerbera. VI. Role of pressure potential. *Scientia Hort.* 12:282–292.

VAN MEETEREN, U. 1981. Role of pressure potential in keeping quality of cut gerbera inflorescences. *Acta Hort.* 113 (in press).

VAN MEETEREN, U. and H. VAN GELDER. 1980. Water relations and keeping quality of cut gerbera. V. Role of endogenous cytokinins. *Scientia Hort.* 12:273–281.

VAN STADEN, J. 1979. The effect of emasculation on the endogenous cytokinin levels of cymbidium flowers. *Scientia Hort.* 10:277–284.

VAN STADEN, J. and G.G. DIMALLA. 1980. The effect of silver thiosulphate preservative on the physiology of cut carnations. II. Influence on endogenous cytokinins. *Z. Pflanzenphysiol.* 99:19–26.

VARNER, J.E. and D.T.H. HO. 1976. Hormones. p. 714–770. *In* J. Bonner and J.E. Varner (eds.) Plant biochemistry. Academic Press, New York.

VEEN, H. 1979a. Effects of silver on ethylene synthesis and action in cut carnations. *Planta* 145:467–470.

VEEN, H. 1979b. Effects of silver salts on ethylene production and respiration of cut carnations. *Acta Hort.* 91:99–103.

VEEN, H. and S.C. VAN DE GEIJN. 1978. Mobility and ionic form of silver as related to longevity of cut carnations. *Planta* 140:93–96.

VEGA, E.A. 1964. Experiments on the transport of cut flowers by ship (Hebrew). Volcani Inst. of Agr. Prelim. Rpt., Beit Dagan, Israel. 474:1–14.

VENKATARAYAPPA, T., M.J. TSUJITA, and D.P. MURR. 1980. Influence of cobaltous ion (Co^{++}) on the postharvest behavior of 'Samantha' roses. *J. Amer. Soc. Hort. Sci.* 105:148–151.

VIGODSKY-HAAS, H. 1979. Control of Botrytis disease in roses by treatments to the cut flowers (Hebrew). *Chamaimot Uprachim* 9(2):39–51.

WAGNER, G.J. 1979. Content and vacuole/extravacuole distribution of neutral sugars, free amino acids, and anthocyanins in protoplasts. *Plant Physiol.* 64:88–93.

WALLNER, S., R. KASSALEN, J. BURGOOD, and R. CRAIG. 1979. Pollination, ethylene production and shattering in geraniums. *HortScience* 14:446.

WANG, C.Y. and J.E. BAKER. 1979. Vase life of cut flowers with rhizobitoxine analogs, sodium benzoate and isopentenyl adenosine. *HortScience* 14:59–60.

WANG, C.Y. and J.E. BAKER. 1980. Extending vase life of carnations with aminooxyacetic acid, polyamines, EDU and CCCP. *HortScience* (in press).

WANG, C.Y., J.E. BAKER, R.E. HARDENBURG, and M. LIEBERMAN. 1977. Effects of two analogs of rhizobitoxine and sodium benzoate on senescence of snapdragons. *J. Amer. Soc. Hort. Sci.* 102:517–520.

WARNE, L.G.G. 1947. Bud and flower dropping in lupine. *J. Roy. Hort. Soc.* 72:193–195.

WATERS, W.E. 1966a. The influence of postharvest handling techniques on vase life of gladiolus flowers. *Proc. Fla. State Hort. Soc.* 79:452–456.

WATERS, W.E. 1966b. Toxicity of certain Florida waters to cut flowers. *Proc. Fla. State Hort. Soc.* 79:456–459.

WATERS, W.E. 1967. Effects of fertilization schedules on flower production, keeping quality, disease susceptibility and chemical composition at different growth stages of *Chrysanthemum morifolium*. *Proc. Amer. Soc. Hort. Sci.* 91:627–632.

WATERS, W.E. 1968a. Relationship of water salinity and fluorides to keeping quality of chrysanthemum and gladiolus cut flowers. *Proc. Amer. Soc. Hort. Sci.* 92:633–640.

WATERS, W.E. 1968b. Influence of well water salinity and fluorides on keeping quality of 'Tropicana' roses. *Proc. Fla. State Hort. Soc.* 81:357–359.

WATSON, D.P. and T. SHIRAKAWA. 1967. Gross morphology related to shelf-life of anthurium flowers. *Hawaii Farm Sci.* 16(3):1–4.

WEBSTER, B.D. and H.W. CHIU. 1975. Ultrastructural studies in *Phaseolus*: characteristics of the floral abscission zone. *J. Amer. Soc. Hort. Sci.* 100:613–618.

WEINSTEIN, L.H. and H.J. LAURENCOT. 1963. Studies on the preservation of cut flowers. *Contrib. Boyce Thomp. Inst.* 22:81–90.

WHITEMAN, T.M. 1949. Sodium alpha-naphtyl acetate tests on life of cut flowers. *Flor. Exchange* 112(12):16.

WHITEMAN, T.M. 1957. Freezing points of fruits, vegetables and florist stocks. USDA Mktg. Res. Rpt. 196.

WHITEMAN, T.M. and W.D. MCCLELLAN. 1946. Tip curvature of cut gladioli. *Proc. Amer. Soc. Hort. Sci.* 47:515–521.

WIATR, S.M. 1978. Physiological and structural investigation of petal abscission in *Linun lewisii*. PhD Thesis, University of California, Davis.

WIDMER, R.E. and L.K. STRUCK. 1973. Prolonging the keeping qualities of cut flowers and greens. Univ. Minn. Agr. Ext. Ser. Bul.

WIEMKEN, A. and E. NURSE. 1973. Isolation and characterization of the amino acid pools located within the cytoplasm and vacuoles of *Candida utilis*. *Planta* 109:293–306.

WIERSMA, O. and W.C. BOER. 1970. Snijbloemen en vacumkoelen. *Vakbl. Bloemist.* 25:1846–1847.

WILBERG, B. 1973. Physiologische Untersuchungen zum Knicker-Problem als Voraussetzung für de Selektion haltbarer Gerbera-Schnittblumen. *Z. Pflanzenzüchtung* 69:107–114.

WILKINS, M.B. 1978. Gravity-sensing guidance mechanisms in roots and shoots. *Bot. Mag. Tokyo Special Issue* 1:255–277.

WILKINS, H.F. and B.T. SWANSON. 1975. The relationship of ethylene to senescence. *Acta Hort.* 41:133–142.

WINKENBACH, F. and P. MATILE. 1970. Evidence for *de novo* synthesis of an invertase inhibitor protein in senescing petals of *Ipomoea*. *Z. Pflanzenphysiol.* 63:292–295.

WINTZ, R.P. 1954. Storage of cut carnations with some fruits and vegetables. *Colorado Flower Growers Assoc. Bul.* 53:1–3.

WOLTZ, S.S. and F.J. MAROUSKY. 1972. Effect of fluoride and a floral preservative on fluoride content and injury to gladiolus florets and injury to poinsettia bracts. *Proc. Fla. State Hort. Soc.* 85:416–418.

WOLTZ, S.S. and W.E. WATERS. 1967. Effects of storage lighting and temperature on metabolism and keeping quality of *Chrysanthemum morifolium* cut flowers relative to nitrogen fertilization. *Proc. Amer. Soc. Hort. Sci.* 91:633–642.

WOLTZ, S.S. and W.E. WATERS. 1976. Effects of light and temperature on keeping quality of cut flowers. *Fla. Flower Grower* 13(5):1–4.

WONG, CH.H. and D.J. OSBORNE. 1978. The ethylene-induced enlargement of target cells in flower buds of *Ecballium elaterium* (L.) A Rich. and their identification by the content of endoreduplicated nuclear DNA. *Planta* 139:103–111.

YANG, S.F. 1974. The biochemistry of ethylene: biogenesis and metabolism. p. 131–164. *In* E. Sondheimer and D.C. Walton (eds.) The chemistry and biochemistry of plant hormones, Vol. 7. Recent advances in phytochemistry. Academic Press, New York.

YANG, S.F. 1980. Regulation of ethylene biosynthesis. *HortScience* 15:238–243.

YONG, H.C. and H.T. ONG. 1979. Effects of chemicals applied to cut stalks on the shelf life on *Oncidium* Goldiana flowers. *Orchid Rev.* 87(1035):292–295.

YU, Y., D.O. ADAMS, and S.F. YANG. 1979. 1-Aminocyclopropanecarboxylate synthase, a key enzyme in ethylene biosynthesis. *Arch. Biochem. Biophys.* 198:280–286.

YVERNEL, D. and A. PAULIN. 1976. Expériences préliminaires sur le transport des fleurs coupées. *Pepin. Hort. March.* 163:15–19.

YVERNEL, D. and D. SOUTER. 1975. Le transport des fleurs coupées. *In* La conservation des fleurs cupées. *CNIH Paris*, Mon. Tech. 75(2):65–82.

ZIESLIN, N. and A.M. KOFRANEK. 1980. Roses can be treated to prevent rose neck droop. *Flor. Rev.* 165(4289):67–69.

ZIESLIN, N., H.C. KOHL, JR., A.M. KOFRANEK, and A.H. HALEVY. 1978. Changes in the water status of cut roses and its relationship to bent-neck phenomenon. *J. Amer. Soc. Hort. Sci.* 103:176–179.

ZIMMERMAN, P.W., A.E. HITCHCOCK, and W. CROCKER. 1931. The effect of ethylene and illuminating gas on roses. *Contrib. Boyce Thomp. Inst.* 3:459–481.

4

Effect of Nutritional Factors on Cold Hardiness of Plants

Harold M. Pellett and John V. Carter
Department of Horticultural Science and Landscape Architecture, University of Minnesota, St. Paul, Minnesota 55108

I. INTRODUCTION

The supposition that a high soil fertility level, especially nitrogen, induces plants to continue growth late into the fall and delay natural maturity development is well entrenched in the literature. This delayed maturity is felt to predispose plants to greater risk of winter injury (Treshow 1970; Childers 1969; Janick 1979).

There have been several reviews on the subject of plant cold hardiness (Levitt 1972; Vasil'yev 1967; Burke *et al.* 1975; Tumanov 1967; George and Burke 1976; Weiser 1970; Mazur 1969), but these have not given detailed attention to the relationship of nutrition to cold hardiness.

Levitt (1956) summarized earlier literature in tabular form that reported effects of various inorganic elements on crop plant cold hardiness. Alden and Hermann (1971) summarized results of a few papers dealing with this topic, but made no attempt to interpret the information.

One purpose of this review is to summarize and discuss what is presently known concerning the relationship(s) between plant nutrition and cold hardiness. Another purpose is to critically examine the methodologies that have been used in past studies, with the hope that such scrutiny may lead to additional research which will contribute to a better understanding of these relationships.

II. EVALUATION OF RESEARCH RELATING NUTRITION TO PLANT COLD HARDINESS

As a result of the general belief that high nutrition predisposes plants to winter injury, it long has been generally recommended that application of high rates of fertilizers, principally N, should be avoided, especially late in the growing season in locations where injury due to cold temperatures might be a problem. An analysis of the literature indicates that acceptance of this concept is not entirely justified. There are also several references that propose beneficial effects on plant cold acclimation from applying potassium (Levitt 1956; Beard 1973), but here again research results do not unequivocally support this practice.

Various studies report the effects of different rates of fertilizers, especially N, P, and K, on the ability of many plant species to withstand winter conditions. Fertilizers have been applied at different periods of the year, and different formulations have been compared. Table 4.1 summarizes the effects of N, P, and K applications on cold tolerance of various species as reported by different authors. Note that the results are inconsistent and sometimes directly conflicting.

TABLE 4.1. INFLUENCE OF VARIOUS FERTILIZERS ON ABILITY OF PLANTS TO TOLERATE COLD TEMPERATURES AS OBSERVED BY VARIOUS RESEARCHERS

		Cold Tolerance Reaction[1]			
Crop	Reference	Complete Fertilizer	N	P	K
GYMNOSPERMS					
Abies grandis (D. Don ex Lamb) Lindl.	Benzian *et al.* 1974		0		
Chamaecyparis obtusa E.	Kawana *et al.* 1964	+			
Cryptomeria japonica D. Don	Kawana *et al.* 1964	+			
Juniperus chinensis L. 'Hetzii'	Pellett and White 1969		0		
Picea A. Dietr.	Aldhous 1972		+0		+
Picea abies (L.) Karst	Kopitke 1941				+
	O'Carrol 1973				−
Picea sitchensis (Bong.) Carriere	Malcolm and Freezaillah 1975			−	
	Benzian 1966		+		+
	Benzian *et al.* 1974		−		0
Pinus L.	Aldhous 1972		+0		+
Pinus resinosa Ait.	Kopitke 1941				+
Pinus strobus L.	Kopitke 1941				+

TABLE 4.1. *(Continued)*

Crop	Reference	Cold Tolerance Reaction[1]			
		Complete Fertilizer	N	P	K
Pinus sylvestris L.	Christersson 1975a				0
Tsuga heterophylla (Raf.) Sarg.	Benzian 1966		+		+
	Benzian *et al.* 1974		0		
ANGIOSPERMS—MONOCOTS					
Agrostis alba L.	Carroll 1943		0		
Agrostis canina L.	Carroll 1943		−		
Agrostis tenuis L.	Carroll 1943		−		
Anthoxanthum odoratum L.	Carroll 1943		−		
Avena sativa L.	Dexter 1935		−		
Cynodon dactylon L.	Adams and Twersky 1960		−		+
	Alexander and Gilbert 1963		−		+
	Gilbert and Davis 1971	+	−	−	+
Cynosurus cristatus L.	Carroll 1943		−		
Dactylis glomerata L.	Howell and Jung 1965		−		
Eremochloa ophiuroides (Munro) Hack.	Palmertree *et al.* 1973		0−	0	0+
Festuca arundinacea Schreb.	Cook and Duff 1976				0
Festuca rubra L.	Carroll 1943		−		
Hordeum vulgare L.	Yasuda 1927				+
	Dexter 1935		−		
	Gamozhina and Sevastyanov 1971			+	
Lolium multiflorum L.	Carroll 1943		−		
Lolium multiflorum L.× *L. perenne* L.	Baker and Davis 1963		−		
Lolium perenne L.	Carroll 1943		−		
Poa annua L.	Carroll 1943		−		
Poa compressa L.	Carroll 1943		−		
Poa nemoralis L.	Carroll 1943		−		
Poa pratensis L.	Carroll and Welton 1939		−		
	Carroll 1943		−		
	Ferguson 1966		+		
	Wilkinson and Duff 1972		−		
Poa trivialis L.	Carroll 1943		−		
Secale cereale L.	Dexter 1935		−		
	Pyiklik 1963			+	+
Triticum aestivum L.	Dexter 1935		−		
	Kuksa 1939		−	+	+
	Pyiklik 1963			+	+
	Slavnyi and Musienko 1972	+	−	+	+
DICOTS—TEMPERATE WOODY					
Carya illinoinensis (Wang.) K. Koch	Skinner and Reed 1925	0			
Cornus sericea L.	Pellett 1973		0	0	
Forsythia × intermedia Zab.	Pellett 1973		−0	0	
	Beattie and Flint 1973				−0
Ilex crenata Thunb	Havis *et al.* 1972		0		0
Liriodendron tulipifera L.	White and Finn 1964				+
Malus domestica Borkh.	Rawlings and Potter 1937		−		
	Havis and Lewis 1938	0			
	Tingley *et al.* 1938		−0		
	Smith and Tingley 1940	−			
	Sudds and Marsh 1943		−		
	Way 1954		−		
	Edgerton 1957		−		
Populus L.	Viart 1965			+	+
Prunus cerasus L.	Kennard 1949		0		
Prunus persica (L.) Batsch	Green and Ballou 1904	+−			
	Chandler 1913		+		

TABLE 4.1. *(Continued)*

Crop	Reference	Cold Tolerance Reaction[1]			
		Complete Fertilizer	N	P	K
	Crane 1924		–		
	Cooper and Wiggans 1929	0			
	Cullinan 1931		+		
	McMunn and Dorsey 1935		0		
	Higgins *et al.* 1943		0+		
	Edgerton and Harris 1950		0		
	Proebsting 1961		+		
Pyracantha coccinea M. J. Roem	Kelley 1972		–		0
Rhododendron L.	Joiner and Ellis 1964		0		0
Ribes uva-crispa L.	Faustov 1965		–		
Rosa L.	Carrier 1953		–		
Rubus allegheniensis T. C. Porter	Lott 1926		+		
Vitis labrusca L.	Gladwin 1917		0	0	0
	Clore and Brummond 1965	–			
DICOTS—SUBTROPICAL PLANTS					
Aleurites Fordii Hemsl.	Brown and Potter 1949		+		+
	Sitton *et al.* 1954		+–	–	–
Citrus Limon (L.) Burm.	Bunina 1957			+	
Citrus × *paradisi* Macfady 'Marsh'	Smith and Rasmussen 1958		+	–	–
Citrus reticulata Blanco.	Bedrikovskaja 1952		+		
Citrus sinensis (L.) Osbeck.	Smith and Rasmussen 1958		+	–	–
	Jackson and Gerber 1963				0
Litchi chinensis Sonn.	Young and Noonan 1959		0		
Sansevieria trifasciata Prain.	Marlatt 1974		–	0	–
	Conover and Poole 1977		–		0
Solanum tuberosum L.	Lacio 1930				+
	Drozdov and Sycheva 1965		+		
DICOTS—TEMPERATE HERBACEOUS PLANTS					
Allium L.	Wallace 1926				+
Brassica oleracea L. var. capitata	Boswell 1925		–		
	Kimball 1927		–		
	Dexter 1935		–		
	Ragan and Nylund 1977		–		
Fragaria L.	Bedard and Therrien 1970				+
	Zurawicz and Stushnoff 1977	+	–		–0
Medicago sativa L.	Jung and Smith 1959			+–	+–
	Calder and Macleod 1966				+–

[1] + = Greater cold tolerance from increasing levels of the element.
– = Less cold tolerance from increasing levels of the element.
0 = No influence on cold tolerance from increasing levels of the element.
Note: More than one symbol indicates that the element was studied at more than two concentrations and that results differed between successive steps and concentration level, or that more than one experiment was reported with different results.

Many conclusions on the relationship of nutrition to plant cold hardiness have been based solely on circumstantial evidence. In addition, differences in amount of injury which may result from small differences in hardiness level have been interpreted as showing a major influence of nutrient levels on plant hardiness. Thus, in studying the various responses reported, the following experimental procedures should be scrutinized closely.

A. Experimental Design

Many early reports were based solely on field observations of survival of crops with past histories of high fertilizer applications preceding severe winters. Although these types of observations can provide some insights, the results of well designed experiments are needed for conclusive results and observations should be weighed accordingly.

B. Fertilizer Application Procedures

Several aspects of application procedures should be evaluated. In comparisons of different forms of an essential element, equal rates of that element must be applied. In a comparison of nutrient rates, tissue content of the element should be determined to characterize the nutrient status of the plant. Growth data are also very useful in determining whether rates given were below, above, or optimum for plant growth. The balance of the various elements essential to the plant should be known to properly interpret results. High levels of one element may result in deficiency levels of other elements within plant tissues. A good example is the study presented by Adams and Twersky (1960) (discussed on p. 159). What might appear as an obvious conclusion that a high N level predisposes coastal bermudagrass to winter injury may, in fact, be an artifact. The lack of proper cold acclimation may have been due to insufficient K levels brought about by removal of plant tissue following rapid growth induced by high N levels and not due to a direct effect of high N. The relation of application time to stage of growth and development is necessary to interpret plant response.

C. Cold Stress Applied or Encountered

Laboratory procedures which allow determination of the hardiness levels of plants at various times during the winter are preferred over field observations where environmental control is not possible. Controlled procedures make it possible to determine the magnitude of hardiness differences at various stages in the acclimation sequence and to assess differences in hardiness levels of specific tissues.

In laboratory freezing tests, samples from each treatment should be frozen at temperatures over a range to determine cold tolerance limits. This provides data on the magnitude of any hardiness difference between treatments as well as knowledge of the maximum hardiness capability. The conclusion of many of the papers cited in Table 4.1—that fertility treatments resulted in hardiness differences—was based on exposing the plants to a single temperature minimum. Although differences in injury were observed, the magnitude of differences in hardiness level and their importance under field situations were not reported. The difficulties inherent in such techniques are demonstrated in the report by

Marcellos (1977) in which injury caused by exposing spring wheat to freezing temperatures ranged from no damage to death over just a 2°C range. Such a narrow killing temperature range is commonly found in cold hardiness research.

D. Injury Assessment

Plant regrowth provides the most reliable method of assessing the extent of injury. However, regrowth analysis is not always possible. In many instances, entire plants cannot be used due to size limitations of the freezing equipment or following applications to trees or other plants under field conditions. Thus, small samples of the tissues under study must be severed from the plant, and these severed tissues may not be capable of regrowth even though not injured. Another limitation of using regrowth for injury analysis is the time period necessary before results can be obtained.

Visual observations of tissue injury are frequently used and can be fairly reliable with many plants. However, when a slight amount of injury is detected, it is often difficult to predict accurately whether the plant could recover under field conditions with equal amount of injury. Thus, visual observations permit only an approximation of the killing temperature.

Injury following freezing tests also may be detected by the conductivity method (Dexter *et al.* 1932), which has the advantage of providing numerical data that easily can be statistically analyzed. However, interpretation of the data can be misleading if one fails to relate the data to what is practically meaningful under field conditions. It is often very difficult to know what level of conductivity reading would parallel death or severe injury in the field, especially if the conductivity method of analysis is used in conjunction with freezing at only a single temperature.

III. RELATIONSHIP OF NUTRITION TO MODE OF ACTION OF COLD ACCLIMATION PROCESSES

Although there have been various studies of the relationship of nutrient levels to cold tolerance, very little effort has been made to determine the nutrients' possible modes of action in their effects on cold tolerance of crop plants. Although certain biochemical changes that frequently parallel cold acclimation are known, the relationship of these metabolic activities to cold acclimation mechanisms has not been well established. Cold hardiness levels long have been positively correlated with total carbohydrate accumulation (Kopitke 1941; Dexter 1935; Weiser 1970).

In a series of experiments, Dexter (1935) examined N's role in cold acclimation. In one study, 3-week-old winter wheat plants growing in solution culture were hardened at 2°C for 2 weeks. Half of the plants were grown in the light and half were grown in the dark. Data were collected, from representative samples at the start and after 2 weeks of cold acclimation, for weight, N content, and hardiness as measured by the freezing exosmosis method. Plants placed in the light had much greater dry weight than those in the dark, even though visual growth was not evident. Plants in the light also had a lower N concentration, but total N per plant was much greater indicating that at 2°C illuminated plants carried on active photosynthesis and N uptake. These plants hardened noticeably, while those plants stored under darkness in the cold did not harden. Reducing sugar content dropped to a very low level in the dark-stored plants. In another experiment, winter wheat plants were established in nutrient solution. Half of the pots then were grown without N until tissue tests indicated that they were practically free of inorganic N. This study was designed to (1) determine if soluble organic N, often reported to increase during cold acclimation, is due to N uptake from the soil or due to breakdown of proteins, and (2) to determine if N-deficient plants can harden at a low temperature in the dark. Similar procedures were followed using cabbage plants. Results of these studies indicated that the N-deficient plants increased in soluble organic N during cold storage as much as did those plants with a high inorganic N content. Thus it appears that there is a breakdown of proteins during cold storage. This accumulation of soluble organic N may not be related to the hardening process since both plants stored in the dark and those stored in the light at 2°C accumulated soluble N. Those that hardened the least had the greatest accumulation. Nitrogen-deficient plants accumulated carbohydrates and were capable of hardening under subsequent cold temperatures in the dark. Thus it appears that wheat and cabbage, at least, do not need an external N source when acclimating.

IV. INFLUENCE OF NUTRIENT LEVELS ON DIFFERENT TYPES OF COLD INJURY

Most of the literature on the relationship of fertility levels to cold injury reports the extent of winter injury observed during the following growing season. Influence on initiation of and rate of cold acclimation generally has not been separated from influence on ability to develop maximum cold tolerance. Reports on the relationships between nutrition and other types of cold temperature-related plant injury, such as chilling injury, bark splitting, sunscald, and winter burn of conifers, are sparse.

Marlatt (1974) reported that field-grown *Sansevieria* with no added N exhibited very little chilling injury following a cold period in mid-February when temperatures dropped to 3°C. Plants grown with increasingly higher levels of N and K exhibited increasingly more chilling injury. Phosphorus levels had no apparent effect. A later report by Conover and Poole (1977) also showed increased sensitivity of *Sansevieria* to chilling injury when grown with high N levels. They also reported that K levels did not affect the amount of chilling injury exhibited by plants following a cold rain.

Sudds and Marsh (1943) reported a severe incidence of frost cracking on the southern side of trunks of young apple trees that had been fertilized with 0.34 kg of sodium nitrate per tree during the first week of November, 1941. Severe cracking of the bark occurred on 19 to 20% of the trees, depending on cultivar. These trees ranged from 4 to 9 cm in trunk diameter. All trees in the block had been fertilized, so no direct comparison to nonfertilized trees was possible. Another block of trees of another cultivar of equal size growing on a site of similar exposure and soil type did not receive injury. Rawlings and Potter (1937) reported similar observations. Tingley *et al.* (1938) reported that they had induced considerable winter trunk injury in 5 of 7 small 'Baldwin' apples with trunk diameter of 4 to 5 cm by applying 0.57 kg of ammonium sulfate per tree on October 14. Only 1 of the 7 trees fertilized with 0.45 kg of cyanamid was injured. One of the control trees and none of those fertilized with cyanamid or ammonium sulfate at three other dates during the fall received any injury. In another study, they applied from 1.8 to 7.3 kg of cyanamid to 37 m^2 of sod around the trunk of each tree in mid-October (1.8 kg/tree was the maximum recommended rate). On various dates during the winter, trees were cut; trunks then were cut into sections and frozen to temperatures ranging from −18° to −39°C. Microscopic examination 1 month after freezing failed to show consistent differences in injury due to fertilizer treatments.

Proebsting (1961) established a split plot soil management experiment with peaches using six management systems and five N levels. Nitrogen applications were 0, 0.11, 0.23, 0.46, and 0.92 kg per tree. Bark splitting occurred following the winter of 1955−1956. Trees fertilized with 0.23 kg or more of N per tree suffered significantly less bark splitting than those trees receiving less N.

V. RELATIONSHIP OF NUTRITION TO COLD HARDINESS OF SPECIFIC PLANTS

A. Temperate Woody Plants

Christersson (1975), analyzing frost hardiness development in *Pinus sylvestris* seedlings, concluded that there was no correlation between frost hardiness development and K or Ca content. Plants were grown in growth chambers and given a nutrient solution with all combinations of 3 levels of K applied as KCl at 9, 96, and 865 mg per liter and 3 levels of Ca applied as $CaCl_2 \cdot 2H_2O$ at 0, 147, and 442 mg per liter. After 4 months of growing with various fertility treatments, plants were acclimated for 3 weeks under 8-hour days at 20°C plus 3 weeks of 8-hour days at 3°C. Tops of plants were frozen to a series of temperatures between −15°C and −23°C, with plants removed at 1° intervals. Plants were slowly thawed, then placed in a warm greenhouse and evaluated for survival after 8 weeks. Foliage was analyzed for N, P, K, Ca, and Mg content at the time of the freezing test. Killing first occurred between −18°C and −21°C, and all seedlings were killed below −21°C. There was no correlation of survival to Ca or K content or to the ratio between these nutrients. Potassium content of needles ranged from 0.12 to 0.50 mmoles per gram of dry weight and Ca content varied from 0.33 to 0.86 mmoles.

In an earlier experiment conducted with similar techniques, Christersson (1973) found no effect of K, Ca, or Mg levels on cold hardiness of *Pinus sylvestris* following 3 weeks of hardening under 8-hour days and 3°C. During the 3 weeks of hardening, water content of shoots decreased with all treatments and K content increased in shoots but decreased in the roots. Thus it appears that K ions are transported from the root to the shoot during hardening.

Benzian (1966) and Benzian *et al.* (1974) examined the effect of high N and K applied in fall to induce "luxury consumption" of N or K in five species of conifers in English forest nurseries. Additional top dressing treatments were applied in early September when top growth had nearly ceased. Plants were dug and stored over winter, and planted out in many forest sites with different soil and climatic conditions in the spring. Tissue analysis indicated that treatment resulted in increased tissue content of the elements applied. Except for *Abies grandis*, the added N advanced bud break of all species during the first season after transplanting. The effects of K were inconsistent. In general, higher N content had no effect on transplant survival. Fall application increased first year growth of Sitka spruce and to a lesser extent of other species at some sites. A severe June frost at one site during the second growing season resulted in severe injury to one provenance planting of Sitka spruce. The extra fall N increased the resulting damage. Potassium treatments had no effect. At another site an early frost at the end of the first growing season resulted in severe injury, but no effect of fertility treatments was apparent. No data were given to indicate if differences in elemental concentrations in plant tissue still existed at the time of the frosts.

Kopitke (1941) analyzed the effect of K levels on development of cold hardiness and summarized that, "a balanced ratio of nutrients and especially an adequate supply of available potash appears to be the most important prerequisite for the production of frost free stock." In this study, white spruce trees were grown from seeds planted April 1 in a greenhouse. Nutrient solutions with four different concentrations of KCl were used. Plants were harvested on July 26 and analyzed for simple sugars. In a field experiment 2-year-old white spruce seedlings growing on a site found to be lacking in K were fertilized with various amounts of K in liquid form in early August. Seedlings were collected on October 27 and analyzed for simple sugar, invert sugar, starch, and protein N content. In another field experiment, the effects of three levels of K on the freezing point, osmotic pressure, content of total solids, and simple sugar content of extracts of white spruce, red pine, and white pine were studied. Plants were collected on June 28 for analysis after two seasons of growth under the potash treatments. Tissue analysis indicated that applications of K fertilizers increased the accumulation of simple and invert sugars and total solids. Increased K rates of up to 337 kg/ha increased osmotic pressure and lowered the freezing point of sap, while a further increase to 560 kg/ha decreased accumulation of sugars. Kopitke concluded that increased K rates increased frost hardiness; we feel that this conclusion is not justified since differences in cold hardiness of plant tissues were not actually tested. The demonstrated biochemical differences due to K treatments are usually an insignificant factor in total hardiness of plant species with capabilities of hardening to withstand extremely low temperatures. In two experiments, Kopitke collected samples for analysis during a period of the year when plants would be expected to be in the least hardy state.

Pellett and White (1969) studied the influence of N treatments on cold acclimation of root and stem tissue of container-grown Hetzi juniper. Plants were established in pots and grown with balanced fertility at a level considered to be sufficient to produce optimum growth. No nitrogen was applied to low N plants after September 18, while high N plants received additional N applications on September 18 and October 18. Levels of cold acclimation, N concentration, and tissue moisture content of root and stem tissues were determined on 6 dates between September 11 and December 2. Fall N application increased N content of both root and stem tissue at all sampling dates in high N-treated plants. Hardiness was not influenced by N treatment as determined by visual observation of either root or stem tissue. Soil N treatments did not consistently affect tissue moisture percentages.

Following an early September frost in Scotland, Malcolm and Freezaillah (1975) noted greater injury to young Sitka spruce which had a high P content. In this study, they planted young spruce in containers of peat and applied differential P treatment (+P and −P). The containers were planted outdoors in spring at several sites differing in elevation and climate. Weekly growth data were determined, and maximum and minimum temperatures were recorded throughout the growing season. A September 8 freeze damaged plants at the 2 lower sites. The +P plants were still in active growth during the week preceding the frost. The −P plants had ceased height extension. The +P treatment resulted in more severe injury at one of the lower sites than the −P treatment, although all but one plant from each treatment exhibited some injury.

There have been numerous reports relating nutrient levels to winter injury suffered by fruit trees. Many of these reports were published in the first half of this century and were based on field observation of orchards that suffered severe injury during winter when very low temperatures occurred. Many of the reports were based on observations of commercial orchards that had been fall-fertilized or that had been fertilized heavily during the preceding growing season. Other reports were based on observation of research plots of ongoing nutrition studies.

Green and Ballou (1904), in assessing possible causes of winter injury of peach trees in the Lake Erie fruit belt, concluded that peach trees with low vigor due to disease or lack of nutrition were more susceptible to cold injury. They also stated that, "with healthy trees, those with high fertility which favored extreme growth of soft poorly ripened wood were more susceptible."

Chandler (1913) examined the influence of heavy pruning and fertilization on survival of peach flower buds. Better survival of flower buds was found on trees that had been pruned severely or had been fertilized heavily the previous spring. Nonfertilized trees lost 91% of their flower buds and trees fertilized with ammonium sulfate lost 45% of buds to low temperature on February 22 following an extended warm period. Chandler attributed this effect to a shift in rest period. He felt that the treatments prolonged the growth in the fall and delayed onset of rest. He assumed that this also delayed satisfaction of the rest requirement and kept buds from deacclimating to the same degree as those trees not fertilized, which entered rest earlier.

McMunn and Dorsey (1935) recorded peach flower bud injury in Illinois over a 7-year period based on injury to trees which were part of a field fertility experiment. They concluded that there was no consistent effect of N on cold hardiness of peach fruit buds. On peach trees fertilized with N, Cullinan (1931) reported an increase in flower bud survival following a mid-December freeze to −20°C. Edgerton and Harris (1950) studied relative cold hardiness of 'Elberta' peach flower buds from trees grown with different N levels. Twigs with flower buds were collected and slowly frozen to a single test temperature on various dates from October to April. After freezing, buds were sectioned and percentage of live buds was determined. In none of the freezing tests were large or consistent differences in flower bud hardiness found among trees fertilized differentially. Twigs from trees grown with high N rates had higher N content and tree growth was greater.

Crane (1924) in West Virginia showed greater winter kill of flower buds on peach trees fertilized with N in comparison to nonfertilized trees. Cooper and Wiggans (1929), working in Arkansas, stated, "the data furnish no evidence that the winter hardiness of fruit buds can be influenced by fertilizers except as vigor is influenced." Higgins *et al.* (1943), in a randomized field study in Georgia, found that N had no effect on flower bud hardiness but resulted in increased wood hardiness.

Because of discrepancies in the literature relating influence of nutrition on peach hardiness, Proebsting (1961) established a detailed study utilizing laboratory techniques to evaluate differences in hardiness level of trees grown under a split plot design with five different N rates and six soil management systems. Increased N application rates resulted in increased leaf N content and increased yield. Flower buds collected from trees grown with high and low N treatments were frozen on different dates between December 10 and March 24 to a series of temperatures spanning the lethal range. Following freezing, buds were sectioned and percentage of survival was determined. Test temperatures that resulted in death of at least 50% of the buds were considered to be the killing temperatures (T_{50}). When averaged over all sampling dates, the T_{50} was −20.1°C for buds from high N treatments and −19.9°C for low N treatments. This difference of 0.2°C was statistically significant. The T_{50} at different sampling dates varied from −16.0°C to −11.1°C. The magnitude of change in T_{50} with the different N treatments was consistent at the different sampling dates. Thus, there was no evidence showing tendency for high N trees to retain hardiness longer due to a delay in completion of rest.

Edgerton (1957) researched influence of methods and timing of N application on cold acclimation of apples during a 3-year period. Treatments included different dates for fall and early winter soil applications of ammonium nitrate at 1.3 kg per tree. Foliar urea sprays were applied in two October applications to another group of trees. The control trees had been fertilized the previous spring. Equal amounts of N were applied in all treatments. Cold hardiness levels were determined at various dates during the fall and winter. Sections from basal stem portions of current season's growth were frozen to a single test temperature and evaluated by the conductivity method. Hardiness levels of bark samples from trunks of treated trees also were determined. Conductivity readings were significantly different among treatments. Edgerton concluded that October and early November applications of N decreased cold hardiness levels of both twigs and bark, especially during November and December. December applications of ammonium N had no effect on cold hardiness. As mentioned previously, differences in the magnitude of conductivity readings such as those reported can be produced easily by a difference of 1° to 2°C in the lowest temperature to which plants are exposed. In comparing readings of frozen samples to unfrozen controls, it appears that the test freezing temperatures used in these experiments had produced significant, perhaps lethal, injury to all treatments in many of the evaluations.

Way (1954) also studied the influence of fall N applications on cold hardiness in apple. Treatments included (1) nontreated trees, (2) 1.4 kg of ammonium nitrate per tree on October 31, and (3) 3.2 kg of ammonium nitrate per tree on November 10. Twig sections were frozen to single test temperatures February 1, March 8, and April 23, and hardiness levels were evaluated by conductivity. Twigs from nontreated trees had significantly less injury than twigs from trees from the fall-fertilized treatments. However, differences in conductivity readings were relatively small. Since conductance values from all treatments were more than twice those of nonfrozen samples, all the samples apparently had been exposed to test temperatures that were lethal.

Kelley (1972) examined the influence of N and K levels on growth and winter survival of *Pyracantha*. Plants grown in containers for one season were given fertilizer treatments of N at 6 rates and of K at 6 rates in a factorial design. Nitrogen rates used were 0, 0.79, 1.58, 2.38, 3.05, and 4.0 g of ammonium nitrate per liter of irrigation water applied every 10 days. Increased N rates resulted in an increase in foliage N, with the greatest increase occurring between the 0 g and 0.79 g rate. Total growth increased greatly at the 0.79 g rate compared to the 0 g rate. There was an additional small increase at the 1.58 g rate before leveling-off occurred. Plant survival recorded in May decreased as N level increased. Survival rates were 85, 60, 40, 30, 25, and 20%, respectively, as N level increased. Increases in K level resulted in no significant differences in growth rate or winter survival.

Havis *et al.* (1972) studied the influence of N rate and form and K rate on cold hardiness of *Ilex* roots. Plants were grown in containers with a complete fertilizer program until the end of August. Combinations of high and low NH_4^+ and NO_3^- with high and low levels of K were applied on September 8 and 30 and October 8 and 15. Plants were covered with white polyethylene on November 4. Hardiness determinations were made on November 20 and January 28 by freezing root segments from each treatment to a series of temperatures at 2°C intervals. Injury was rated by using the triphenyl tetrazolium chloride method. Additional hardiness determinations were made in mid-December by freezing root balls of intact plants and rating injury by visual evaluation after a recovery period. Leaves were collected January 25 for analysis of N and K content. The NH_4^+ treatment resulted in greater leaf N than did the NO_3^- treatment. Also, greater K levels were found in leaves where K was applied alone or in combination with NO_3^- than when K was applied with NH_4^+. There were no significant differences in root cold hardiness due to nutrient levels or forms studied.

Beattie and Flint (1973), using *Forsythia*, attempted to study the relationship between optimum K levels for plant growth and cold hardiness development. Plants were grown in pots and fertilized uniformly with complete fertilizers except for K, which was applied at 4 different rates (25, 75, 225, and 675 ppm in irrigation water) to provide a suboptimum rate, 2 rates within the optimum range, and a rate well above the optimum range. Representative plants were sampled in early September for dry weight measurements of growth and analysis of foliage nutrient concentrations. Remaining plants were moved into a cold greenhouse in mid-October. On December 7 and January 11, samples from the middle third of the current season's stem growth were used for tissue analysis and cold hardiness determination. Stem segments were frozen at different temperatures and injury analyzed by a modification of Dexter's electrolytic method. Potassium concentration in plant tissue increased at each higher fertilization rate. Maximum growth was reached with the 75 ppm rate and decreased significantly at the other rates. The highest K level resulted in significantly greater freezing injury (on both testing dates than the lower rates). The optimum K rate for development of cold hardiness was at or lower than that producing optimum growth.

Pellett (1973) examined the influence of N and P on cold acclimation of roots and stems of *Forsythia* and *Cornus*. In one experiment, *Forsythia* plants were grown in pots with 4 different rates of N (25, 100, 200, and 300 ppm N) applied as calcium nitrate at each irrigation. In a second experiment, potted plants of *Forsythia* and *Cornus* were given factorial combinations of two N and three P levels. Nitrogen treatments were 25 ppm and 200 ppm N at each irrigation and P treatments consisted of 0, 0.5, and 2.0 tsp of 20% superphosphate applied to each pot on July 1, July 30, August 18, and September 21. Hardiness levels were determined for *Cornus* stems on October 27 and *Cornus* roots on November 17. Hardiness of *Forsythia* stems was determined on November 9 and roots were tested on November 11. Hardiness was determined by visual rating of injury following controlled freezing of plant tissues to a series of temperatures. Growth was measured and tissue content of N, P, K, Ca, and Mg was determined. Increases in nutrient rates resulted in increased concentration of the applied elements in plant tissue and in increased growth. In the N study with *Forsythia*, plants which received the higher

N treatments had slightly greater injury of roots and stems than plants fertilized at lower rates. These differences in injury ratings could be equalled by temperature differences of about 1.4° to 2.0°C. In the second study, there was little obvious effect of N level alone on injury resulting from freezing. Changing P levels resulted in slight changes in hardiness. Following freezing, *Cornus* stems from the lowest P treatment and roots from the lower two treatments had least injury, and *Forsythia* stems and roots from the intermediate P level had less injury than tissues from plants given either more or less P. The high N, high P treatment resulted in greatest freezing injury. Differences in hardiness were, however, very small and usually not statistically significant. Pellett concluded that the differences in nutrient levels tested, "resulted in small or no significant inhibition of the cold acclimation of these plants as measured in mid-autumn." Weather records were used to illustrate that the freezing temperature required to injure the plants was about 17°C lower than the minimum temperature that might be encountered with a 10% probability on the dates hardiness was studied. Thus, Pellett concluded that the 2° or 3° difference in cold acclimation that might result from utilization of high fertility practices is probably not important if plants (grown) are well within their geographical hardiness zone.

Skinner and Reed (1925) observed no difference in winter injury between pecan trees in fertilized and nonfertilized blocks. Fertilizer treatments studied did induce noticeable differences in growth.

B. Temperate Herbaceous Plants

In an examination of the effect of level and timing of nitrogenous fertilizers on cold resistance of Kentucky bluegrass, Carroll and Welton (1939) determined tissue N, water status, and total sugar levels at different times of the year in both fertilized and nonfertilized plants. On every sampling date, the high N-treated plants had higher tissue N than did the unfertilized controls, indicating that the fertilization supplied an excess of N above the amount necessary for optimum growth. In almost all of the samples analyzed, total sugar content was lower in the high N than in the control plants. Freezing stress tests were performed on greenhouse-grown grass that had been cold-acclimated by storage for 10 hours at 0°C and on grass grown out-of-doors that had cold-acclimated naturally. Both populations showed a higher level of damage in the high N-treated plants than in the controls. For example, greenhouse-grown plants exposed to −12°C air temperature for 8 hours (final soil temperature was −6°C) showed 95% survival in the control group and 50% survival in the high N group. Survival was checked two weeks after the test. When tested on October 20, 1936, plants grown out-of-doors showed no effect of N treatment on cold resistance. A freezing stress given on November 11, 1936, after the plants had cold acclimated, resulted in no injury of control plants, while only 10% of the high N group survived exposure to −25°C for 3 hours. In both greenhouse- and out-of-door-grown grass, the level of N applied made no difference in survival of freezing tests if the grass was not subjected to a hardening treatment.

Wilkinson and Duff (1972), in their study of Kentucky bluegrass cold resistance as influenced by N fertilization, found essentially the same result as that of Carroll and Welton (1939). Although they did not measure tissue N concomitant with cold resistance determination, Wilkinson and Duff (1972) did measure chlorophyll content and color changes. Treatment leading to darker color reduced cold resistance. Early fall fertilization produced the greatest reduction of cold resistance in late fall while late fall N fertilization caused a larger reduction in cold resistance the following spring. Cold resistance was determined by comparing the electrolytic content lost following exposure to −7°C for 4 hours with the total electrolytic content as determined by heating the tissue at 100°C for 15 minutes. Plants fertilized with N, although less hardy during fall than nonfertilized plants, nevertheless did cold acclimate. By midwinter they all had hardened significantly, with most achieving the same hardiness level as the control plants. This indicates that N fertilization acts to delay cold acclimation, as other treatments which prolong growth have been shown to do (Proebsting and Mills 1974).

Carroll (1943) published results of an extensive study on N fertilization's effect on drought, heat, and cold resistance of 16 grasses using earlier developed techniques (Carroll and Welton 1939) for determining cold resistance. Nitrogen fertilization consisted of 245 kg/ha of ammonium sulfate applied in April, July, and September. All 16 grass species were found to be much less resistant to freezing temperatures when they received N. The deleterious effect of N fertilization also was manifested in a loss of drought resistance and heat resistance: those plants receiving N were less resistant to high soil or air temperatures and to water stress than were unfertilized plants of the same species. The tests were performed on 11 cm-diameter samples of sod taken from plots during active growth and maintained in earthenware jars.

A field study on natural stands of coastal bermudagrass in Georgia and Alabama which had received differential N, P, K treatments during 1955 to 1957 revealed two relationships between nutrition and cold resistance of this plant species (Adams and Twersky 1960). A low temperature of −15°C on February 7, 1958 provided the cold resistance test. Those plants which had received no N suffered least. It also was found that, for a given level of N applied, the degree of frost damage decreased as the level of K increased. The maximum K level was 225 kg/ha, but extrapolation of this relationship suggests that K might completely counteract the deleterious effects of N if applied at sufficiently high levels. Soil tests showed that available K was depleted by high N fertilization. A similar study (Gilbert and Davis 1971) carried out on two golf-green bermudagrass cultivars, which were field-established and then transferred to a greenhouse, corroborated the relationship between N and K levels. P levels were varied also. The optimum ratio of N-P-K for winter hardiness in both 'Tifdwarf' and 'Tifgreen' bermudagrass was found to be 4:1:6. Doubling the P level decreased the hardiness level by about 1°C. Once again, N in the absence of P and K produced plants with the lowest cold resistance.

Potassium fertilization of tall fescue at levels of 0, 50, and 200 ppm for sand cultures and 1, 10, 20, and 40 kg/ha for field plots had no measureable effect on cold hardiness levels (Cook and Duff 1976). Test plants again were frozen to a single temperature, −8°C, for 4 hours, and relative injury determined by comparing conductivities of soak solutions from the frozen plant tissue and from the same tissue after boiling.

Palmertree *et al.* (1973) examined the effects of different N, P, and K levels on the cold hardiness of both field-grown and greenhouse-grown centipede grass. Cold hardiness levels were determined by visual rating following freezing to different temperatures. They found that varying P and K levels had no detectable effect on killing temperature: the hardened field sod survived to −11°C and the unhardened greenhouse sod survived to about −10.5°C. Sod with N levels of 0.98 kg/ha and 1.96 kg/ha had no difference in killing temperature, but when N was increased to 2.94 kg/ha, the grass was less cold hardy by 3°C. They also examined longer term effects by determining the influence of N, P, and K applied during summer on survival the following spring. Live sprigs per unit area were counted and again the 2.94 kg/ha N treatment was found to be harmful. Less than half the number of sprigs that survived in the sod fertilized at the lower levels—0.98 kg/ha and 1.96 kg/ha—survived in the plots receiving 2.94 kg/ha. Phosphorus levels showed no effect on hardiness by this rating method while stolon viability was significantly increased at the two higher K levels. Their study indicates that too little N produces centipede grass of inferior quality while excessive N levels lead to long-term deterioration due to enhanced winter damage.

Howell and Jung (1965) analyzed cold hardiness of 'Potomac' orchardgrass at two levels of N fertilization. Following hardiness changes throughout the year by measuring the percentage of total electrolytes exosmosed after freezing tests, they found that the higher N level produced less cold hardy plants than did the lower N level. However, since killing temperatures were not determined, their study does not show whether the protection afforded by withholding N was large enough to be of practical importance.

When winter wheat and winter rye were grown on soils of low nutrition, K and P applications increased the plants' sugar content, dry weight, and winter survival (Pyiklik 1963). High N applications (100 to 150 kg of ammonium nitrate per hectare) applied in autumn decreased the amount of sugars and percentage of dry matter. However, a smaller amount (30 to 50 kg/ha) applied to poorer soils increased sugar content. Pyiklik related ability of plants to overwinter to their sugar accumulation and rate of winter utilization of those reserves. He noted that stands of plants under deep snow when soil was thawed utilized their sugar reserves faster and suffered more winter kill than plants in frozen soil.

Application of K to alfalfa can increase its cold hardiness as well as yield (Calder and MacLeod 1966). However, the level of K must not be too high or the beneficial effects will be lost. The optimum level found by Calder and MacLeod (1966) was 225 kg/ha, with 337 kg/ha being too high. In this study, relative cold hardiness was estimated by determining the loss of stand caused by 7-day exposure to −3.9°C, a regime which in a prior study with 3 cultivars had caused 63% loss of stand. The previous criticism of single temperature freezing tests is applicable to this study. No attempt to correlate K and N levels was made, but P and K tissue levels were determined. Application of K reduced tissue P and increased tissue K levels. The effect of K on alfalfa cold hardiness depended on the number of harvests per season. Calder and MacLeod (1966) felt that the interactive effects of the two variables (K level and cropping frequency) were related to total available carbohydrate content of the roots, which they also measured. Potassium nutrition appears to be related to several aspects of carbohydrate metabolism so that this conclusion is well-founded.

Drozdov and Sycheva (1965) examined the effect of nutrition on cold hardiness of potato, presumably *Solanum tuberosum*. They were specifically concerned with cold resistance of the potato vines. When the standard N level was doubled, vine damage resulting from a 4-hour exposure to −2.6°C was decreased from 59 to 42%. They also found a small increase in cold resistance of vines when N was applied as ammonium rather than as nitrate ion, which they ascribed to a change in growth rate due to soil acidity changes caused by ammonium sulfate and calcium nitrate. However, because the method of evaluating the freezing injury was not mentioned, it is difficult to assess the significance of their results.

In an examination of strawberry cold tolerance as affected by N, P, and K nutrition, Zurawicz and Stushnoff (1977) found that the least tolerant plants at the onset of acclimation were those with nutrient deficiencies, while the most resistant had received 1:1:1 or 1:2:1 ratios of N, P, and K.

This was also true of fully hardy plants examined in January. The freezing tests consisted of exposure to −2.2°C for the tender plants and to −9°, −11°, and −13°C for the fully hardy plants. Damage was determined by looking at recovery 30 days after the freezing test. From their work, in which nine fall fertilizer treatments were evaluated, it seems clear that the N: P: K ratio is more important than the level of any individual element. The 2 treatments which resulted in no plants with flowers 30 days after the first freezing test were 1:1:4, with N at 112 kg/ha, and 2:0:0, with N at 224 kg/ha. The average recovery score 30 days after the second freezing test (midwinter) was worst for the 1:1:4 treatment. In a conflicting report, Bedard and Therrien (1970) found that K increased cold resistance of strawberries in sand culture, but they did not report on the role of other nutrients or on the interrelationships of nutrients. Ragan and Nylund (1977) found that cabbage seedlings fertilized with N, P, and K and N and P were less cold resistant than those receiving P and K or no fertilizer. The treatment containing N but no K reduced cold tolerance more than the N, P, and K treatment.

C. Subtropical Plants

Following a mid-December freeze in Florida when temperatures dropped to −5.6° to −7.8°C, Smith and Rasmussen (1958) reported observations of freeze injury of citrus trees grown under different experimental nutrition regimes. In a study in which different N rates resulted in differences in leaf N content but had not affected yield significantly for 5 years, they observed that high N, which was probably at a luxury level, slightly reduced cold injury of foliage. In another study with grapefruit, trees fertilized with ammonium sulfate received greater injury than trees fertilized with calcium nitrate or ammonium nitrate (Smith and Rasmussen 1958). In a study of K fertilization, rates of 0.23, 0.91, and 2.3 kg potassium oxide per tree per year applied for 5 years resulted in differences in cold injury. Less than 10% defoliation occurred on the trees fertilized with low K, while the trees given the high rate were about 30% defoliated. In studies with young trees, trees planted only for 1 year were injured less when the last N application was applied in August than when an October application was given. This effect was not noted on 2-year-old trees. Smith and Rasmussen concluded, "All the observations reported here indicate that fertilization practices have a relatively small effect on cold tolerance of citrus when obvious deficiencies are absent. Fertilization practices that produce a healthy vigorous tree seem to offer about the maximum protection that can be attained within the tree."

Orange seedlings grown on solution culture for 5 weeks with K rates of 0, 8, 32, and 128 ppm showed, on foliar analysis, leaf K contents of 0.59,

0.70, 0.89, and 1.60%, respectively (Jackson and Gerber 1963). Detached leaves were frozen, and temperature recorded at the freezing point indicated that the leaves supercooled to −6.9°C before freezing. Whole plants were slowly frozen to −5.5°, −6.9°, and −8.3°C, then slowly thawed and held at warm temperatures for 15 days. Plants cooled to −5.5°C showed little injury, while those frozen to −6.9°C were completely defoliated. Temperatures of 8.3°C killed the plants. There were no differences in leaf freezing point or in injury of intact seedlings due to K nutritional differences.

Sitton *et al.* (1954) observed response of tung trees to factorial combinations of three levels each of N, P, and K. Increased tree growth and yield of nuts were associated with increasing levels of N and K, but little difference was found due to P levels. Leaf analysis showed an increase in concentration of N, P, and K as soil applications of those elements were increased. Freezes on November 23, 1950 and November 3, 1951 resulted in considerable cambium injury to tree trunks. Medium and high levels of N and P increased amount of injury received. Combination of high N and high P resulted in greater injury than could be accounted for by additive effects of the two elements. On March 9, 1950 and March 14, 1951, freezing temperatures destroyed many flower buds. In both years, injury was less on trees fertilized with high N. Trees given high and medium levels of K were injured less than trees fertilized at the low K level. Brown and Potter (1949) in Mississippi also reported less injury of tung trees following a spring freeze if trees had been fertilized with high levels of N or K.

Based on observation of trees growing in a field fertility study, Young and Noonan (1959) reported that lychee trees fertilized with sodium nitrate received more cold injury than did those fertilized with sludge or ammonium forms at an equal N rate. There was no consistent relationship between N rate and corresponding cold injury, although there was a slight trend toward less damage at higher rates used in the study. Rates studied produced no obvious differences in growth or maturity development.

D. Effect of Nutrients Other than N, P and K on Cold Hardiness

Most reports of nutritional influence on plant cold hardiness are limited to N, P, and K. However, there are a few reports on relationship of other essential elements to cold hardiness. Treatment of corn seed with zinc sulfate prior to sowing improved cold resistance of young corn plants (Pankratova 1963), and a high zinc content was shown to increase chilling tolerance and increase water permeability of tomato cell membranes (Rhoton 1967). Cold resistance of *Eucalyptus* was enhanced by boron

(Cooling 1967). Atterson (1967) reported that a magnesium deficiency increased amount of frost injury suffered by Lodgepole pine and Sitka spruce. Applications of calcium chloride, sodium chloride, and sodium sulfate to grapefruit trees resulted in boron, magnesium, and potassium deficiency and increased severity of frost injury (Peynado and Young 1963). Calcium and magnesium content of plant tissue were not correlated with cold hardiness of Norway spruce (Christersson 1975b).

VI. CONCLUSIONS

The widely held belief that high fertility may predispose plants to severe winter injury is not universally supported by research data. After critical review of the literature and based on our own experiences, we feel that most plants fertilized at levels which promote optimum growth will cold acclimate at a similar rate and to the same degree as plants grown under a lower fertility regime and may even exceed cold hardiness development of plants grown under severe nutrient deficiencies. Supra-optimum fertility levels can retard cold acclimation.

In reviewing studies in which comparisons were made between two or more nutrient levels, we tried to determine whether levels used limited, supported, or were in excess of optimum plant growth. In most reports this was difficult because of limited data. Foliar nutrient concentration data provide insight into the level of nutrition. Approximations of optimum foliar concentration for a particular plant can be made by examining levels that have been reported for closely related plants or for plants of similar growth characteristics. An examination of conflicting reports of the relationship of nutrition to cold hardiness shows that, in woody plants, decreasing hardiness with high nutrient levels may be due to rates higher than necessary for optimum growth. However, in studies reporting a benefit from higher rates, deficient levels often were used for the lowest rate. Beattie and Flint (1973), working with *Forsythia*, proposed that, "Potassium levels within or below the optimum range for growth are necessary for full development of frost hardiness, but higher levels are of no benefit, and may reduce hardiness." This relationship may hold for other essential elements with woody plants. However, additional work is needed to critically test this concept.

Of the processes currently understood, the main way in which nutrition may influence cold hardiness is through its relationship to growth. For nutrient levels to be of practical concern in plants' winter survival, several conditions must be met. First, growth cessation and cold acclimation of the plant of concern must not be influenced by photoperiod, or the plant must be growing in an area of marginal adaptation in which the critical photoperiod for triggering acclimation is not reached until very late in the season. Second, temperatures must be high enough to support continual growth of the plant, preventing carbohydrate accumulation and initiation of acclimation. Third, an early severe freeze must be encountered before acclimation proceeds sufficiently to allow survival. Under this set of circumstances, a nutrient level insufficient to support rapid growth may result in carbohydrate accumulation and an increase in cold hardiness development, in comparison to the same plant grown at a higher level which may not have sufficient hardiness to survive the early freeze. However, the potential difference in hardiness levels that may occur when these circumstances are met would be at most only a few degrees.

Certain non-perennials, such as cabbage, that have some capacity to harden may not be triggered into acclimation by photoperiod and thus could be influenced by nutrient levels. Also, cool season grasses commonly grow late into the fall and acclimation of these species may not be strongly influenced by photoperiod. Studies with such plants indicate that low nutrient levels usually favor increased hardiness.

Another possible relationship between nutrition and cold hardiness is the effect nutrient levels may have on specific metabolic processes accompanying changes in cold hardiness. However, the role of nutrition in this area can not be evaluated until the relationship of these metabolic processes to cold hardiness is understood.

VII. LITERATURE CITED

ADAMS, W.E. and M. TWERSKY. 1960. Effect of soil fertility on winter killing of coastal bermudagrass. *Agron. J.* 52:325–326.

ALDEN, J. and R.K. HERMANN. 1971. Aspects of the cold hardiness mechanisms in plants. *Bot. Rev.* 37:37–142.

ALDHOUS, J.R. 1972. Nursery practice. *Bul. For. Commn.* No. 43, p. 38. London.

ALEXANDER, P.M. and W.B. GILBERT. 1963. Winter damage to bermuda greens. *Golf Course Rptr.* 31:50–53.

ATTERSON, J. 1967. Slow-acting inorganic fertilizers. p. 21–23. *In* Report on forestry research 1965–66. Forestry Commission, London. (*For. Abstr.* 28: 5523).

BAKER, H.K. and G.L. DAVIS. 1963. Winter damage to grass. *Agriculture* 70:380–382.

BEARD, J.B. 1973. Turfgrass science and culture. Prentice-Hall, Englewood Cliffs, N.J.

BEATTIE, D.J. and H.L. FLINT. 1973. Effect of K level on frost hardiness of stems of *Forsythia* × *intermedia* Zab. 'Lynwood'. *J. Amer. Soc. Hort. Sci.* 98:539–541.

BEDARD, R. and H.P. THERRIEN. 1970. Influence of potassium nutrition on the cold resistance of the strawberry and changes in the mineral composition during hardening (in French). *Natur. Can.* 97:307–313.

BEDRIKOVSKAJA, N.P. 1952. Nutrition and frost resistance of the mandarin tree. *Pokladyvsesojuz Akad. Seljsk. Nauk*, 17:30–34. (*Hort. Abstr.* 23: 166).

BENZIAN, B. 1966. Effects of nitrogen and potassium concentrations in conifer seedlings on frost damage. Extracted from Report on Rothamsted Expt. Sta., Harpenden, Herts, England, p. 58–59. *(For. Abstr.* 28:510).

BENZIAN, B., R.M. BROWN, and S.C.R. FREEMAN. 1974. Effect of late season top dressing of N (and K) applied to conifer transplants in the nursery on their survival and growth on British forest sites. *Forestry* 47:153–184.

BOSWELL, V.R. 1925. A study of some environmental factors influencing the shooting to seed of wintered-over cabbage. *Proc. Amer. Soc. Hort. Sci.* 22: 380–393.

BROWN, R.T. and G.F. POTTER. 1949. Relation of fertilizers to cold injury of tung trees occurring at Lucedale, Mississippi in March 1948. *Proc. Amer. Soc. Hort. Sci.* 53:109–113.

BUNINA, N.N. 1957. The effect of phosphatic fertilizers on the growth, development and frost resistance of lemon seedlings (in Georgian). Chaisa da subtropikul kulturata sruliad sakavehiro sametsnierokoleveti institutis biuleteni 4:93–103. (*Biol. Abstr.* 46:26902).

BURKE, M.J., M.F. GEORGE, and R.G. BRYANT. 1975. Water in plant tissues and frost hardiness. p. 111. *In* R.B. Duckworth (ed.) Water relations of foods. Academic Press, London.

CALDER, F.W. and L.B. MACLEOD. 1966. Effect of cold treatment on alfalfa as influenced by harvesting system and rate of potassium application. *Can. J. Plant Sci.* 46:17–26.

CARRIER, L.E. 1953. Environment and rose hardiness. *Proc. Amer. Soc. Hort. Sci.* 61:573–580.

CARROLL, J.C. 1943. Effects of drought, temperature and nitrogen on turfgrasses. *Plant Physiol.* 18:19–36.

CARROLL, J.C. and F.A. WELTON. 1939. Effect of heavy and late applications of nitrogen fertilizer on the cold resistance of Kentucky bluegrass. *Plant Physiol.* 14:297–308.

CHANDLER, W.H. 1913. The killing of plant tissues by low temperature. *Mo. Agr. Expt. Sta. Res. Bul.* 8.

CHILDERS, N.F. 1969. Modern fruit science, 4th ed. Horticultural Publications, New Brunswick, N.J.

CHRISTERSSON, L. 1973. The effect of inorganic nutrients on water economy and hardiness of conifers. *Studia Forest. Suecica* 103.

CHRISTERSSON, L. 1975a. Frost-hardiness development in *Pinus silvestris* L. seedlings at different levels of potassium and calcium fertilization. *Can. J. For. Res.* 5:738–740.

CHRISTERSSON, L. 1975b. Frost hardiness development in rapid and slow growing Norway spruce seedlings. *Can. J. For. Res.* 5:340–343.

CLORE, W.J. and V.P. BRUMMOND. 1965. Cold injury to grapes in eastern Washington 1964–1965. Proc. 61st Annu. Meeting. Washington State Hort. Assoc. p. 143–144.

CONOVER, C.A. and R.T. POOLE. 1977. Influence of nutrition on yield and chilling injury of *Sansevieria trifasciata*. *Proc. Fla. State Hort. Soc.* 89:305–307.

COOK, T.W. and D.T. DUFF. 1976. Effects of K fertilization on freezing tolerance and carbohydrate content of *Festuca arundinacea* Schreb. maintained as turf. *Agron. J.* 68:116–119.

COOLING, E.N. 1967. Frost resistance in *Eucalyptus grandis* following the application of fertilizer borate. *Rhodesia Zambia Malawi J. Agr. Res.* 5:97–100. (*For. Abstr.* 28:5770).

COOPER, J.R. and C.B. WIGGANS. 1929. A study of the effect of commercial fertilizers on the performance of peach trees. *Ark. Agr. Expt. Sta. Bul.* 239.

CRANE, H.L. 1924. Experiments in fertilizing peach trees. *West Va. Agr. Expt. Sta. Bul.* 183.

CULLINAN, F.P. 1931. Some relationships between tree response and internal composition of shoots of the peach. *Proc. Amer. Soc. Hort. Sci.* 28:1–5.

DEXTER, S.T. 1935. Growth, organic nitrogen fractions and buffer capacity in relation to hardiness of plants. *Plant Physiol.* 10:149–158.

DEXTER, S.T., W.E. TOTTINGHAM, and L.F. GRABER. 1932. Investigations of hardiness of plants by measurement of electrical conductivity. *Plant Physiol.* 7:63.

DROZDOV, S.N. and Z.F. SYCHEVA. 1965. Relationship of the frost resistance of potato haulms with the level of nitrogen metabolism. *Soviet Plant Physiol.* 12:274–279.

EDGERTON, L.J. 1957. Effect of nitrogen fertilization on cold hardiness of apple trees. *Proc. Amer. Soc. Hort. Sci.* 70:40–45.

EDGERTON, L.J. and R.W. HARRIS. 1950. Effect of nitrogen and cultural treatment on Elberta peach fruit bud hardiness. *Proc. Amer. Soc. Hort. Sci.* 55:51–55.

FAUSTOV, V.V. 1965. The effect of mineral nutrition on the winter hardiness of gooseberries. *Agrobiologia* 2:289–290.

FERGUSON, A.C. 1966. Winter injury north of the 49th. *Golf Supt.* 39:38–39.

GAMOZHINA, S.T. and V.T. SEVASTYANOV. 1971. Phosphorus containing ingredients in winter-barley under different growing conditions. *SKH Biol.* 6:256–269.

GEORGE, M.F. and M.J. BURKE. 1976. The occurrence of deep super-cooling in cold hardy plants. *Curr. Adv. Plant Sci.* 22:349.

GILBERT, W.B. and D.L. DAVIS. 1971. Influence of fertility ratios on winter hardiness of bermudagrass. *Agron. J.* 63:593–595.

GLADWIN, F.E. 1917. Winter injury of grapes. *New York (Geneva) Agr. Expt. Sta. Bul.* 433.

GREEN, W.J. and F.H. BALLOU. 1904. Winter killing of peach trees. Report of investigations in the Lake Erie fruit belt. *Ohio Agr. Expt. Sta. Bul.* 157: 115–134.

HAVIS, J.R., R.D. FITZGERALD, and D.N. MAYNARD. 1972. Cold hardiness response of *Ilex crenata* Thunb. cv Hetzi roots to nitrogen source and potassium. *HortScience* 7:195–196.

HAVIS, L. and I.P. LEWIS. 1938. Winter injury of fruit trees in Ohio. *Ohio Agr. Expt. Sta. Bul.* 596.

HIGGINS, B.B., G.P. WALTON, and J.J. SKINNER. 1943. The effect of nitrogen fertilizer on cold injury of peach trees. *Ga. Agr. Expt. Sta. Bul.* 226.

HOWELL, J.H. and G.A. JUNG. 1965. Cold resistance of Potomac orchardgrass as related to cutting management, nitrogen fertilization and mineral levels in the plant sap. *Agron. J.* 57:525–529.

JACKSON, L.K. and J.F. GERBER. 1963. Cold tolerance and freezing point of citrus seedlings. *Proc. Fla. State Hort. Soc.* 76:70–74.

JANICK, J. 1979. Horticultural science, 3rd ed. W.H. Freemen and Co., San Francisco, Calif.

JOINER, J.N. and E.R. ELLIS, JR. 1964. Effects of varying levels of nitrogen and potassium on cold hardiness of 'George Tabor' azaleas. *Proc. Fla. State Hort. Soc.* 77:525–527.

JUNG, G.A. and D. SMITH. 1959. Influence of soil potassium and phosphorus content on cold resistance of alfalfa. *Agron. J.* 51:585–587.

KAWANA, A., M. NAKAHARA, B. SUGIMOTO, and H. HURUHATA. 1964. The effect of fertilization on the growth and frost damage of sugi (*Cryptomeria japonica* D. Don) and hinoki (*Chamaecyparis obtusa* Endl.). *J. Jap. For. Soc.* 46:355–363.

KELLEY, J.D. 1972. Nitrogen and potassium rate effects on growth, leaf nitrogen and winter hardiness of *Pyracantha coccinea* 'Lalandi' and *Ilex crenata* 'Rotundifolia'. *J. Amer. Soc. Hort. Sci.* 97:446–448.

KENNARD, W.D. 1949. Defoliation of Montmorency sour cherry trees in relation to winter hardiness. *Proc. Amer. Soc. Hort. Sci.* 53:129–133.

KIMBALL, D.A. 1927. Effect of the hardening process on the water content of some herbaceous plants. *Proc. Amer. Soc. Hort. Sci.* 24:64–69.

KOPITKE, J.C. 1941. The effect of potash salts upon the hardening of conifer seedlings. *J. For.* 39:555–558.

KUKSA, I.N. 1939. The effect of mineral nutrition on winter hardiness and yield of winter wheat. *Himiz. Soc. Zemled.* 1:70–79. (*Herb. Abstr.* 9:635).

LACIO, A.H. 1930. Minerals leicits ka kalja mesolsauas lidzeklis. *Lauksaimniecibas Menesraksts.* (Riga) 8:506–513. (*Biol. Abstr.* 5:24571).

LEVITT, J. 1956. The hardiness of plants. Academic Press, New York.

LEVITT, J. 1972. Responses of plants to environmental stresses. Academic Press, New York.

LOTT, R.V. 1926. Correlation of chemical composition with hardiness in brambles. *Mo. Agr. Expt. Sta. Res. Bul.* 95.

MALCOLM, D.C. and B.C.Y. FREEZAILLAH. 1975. Early frost damage on Sitka spruce seedlings and the influence of phosphorus nutrition. *Forestry* 48:139–145.

MARCELLOS, H. 1977. Wheat frost injury—freezing stress and photosynthesis. *Austral. J. Agr. Res.* 28:557–564.

MARLATT, R.B. 1974. Chilling injury in sansevieria. *HortScience* 9:539–540.

MAZUR, P. 1969. Freezing injury in plants. *Annu. Rev. Plant Physiol.* 20: 419.

MCMUNN, R.L. and M.J. DORSEY. 1935. Seven years results of the hardiness of Elberta fruit buds in a fertilizer experiment. *Proc. Amer. Soc. Hort. Sci.* 32:239–243.

O'CARROL, N. 1973. Chemical weed control and its effect on the response to K fertilization. *Irish For.* 29:20–31.

PALMERTREE, H.D., C.Y. WARD, and R.H. PLUENNEKE. 1973. Influence of mineral nutrition on the cold tolerance and soluble protein fraction of centipedegrass *(Eremochloa ophiuroides).* p. 500–507. *In* Proc. Second Intern. Turfgrass Res. Conf., Blacksburg, Va. American Society of Agronomy and Crop Science Society of America, Madison, Wisc.

PANKRATOVA, E.M. 1963. Effect of variable temperature hardening and trace element seed treatments on frost hardiness and yield of corn. *Agrobiologiya* 4:553–557.

PELLETT, N.E. 1973. Influence of nitrogen and phosphorus fertility on cold acclimation of roots and stems of two container-grown woody plant species. *J. Amer. Soc. Hort. Sci.* 98:82–86.

PELLETT, N.E. and D.B. WHITE. 1969. Effect of soil nitrogen and soil moisture levels on the cold acclimation of container grown *Juniperus chinensis* 'Hetzi'. *J. Amer. Soc. Hort. Sci.* 94:457–459.

PEYNADO, A. and R.H. YOUNG. 1963. Toxicity of three salts to greenhouse-grown grapefruit trees and their effects on ion accumulation and cold hardiness. *J. Rio Grande Valley Hort. Soc.* 17:60–67. (*Biol. Abstr.* 46:100239).

PROEBSTING, E.L. and H.H. MILLS. 1974. Time of gibberellin application determines hardiness response of 'Bing' cherry buds and wood. *J. Amer. Soc. Hort. Sci.* 99:464–466.

PROEBSTING, E.L., JR. 1961. Cold hardiness of Elberta peach fruit buds as influenced by nitrogen level and cover crops. *Proc. Amer. Soc. Hort. Sci.* 77:97–106.

PYIKLIK, K.M. 1963. Accumulation and utilization of sugars in relation to overwintering in winter crops. *Soviet Plant Physiol.* 10:104–108.

RAGAN, P. and R.E. NYLUND. 1977. Influence of N-P-K fertilizers on low temperature tolerance of cabbage seedlings. *HortScience* 12:320–321.

RAWLINGS, C.O. and G.F. POTTER. 1937. Unusual and severe winter injury to the trunks of McIntosh apple trees in New Hampshire. *Proc. Amer. Soc. Hort. Sci.* 34:44–48.

RHOTON, V.D. 1967. The relation of zinc and cell organelle to low temperature tolerance of tomato plants. PhD Thesis, Arizona State University, Tuscon. (*Diss. Abstr.* 28:803-B).

SITTON, B.G., W. LEWIS, M. DROSDORF, and H. BARROWS. 1954. Trends in response of bearing tung trees to N, P and K fertilizers. *Proc. Amer. Soc. Hort. Sci.* 64:29–46.

SKINNER, J.J. and C.A. REED. 1925. Fertilizers, cover crops, soil conditions. *Amer. Nut. J.* 22:90–93.

SLAVNYI, P.S. and N.N. MUSIENKO. 1972. Formation of wheat root systems and its frost resistance in dependence on nutrition. *Fiziologiya I Biokhimiya Kulturnykh. Rastenii* 4:68–73.

SMITH, D.F. and G.K. RASMUSSEN. 1958. Relation of fertilization to winter injury of citrus trees. *Proc. Fla. State Hort. Soc.* 71:170–175.

SMITH, W.W. and M.A. TINGLEY. 1940. Frost rings in fall fertilized McIntosh apple trees. *Proc. Amer. Soc. Hort. Sci.* 37:110–112.

SUDDS, R.H. and R.S. MARSH. 1943. Winter injury to trunks of young bearing apple trees in West Virginia following a fall application of nitrate of soda. *Proc. Amer. Soc. Hort. Sci.* 42:293–297.

TINGLEY, M.A., W.W. SMITH, T.G. PHILLIPS, and G.F. POTTER. 1938. Experimental production of winter injury to the trunks of apple trees by applying nitrogenous fertilizers in the autumn. *Proc. Amer. Soc. Hort. Sci.* 36:177–180.

TRESHOW, M. 1970. Environment and plant response. McGraw-Hill Book Co., New York.

TUMANOV, I.I. 1967. Physiological mechanism of frost resistance of plants. *Soviet Plant Physiol.* 14:440.

VASIL'YEV, I.M. 1967. Wintering of plants. American Institute of Biological Sciences, Washington, D.C.

VIART, M. 1965. Fertilizing poplars. *Bulletin du service de Culture et d'Etudes du Peuplier et du Saule.* Paris. p. 1–37. (*For. Abstr.* 27:5934).

WALLACE, T. 1926. An experiment on the winter killing of vegetable crops in market gardens. *J. Pomol. Hort. Sci.* 5:205–209. (*Biol. Abstr.* 2:2679).

WAY, R.D. 1954. The effect of some cultural practices and of size of crop on the subsequent winter hardiness of apple trees. *Proc. Amer. Soc. Hort. Sci.* 63:163–166.

WEISER, C.J. 1970. Cold resistance and injury in woody plants. *Science* 169: 1269.

WHITE, D.P. and R.F. FINN. 1964. Frost damage in a tulip poplar plantation as related to foliar potassium content. *Pap. Mich. Acad. Sci., Arts, & Letters* 49:75–80.

WILKINSON, J.F. and D.T. DUFF. 1972. Effects of fall fertilization on cold resistance, color and growth of Kentucky bluegrass. *Agron. J.* 64:345–348.

YASUDA, S. 1927. On the winter hardiness of barley. II. Effect of potassium salts. *J. Sci. Agr. Soc.* 295:273–281. (*Biol. Abstr.* 4:3833).

YOUNG, T.W. and J.C. NOONAN. 1959. Influence of nitrogen source on cold hardiness of lychees. *Proc. Amer. Soc. Hort. Sci.* 73:229–233.

ZURAWICZ, E. and C. STUSHNOFF. 1977. Influence of nutrition on cold tolerance of 'Redcoat' strawberries. *J. Amer. Soc. Hort. Sci.* 102:342–345.

5

Mycorrhizal Fungi and Their Importance in Horticultural Crop Production[1]

Dale M. Maronek
Department of Horticulture and Landscape Architecture,
University of Kentucky, Lexington, Kentucky 40546
James W. Hendrix and Jennifer Kiernan
Department of Plant Pathology,
University of Kentucky, Lexington, Kentucky 40546

I. INTRODUCTION

Production of horticultural crops has undergone enormous change during the past several decades due to the development and use of

[1] Paper 80-10-11-22, Kentucky Agricultural Experiment Station. Previously unpublished research reported herein was supported by grants from USDA-SEA (706-15-16), the University of Kentucky Institute of Mining and Minerals Research (S4187, S4322), and the Office of Surface Mining, U.S. Department of Interior (G5105004).

fertilizers, pesticides, mechanization, soil fumigation, container production in soilless media, breeding, tissue culture methods for propagation and pathogen elimination, and a number of other innovations. Recently there have been indications that advances in productivity due to technology are peaking out. Furthermore, in the near future, we face severe problems in one of the most important factors involved in the technological revolution: fertilizer production. Production of nitrogen fertilizers is energy-intensive, and reserves of some fertilizer components, especially phosphate, are becoming limiting.

Scientists in recent years have demonstrated the beneficial or essential role of mycorrhizal fungi in the lives of many plants of horticultural interest, especially when growing in natural, unmanaged situations. We suspect that organic techniques practiced decades ago, such as use of manures, crop rotation, and green manure crops, were effective in large part because they influenced mycorrhizal fungi. We are not proposing to discard the advances that led to our present state of horticultural production. However, in evolving during this decade to the point where our production is both in equilibrium with our resources and environmentally compatible, we feel that mycorrhizal fungi will find significant roles.

In this review, we present experimental evidence suggesting uses of mycorrhizal fungi in horticultural production and point out some research voids which must be filled.

II. CONCEPTS OF MYCORRHIZAE AND SYMBIOSIS

Mycorrhizae have been known for quite a long time. In 1885 Frank coined the term mycorrhizae to describe the symbiotic association of plant roots and fungi. Mycorrhiza literally means "fungus root." About the same time, in 1879, Anton de Bary introduced the concept of symbiosis, or "living together." This broad definition of symbiosis covers a number of situations, such as one partner causing harm to the other (pathogenesis), neither partner benefiting or being damaged from the association (commensalism), and both partners benefiting from the association (mutualism). Today, symbiosis and mutualism are often used synonymously, and indeed it is mutualism which we want to utilize in horticultural production. Following Frank's observation, it became apparent that with certain kinds of plants, health and vigor were associated with abundant mycorrhizae. This gradually was thought of as a mutually beneficial association, in contrast to the root-invading parasites which cause damage.

Thus, to avoid becoming ensnared in terminology, "mycorrhizal fungi" in this review refers to the types of fungi that usually form beneficial associations with plant roots. This concept of mycorrhizal fungi excludes mutualism as a requirement and allows us to deal with situations in which (1) a fungus benefits its host in one set of conditions but harms it in another, (2) the fungus benefits one host but harms another, or (3) an occasional rogue apparently does not benefit any plant but taxonomically is associated with the groups of fungi known to form mutualistic associations with plants.

Mycorrhizal associations are so prevalent that the nonmycorrhizal plant is more the exception than the rule (Gerdemann 1968). Consequently, it is easier to name the plant groups in which the associations do not occur or have yet to be reported: the order Centrospermae and the families Cruciferae, Cyperaceae, Fumariaceae, Commelinaceae, Urticaceae, and Polygonaceae (Gerdemann 1968). However, Gerdemann (1975) has cited exceptions—where endomycorrhizae were found for several members of the order Centrospermae, family Chenopodiaceae (Ross and Harper 1973; Kruckleman 1973, 1975; Williams *et al.* 1974), and several species in the Cyperaceae (Mejstrik 1972) and Cruciferae (Ross and Harper 1973; Kruckelman 1973, 1975). Hirrel *et al.* (1978) and Ocampo *et al.* (1980) found that several species of Chenopodiaceae and Cruciferae become infected to a limited extent if grown in the presence of a mycorrhizal plant, but not if they are grown alone.

Anatomically mycorrhizae can be divided into three classes: ecto-, endo-, and ectendomycorrhizae. Of these, ecto- and endomycorrhizae are the two major classes of mycorrhizae. The three types of mycorrhizal relationships are typified by the fungal intrusion being limited to the cortical region of unsuberized roots. The association involving only unsuberized roots may in part be the result of root expansion and subsequent loss of the primary cortex in mature roots. Attempts by the fungus to penetrate further into the inner layers of the root may be overcome by defense mechanisms of the host (Meyer 1974).

A. Ectomycorrhizae

Ectomycorrhizae are most common among forest and ornamental tree species in the families Pinaceae, Salicaceae, Betulaceae, Fagaceae, and Tiliaceae, as well as in some members of the Rosaceae, Leguminosae, Ericaceae, and Juglandaceae (Trappe 1962a; Meyer 1973). The fungal partner in an ectomycorrhiza most frequently belongs to the Basidiomycetes, primarily in Amanitaceae, Boletaceae, Cortinariaceae, Russiclaceae, Tricholomataceae, Rhizopoganaceae, and Sclerodermataceae (Marx 1972). Some orders of Ascomycetes, Eurotiales, Tuberales, Pezizales, and Helotiales contain species that form ectomycorrhizae. Basidiomycetes include those fungi that produce mushrooms and puffballs, while the major Ascomycetes that form ectomycorrhizae are the truffles, some of which are gourmet delights in many European countries.

Structurally, ectomycorrhizae can be distinguished by the presence of hyphal strands coalescing to form a thick weft or sheath around the feeder roots known as a mantle (Fig. 5.1E). The mantle can range from thin to profuse and the texture can vary from smooth to cottony or granular (Zak 1973). The mantle replaces the root hairs with fungal strands, greatly enhancing root surface absorptive area. These hyphal strands are capable of permeating outward from the root surface several meters or more and exploring regions not accessible by root hairs. Hyphae also penetrate through the epidermis into the intercellular spaces of the cortical cells apparently replacing the middle lamella and forming an interconnecting network known as the 'Hartig net'. Ectomycorrhizal roots are generally recognizable by their short, swollen appearance and distinctive colors of either white, black, orange, yellow, or olive green which are contingent on interactions between the plant species and its fungal associate (Fig. 5.1F). Ectomycorrhizae are also characterized by specific branching patterns ranging from monopodial to multi-forked (ramiform) or coralloid.

Ectomycorrhizal fungi can exist in the soil as spores, sclerotia, and rhizomorphs. A rhizomorph is a coalesced group of hyphal strands which may extend from the ectomycorrhiza to or near the surface of the soil, forming fruiting bodies containing spores that can be disseminated by wind, surface water, or animals.

Ectomycorrhizal fungi are dependent upon their hosts in their natural habitats for reduced carbon compounds, but many species can be cultured without a host using pure culture techniques (Mikola 1970; Göbl 1975; Marx *et al.* 1970, 1976). With perhaps a few exceptions, these fungi differ from lignicolous and litter-decaying fungi which use the complex polysaccharides, cellulose, and lignin as a carbon source.

B. Endomycorrhizae

Most endomycorrhizal fungi belong to the order Phycomycetes, genera *Glomus, Sclerocystis, Endogone, Gigaspora,* and *Acaulospora* (Gerdemann and Trappe 1975). These form what is known as vesicular-arbuscular (VA) mycorrhizae with plants. Endomycorrhizae are the most widely distributed of any of the mycorrhizae, being found in many herbaceous, shrub, and tree species, including most annual agronomic and horticultural crops (Gerdemann 1975). A few endomycorrhizae are formed by an association with Basidiomycetes, which differ somewhat from endomycorrhizae in cellular morphological structure by having septate hyphae. Those endomycorrhizae occur primarily in the Orchidaceae, Gentianaceae, and Ericaceae.

Endomycorrhizae differ from ectomycorrhizae in that there are essentially no discernible morphological changes in the external root struc-

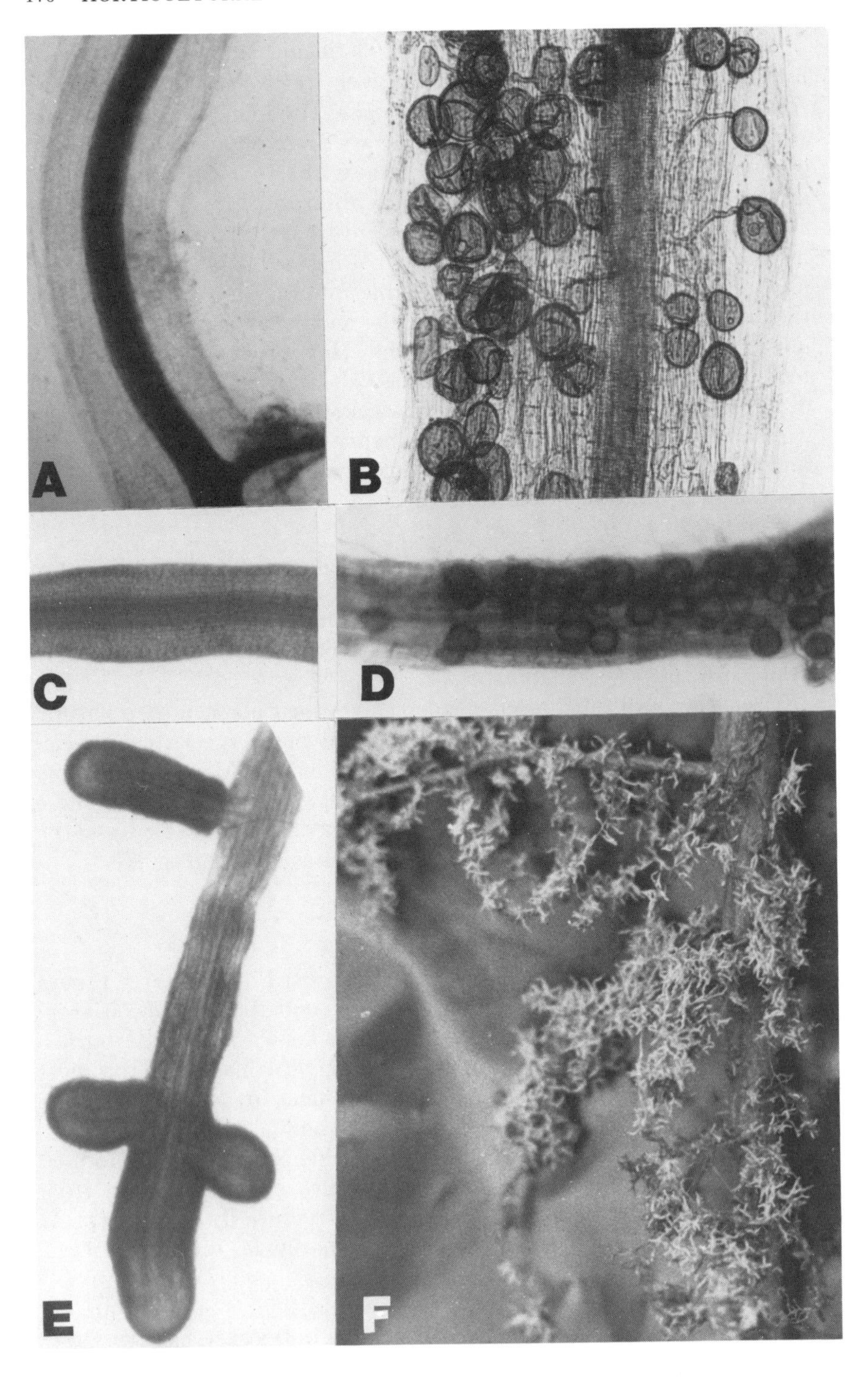
A
B
C
D
E
F

ture of the host. A fungal mantle is not present, but a loose network of hyphae which radiate outward several centimeters or more from the root may be seen on feeder roots (Rhodes and Gerdemann 1975). Hyphae of endomycorrhizal fungi generally penetrate through the epidermis or root hair into the cortical cells; hence, the prefix "endo" is used. The penetrating hyphae often form specialized structures called vesicles and arbuscules (Fig. 5.1A–D). The longevity of arbuscules is short, 5 to 15 days (Bevege and Bowen 1975), as they continually disintegrate (Kaspari 1973), leading to the hypothesis that this process may be a valuable source of mineral nutrients to the host (Gerdemann 1968). Round or oval bodies known as vesicles are also formed in the cortex and in some instances outside the root. Little is known about vesicles; however, it has been suggested that they may function in some storage capacity (Gerdemann and Nicholson 1963).

Spores of endomycorrhizae are borne in small sporocarps or occur individually in the soil or on or in the root (Fig. 5.1A–D). The sporocarps are often not easily observed, as in the Basidiomycetes, and must be extracted from the soil using wet sieving techniques (Gerdemann and Nicholson 1963).

Some plants are capable of forming both ecto- and endomycorrhizal associations. Those include the families Salicaceae, Juglandaceae, Tiliaceae, and Myrtaceae as well as some species of *Juniperus* and *Chamaecyparis* (Gerdemann 1975).

C. Ectendomycorrhizae

A third type of mycorrhizae exhibiting both ecto- and endomycorrhizal characteristics has been classified as ectendomycorrhizae. Pure culture synthesis of ectendomycorrhiza in aseptic conditions has been successful (Wilcox and Ganmore-Neuman 1974). However, very little is known about the species of fungi involved. They appear to have a limited distribution in forest soils and have been associated with species in nursery seedbeds that are normally ectomycorrhizal (Mikola 1965). The ectendomycorrhizal characteristics may be determined by the host. An ericaceous plant produced ectendomycorrhizae when it was inoculated

FIG. 5.1. MYCORRHIZAE
A. Black cherry (*Prunus serotina* J.F. Ehrh.), as a control. B. Black cherry infected with endomycorrhizae, formed by *Glomus fasciculatus*. (Enlargement B is approximately 2× that of A.) Note the presence of fungal spores and vesicles. C. Sweetgum *(Liquidambar styraciflua)*, as a control. D. Sweetgum *(Liquidambar styraciflua)* infected with endomycorrhizae, formed by artificial inoculation of container-grown seedlings. Note the presence of fungal spores and vesicles. E. Ectomycorrhizae formed on Norway spruce (*Picea abies* (L.) Karst.) with *Pisolithus tinctorius* by artificial inoculation of container-grown seedlings. Note the presence of fungal mantle. F. Ectomycorrhizal on pin oak *(Quercus palustris)* formed by unidentified bolete (Isolate M22) through artificial inoculation in nursery seedbeds. Note the presence of abundant short roots.

with fungi that induce ectomycorrhizae in other hosts (Zak 1976).

III. MYCORRHIZAL DEVELOPMENT AND SOME BIOCHEMICAL INTERACTIONS BETWEEN HOST AND FUNGUS

The interactions between the fungus and host are complex and appear to be influenced by a myriad of interrelated biochemical, physiological, and environmental processes. Furthermore, there also appear to be some reciprocal relationships between the fungus and the plant host, but it is difficult to interpret the exact contribution either organism lends to the association (Meyer 1974). We do know that the fungal symbiont must enter and maintain a parasitic relationship with its host for procurement of organic compounds required for its growth and reproduction. It is during mycorrhizal formation that we can begin to understand how the biochemical interactions between the partners develop and how these interactions may influence plant growth and development of horticulture crops. Most of the work on mycorrhizal formation has centered around ectomycorrhizae, where the morphological changes brought about by various biochemical influences are more easily discernible.

A. Growth Regulating Substances

Plant and fungal hormones are suspected of being involved in mycorrhizal development. Auxins, cytokinins, gibberellins, and vitamins have been shown to be produced by mycorrhizal fungi in pure culture (Miller 1971; Crafts and Miller 1974; Slankis 1975), and the effects of these compounds on plant growth and development are well documented (Torrey 1976). Also, the type and quantity of growth substances produced by both partners could affect their association and subsequent overall plant responses. However, no one has yet shown that any mycorrhizal fungus produces a growth hormone while in association with the root.

Mycelial growth of symbiotic fungi can be regulated by many plant growth hormones. Melin (1963) reported that an unidentified substance which he termed "M factor" was exuded from Scotch pine (*Pinus sylvestris* L.) roots which stimulated mycelial growth of symbiotic fungi. The identity of the "M factor" is still subject to debate, but the "M factor" has been shown to be replaceable with diphosphopyridine nucleotide (Melin 1962).

However, Gogala (1970) suggested that the "M factor" may be related to cytokinins. She found that cytokinins extracted from pine seedlings can stimulate mycelial growth of *Boletus edulis* var. piniculus. Similarly, Maronek and Hendrix have found (unpublished data) that mycelial growth of the ectomycorrhizal fungus *Pisolithus tinctorius* (Pers.) Coker and Couch can be altered also by auxin concentrations in pure culture.

Nielson (1930) first showed that an auxin-like compound was produced by cultured ectomycorrhizal fungi, and several years later Thimann (1935) identified the compound as IAA. Slankis (1949, 1950, 1951, 1958) was able to show that synthetic auxins can influence the induction of mycorrhizal structures on the roots of Scotch and white pine (*Pinus sylvestris* and *P. strobus* L.). When the auxin treatment was discontinued, mycorrhizal characteristics of the roots disappeared, suggesting that auxin and perhaps other related compounds in the root are responsible for the typical mycorrhizal formation and morphology. Although auxin appears to be the most common substance inducing mycorrhiza-like structures, others have found activity with colchicine, kinetin, and various vitamins (Slankis 1975).

The role that cytokinin may play in the mycorrhizal association and, for that matter, the role that fungus-produced cytokinins play in overall plant growth and development are subject to speculation. The development of short, swollen roots (Fig. 5.1E and F) on some plants infected with a mycorrhizal fungus suggests that cytokinins may be involved in the symbiotic relationship. Cytokinins have been synthesized by mycorrhizal fungi in pure culture (Miller 1971; Crafts and Miller 1974). However, not all mycorrhizal fungi may be capable of synthesizing cytokinins or of synthesizing them in sufficient quantities. Miller (1971) and Crafts and Miller (1974) have screened over 25 species of mycorrhizal or suspected mycorrhizal fungi. They have found only six species that definitely produce cytokinins in detectable quantities. They further point out that despite the lack of complete correlation between cytokinin production and fungi which form mycorrhizae, the role of cytokinin in associations where a correlation does exist should not be overlooked. Perhaps the fungus takes over cytokinin synthesis or high indigenous concentrations of cytokinins in the root, modifies growth, and predisposes the root to mycorrhizal infection (Crafts and Miller 1974).

B. Carbohydrates

The involvement of plant growth hormones such as auxin in mycorrhizal formation is further complicated by the need for simple carbohydrates for fungal growth. Auxin has been shown to influence translocation of sugar from starch reservoirs of the plant (Thimann 1972) as well as the hydrolysis of starch into sugar (Borthwick *et al.* 1937; Alexander 1938; Bausor 1942). The quantity of soluble sugar in the roots also may have a direct relationship to the degree of ectomycorrhizal development (Hacskaylo 1971). Björkman (1942, 1970) suggested that, since mycorrhizal fungi generally assimilate soluble carbohydrates, their absence or presence in small quantities could influence mycorrhizal formation. Marx *et al.* (1977) reported that a decrease in sucrose content of loblolly pine (*Pinus taeda* L.) short roots reduced their susceptibility to colonization by the ectomycorrhizal fungus *Pisolithus tinctorius*. Ratnayake *et al.* (1978) proposed that a reduction in root exudation of soluble amino acids and reducing sugars also may influence endomycorrhizal infection.

Another interesting aspect in the biochemical interaction between the fungus and the plant is the possibility that the fungus is capable of converting absorbed soluble sugars into storage compounds that are not readily available to the plant, such as glycogen, mannitol, or trehalose. Lewis and Harley (1965a,b,c) found both mannitol and trehalose in the soluble fraction of beech ectomycorrhizae, but none of the compounds were reportedly found in the host. Smith *et al.* (1969) further substantiated these results by reporting that glucose derived from the host was converted into fungal carbohydrates (trehalose and mannitol) which could not be metabolized by the host plant. Meyer (1974) suggests that the fungus intervenes in the carbohydrate metabolism of the host by acting as a sink. A withdrawal of soluble carbohydrates from roots would tend to suggest that a concentration gradient of soluble carbohydrates is induced in the roots by the mycorrhizal fungus. This source to sink relationship could be reciprocal and may even exist at certain times during the year, based on plant and fungal needs. For instance, Harley (1971) suggests that an estimated 10% of the annual potential wood production in a spruce (*Picea* sp.) forest was diverted from the host plant for production of mycorrhizal fruiting bodies. Fruiting of many mycorrhizal fungi in native habitats is periodic. The possibility that the fungus acts like a sink for carbohydrates and other plant compounds is analogous to the seasonal source to sink relationship found in plants during fruit and seed production.

In endomycorrhizae the carbohydrate storage process may be different. Trehalose and mannitol have not been found in endomycorrhizal fungi (Hepper and Mosse 1972; Bevege *et al.* 1975), and it has been hypothesized that lipid may serve an important storage function in endomycorrhizal fungi (Cox *et al.* 1975). Ho (1977) also reported that roots of corn (*Zea mays* L.) inoculated with endomycorrhizal fungi contained significantly greater concentrations of total sterols, cholesterol, and campesterol than nonmycorrhizal roots. Endomycorrhizal roots of onion (*Allium cepa* L.), clover (*Trifolium repens* L.), and rye grass (*Lolium perenne* L.) contained more total lipid than nonmycorrhizal roots (Cooper and Losel 1978). Abundant oil droplets were observed in the internal and external hyphae and in the vesicles and chlamydospores. The abundance of lipid globules could be the result of the fungus acting as an alternative sink for photosynthate (Ho and Trappe 1975). Ho and Trappe also suggested that lipid globules could be involved in carbon translocation moving via cytoplasmic streaming.

C. Environmental Influences

The available soluble carbohydrates and the presence of plant growth hormones or other substances found in the root are not considered to be solely responsible for mycorrhizal formation and maintenance of the symbiosis. Some environmental factors affecting mycorrhizal development as well as plant growth and development include light, soil conditions (moisture, mineral nutrients, pollutants), and interaction with other soil organisms.

Ectomycorrhizal formation generally is greater when the nitrogen (N) content of the soil is low (Björkman 1942; Marx *et al.* 1977). A possible explanation for this is that at high soil N levels, the host plant utilizes the nitrogen for making protein, resulting in a reduction in available carbohydrates. At high N levels, there is also an increase in plant auxin levels, making the fungus less competitive with the plant for the carbohydrate supply (Björkman 1942). However, mycorrhizal development is not always indicative of nutritionally poor soils, but also may be dependent upon rooting behavior of trees, soil types, and balance of nutrients (Meyer 1974). In addition, the rate of fertilizer release may influence mycorrhizal development. In greenhouse studies with pin oak (*Quercus palustris* Muenchh.) seedlings inoculated with the ectomycorrhizal fungus *Pisolithus tinctorius*, a greater percentage of the root system was mycorrhizal on seedlings fertilized with the manufacturer's recommended rate of 4.5 kg/m^3 of a slow release fertilizer than on those fertilized at 1.9 kg/m^3 rate (Maronek and Hendrix 1979a). Also, atmospheric pollutants may influence the population of mycorrhizal fungi inhabiting an area. For example, industrial ash and dust fall out over a long period were reported to change the mycorrhizal fungus flora of a forest community (Sabotka 1974).

Research on soil fertility's influence on endomycorrhizal formation is limited. Endomycorrhizae tend to be more prevalent in low fertility soils, especially when nitrogen and phosphorus are low (Gerdemann 1975). The interaction of mycorrhizal fungus and soil fertility will be discussed later.

Mycorrhizal development is greatest at higher light intensities (Björkman 1970). This would appear to be linked with increased photosynthetic activity and subsequent carbohydrate translocation to the roots. Björkman (1942) reported that a reduction in light intensity to below 25% of full daylight reduced ectomycorrhizae, and at 6% light intensity no infection occurred. Unfortunately, conflicting research data suggest that this relationship is not so clear-cut. For example, Harley and Waid (1955) found that while a general correlation between light and mycorrhizal development exists, a degree of complexity arises in that ectomycorrhizae formed by the association of *Cenococcum graniforme* (Sow.) Ferd. and Winge. with beech (*Fagus* sp.) trees were more plentiful at lower light intensities. Mikola (1948) also reported better mycorrhizae on birch (*Betula pubescens* J.F. Ehrh.) with *C. graniforme* at 10% than at 39% of daylight.

The effect of light on endomycorrhizae is similar to that on ectomycorrhizae. Hayman (1974) reported that mycorrhizal development is best in onions *(Allium cepa)* infected with *Endogone* (yellow vacuolate spore type) at high light intensity, particularly in the number of arbuscules. Furlan and Fortin (1977), however, found that onion plants inoculated with *Gigaspora calospora* (Nicol. and Gerd.) showed more extensive and rapid mycorrhizal infection at low light intensities than at high ones. They concluded that it cannot be assumed that high light intensity favors mycorrhizal infection.

Most mycorrhizal fungi have an optimum temperature for establishment of the symbiotic relationship and the survival of the mycorrhizal condition. There is, however, considerable variation in the temperature range tolerance of individual fungi. Marx *et al.* (1970) found that the fungus *Thelephora terrestris* Ehrh. ex. Fr. had a considerably lower tolerance towards a wide temperature range than did *Pisolithus tinctorius*. *Thelephora terrestris* formed ectomycorrhizae on 45% of the feeder roots of loblolly pine *(Pinus taeda)* at 14°, 19°, and 24°C, but the percentage dropped to 30% at 29°C and none at all at 34°C. *Pisolithus tinctorius* formed mycorrhizae in increasing numbers up to a maximum of over 80% at 34°C. *Pisolithus tinctorius* has been shown to have a good survival rate even at 40°C (Marx and Bryan 1971). Schenck and Schroder (1974) found that there was also a temperature response by an 'Endogone' mycorrhiza on soybean roots with arbuscular development being favored at 30°C, mycelial development greatest at 28° to 34°C, and spore and vesicle production being greatest at 35°C.

IV. DISEASE RESISTANCE

Wilt, root, and stem diseases in horticultural crops cause large economic losses annually. These root-invading pathogens are responsive to soil type, pH, temperature, fertility, moisture, and the presence of other soil-borne organisms. Production systems which advocate the use of high density planting and monocropping must maintain continuous disease prevention programs. Generally, once these diseases have been identified it is too late to implement successful control. An important contribution of some mycorrhizal fungi in horticulture crop production may be their ability to increase disease resistance.

Zak (1964) postulated four disease protection mechanisms afforded to plants by ectomycorrhizal fungi: (1) the utilization of surplus carbohydrates in the roots which reduces the roots' attraction to pathogens; (2) the fungal mantle, serving as a physical barrier to infection; (3) secretion of antibiotics; and (4) favoring a protective rhizosphere of organisms. Schenck and Kellam (1978) suggested that all these phenomena, except that of providing a physical barrier to infection, also could exist with endomycorrhizal fungi, since profuse fungal mantles are lacking in endomycorrhizae. They also reported that disease severity can be increased, decreased, or not affected by the presence of endomycorrhizal fungi. Furthermore, the influence of endo- or ectomycorrhizae on a root may vary with plant species and type of mycorrhizal and pathogenic fungi involved (Marx 1972; Chou and Schmitthenner 1974; Richard and Fortin 1975; Schonbeck and Dehne 1977; Davis *et al.* 1978).

The simplest disease protection offered by an ectomycorrhizal fungus may be fungal mantle providing a physical barrier against infection by pathogenic fungi. The mantle of a mycorrhiza can range from none or barely perceptible to as much as 70 to 100 μm thick, which has been recorded for red pine (Wilcox 1971).

In experiments to determine the effect of ectomycorrhizal fungi on the resistance of shortleaf pine (*Pinus enchinata* Mill.) to *Phytophthora cinnamomi* (Rands), Marx (1970) found that all nonmycorrhizal roots inoculated with *P. cinnamomi* became infected, but those with well-formed mantles of either *Thelephora terrestris* (mantle thickness 29 μm) or *Pisolithus tinctorius* (mantle thickness 76 μm) were resistant to infection. A few ectomycorrhizal roots with incomplete mantles had some infection by zoospores or vegetative mycelium of *P. cinnamomi*, but infection did not extend beyond the meristematic tissue, suggesting that the Hartig net also acted as a barrier to infection. Also, Marx and Davey (1969a) showed that 5 types of naturally occurring ectomycorrhizae of shortleaf pine with mantles ranging from very thin (5 to 8 μm) to thick (47 to 52 μm) were resistant to pathogenic infections, while nonmycorrhizal roots were 100% infected with *P. cinnamomi*.

Since endomycorrhizae generally do not possess a profuse fungal mantle, they may be dependent on one or more of the other mechanisms of disease resistance rather than on the physical protection of the mantle itself. The physical protection of a fungal mantle may not be sufficient to prevent pathogenic infection since the presence of a few mycorrhizae does not prevent infection of the whole plant. Subsequently, antibiotic (antifungal, antibacterial, or antiviral) production by mycorrhizal fungi also has been hypothesized as a mechanism for protecting the roots of mycorrhizal plants. Production of antibiotic substances has been demonstrated for a large number of fungi, many of which form mycorrhizal associations. Marx (1972) has compiled a list of over 90 species of mycorrhizal fungi which have been reported to produce one or more types of antibiotic-like substances in pure culture or in basidiocarps. Many of the antibiotic substances could have a selective influence on the population of microorganisms around the root. However, the production and identification of antibiotic-like substances in mycorrhizal associations are not well documented. Marx and Davey (1969b) have identified the antibiotic diatretyne nitrile in the ectomycorrhizal association between *Leucopaxillus cerealis* var. piceina (Peck) and shortleaf pine *(Pinus enchinata)* seedlings.

Besides the antibiotic effect, there is a striking difference between the rhizosphere populations of mycorrhizal and nonmycorrhizal roots of a plant species. The rhizosphere is the region of soil around a root in which root influence on microflora occurs, and the area immediately surrounding a mycorrhiza in which its influence is felt is sometimes referred to as the "mycorrhizosphere." The rhizosphere of a plant supports many more microorganisms than the nonrhizospheric areas, and a close interaction between the root system of a plant and its surrounding microflora exists (Zak 1964; Marx 1972). The mycorrhizosphere has been shown to possess bacteria which are found only in association with mycorrhizae. There are also other species of microorganisms which are found only in association with nonmycorrhizal roots, indicating the possibility of selectivity (Oswald and Ferchau 1968). Katznelson *et al.* (1962) found that bacteria growing on mycorrhizal roots tended to have complex nutritional requirements, while those growing on nonmycorrhizal roots grew optimally on a simple synthetic medium containing minerals and glucose. Populations of Actinomycetes, methylene blue-reducing bacteria, and ammonifying bacteria were all considerably higher on mycorrhizal than on nonmycorrhizal roots, but the fungal population was lower on mycorrhizal roots. *Pythium* and *Fusarium* spp. were the predominant fungi in nonmycorrhizal roots, but were absent in mycorrhizal roots. Also, the incidence of *Cylindrocarpon* sp. was reduced from 38% in nonmycorrhizal roots to 21% in mycorrhizal roots.

A fifth mechanism of disease resistance in which chemical inhibitors produced by symbiotically-infected host cortical cells may function as inhibitors to disease infection and subsequent spread in ectomycorrhizae has been postulated by Marx (1972). Wilhelm (1973) suggested that prior colonization of tissue before invasion by a pathogen could be operative in preventing or diminishing the invasion of a pathogen. Endomycorrhizae are capable of inducing morphogenic and biochemical changes within host tissue that may be unfavorable to pathogen invasion or development (Baltruschat *et al.* 1973; Becker 1976). Plant production of inhibitory substances against its fungal symbionts may maintain the mycorrhizal relationship in a balanced state, which at the same time provides protection against pathogenic fungi.

Nematodes in the soil can modify mycorrhizal roots' resistance to disease. With ectomycorrhizae, the majority of work suggests that nematodes can easily penetrate the fungal mantle, opening up pathways to pathogenic infection (Ruehle and Marx 1971; Barham *et al.* 1974). However, the interaction between plant-parasitic nematodes and endomycorrhizae is contradictory. Bird *et al.* (1974) found that treating the soil with nematicides increased endomycorrhizal development in cotton (*Gossypium* sp.), while others reported that endomycorrhizae were unaffected by the presence of nematodes (Roncadori and Hussey 1977). Hussey and Roncadori (1978) suggested that fungal structures formed by the endomycorrhizal fungus *Gigaspora margarita* Becker and Hall inside of cotton cortical cells could compete with nematodes for space and that cells colonized by the fungal symbiont also could be an unfavorable food source for a parasitic nematode.

V. NUTRIENT UPTAKE

One of the functions of mycorrhizal fungi particularly relevant to world wide production of field and greenhouse horticultural crops is their role in the mineral nutrition of plants. Mycorrhizae appear to be extremely advantageous to crops grown in low fertility soils which are characteristic of poorly managed, continuously cropped agricultural lands as well as drastically disturbed landscape and mined reclamation sites. Increases in mineral uptake as the result of mycorrhizal associations are often reflected in increased plant survival, growth, and yield as well as nutrition. While many horticultural production practices tend to reduce or eliminate mycorrhizal fungi, Marx (1975a) also points out that ectomycorrhizal fungi and their hosts are not indigenous in certain areas: former treeless regions of the United States (Hatch 1937), and portions of Asia (Oliveros 1932), Austria (Moser 1963), Australia (Bowen *et al.* 1973), Africa (Gibson 1963), Peru (Marx 1975b), Poland (Dominik 1961), Puerto Rico (Vozzo and Hacskaylo 1971), and Russia (Imshenetskii 1955).

Ectomycorrhizal fungi's beneficial effects on plant nutrition have been known for some time. Hatch (1937) reported that mycorrhizal white pine *(Pinus strobus)* seedlings weighed significantly more and contained more nitrogen (N), phosphorus (P), and potassium (K) than did nonmycorrhizal plants. Later work, reviewed by Trappe (1977) and Meyer (1974), with a variety of woody and herbaceous plant species has substantiated Hatch's findings.

Although endomycorrhizal fungi were identified on plant roots over 75 years ago (Gerdemann 1975), only within the last 20 years have the effects of endomycorrhizal fungi on plant nutrition been as intensively investigated as those of ectomycorrhizal fungi. Until this time, techniques for maintaining pure fungal endomycorrhizal cultures and for spore extraction from the soil had not been developed (Schenck and Kellam 1978).

Both endo- and ectomycorrhizae are capable of increasing P uptake, especially in low fertility soils. In experiments using ^{32}P, it was found that the mantle of ectotrophic mycorrhizae accumulates and stores P (Harley and Brierley 1955). This accumulation remains consistent unless there is a dramatic change in the plant's P status. A P deficiency in the plant may stimulate the release by the mantle to the plant some of its accumulated P. However, whether or not an ectotrophic mycorrhiza can store other nutrients is questionable. Morrison (1962) found that beech (*Fagus* sp.) and pine (*Pinus* sp.) mycorrhizae do not store sulfate, while Harley and Wilson (1959) suggested that beech mycorrhizae also may not store K. In many cases N uptake is also unaffected by mycorrhizae (Bond and Scott 1955; Schenck and Hinson 1973; Carling *et al.* 1978).

It is now recognized that plant growth responses to endomycorrhizal fungi are often the result of increased P nutrition (Mosse 1973a). Increases in P concentrations of plant tissues as the result of endomycorrhizal inoculation have been reported for such horticultural crops as raspberry *(Rubus idaeus)* (Hughes *et al.* 1979), strawberry (*Fragaria* sp.) (Holvas 1966), citrus (*Citrus* sp.) (Kleinschmidt and Gerdemann 1972), and corn *(Zea mays)* (Gerdemann 1964). However, reports of simultaneous increases in N and K have been variable. Ross and Harper (1970) found that soybean (*Glycine max* (L.) Merrill) growth, yield, and nutritional status were increased when fumigated plots were inoculated with two different inoculum levels of *Endogone* spp. Yield and foliar N concentration also were correlated positively with the inoculum levels, while foliar P concentration at both inoculum levels was nearly doubled. But in non-fumigated plots, plant growth and yield were similar in both inoculated and non-inoculated soil. Hughes *et al.* (1979) also reported that presence of endomycorrhizae did not influence N concentration in raspberry plants grown at 0, 22, and 44 ppm soil P levels, and at the 22 ppm P level the K concentration was actually lower than controls, but total K uptake was the same.

Mycorrhizal fungi also may facilitate increased or selective uptake of minor elements, but the ultimate effects appear to be dependent on individual host-fungus interactions. Gilmore (1971) found that a zinc (Zn) deficiency could be corrected by inoculating peach (*Prunus* sp.) plants growing in a sterilized Zn-deficient medium with an endomycorrhizal fungus. However, the effectiveness in correcting Zn-deficiency varied among the fungal species used. Only two of six endomycorrhizal cultures tested completely eliminated the Zn-deficiency symptoms while increasing the foliar Zn levels two to three times those of the controls. Unfortunately, the cultures were not completely identified. LaRue *et al.* (1975) also noted beneficial effects in peach seedlings from inoculations with *Glomus fasciculatus* (Thaxter sensu Gerdemann) Gerdemann and Trappe. These included increased Zn and decreased Mn levels. Benson and Covey (1976) found that inoculating apple (*Malus* sp.) seedlings with several species of Glomus endomycorrhizal fungi significantly increased shoot weight but not Zn concentrations. However, Zn fertilization plus mycorrhizal inoculation significantly increased growth two- to three-fold and Zn uptake six- to seven-fold over non-inoculated controls. Ca, Cu, and Mn also have been found at higher foliar concentrations in soybean plants grown in fumigated soil and inoculated with *Endogone* spp. (Ross and Harper 1970).

The involvement of mycorrhizae in plant nutrition is somewhat paradoxical and not clearly understood. There exist variations in mycorrhizal dependency levels among plant species and/or fertility regimes. Some plants cannot grow normally without the presence of mycorrhizae (Marx 1975b). Kleinschmidt and Gerdemann (1972), examining citrus (*Citrus* spp.) seedlings grown in fumigated or heat-treated soils, found them consistently stunted and chlorotic. These plants were found to be nonmycorrhizal. When inoculated with an endomycorrhizal fungus, the plants recovered and resumed normal growth. Some signs of recovery also were seen in nonmycorrhizal plants when they were treated with P fertilizer, but growth was still not as satisfactory as in endomycorrhizal plants. They suggest that lack of sufficient P uptake in nonmycorrhizal plants may account for the stunting and chlorosis since significantly higher concentrations of P were found in all the mycorrhizal plants when compared to the nonmycorrhizal ones. There also exist such anomalies as those found in papaya (*Carica papaya* var. solo) where three different species of endomycorrhizal fungi produced different degrees of increased plant height in fertilized autoclaved soil, while the same three fungi had no effect in increasing growth over nonmycorrhizal plants in unfertilized autoclaved soil. Sparling and Tinker (1978) also reported a reduction in growth of some grassland species inoculated with endomycorrhizae, but attributed the negative effects to greater availability of soil nutrients. Apparently, mycorrhizal uptake of P in grasses is not significant until the soil is severely deficient.

Numerous experiments using labeled ^{32}P have confirmed that ecto- and endomycorrhizal fungi are capable of increasing P uptake, but the exact mechanism by which mycorrhizae are able to enhance nutrient absorption—especially in low fertility soils—is not completely known. Sanders and Tinker (1971), growing mycorrhizal and nonmycorrhizal onions *(Allium cepa)* in soil of low P status labelled with ^{32}P, noted that the *Endogone mosseae* (*Glomus mosseae* (Nicol. and Gerd.) Gerdemann and Trappe) mycorrhizal plants were clearly larger and had 5 times the dry weight and twice the P content of the nonmycorrhizal plants after 4½ weeks. Specific activity of P in both mycorrhizal and nonmycorrhizal plants was about the same. But mycorrhizal plants showed influx rates 3 to 16 times greater, indicating that although both types of association utilized the same sources of P, there is increased efficiency of uptake in the mycorrhizal system. Gray and Gerdemann (1969) also found that endomycorrhizal fungi increased P uptake by nonmycorrhizal root segments present on mycorrhizal plants over that of nonmycorrhizal roots in uninfected plants. They also found that application of a fungitoxicant (PCNB) decreased ^{32}P accumulation of mycorrhizal roots to approximately the same level as that of nonmycorrhizal plants' roots. Increased P uptake via mycorrhizal fungi also may be detrimental. Mosse (1973a) found that both mycorrhizal and nonmycorrhizal onion plants fed increasing amounts of $Ca(H_2PO_4)_2$ showed a decrease in growth, but mycorrhizal plants responded to the effects of P toxicity at lower levels of P than nonmycorrhizal plants, probably due to the greater efficiency of P uptake in mycorrhizae. With more than 1 g $Ca(H_2PO_4)_2$/kg soil, mycorrhizae declined. Large amounts of P caused arbuscules to disappear along with all signs of intracellular infection. Bowen (1975) suggested that a prime consideration for nutrient uptake (especially P) is the selection of a fungus capable of producing mycelial strands over a wide range of conditions. Increased nutrient uptake by mycorrhizal plants could be attributed to solubilization of elemental compounds, use of one or more elemental sources, rate of translocation to the plant via hyphal strands, increased root surface to volume ratios, and permeation by hyphal strands into soil regions inaccessible by root hairs.

The typical root hair is approximately 0.7 mm long with 50 hairs per millimeter of root. Although penetration of mycorrhizal hyphae occurs at only 1 to 20 connections per millimeter of root, they are capable of extending 0.1 to 7 cm out into the soil (Bieleski 1973; Rhodes and Gerdemann 1975). This means that plants are able to obtain P from an area outside the P-depletion zone 1 to 2 mm wide which generally surrounds a root. We have found mycelial strands of the ectomycorrhizal fungus *Pisolithus tinctorius* extending 5 to 10 m from naturally infected *Pinus* spp.

Onion *(Allium cepa)* roots inoculated with *Endogone mosseae (Glomus mosseae)* and *E. fasciculatus (G. fasciculatus)* were found to have considerably higher uptakes of ^{32}P introduced 27 mm away from the root surface than those of nonmycorrhizal plants (Hattingh *et al.* 1973). Diffusion could not account for the increased uptake as its effect was found to extend only 7.5 mm from the point of injection and the mycorrhizal root surface was at a minimum, 15 mm away. This suggested an active transport mechanism in the fungus extending beyond the immediate vicinity of the root surface. When hyphae growing from the mycorrhizae were severed, the ^{32}P uptake differed little between mycorrhizal and nonmycorrhizal plants, emphasizing the active part played by fungal hyphae in P uptake. In roots, inorganic P appears to move at about 2 mm/hour (Crosset and Loughman 1966); in fungal hyphae the movement is 2 cm/hour (Littlefield 1966), indicating a greater efficiency in uptake and possibly the presence of an active transport mechanism such as cytoplasmic streaming in the mycorrhizal fungus hyphae. Evidence suggests that mycorrhizal transport of P to the plant via cytoplasmic streaming occurs with the P being contained in the vacuoles. Cox *et al.* (1975), using electron microscopic techniques, found that *Glomus mosseae*-infected onion roots had fungal vacuoles containing metachromatic lead-staining bodies which appeared to be polyphosphate granules. Later, Callow *et al.* (1978) confirmed that polyphosphate comprises at least 40% of the total P present in fungal components of the mycorrhizal onion root. This evidence, along with these authors' earlier observations of bi-directional cytoplasmic streaming at the rate of several cm $hour^{-1}$ in the germ tubes of various endomycorrhizal fungi, suggests a possible transport mechanism. Photomicrographs also showed that the vacuoles lost their polyphosphate granules on reaching the finer hyphal branches. They suggested that the hydrolysis of the polyphosphate granules and the site of phosphate transfer to the host may occur in these branches. Ling-Lee *et al.* (1975), working with three mycorrhizal fungi (ecto-, endo-, and ectendomycorrhizal fungi), found polyphosphate granules in all three fungi, further substantiating polyphosphate granules as a major form of P storage. Schramm (1966) suggested that the movement of nutrients by mycorrhizal fungi could be via retraction of protoplasm from proliferating hyphae. Although possible, this is probably not the primary means of P translocation.

Evidence as to whether mycorrhizal fungi are capable of solubilizing unavailable forms of P in the soil is conflicting. Sanders and Tinker (1971) and Mosse *et al.* (1973) found no differences in specific activity of mycorrhizal and nonmycorrhizal plants provided with ^{32}P, suggesting that some endomycorrhizae do not utilize P sources unavailable to the nonmycorrhizal plant. Nevertheless, some studies with onion *(Allium cepa)* and tomato (*Lycopersicon esculentum* Mill.) (Daft and Nicolson 1972), pine (*Pinus* sp.) and beech (*Fagus* sp.) (Bartlett and Lewis 1973) mycorrhizae have indicated the ability to absorb normally insoluble P sources through the presence of phosphate enzymes on mycorrhizal surfaces.

Although the benefits of mycorrhizae are more evident in low fertility soils (Gerdemann 1975), some contradictory evidence exists between mycorrhizal development and fertility regimes. High fertility levels have been found to suppress mycorrhizal development (Khan 1972; Mosse 1973a; Marx *et al.* 1977), while there have been reported instances where plants were heavily mycorrhizal in very fertile soils (Hayman *et al.* 1976; Gerdemann 1970). Mycorrhizal development in relation to fertility levels appears to be influenced by host specificity, ecotypes, soil characteristics, initial inoculum density, pH, season changes, and moisture levels. Sanders and Tinker (1973) reported that P may control endomycorrhizal infections since they found that root infection was negatively correlated with P levels in the soil. These results did not completely establish whether P directly inhibited endomycorrhizal spore germination or disrupted fungal-host compatibility. Sanders (1975) and Menge *et al.* (1978) later reported evidence of a feedback system in mycorrhizal plants. Sanders (1975), using foliar applied P given to mycorrhizal onion *(Allium cepa)* plants, found that P applied to the leaves (1) reduced the rate of spread and intensity of the mycorrhizal infection, (2) reduced the weight of external mycelium associated with each centimeter of infected root, and (3) depressed the supply of P provided to the host via the mycorrhizae. This is indicative of an interaction between fungus and plant, but the mechanism of regulation is not known. Ratnayake *et al.* (1978) proposed that P inhibition of a mycorrhizal symbiosis is associated with a membrane-mediated decrease in root exudation of soluble amino acids and reducing sugars rather than changes in root concentrations of these substances.

Evidence that mycorrhizae are actually fixing N is inconclusive. Nitrate reductase activity has been found in certain ecto- and endomycorrhizal fungi produced during the plant endophyte symbiosis (Trappe 1967; Ho and Trappe 1975). The ability to reduce nitrate could facilitate increased symbiotic effectiveness in N assimilation and translocation to the host. Mycorrhizal associations have been found with plants capable of fixing N and are also known to stimulate nodulation (Crush 1974; Carling *et al.* 1978; Bagyaraj *et al.* 1978). Asai (1944) demonstrated that several legumes grew poorly and failed to nodulate in autoclaved soils unless they were mycorrhizal. Daft and El-Giahmi (1975) attributed better nodulation on three legume species to endomycorrhizae and increased nutrition. In addition to P, mycorrhizal plants also have been found to contain higher concentrations of Zn (Gilmore 1971) and Cu (Hughes *et al.* 1979; Ross and Harper 1970) which are also known to influence nodulation (Demeterio *et al.* 1972). However, Mosse (1977) observed that *Stylo-*

santhes guyanensis Aubl. and corn (*Zea* sp.) grown in nonsterile soils exhibited no clear relationship between plant P and nodulation, although an endomycorrhizal inoculation often stimulated nodulation by Rhizobia.

The relationship between mycorrhizal and N-fixation endophytes needs to be explored further. There are structures produced during the plant-N-fixation-endophyte symbiosis which are somewhat similar to endomycorrhizal structures, e.g., vesicles (Quispel 1974). Environmental pressures may have resulted in a segregation of an endophytic population into separate groups. Any possible compatibility between the endophytic groups could have far reaching implications in genetic engineering for N-fixation and extension of N-fixation process to nonlegumenous horticultural crops. Zobel and Wallace (1978) reported that insertion of an N-fixing bacterium into mycelium of an ectomycorrhizal fungus was successful. The fungus was shown to fix N, and recently a mycorrhizal association has been synthesized on pine seedlings (*Pinus* sp.) with the fungus.

VI. PLANT STRESS

One of the distinct advantages afforded to mycorrhizal plants is their ability to produce better than nonmycorrhizal plants under a variety of stress situations. Many mycorrhizal fungi possess specific individual traits with respect to tolerance to soil temperature extremes, pH, moisture, low fertility, salinity, toxicants, etc., which may provide the host plant with an ecological competitive advantage facilitating increased plant survival, growth, nutrition and/or yield under stress conditions (Trappe 1977). Utilization of these mycorrhizal traits in horticultural cropping systems is contingent upon understanding fungal-host interactions and environmental influences on mycorrhizal associations.

Basic and applied research pertaining to stress tolerances afforded to horticulture and agronomic crops by mycorrhizal fungi are limited. The vast majority of work on stress physiology has centered on applications in which mycorrhizal tree and shrub species are being produced and utilized for revegetation of drastically disturbed lands, e.g., severely eroded land, barrow pits, and coal spoils (Aldon 1975; Marx 1975a; Harris and Jurgensen 1977; Marx and Artman 1979). These revegetation studies as well as those few pertaining to horticulture and agronomic crops have implications as to how mycorrhizal fungi may be utilized in cropping systems.

The alteration of root morphology and/or the presence of fungal hyphae could markedly affect plant tolerance to moisture extremes. Safir *et al.* (1971), working with mycorrhizal and nonmycorrhizal soybeans, found that the presence of mycorrhizae formed with *Glomus mosseae* decreased the plants' resistance to water transport considerably compared to nonmycorrhizal plants. In a subsequent paper (1972) these authors confirmed decreases in resistance to water transport of about 40% in mycorrhizal soybeans compared to the nonmycorrhizal controls. Since there were no appreciable differences in stem and leaf resistances between nonmycorrhizal and mycorrhizal plants, the differences in resistance to water transport were concluded to occur in the roots. Addition of nutrients to nonmycorrhizal plants growing in low nutrient-status soils lowered the resistance to water transport to the same level as mycorrhizal plants, suggesting that the lowered resistance in mycorrhizal plants was due to the enhanced nutrient status brought about by the fungus. Survival under drought conditions may be contingent on fungal-host specificity as exemplified by studies involving inoculation of fourwing saltbush (*Atriplex canescens* (Pursh) Nutt.) with different species of the endomycorrhizal genus *Glomus*. Fourwing saltbush plants grown in a semi-arid area of New Mexico (precipitation less than 250 mm per annum) and inoculated with *G. mosseae* showed increased survival rates over nonmycorrhizal plants (Aldon 1975). In contrast, inoculation of fourwing saltbush and rabbit brush (*Chrysothamnus nauseosus* (Pall.) Britt.) with *G. fasciculatus* resulted in increased growth and survival of rabbit brush while saltbush failed to form mycorrhizae and exhibited no increase in height (Lindsey *et al.* 1977).

Among different species of ectomycorrhizal fungi there also exist varying degrees of drought tolerance. The ectomycorrhizal fungus *Cenococcum graniforme* grew well and formed mycorrhizae under severe moisture stress conditions (Worley and Hacskaylo 1959; Trappe 1962b). *Cenococcum graniforme* is considerably more tolerant of low water potential than is either *Suillus luteus* Gray or *Thelephora terrestris* (Mexal and Reid 1973). However, Theodorou (1978) found *C. graniforme* to be more effective in colonizing Monterey pine (*Pinus radiata* D. Don) roots at higher moisture levels than *Rhizopogon luteolus* Fr. and Nordholm. Monterey pine inoculated with *R. luteolus* or *S. granulatus* showed increased survival over uninoculated plants during a very dry summer following planting. *Pisolithus tinctorius* also appears to be fairly drought tolerant since it is often found on droughty coal spoils in Kentucky. Selection of some mycorrhizal fungi with drought tolerant characteristics could be of commercial importance to horticulturists during plant production as well as after outplanting to the field. For example, during high density container greenhouse production of oak seedlings (white, *Quercus alba* L.; pin, *Q. palustris*; red, *Q. rubra* L.; swamp chestnut, *Q. michauxii* (Nutt.)), Maronek and Hendrix (unpublished data) found that nonmycorrhizal seedlings exhibit wilt sooner and grow less than mycorrhizal *P. tinctorius*-infected seedlings. Furthermore, when these container-grown greenhouse pin oak seedlings were transplanted into 15-cm diameter pots and grown in a nursery, these differences in growth and moisture tolerances persisted.

Generally, high moisture levels tend to retard mycorrhizal formation. This is especially true during the winter months when many soils become waterlogged (Powell and Sithamparanathan 1977). Khan (1974) reported a reduction in spore populations of *Endogone* species in waterlogged soils. We have consistently found poorer mycorrhizal development on conifer and hardwood species inoculated with *Pisolithus tinctorius* growing in poorly drained container media. However, certain ectomycorrhizal fungi appear to be adapted to wet soils. Trappe (1977) reported that *Hymenogaster alnicola* Vitt. and *Lactarius obscuratus* Pers. form alder (*Alnus* sp.) mycorrhizae in continuously wet soils.

Mycorrhizae also exhibit differences in tolerance to temperature extremes. *Pisolithus tinctorius* is more tolerant to high temperatures than is *Thelephora terrestris*. *Pisolithus tinctorius* mycorrhizal loblolly pine *(Pinus taeda)* seedlings survived as did most of their mycorrhizae at temperatures of 40°C for 5 weeks, while either *T. terrestris* seedlings died or mycorrhizae declined drastically (Marx and Bryan 1971). *Pisolithus tinctorius* is also capable of surviving at temperatures to at least −70°C (Hendrix and Maronek 1978). Low temperature tolerance by mycorrhizal fungi also may be of critical importance in plant cold hardiness. Conifers naturally infected with *P. tinctorius* and growing on abandoned strip mine land in Kentucky appear to exhibit superior growth and cold hardiness to nonmycorrhizal plants or those infected with the ectomycorrhizal fungus *Thelephora terrestris* (Marx and Artman 1979; Maronek and Hendrix, unpublished data). In addition, container grown 2-0 white pine *(P. strobus)* seedlings inoculated with *P. tinctorius* in early spring and overwintered in an unheated greenhouse exhibited an 80% survival rate compared to a 20% rate for noninoculated controls (Maronek and Hendrix 1978). Greenhouse temperatures during the study dropped to below −26°C.

Plant tolerance to soils with specific toxicity problems also may be increased via mycorrhizal fungi. However, mycorrhizal fungi can vary with respect to soil toxicity problems. Marx (1975c) reported that performance of tree seedlings on very toxic coal spoils of low pH (3.0 to 3.8) is contingent on seedlings having specific mycorrhizae. For example, Virginia (*Pinus virginiana* Mill.) and red (*P. resinosa* Ait.) pine seedlings with *Pisolithus tinctorius* ectomycorrhizae exhibited a greater survival rate and grew significantly more than seedlings with *Thelephora terrestris* ectomycorrhizae growing on a Kentucky coal spoil of pH 3.0. Often associated with low pH mine or coal spoils are high levels of mineral toxicants such as Al, S, Mn, Cu, and Fe. Heavy metals such as these appear to affect the population of indigenous mycorrhizal fungi.

Harris and Jurgensen (1977) found a direct relationship between plant growth and presence of natural ectomycorrhizal formation on willow (*Salix* spp.) and poplar (*Populus* spp.) seedlings grown in iron mine tailings. The same species grown in copper tailings had no natural mycorrhizal formation. Furthermore, tree inoculations with natural forest soils also failed to promote ectomycorrhiza formation on trees growing on the copper tailings. The susceptibility of mycorrhizal fungi to heavy metal toxicity becomes extremely relevant when one considers the use of pesticides that contain heavy metals on horticulture crops. In a recent listing of 280 pesticides by McGrath (1964), 112 pesticides contained heavy metals in their formulations. Göbl and Pumpel (1973) reported that copper fungicides can affect mycorrhizal associations with plants, while Trappe *et al.* (1973) found that mycorrhizae formation declined on apple trees growing on soils with high arsenic levels. The high residual arsenic concentrations on the soil were attributed to the use of lead arsenate as an insecticide.

In some instances, mycorrhizal fungi also may function in detoxification of soils high in metal toxicants. Heavy metal tolerance mechanisms are known to be present in fungi (Ashida 1965). Accumulation and storage of toxic minerals such as sulfur have been detected on *Pisolithus tinctorius* sporocarps (Muncie *et al.* 1975). Sequestering and storage of heavy metals by sporocarp tissue followed by wind dissemination of spores could be a functional biological transformation system for detoxification of soils with metal toxicants. However, more work on sensitivity of mycorrhizal fungi to heavy metals is needed, since tolerance to one element does not necessarily mean tolerance to other elements (Huisingh 1974).

Some endomycorrhizal fungi also may appear to be ecologically adapted to adverse sites. Some *Glomus* spp. have been reported to be adapted to higher soil temperatures than others (Schenck and Shroder 1974; Schenck *et al.* 1975). In fact, endomycorrhizal fungi may have even more impact on horticultural crop production and revegetation of adverse sites than ectomycorrhizae fungi because of their broader host range.

VII. MYCORRHIZAL FUNGI IN HORTICULTURAL CROP PRODUCTION

The potential for utilization of mycorrhizal fungi in crop production has been demonstrated primarily with woody species grown in media devoid of or low in mycorrhizal fungi, either naturally or because of some sterilization treatment. These include citrus (*Citrus* spp.) (Kleinschmidt and Gerdemann 1972; Schenck and Tucker 1974; Timmer and Leyden 1978), peach (*Prunus* sp.) (LaRue *et al.* 1975; Lambert *et al.* 1979), sweetgum (*Liquidambar styraciflua* L.) (Bryan and Ruehle 1976; Bryan and Kormanik 1977), avocado (Menge *et al.* 1980), and redbud (*Cercis canadensis* L.) (Maronek and Hendrix 1978), grown in fumigated soil. Growth enhancement of loblolly pine (*Pinus taeda*) also was obtained with specific ectomycorrhizal fungi in a fumigated field nursery (Marx *et al.* 1978). This nursery was located in a plains area in which deficiency in mycorrhizal fungi was corrected a few years previously by planting mycorrhizal trees in the nursery as a source of basidiospore inoculum; apparently this inoculum level was insufficient for optimal seedling growth.

Growth benefits also have been obtained in production systems using soilless media which also may have been fumigated or heat-sterilized. Crews *et al.* (1978) obtained enhanced growth of three woody ornamentals, *Viburnum suspensum* Lindl., *Podocarpus macrophylla* (Thunb.) D. Don, and *Pittosporum tobira* (Thunb.) Ait., with inoculation with two endomycorrhizal fungi. Growth of southern magnolia (*Magnolia grandiflora* L.) but not Bar Harbor juniper (*Juniperus horizontalis* Moench) was increased by inoculation with *Glomus fasciculatus* (Maronek *et al.* 1980). Vanderploeg *et al.* (1974) obtained superior growth of two lilies (*Lilium* sp.) in a medium containing apparently unsterilized soil mixed with soil from beneath a naturally inoculated garden lily. The poor growth of the control lilies was apparently due to either the low populations of endomycorrhizal fungi in the unsterilized soil component of the potting medium or that lilies require specific mycorrhizal fungi not present in the soil component. Ames and Linderman (1978) obtained enhancement of growth of Easter lily (*Lilium longiflorum* Thunb.) through inoculation with a pathogen-free culture of the endomycorrhizal fungus *Acaulospora trappei*; but if *Fusarium oxysporum* and perhaps other pathogens were present, inoculation with a mixture of endomycorrhizal fungi was deleterious. Inoculation with *Gigaspora margarita* also was found to be beneficial in poinsettia (*Euphorbia pulcherrima* Willd. ex Klotzsch) (Barrows and Roncadori 1977). Enhanced growth of pin oak (*Quercus palustris*) seedlings grown in 165-ml tubes containing peat-perlite was obtained with one ectomycorrhizal isolate (M3) of *Pisolithus tinctorius*, but not another (M1) (Maronek and Hendrix 1979a). Conversely, when pin oak seedlings were grown in an outdoor nursery in 15-cm diameter pots containing a composted hardwood bark-shale medium, growth enhancement was obtained with isolate M1, but not isolate M3 (Maronek and Hendrix 1978). Isolate M1 also increased growth of red oak (*Q. rubra*) and swamp chestnut oak (*Q. michauxii*) seedlings grown under the latter conditions. Molina (1979) reported variability among six different *P. tinctorius* isolates during culture as well as when they were used to inoculate container-grown Douglas fir (*Pseudotsuga menziesii* (Mirb.) Franco) and lodgepole (*Pinus contorta* Dougl. ex

Loud.) pine. He found that the percentage of the seedlings that formed *P. tinctorius* ectomycorrhizae and the percentage of the root system colonized differed significantly, but there were no differences among inoculated and control seedlings in height, stem diameter, stem and root dry weights, and shoot/root ratios. Enhanced rooting of cuttings also was obtained with some mycorrhizal fungi (Linderman and Call 1977).

Woody plants are usually high value crops, and their production often involves expensive practices such as soil fumigation, sophisticated fertilization, high labor requirements, and expensive facilities such as greenhouses and specialized equipment. Enhanced growth due to inoculation with mycorrhizal fungi is desirable because of more rapid turnover and thus greater productivity of facilities; decreased labor, fertilizer, and water per plant unit; and greater appeal to consumers because mycorrhizal plants are usually more vigorous and have better color than nonmycorrhizal plants.

Similar benefits would be desirable for other horticultural crops, such as field-grown annual crops. However, benefits to such plants usually have been demonstrated under adverse conditions—such as deficient soil phosphorus (Gerdemann 1968; Mosse 1973b)—not under normal production conditions. The nature of the root system apparently influences the response of a plant species to mycorrhizal fungi. Baylis (1975) separates root types into three groups: magnolioid, graminoid, and intermediate. Magnolioid roots are coarsely branched, the finest roots usually being over 0.5 mm in diameter. They have few or no root hairs; when present, root hairs are wide and short, less than 0.1 mm in length. Graminoid roots, in contrast, are smaller (0.1 mm or less in diameter) and densely covered with slender root hairs 1 to 2 mm long. At the extreme in this group, rushes and sedges grow well in the nonmycorrhizal state, whereas grasses may grow better in the mycorrhizal than in the nonmycorrhizal state. Baylis considers that most plants have root systems that place them in the intermediate group, with the smallest roots between 0.1 m and 0.5 mm diameter and abundant root hairs less than 1 mm long. He considers the length and frequency of root hairs to be the best index of the capacity of a plant to grow without mycorrhizal fungi. Many of the plants shown to be strongly dependent on endomycorrhizal fungi for growth, e.g., southern magnolia *(Magnolia grandiflora)* and yellow poplar (*Liriodendron tulipifera* L.), have magnolioid roots. Most or all annual crops have root systems with abundant root hairs which place them in the graminoid or intermediate groups. These plants, according to Baylis, would be expected to be capable of adequate growth without mycorrhizal fungi provided fertilizer and water are adequate. St. John (1980) found a significant positive correlation between magnolioid root characteristics and endomycorrhizal infection.

Since economic production is usually done under near-optimal fertility and water, the role of mycorrhizal fungi in the present production of field-grown crops is questionable. Endomycorrhizal fungi are present in virtually all soils and may be greater in both diversity and populations in cultivated soils than in noncultivated soils (Ames and Linderman 1977). Endomycorrhizal fungi are part of the plant production system, but we have little information on how they affect the system. Although research on the ecology of the fungi is intensifying, we presently have little information on the species found in a particular cropping system (Ames and Linderman 1977) or the influence of such factors as monocropping, crop rotation systems, winter cover crops, fertilization practices, and the use of soil fumigants, herbicides, and fungicides. The taxonomy of these fungi is currently under intensive development. While Gerdemann and Trappe (1974) have provided a workable taxonomic framework, many of the fungi currently encountered are apparently undescribed species. Development of the taxonomy is hampered by the fact that these fungi are obligate root parasites and our single culture methods do not work well for all species encountered. Consequently, much of our research deals with easily cultured fungi, such as certain isolates of *Glomus fasciculatus* and *G. mosseae*, but insight into the role of mycorrhizal fungi in a particular cropping system requires knowledge of and control over the entire spectrum of fungi.

The ubiquitous nature of endomycorrhizal fungi in soils complicates research on their role in field production of crops. Problems associated with soil fumigation with methyl bromide or methyl bromide-chloropicrin mixtures under plastic have been reported for production of seedlings such as citrus (*Citrus* sp.) (Nemec and O'Bannon 1979; Schenck and Tucker 1974; Kleinschmidt and Gerdemann 1972), peach (*Prunus* sp.) (Lambert *et al.* 1979), sweet gum *(Liquidambar styraciflua)* (Bryan and Kormanik 1977), and yellow poplar *(Liriodendron tulipifera)* (Kormanik *et al.* 1977). Indeed, these problems stimulated research on the role of endomycorrhizal fungi in growth of these plants. However, soil fumigation is routinely used in production of most of these seedlings without growth problems in most years. With peach, the occasional problems are usually associated with growth inhibition early in the season, so that plants do not reach sufficient size to be grafted by a critical cutoff date. We have observed similar behavior with production of yellow poplar in methyl bromide-fumigated soil. Beds were seeded in the fall, and in the spring seedlings emerged, grew to a few inches in height, then failed to grow for several weeks. Then in isolated spots, plants began to grow rapidly (Fig. 5.2A). These spots quickly enlarged (Fig. 5.2B). Plants in the same beds inoculated with *Glomus fasciculatus* before planting in the fall grew rapidly and uniformly early in the season (Fig. 5.2C, Table 5.1). However, by the time of fall lifting of the seedlings, plants in uninoculated portions of the beds were equal in size to inoculated ones; consequently,

FIG. 5.2. FUMIGATED SEEDBEDS OF YELLOW POPLAR *(LIRIODENDRON TULIPIFERA)* PLANTED LATE FALL, 1977
A. Spring, 1978, rapidly growing seedlings in isolated portions of the seedbed. B. Continued enlargement of seedlings one month later. C. Portion of the same seedbed inoculated with *Glomus fasciculatus*. Note uniform growth and increased size of inoculated seedlings (left) compared to that of uninoculated seedlings (right). Photos B and C were taken at the same time. We acknowledge the cooperation of Mr. David Fisher, Nurseries Supervisor, Division of Forestry, Department of Natural Resources and Environmental Protection, Gilbertsville, Kentucky.

in this nursery soil fumigation is not considered to be deleterious to production. Apparently, plants became mycorrhizal and served as inoculum points, perhaps because (1) occasional spores were not killed by fumigation, (2) roots reached inoculum below the fumigation zone (usually considered to be the plow layer), or (3) because of reintroduction of spores by various means. These kinds of observations raise a number of questions about the influence of soil fumigation on endomycorrhizal fungi. Are mycorrhizal fungi more resistant to fumigation than the target weeds and

TABLE 5.1. EFFECT OF *GLOMUS FASCICULATUS* INCORPORATED INTO NURSERY SEEDBEDS ON HEIGHT AND STEM DIAMETER OF YELLOW POPLAR *(LIRIODENDRON TULIPIFERA)* SEEDLINGS

Treatment	Height (cm)[1]	Stem Diameter (mm)[1]
Uninoculated control	14.7±6.4	2.3±0.04
Inoculated	37.5±3.2	4.0±0.03

[1]Significant at 1% level; mean of 150 trees.

pathogens? Do species or genera of mycorrhizal fungi vary in susceptibility to fumigation? Menge *et al.* (1978) found two endomycorrhizal fungi to be more susceptible to methyl bromide than many common pathogenic fungi, and they concluded that current fumigation practices should destroy mycorrhizal fungi in the top 1 m of soil. However, they also found survival of small amounts of inoculum at lethal dosages and concluded that it is difficult to eliminate completely the population of mycorrhizal fungi in soil. What is the reason for the occasional stunting problems we observe with fumigation? Reduction of mycorrhizal fungi populations to levels lower than usual with fumigation due to unknown conditions at the time of fumigation? Some effect of an environmental factor so that the normal populations following fumigation are inadequate for normal growth? Or failure in those years of reintroduction of mycorrhizal fungi by such factors as animals or equipment if these are of consequence? Large acreages of tomatoes *(Lycopersicon esculentum)* and strawberries (*Fragaria* spp.) are planted on methyl bromide-fumigated land. While pathogen control by fumigation undoubtedly is markedly beneficial to production, mycorrhizal deficiency induced by fumigation is not reported. Yet, both tomato (Hall 1975) and strawberry (Holvas 1966) are reported to respond favorably to mycorrhizal fungi. Mycorrhizal fungi may be reintroduced routinely on strawberry plants, but tomato plants are often produced on fumigated soil or in soilless mix. Are these plants capable of reaching and proliferating inoculum more rapidly than plants such as peach? Are they capable of benefiting from a more diverse group of fungi, some of which may be more resistant to fumigation? Or are mycorrhizal fungi inconsequential to production of tomatoes and strawberries? Soil fumigation needs further evaluation from the standpoint of mycorrhizal fungi rather than plant response or control of weeds and pathogens. We can evaluate realistically the role of mycorrhizal fungi in crop production in the field only by totally eliminating these fungi from land that is normally highly productive.

Ectomycorrhizal fungi exhibit ecological selectivity. *Pisolithus tinctorius*, for example, appears to be much superior to *Thelephora terrestris* as a symbiont for southern pines transplanted onto such disturbed sites as strip mine spoils (Marx 1975c; Marx and Artman 1979). Trappe (1977) summarized ecological adaptability of ectomycorrhizal fungi and gave insight into ways in which these fungi may be used in production. Similar knowledge of endomycorrhizal fungi is generally lacking, but it seems probable that for a particular purpose some fungi will prove to be superior to others. At present, an empirical approach is the only one possible. This will involve locating superior plants growing under the particular conditions for which superior plant performance is desired, isolating mycorrhizal fungi from these plants, producing inoculum, inoculating seedlings, and evaluating the specifically-infected seedlings under the conditions originally chosen, i.e., growing a plant species in a particular container mix, or growing on acid coal spoils.

The superiority of a particular symbiont for plant growth under special adverse conditions, such as *Pisolithus tinctorius* for growth of pines on acid mine spoils, may appear to offer only limited application of mycorrhizal technology, but drastically disturbed lands are much more common than one may think. Probably most ornamental and vegetable plants sold for home use are planted on disturbed land, because extensive grading is done in most suburban lands before and during building. Often top soil is pushed into areas as fill, requiring the homeowner to grow plants on the B or C horizon. The production of plants that will grow well under any conditions may offer the greatest challenge for the use of mycorrhizal technology in the horticultural industry. Since it seems unlikely that a single fungus will produce superior results under all conditions, it may be necessary to produce plants infected with several symbionts, only one of which may continue to develop after the plants are transplanted. The objective would be to produce a plant that would grow well in any situation, whether or not adequate populations of ecologically-adapted mycorrhizal fungi were naturally present.

Production of specifically-infected plants may not be such a radical departure from many current production systems, such as those making use of tissue culture, rooting of cuttings, or production from seed in sterilized or soilless mixes. These plants probably would become mycorrhizal if planted into normal soil, and many might need to be mycorrhizal before transplanting to survive transplanting shock and to get a rapid start. Johnson and Crews (1979) obtained better survival and growth four months after transplanting with *Glomus mosseae*-inoculated azalea (*Rhododendron simsii* Planch.) plants than with nonmycorrhizal plants transplanted to a normal (nonsterilized) site. If mycorrhizal plants are superior to nonmycorrhizal plants in the hands of the consumer, why not inoculate them with mycorrhizal fungi selected for superior performance?

Production of specifically-infected mycorrhizal plants probably will have unique problems; simply adding inoculum often will not work. Pesticides must be evaluated for their effects on mycorrhizal fungi; some fungicides and fumigants affect mycorrhizal fungi and plant responses to mycorrhizal fungi more than others (see Nemec 1980 and Nemec and O'Bannon 1979, and references cited therein). Another factor that will require modification is fertilization. Current production practices often employ extremely heavy fertilization, often, perhaps, to force growth of plants with heavy dependency for mycorrhizal fungi. Normal fertilization practices often inhibit mycorrhizal development (Marx *et al.* 1977; Gerdemann 1968). We have found it possible to produce mycorrhizal plants at an adequate growth rate by using slow-release fertilizers (Maronek and Hendrix 1979a,b, 1980; Maronek *et al.* 1980). Slow release of fertilizer over an extended time apparently is in keeping with the evolutionary association of the plant and fungus, whereas the cyclic nutrient availability associated with periodic application of soluble fertilizers is not.

Production of many fertilizers, especially nitrogen N, is energy-intensive, and supplies of other fertilizer components are limited. Phosphate mining, for example, is expected to peak in the 1980s. It seems clear that limited phosphate for fertilizers and limited energy for production of other types of fertilizers will end the gains in agricultural productivity that we have become accustomed to in recent decades. Recent fertilization practices have resulted in other problems, such as pollution of groundwater and streams, and these problems are under attack on environmental grounds. It is obvious that production practices will change in order to conserve fertilizer and to prevent waste. Under future production conditions, mycorrhizal fungi may play a central role, even with crops not benefited by mycorrhizae in present production systems. We may find it necessary to monitor and control mycorrhizal populations to maximize production, and controlling these populations may involve inexpensive management procedures such as using certain crop rotations or controlling pathogens of mycorrhizal fungi. Because of the time lag involved in developing commercially-feasible practices, basic research on a broader front of mycorrhizal issues needs to be initiated now.

VIII. CONCLUSION

New resources for increasing crop productivity need to be explored. We cannot rely on many of our diminishing resources (oil and gas) to meet food and fiber demands of the world. We feel that a more intensive investigation into the plant-soil microflora complex is needed. In the past decade, we have seen a greater emphasis on nitrogen fixation research, but this is only part of the much broader research base needed in plant-soil microflora interactions. The information discussed in this text has emphasized the potential benefits of mycorrhizal fungi, but there is limited information on the role of mycorrhizal fungi in horticultural crop production. To use mycorrhizal fungi in cropping systems, we must await much research on applied and basic levels.

The absence of mycorrhizal fungi as well as most other soil microflora from nearly all nutritional and physiological experiments done in greenhouses or growth chambers poses serious questions concerning some of our theories of plant growth and development. Mycorrhizal fungi appear to be involved in some physiological plant functions. We must establish the contribution that mycorrhizal fungi are making to horticultural crops, especially in those areas where laboratory and greenhouse studies do not correlate with field tests. If the mycorrhizal state is the normal one for most horticultural crops, then most of our information is based on research on abnormal plants.

IX. LITERATURE CITED

ALDON, E.F. 1975. Endomycorrhizae enhance survival and growth of four-wing saltbush on coal mine spoils. USDA For. Serv. Res. Note RM-294, Rocky Mountain Forest and Range Expt. Sta., Fort Collins, Colo.

ALEXANDER, T.R. 1938. Carbohydrates of bean plants after treatment with indole-3-acetic acid. *Plant Physiol.* 13:845–858.

AMES, R.N. and R.G. LINDERMAN. 1977. Vesicular-arbuscular mycorrhizae of Easter lily in the Northwestern United States. *Can. J. Microbiol.* 23:1663–1668.

AMES, R.N. and R.G. LINDERMAN. 1978. The growth of Easter lily *(Lilium longiflorum)* as influenced by vesicular-arbuscular mycorrhizal fungi, *Fusarium oxysporum*, and fertility level. *Can. J. Bot.* 56:2773–2780.

ASAI, T. 1944. Uber die Mykorrhizenbilding der Legumenosen Pflanzen. *Jap. J. Bot.* 13:463–465.

ASHIDA, J. 1965. Adaptation of fungi to metal toxicants. *Annu. Rev. Phytopathol.* 3:153–174.

BAGYARAJ, D.J. and J.A. MENGE. 1978. Interaction between a VA mycorrhiza and azotobacter and their effects on rhizosphere microflora and plant growth. *New Phytol.* 80:567–573.

BALTRUSCHAT, H., R.A. SIKORA, and F. SCHONBECK. 1973. Effect of VA mycorrhizae *(Endogone mosseae)* on the establishment of *Thielaviopsis basicola* and *Meloidogyne incognita* in tobacco. 2nd Intern. Congr. Plant Pathol. Sept. 5–12, 1973, Minneapolis. The American Phytopathological Society, St. Paul, Minn. (*Abstr.* 0661.)

BARHAM, R.O., D. MARX, and J. RUEHLE. 1974. Infection of ectomycorrhizal and nonmycorrhizal roots of shortleaf pine by nematodes and *Phytophthora cinnamomi. Phytopathology* 64:1260–1264.

BARROWS, J.B. and R.W. RONCADORI. 1977. Endomycorrhizal synthesis by *Gigaspora margarita* in poinsettia. *Mycologia* 69:1173–1184.

BARTLETT, E.M. and D.H. LEWIS. 1973. Surface phosphatase activity of mycorrhizal roots of beech. *Soil Biol. Biochem.* 5:249–257.

BAUSOR, S.S. 1942. Effect of growth substances on reserve starch. *Bot. Gaz.* 104:115–121.

BAYLIS, G.T.S. 1975. The magnolioid mycorrhiza and mycotrophy in root systems derived from it. p. 373–389. *In* F.L. Sanders, B. Mosse and P.B. Tinker (eds.) Endomycorrhizas. Academic Press, New York.

BECKER, W.N. 1976. Quantification of onion vesicular-arbuscular mycorrhizae and their resistance to *Pyrenochaeta terrestris.* PhD Dissertation, University of Illinois, Urbana.

BENSON, N.R. and R.P. COVEY, JR. 1976. Response of apple seedlings to zinc fertilization and mycorrhizal inoculation. *HortScience* 11:252–253.

BEVEGE, D.I. and G.E. BOWEN. 1975. Endogone strain and host plant differences in development of vesicular-arbuscular mycorrhizae. p. 77–86. *In* F.E. Sanders, B. Mosse and P.B. Tinker (eds.) Endomycorrhizas. Academic Press, New York.

BEVEGE, D.I., G.D. BOWEN, and M.F. SKINNER. 1975. Comparative carbohydrate physiology of ecto- and endomycorrhizas. p. 149–174. *In* F.E. Sanders, B. Mosse and P.B. Tinker (eds.) Endomycorrhizas. Academic Press, New York.

BIELESKI, R.L. 1973. Phosphate pools, phosphate transport and phosphate availability. *Annu. Rev. Plant Physiol.* 24:225–252.

BIRD, G.W., J.R. RICH, and S.V. GLOVER. 1974. Increased endomycorrhizae of cotton roots in soil treated with nematicides. *Phytopathology* 64:48–51.

BJÖRKMAN, E. 1942. Über die Bedingungen der Mykorrhizabildung bei Kiefer und Fichte. *Symb. Bot. Upsal.* 6:1–191.

BJÖRKMAN, E. 1970. Mycorrhiza and tree nutrition in poor forest soils. *Stud. For. Suec.* 83:1–24.

BOND, G. and G.D. SCOTT. 1955. An examination of some symbiotic systems for fixation of nitrogen. *Ann. Bot.* 19:67–77.

BORTHWICK, H.A., K.C. HAMMER, and M.W. PARKER. 1937. Histological and microchemical studies of the relations of tomato plants to indoleacetic acid. *Bot. Gaz.* 98:491–519.

BOWEN, G.D. 1975. Mineral nutrition of ectomycorrhizae. p. 151–205. *In* G.C. Marx and T.T. Kozlowski (eds.) Ectomycorrhizae. Academic Press, New York.

BOWEN, G.D., C. THEODOROU, and M.F. SKINNER. 1973. Towards a mycorrhizal inoculation programme. Proc. American-Australian Forest Nutrition Conference, Canberra. 1971.

BRYAN, W.C. and P.P. KORMANIK. 1977. Mycorrhizae benefit survival and growth of sweetgum seedlings in the nursery. *South. J. Appl. For.* 1:21–23.

BRYAN, W.C. and J.L. RUEHLE. 1976. Growth stimulation of sweetgum seedlings induced by the endomycorrhizal fungus *Glomus mosseae*. *Tree Planters Notes* 27:9, 24.

CALLOW, J.A., L.C.M. CAPACCIO, G. PARISH, and P.B. TINKER. 1978. Detection and estimation of polyphosphate in vesicular-arbuscular mycorrhizas. *New Phytol.* 80:125–134.

CARLING, J.E., W.G. RIEHLE, M.F. BROWN, and D.R. JOHNSON. 1978. Effects of a vesicular-arbuscular mycorrhizal fungus on nitrate reductase and nitrogenase activities in nodulating and non-nodulating soybeans. *Phytopathology* 68:1590–1596.

CHOU, L.G. and A.F. SCHMITTHENNER. 1974. Effect of *Rhizobium japonicum* and *Endogone mosseae* on soybean root rot caused by *Pythium ultimum* and *Phytophthora megasperma* var. sojae. *Plant Dis. Rptr.* 58:221–225.

COOPER, K.M. and D.M. LOSEL. 1978. Lipid physiology of vesicular-arbuscular mycorrhizae I. Composition of lipids in roots of onions, clover and ryegrass infected with *Glomus mosseae*. *New Phytol.* 80:143–151.

COX, G., F.E. SANDERS, P.B. TINKER, and J.A. WILD. 1975. Ultrastructural evidence relating to host-endophyte transfer in a vesicular-arbuscular mycorrhizae. *In* F.E. Sanders, B. Mosse and P.B. Tinker (eds.) Endomycorrhizas. Academic Press, New York.

CRAFTS, C.B. and C.O. MILLER. 1974. Detection and identification of cytokinins produced by mycorrhizal fungi. *Plant Physiol.* 54:586–588.

CREWS, C.L., C.R. JOHNSON, and J.N. JOINER. 1978. Benefits of mycorrhizae on growth and development of three woody ornamentals. *HortScience* 13:429–430.

CROSSET, R.N. and B.C. LOUGHMAN. 1966. The absorption and translocation of phosphorus by seedlings of *Hordeum vulgare* (L). *New Phytol.* 65: 459–468.

CRUSH, J.R. 1974. Plant growth responses to vesicular-arbuscular mycorrhizae. VII. Growth and nodulation of some herbage legumes. *New Phytol.* 73: 743–749.

DAFT, M.J. and A.A. EL-GIAHMI. 1975. Effects of *Glomus* infections on three legumes. p. 581–592. *In* E.F. Sanders, B. Mosse and P.B. Tinker (eds.) Endomycorrhizas. Academic Press, New York.

DAFT, M.J. and T.H. NICOLSON. 1972. Effect of Endogone mycorrhizae on plant growth. IV. Quantitative relationships between the growth of the host and the development of the endophyte in tomato and maize. *New Phytol.* 71:287–295.

DAVIS, R.M., J.A. MENGE, and G.A. ZENTMYER. 1978. Influence of vesicular-arbuscular mycorrhizae on phytophthora root rot of three crop plants. *Phytopathology* 68:1614–1617.

DEBARY, A. 1879. Die Erscheinungen der symbiose Tagebl. 51. Vers. Deut. Naturforscher und Aerzle zu Cassel, 1878, Strassburg, Germany.

DEMETERIO, J.L., R. ELLIS, and G.M. PAULSEN. 1972. Nodulation and nitrogen fixation by two soybean varieties as affected by phosphorous and zinc nutrition. *Agron. J.* 64:566–568.

DOMINIK, T. 1961. Experiments with inoculation of agricultural land with microbial cenosis from forest soils. *Pr. Inst. Badaw. Lesn.* 210:103–105.

FRANK, A.B. 1885. Uber die auf Wurzelsymbiose beruhende Ernährung gewisser Baume durch unterirdische Pilze. *Ber. Deut. Bot. Ges.* 3:128–145.

FURLAN, V. and J.A. FORTIN. 1977. Effect of light intensity on the formation of vesicular-arbuscular endomycorrhizas on *Allium cepa* by *Gigaspora calospora*. *New Phytol.* 79:335–340.

GERDEMANN, J.W. 1964. The effect of mycorrhiza on growth of maize. *Mycologia* 56:342–349.

GERDEMANN, J.W. 1968. Vesicular-arbuscular mycorrhiza and plant growth. *Annu. Rev. Phytopathol.* 6:397–418.

GERDEMANN, J.W. 1970. The significance of vesicular-arbuscular mycorrhizae on plant nutrition. p. 125–129. *In* T.A. Toussoun, R.V. Bega and P.E. Nelson (eds.) Root diseases of soil-borne plant pathogens. University of California Press, Berkeley.

GERDEMANN, J.W. 1975. Vesicular-arbuscular mycorrhizae. p. 575–591. *In* J.G. Torrey and D.T. Clarkson (eds.) The development and function of roots. Academic Press, New York.

GERDEMANN, J.W. and T.H. NICHOLSON. 1963. Spores of mycorrhizae Endogone species extracted from soil by wet sieving and decanting. *Trans. Brit. Mycol. Soc.* 46:235–244.

GERDEMANN, J.W. and J.M. TRAPPE. 1974. The Endogonaceae in the Pacific Northwest. Mycol. Mem. 5, Mycol. Soc. Am. The New York Botanical Garden, Bronx, N.Y.

GERDEMANN, J.W. and J.M. TRAPPE. 1975. Taxonomy of Endogonaceae. p. 35–51. *In* F.E. Sanders, B. Mosse and P.B. Tinker (eds.) Endomycorrhizas. Academic Press, New York.

GIBSON, I.A.S. 1963. Eine Mitteilung über die Kiefernmykorrhiza in den Wäldern Kenias. p. 49. *In* W. Rawald and H. Lyr (eds.) Mykorrhiza. Fisher, Jena.

GILMORE, A.E. 1971. The influence of endotrophic mycorrhizae on growth of peach seedlings. *J. Amer. Soc. Hort. Sci.* 96:35–37.

GÖBL, F. 1975. Erfahrungen bei der Anzucht von Mykorrhiza-Inpfmaterial Centralbl. *Gesamte Forstwes.* 92:227–237.

GÖBL, F. and B. PÜMPEL. 1973. Einfluss von 'Grünkupfer Linz' auf Pflanzenausbildung, Mykorrhizabesatz sowie Frosthärte von Zirbenjungpflanzen. *Eur. J. For. Pathol.* 3:242–245.

GOGALA, N. 1970. Einfluss der natürlichen Cytokinine von *Pinus silvestris* L. und anderer Wuchsstoffe auf das Mycelwachstum von *Boletus edulis* var. pinicolus Vitt. Oesterr. *Bot. Z.* 118:321–328.

GRAY, L.E. and J.W. GERDEMANN. 1969. Uptake of phosphorous -32 by vesicular-arbuscular mycorrhizae. *Plant & Soil* 30:415–422.

HACSKAYLO, E. 1971. Metabolite exchanges in ectomycorrhizae. p. 175–182. *In* E. Hacskaylo (ed.) Mycorrhizae. USDA For. Serv. Misc. Publ. 1189. GPO, Washington, D.C.

HALL, I.R. 1975. Endomycorrhizas of *Metrosideros umbellata* and *Weinmannia racemosa. N. Z. J. Bot.* 13:463–472.

HARLEY, J.L. 1971. Fungi in ecosystems. *J. Ecol.* 59:563–668.

HARLEY, J.L. and J.K. BRIERLEY. 1954. Uptake of phosphate by excised mycorrhizal roots of the beech. VI. Active transport of phosphorus from the fungal sheath into host tissue. *New Phytol.* 53:240–252.

HARLEY, J.L. and J.K. BRIERLEY. 1955. Uptake of phosphate by excised mycorrhizal roots of the beech VII. Active transport of P-32 from fungus to host during uptake of phosphate from solution. *New Phytol.* 54:296–301.

HARLEY, J.L. and J.S. WAID. 1955. The effect of light on the roots of beech and its surface population. *Plant & Soil* 7:96–112.

HARLEY, J.L. and J.M. WILSON. 1959. The absorption of potassium by beech mycorrhizas. *New Phytol.* 58:281–298.

HARRIS, M.M. and M.F. JURGENSEN. 1977. Development of *Salix* and *Populus* mycorrhizae in metallic mine tailings. *Plant & Soil* 47:509–517.

HATCH, A.B. 1937. The physical basis of mycotrophy in the genus *Pinus*. *Black Rock For. Bul.* 6:1–168.

HATTINGH, M.J., L.E. GRAY, and J.W. GERDEMANN. 1973. Uptake and translocation of P-32 labeled phosphate to onion roots by endomycorrhizal fungi. *Soil Sci.* 116:383–387.

HAYMAN, D.S. 1974. Plant growth responses to vesicular-arbuscular mycorrhiza VI. Effect of light and temperature. *New Phytol.* 73:71–80.

HAYMAN, D.S., J.M. BAREA, and R. AZCON. 1976. Vesicular-arbuscular mycorrhiza in southern Spain: its distribution in crops growing in soil of different fertility. *Phytopathol. Med.* 15:1–6.

HENDRIX, J.W. and D.M. MARONEK. 1978. Survival of the ectomycorrhizal fungus *Pisolithus tinctorius* at ultralow temperatures and its implications in cold hardiness and production of specifically infected seedlings. Proc. South. Nurserymen's Res. Conf. 23:49–50. Atlanta, Ga. Southern Nurserymen's Assoc., Nashville, Tenn.

HEPPER, C.M. and B. MOSSE. 1972. Trehalose and mannitol in vesicular-arbuscular mycorrhizae. Rpt. Rothamsted. Expt. Sta., Harpenden, Herts, U.K.

HIRREL, M.C., H. MEHRAVARAN, and J.W. GERDEMANN. 1978. Vesicular-arbuscular mycorrhizae in the Chenopodiaceae and Cruciferae: do they occur. *Can. J. Bot.* 56:2813–2817.

HO, I. 1977. Phytosterols in root systems of mycorrhizal and nonmycorrhizal *Zea mays* L. *Lloydia* 40:476–478.

HO, I. and J.M. TRAPPE. 1975. Nitrate reducing capacity of two vesicular-aruscular mycorrhizal fungi. *Mycologia* 67:886–888.

HOLVAS, C.D. 1966. The effect of a vesicular-arbuscular mycorrhizae on the uptake of soil phosphorous by strawberry (*Fragaria* sp. var. Cambridge Favourite). *J. Hort. Sci.* 41:57–64.

HUGHES, M., M.H. CHAPLIN, and L.W. MARTIN. 1979. Influence of mycorrhiza on nutrition of red raspberries. *HortScience* 14:521–523.

HUISINGH, D. 1974. Heavy metals: implications for agriculture. *Annu. Rev. Phytopathol.* 12:375–388.

HUSSEY, R.S. and R.W. RONCADORI. 1978. Interaction of *Pratylenchus brachyurus* and *Gigaspora margarita* on cotton. *J. Nematol.* 10:16–20.

IMSHENETSKII, A.A. (ed.) 1955. Mycotrophy in plants (in Russian). Izd. Akad., Nauk USSR, Moscow. Transl. by Isr. Prog. Sci. Transl., Jerusalem, 1967.

JOHNSON, C.R. and C.F. CREWS, JR. 1979. Survival of mycorrhizal plants in the landscape. *Amer. Nurseryman* 150(1):15, 59.

KASPARI, H. 1973. Elektronenmikroskopische Untersuchung zur Feinstruktur der endotrophen Tabakmykorrhiza. *Arch. Mikrobiol.* 92:201–207.

KATZNELSON, H., J.W. ROUATT, and E.A. PETERSON. 1962. The rhizosphere effect of mycorrhizal and nonmycorrhizal roots of yellow birch seedlings. *Can. J. Bot.* 40:377–382.

KHAN, A.G. 1972. The effect of vesicular-arbuscular mycorrhizal associations on growth of cereals. *New Phytol.* 71:613–619.

KHAN, A.G. 1974. The occurrence of mycorrhizas in halophytes, hydrophytes and xerophytes and of *Endogone* spores in adjacent soils. *J. Gen. Microbiol.* 81:7–14.

KLEINSCHMIDT, G.D. and J.W. GERDEMANN. 1972. Stunting of citrus seedlings in fumigated nursery soils related to the absence of endomycorrhizae. *Phytopathology* 62:1447–1452.

KORMANIK, P.P., W.C. BRYAN, and R.C. SCHULTZ. 1977. Endomycorrhizal inoculation during transplanting improves growth of vegetatively propagated yellow poplar. *Plant Propagator* 23(4):4–5.

KRUCKLEMAN, H.W. 1973. Die vesikulär-arbuskuläre mykorrhiza und ihre beeinflussung in landwirtschaftlichen Kulturen. Diss. Naturwiss. Fakultat Tech. Universitat. Carolo-Wilhelmina, Braunschweig.

KRUCKLEMAN, H.W. 1975. Effects of fertilizers, soils, soil tillage, and plant species on the frequency of *Endogone* chlamydospores and mycorrhizal infection in arable soil. p. 511–525. *In* F.E. Sanders, B. Mosse and P.B. Tinker (eds.) Endomycorrhizas. Academic Press, New York.

LAGERWERFF, J.V. 1967. Heavy metal contamination of soils. p. 343–364. *In* N.C. Brady (ed.) Agriculture and the quality of our environment. American Assoc. for the Advancement of Science, Washington, D.C.

LAMBERT, D.H., R.F. STOUFFER, and H. COLE, JR. 1979. Stunting of peach seedlings following soil fumigation. *J. Amer. Soc. Hort. Sci.* 104:433–435.

LARUE, J.H., W.D. MCCLELLAN, and W.L. PEACOCK. 1975. Mycorrhizal fungi and peach nursery nutrition. *Calif. Agr.* 29(5):6–7.

LEWIS, D.H. and J.L. HARLEY. 1965a. Carbohydrate physiology of mycorrhizal roots of beech. I. Identity of endogenous sugars and utilization of exogenous sugars. *New Phytol.* 64:224–237.

LEWIS, D.H. and J.L. HARLEY. 1965b. II. Utilization of exogenous sugars by uninfected and mycorrhizal roots. *New Phytol.* 64:238–255.

LEWIS, D.H. and J.L. HARLEY. 1965c. III. Movement of sugars between host and fungus. *New Phytol.* 64:256–269.

LINDERMAN, R.G. and G.A. CALL. 1977. Enhanced rooting of woody plant cuttings by mycorrhizal fungi. *J. Amer. Soc. Hort. Sci.* 102:529–532.

LINDSEY, D.L., W.A. CRESS, and E.F. ALDON. 1977. The effects of endomycorrhizae on growth of rabbitbrush, fourwing saltbush, and corn in coal mine spoil material. USDA For. Serv. Res. Note RM-343. Rocky Mountain Forest and Range Expt. Sta., Fort Collins, Colo.

LING-LEE, M., G.A. CHILVERS, and A.E. ASHFORD. 1975. Polyphosphate granules in three different kinds of tree mycorrhiza. *New Phytol.* 75:551–554.

LITTLEFIELD, L.J. 1966. Translocation of phosphorous-32 in sporophores of *Collybia velutipes*. *Physiol. Plant.* 19:264–270.

MARONEK, D.M. and J.W. HENDRIX. 1978. Mycorrhizal fungi in relation to some aspects of plant propagation. *Proc. Intern. Plant Prop. Soc.* 28:506–514.

MARONEK, D.M. and J.W. HENDRIX. 1979a. Growth acceleration of pin oak seedlings with a mycorrhizal fungus. *HortScience* 14:627–628.

MARONEK, D.M. and J.W. HENDRIX. 1979b. Slow release fertilizer for optimizing mycorrhizal production in pine seedlings by *P. tinctorius*. Abstr. 4th North American Conference on Mycorrhiza. June 24–28, 1979. Colorado State University, Fort Collins, Colo.

MARONEK, D.M. and J.W. HENDRIX. 1980. Synthesis of *Pisolithus tinctorius* ectomycorrhizae on seedlings of four woody species. *J. Amer. Soc. Hort. Sci.* 105:823–825.

MARONEK, D.M., J.W. HENDRIX, and J. KIERNAN. 1980. Differential growth response to the mycorrhizal fungus *Glomus fasciculatus* of southern magnolia and Bar Harbor juniper grown in containers in composted hardwood bark-shale. *J. Amer. Soc. Hort. Sci.* 105(2):206–208.

MARX, D.H. 1970. The influence of ectotrophic mycorrhizal fungi on the resistance of pine roots to pathogenic infections. V. Resistance of mycorrhizae to infection by vegetative mycelium of *Phytophthora cinnamomi*. *Phytopathology* 60:1472–1473.

MARX, D.H. 1972. Ectomycorrhizae as biological deterrents to pathogenic root infections. *Annu. Rev. Phytopathol.* 10:429–454.

MARX, D.H. 1975a. Mycorrhizae and establishment of trees on strip mined land. *Ohio J. Sci.* 75:288–297.

MARX, D.H. 1975b. Mycorrhizae of exotic trees in the Peruvian Andes and synthesis of ectomycorrhiza on Mexican pines. *For. Sci.* 21:353–358.

MARX, D.H. 1975c. Use of specific mycorrhizal fungi on tree roots for forestation of disturbed lands. Proc. Symp. Forestation of Disturbed Surface Areas, May, 1976, Birmingham, Ala. USDA For. Serv., State and Private Forestry, Atlanta, Ga.

MARX, D.H. and J.D. ARTMAN. 1979. The significance of *Pisolithus tinctorius* ectomycorrhizae to survival and growth of pine seedlings on coal spoils in Kentucky and Virginia. *Reclam. Rev.* 2:23–31.

MARX, D.H. and W.C. BRYAN. 1971. Influence of ectomycorrhizae on survival and growth of aseptic seedlings of loblolly pine at high temperatures. *For. Sci.* 17:37–41.

MARX, D.H., W.C. BRYAN, and C.E. CORDELL. 1976. Growth and ectomycorrhizal development of pine seedlings in nursery soils infested with the fungal symbiont *Pisolithus tinctorius*. *For. Sci.* 22:91–100.

MARX, D.H., W.C. BRYAN, and C.B. DAVEY. 1970. Influence of temperature on aseptic synthesis of ectomycorrhizae by *Thelephora terrestris* and *Pisolithus tinctorius* on loblolly pine. *For. Sci.* 16:424–431.

MARX, D.H. and C.B. DAVEY. 1969a. The influence of ectotrophic mycorrhizal fungi on the resistance of pine roots to pathogenic infections. IV. Resistance of naturally occurring mycorrhizae to infections by *Phytophthora cinnamomi*. *Phytopathology* 59:559–565.

MARX, D.H. and C.B. DAVEY. 1969b. The influence of ectotrophic mycorrhizal fungi on the resistance of pine roots to pathogenic infections. III. Resistance of aseptically formed mycorrhizae to infection by *Phytophthora cinnamomi*. *Phytopathology* 59:549–558.

MARX, D.H., A.B. HATCH, and J.F. MENDICINO. 1977. High soil fertility decreases sucrose content and susceptibility of loblolly pine roots to ectomycorrhizal infection by *Pisolithus tinctorius*. *Can. J. Bot.* 55:1569–1574.

MARX, D.H., W.G. MORRIS, and J.G. MEXAL. 1978. Growth and ectomycorrhizal development of loblolly pine seedlings in fumigated and nonfumigated nursery soil infested with different fungal symbionts. *For. Sci.* 24:193–203.

MCGRATH, H. 1964. Chemicals for plant disease control. *Agr. Chem.* 19:18–37.

MEJSTRIK, V.K. 1972. Vesicular-arbuscular mycorrhizas of the species of a *Molinietum coerulae*. L.I. association: The ecology. *New Phytol.* 71:883–890.

MELIN, E. 1962. Physiological aspects of mycorrhizae of forest trees. p. 247–263. *In* T.T. Kozlowski (ed.) Tree growth. Ronald Press, New York.

MELIN, E. 1963. Some effects of forest tree roots on mycorrhizae basidiomycetes. *In* Symp. Soc. Gen. Microbiol. 13:125–145.

MENGE, J.A., J. LARUE, C.K. LABANAUSKAS, and E.L.V. JOHNSON. 1980. The effect of two mycorrhizal fungi upon growth and nutrition of avocado seedlings grown with six fertilizer treatments. *J. Amer. Soc. Hort. Sci.* 105:400–404.

MENGE, J.A., D.E. MUNNECKE, E.L.V. JOHNSON, and D.W. CARNES. 1978. Dosage response of the vesicular-arbuscular mycorrhizal fungi *Glomus fasciculatus* and *G. constrictus* to methyl bromide. *Phytopathology* 68:1368–1372.

MENGE, J.A., D. STEIRLE, D.J. BAGYARAJ, E.L.V. JOHNSON, and R.T. LEONARD. 1978. Phosphorous concentration in plant responsible for inhibition of mycorrhizal infection. *New Phytol.* 80:575–578.

MEXAL, J. and C.P.P. REID. 1973. The growth of selected mycorrhizal fungi in response to induced water stress. *Can. J. Bot.* 51:1579–1588.

MEYER, F.H. 1973. Distribution of ectomycorrhizae in native and man-made forests. p. 79–105. *In* G.C. Marx and T.T. Kozlowski (eds.) Ectomycorrhizae. Academic Press, New York.

MEYER, F.H. 1974. Physiology of mycorrhizae. *Annu. Rev. Plant Physiol.* 25:567–586.

MIKOLA, P. 1948. On the physiology and ecology of *Cenococcum graniforme*. *Commun. Inst. For. Fenn.* 36:1–104.

MIKOLA, P. 1965. Studies on the ectendotrophic mycorrhiza of pine. *Acta. For. Fenn.* 79:1–56.

MIKOLA, P. 1970. Mycorrhizal inoculation in afforestation. *Intern. Rev. For. Res.* 3:123–196.

MILLER, C.O. 1971. Cytokinin production by mycorrhizal fungi. p. 168–174. *In* E. Hacskaylo (ed.) Mycorrhizae. GPO, Washington, D.C.

MITCHELL, H.L., R.F. FINN, and R.O. ROSENDAHL. 1937. The relation between mycorrhizae and the growth and nutrient absorption of coniferous seedlings in nursery beds. *Black Rock For. Pap.* 1:57–73.

MOLINA, R. 1979. Ectomycorrhizal inoculation of containerized Douglas-fir and lodgepole pine seedlings with six isolates of *Pisolithus tinctorius*. *For. Sci.* 25:585–590.

MORRISON, T.M. 1962. Uptake of sulphur by mycorrhizal plants. *New Phytol.* 61:21–27.

MOSER, M. 1963. Die Bedeutung der Mykorrhiza bei Aufforstungen unter besonderer Berücksichtigung von Hochlagen. p. 407–422. *In* W. Rawald and H. Lyr (eds.) Mykorrhiza. Fisher, Jena.

MOSSE, B. 1973a. Plant growth responses to VA mycorrhiza. IV. In soil given additional phosphate. *New Phytol.* 72:127–136.

MOSSE, B. 1973b. Advances in the study of vesicular-arbuscular mycorrhiza. *Annu. Rev. Phytopathol.* 11:171–196.

MOSSE, B. 1977. Plant growth responses to vesicular-arbuscular mycorrhiza. X. Responses of Stylosanthes and maize to inoculation in unsterile soils. *New Phytol.* 78:277–388.

MOSSE, B., D.S. HAYMAN, and D.J. ARNOLD. 1973. Plant growth responses to vesicular-arbuscular mycorrhiza. V. Phosphate uptake by three plant species from P-deficient soils labeled with P-32. *New Phytol.* 72:809–815.

MUNCIE, J.G., F.M. ROTHWELL, and W.G. KESSEL. 1975. Elemental sulfur accumulation in *Pisolithus*. *Mycopathologia* 55:95–96.

NEMEC, S. 1980. Effects of 11 fungicides on endomycorrhizal development in sour orange. *Can. J. Bot.* 58:522–526.

NEMEC, S. and J.H. O'BANNON. 1979. Response of *Citrus aurantium* to *Glomus etunicatus* and *G. mosseae* after soil treatment with selected fumigants. *Plant & Soil* 52:351–359.

NIELSON, N. 1930. Untersuchungen über einen neuen wachstumregulierenden. Stoff: Rhizopin. *Jahrb. Wiss Bot.* 73:125–130.

OCAMPO, J.A., J. MARTIN, and D.S. HAYMAN. 1980. Influence of plant interactions on vesicular-arbuscular mycorrhizal infections. I. Host and nonhost plants grown together. *New Phytol.* 84:27–35.

OLIVEROS, S. 1932. Effect of soil inoculations on the growth of Benguet pine. *Makiling Echo.* 11:205–207.

OSWALD, E.T. and H.A. FERCHAU. 1968. Bacterial associations of coniferous mycorrhizae. *Plant & Soil* 28:187–192.

POWELL, C.L. and J. SITHAMPARANATHAN. 1977. Mycorrhizae in hill country soils. IV. Infection rate in grass and legume species by indigenous mycorrhizal fungi under field conditions. *N. Z. J. Agr. Res.* 20:489–494.

QUISPEL, A. 1974. The endophytes of the root nodules in non-legumenous plants. p. 499–520. *In* A. Quispel (ed.) The biology of nitrogen fixation. American-Elsevier Publishing Co., New York.

RAMIREZ, B.N., D.J. MITCHELL, and N.C. SCHENCK. 1975. Establishment and growth effects of three vesicular-arbuscular mycorrhizal fungi on papaya. *Mycologia* 5:1039–1041.

RATNAYAKE, R.T., R.T. LEONARD, and J.A. MENGE. 1978. Root exudation in relation to supply of phosphorus and its possible relevance to mycorrhizal formation. *New Phytol.* 543–552.

RHODES, L.H. and J.W. GERDEMANN. 1975. Phosphate uptake zones of mycorrhizal and non-mycorrhizal onions. *New Phytol.* 75:555–561.

RICHARD, C. and J.A. FORTIN. 1975. Rôle protecteur du *Suillus granulatus* contre le *Mycelium radicis atrovirens* sur des semis de *Pinus resinosa*. *J. For. Res.* 5:452–456.

RONCADORI, R.W. and R.S. HUSSEY. 1977. Interaction of the endomycorrhizal fungus *Gigaspora margarita* and root-knot nematode on cotton. *Phytopathology* 67:1507–1511.

ROSS, J.P. and J.A. HARPER. 1970. Effect of Endogone mycorrhiza on soybean yields. *Phytopathology* 60:1552–1556.

ROSS, J.P. and J.A. HARPER. 1973. Hosts of a vesicular-arbuscular *Endogone* species. *J. Elisha Mitchell Sci. Soc.* 89:1–3.

RUEHLE, J. and D. MARX. 1971. Parasitism of ectomycorrhizae of pine by lance nematode. *For. Sci.* 17:31–34.

SABOTKA, A. 1974. Ein Fluss von Immissionen auf die Wurzeinbildung mit begranztem wuchs beider fichte. Int. Tag. Luftverunrein. Forstwirtsch. Oct. 1974. *Mariamsle. Czech.* p. 283–287.

SAFIR, G.R., J.S. BOYER, and J.W. GERDEMANN. 1971. Mycorrhizal enhancement of water transport in soybean. *Science* 172:581–583.

SAFIR, G.R., J.S. BOYER, and J.W. GERDEMANN. 1972. Nutrient status and mycorrhizal enhancement of water transport in soybean. *Plant Physiol.* 49:700–703.

SANDERS, F.E. 1975. The effect of foliar-applied phosphate on the mycorrhizal infections of onion roots. p. 261–276. *In* F.E. Sanders, B. Mosse and P.B. Tinker (eds.) Endomycorrhizas. Academic Press, New York.

SANDERS, F.E. and P.B. TINKER. 1971. Mechanism of absorption of phosphate from soil by Endogone mycorrhizas. *Nature* 233:278–279.

SANDERS, F.E. and P.B. TINKER. 1973. Phosphate flow into mycorrhizal roots. *Pestic. Sci.* 4:385–395.

SCHENCK, N.C., S.O. GRAHAM, and N.E. GREEN. 1975. Temperature and light effect on contamination and spore germination of vesicular-arbuscular mycorrhizal fungi. *Mycologia* 67:1189–1192.

SCHENCK, N.C. and K. HINSON. 1973. Response of nodulating and non-nodulating soybeans to Endogone mycorrhiza. *Agron. J.* 65:849–850.

SCHENCK, N.C. and M.K. KELLAM. 1978. The influence of vesicular-arbuscular mycorrhizae on disease development. *Univ. of Fla., Agr. Expt. Sta. Tech. Bull.* 798.

SCHENCK, N.C. and V.N. SCHRODER. 1974. Temperature response of Endogone mycorrhiza on soybean roots. *Mycologia* 66:600–605.

SCHENCK, N.C. and D.P.H. TUCKER. 1974. Endomycorrhizal fungi and the development of citrus seedlings in Florida fumigated soils. *J. Amer. Soc. Hort. Sci.* 99:284–287.

SCHONBECK, F. and H.W. DEHNE. 1977. Damage to mycorrhizal and nonmycorrhizal cotton seedlings by *Thielaviopsis basicola*. *Plant Dis. Rptr.* 61: 266–267.

SCHRAMM, J.E. 1966. Plant colonization studies on black wastes from anthracite mining in Pennsylvania. *Amer. Phil. Soc.* 56:1–194.

SLANKIS, V. 1949. Wirkung von B-Indolylessigsäure auf die dichotomische Verzweigung isolierter Wurzeln von *Pinus sylvestris*. *Svensk Bot. Tetskr.* 43: 603–607.

SLANKIS, V. 1950. Effect of α-napthalene acetic acid on dichotomous branching of isolated roots of *Pinus sylvestris*. *Physiol. Plant.* 3:40–44.

SLANKIS, V. 1951. Über den Einfluss von V-Indolylessigsäure und anderen Wuchsstoffen auf das Wachstum von Kiefernwurzeln. I. *Symb. Bot. Upsal.* 11:1–63.

SLANKIS, V. 1958. The role of auxin and other exudates in mycorrhizal symbiosis of forest trees. p. 427–443. *In* K.V. Thimann (ed.) The physiology of forest trees. Ronald Press, New York.

SLANKIS, V. 1975. Hormonal relationships in mycorrhizal development. p. 231–298. *In* G.C. Marx and T.T. Kozlowski (eds.) Ectomycorrhizae. Academic Press, New York.

SMITH, D., L. MUSCATINE, and D. LEWIS. 1969. Carbohydrate movement from autotrophs to heterotrophs in parasitic and mutualistic symbiosis. *Biol. Rev.* 44:17–40.

SPARLING, G.P. and P.B. TINKER. 1978. Mycorrhizal infection in Pennine grassland. II. Effects of mycorrhizal infection on growth of some upland grasses on α-irradiated soils. *J. Appl. Ecol.* 15:951–958.

ST. JOHN, T.V. 1980. Root size, root hairs, and mycorrhizal infection: a reexamination of Baylis's hypothesis with tropical trees. *New Phytol.* 84:483–487.

THEODOROU, C. 1978. Soil moisture and the mycorrhizal association of *Pinus radiata* D. Don. *Soil Biol. Biochem.* 10:33–37.

THIMANN, K.V. 1935. On the plant growth hormone produced by *Rhizopus suinus*. *J. Biol. Chem.* 109:279.

THIMANN, K.V. 1972. The natural plant hormones. p. 1–365. *In* F.C. Steward (ed.) Plant physiology. Academic Press, New York.

TIMMER, L.W. and R.F. LEYDEN. 1978. Stunting of citrus seedlings in fumigated soils in Texas and its correction by phosphorous fertilization and inoculation with mycorrhizal fungi. *J. Amer. Soc. Hort. Sci.* 103:533–537.

TORREY, J.A. 1976. Root hormones and plant growth. *Annu. Rev. Plant Physiol.* 27:435–459.

TRAPPE, J.M. 1962a. Fungus associates of ectotrophic mycorrhizae. *Bot. Rev.* 28:538–605.

TRAPPE, J.M. 1962b. *Cenococcum graniforme*—its distribution, ecology, mycorrhiza formation and inherent variation. PhD Dissertation, University of Washington, Seattle.

TRAPPE, J.M. 1967. Principles of classifying ectotrophic mycorrhizae for identification of fungal symbionts. Proc. Intern. Union For. Res. Organ. 5:46–59. München, Germany.

TRAPPE, J.M. 1977. Selection of fungi for ectomycorrhizal inoculation in nurseries. *Annu. Rev. Phytopathol.* 15:203–222.

TRAPPE, J.M., E.A. STAHLY, N.R. BENSON, and D.A. DUFF. 1973. Mycorrhizal deficiency of apple trees in high arsenic soils. *HortScience* 8:52–53.

VANDERPLOEG, J.F., R.W. LIGHTLY, and M. SASSER. 1974. Mycorrhizal association between *Lilium* taxa and Endogone. *HortScience* 9:383–384.

VOZZO, J.A. and E. HACSKAYLO. 1971. Inoculation of *Pinus carribaea* with ectomycorrhizal fungi in Puerto Rico. *For. Sci.* 17:239–245.

WILCOX, H.E. 1971. Morphology of ectendomycorrhizae in *Pinus resinosa*. p. 54–68. *In* E. Hacskaylo (ed.) Mycorrhizae. GPO, Washington, D.C.

WILCOX, H.E. and R. GUNMORE-NEUMAN. 1974. Ectendomycorrhizae in *Pinus resinosa* seedlings. I. Characteristics of mycorrhizae produced by black imperfecti fungus. *Can. J. Bot.* 52:2145–2155.

WILHELM, S. 1973. Principles of biological control of soil-borne plant diseases. *Soil Biol. Biochem.* 5:729–737.

WILLIAMS, S.E., A.G. WOLLUM, II, and E.F. ALDON. 1974. Growth of *Atriplex canescens* (pursh) Nutt. improved by formation of vesicular-arbuscular mycorrhizae. *Soil Sci. Soc. Amer. Proc.* 38:362–365.

WORLEY, J.F. and E. HACSKAYLO. 1959. The effect of available soil moisture on mycorrhiza associations of Virginia pine. *For. Sci.* 5:267–268.

ZAK, B. 1964. Role of mycorrhizae in root disease. *Annu. Rev. Phytopathol.* 2:377–392.

ZAK, B. 1973. Classification of Ectomycorrhizae. p. 43–78. *In* G.C. Marks and T.T. Kozlowski (eds.) Ectomycorrhizae. Academic Press, New York.

ZAK, B. 1976. Pure culture synthesis of Pacific Madrone ectendomycorrhizae. *Mycologia* 68:362–369.

ZOBEL, R.W. and D.H. WALLACE. 1978. Potential for nitrogen fixation in vegetables. *HortScience* 13:679–686.

6

Plant Regeneration from Cell Cultures[1]

D.A. Evans and W.R. Sharp
Campbell Institute for Research and Technology,
Cinnaminson, New Jersey 08077
C.E. Flick
State University of New York,
Center for Somatic-Cell Genetics and Biochemistry,
Binghamton, New York 13901

[1] Literature search for this review was completed on April 1, 1980.

INTRODUCTION

Before cell culture techniques can be applied to either propagation or crop improvement, efficient protocols for plant regeneration must be established. Unfortunately most agriculturally important crop species cannot be regenerated with the ease of model systems such as tobacco or carrot.

There have been several reviews of plant regeneration and *in vitro* propagation (e.g., Murashige 1974; Narayanaswamy 1977; Vasil *et al.* 1979; Pierik 1979). From these reviews it has been difficult to derive the necessary information for reproducing the plant regeneration experiments. Typically, no information about explant source, media, hormone concentrations, or frequency of regeneration has been included. In this review we have not attempted to list all species that can be regenerated *in vitro*, but have attempted to emphasize the techniques of plant regeneration. Consequently, we have consciously eliminated some citations with insufficient details or in which complete plants were not obtained. In addition, we have emphasized species or families that contain agriculturally important plants. For a recent complete list of species that will undergo plant regeneration, we suggest the review by Vasil *et al.* (1979). We have grouped regeneration protocols by plant families when possible in the hope that this review can be used as a guide to techniques for researchers attempting to regenerate untested plant species *in vitro*.

This manuscript will emphasize examples of somatic organogenesis and embryogenesis. Organ culture, including the culture of anthers and meristems, is discussed only briefly.

ABBREVIATIONS USED IN THIS REVIEW

Growth Regulators

Auxin		Cytokinin	
IAA	indoleacetic acid	KIN	kinetin
IBA	indolebutyric acid	6BA	6-benzyladenine
NAA	napthalene acetic acid	ZEA	zeatin
NOA	β-napthoxyacetic acid	2iP	2-isopentyladenine
CPA	ρ-chlorophenoxyacetic acid	PBA	tetrahydropyranyl
2,4-D	2,4 dichlorophenoxyacetic acid	ADE	adenine

Others

GA_3	gibberellic acid
BTOA	2-benzothiazole acetic acid
2,4,5-T	2,4,5 trichlorophenoxyacetic acid

Media

MS	Murashige and Skoog 1962	White	White 1963
SH	Schenk and Hildebrandt 1972	B5	Gamborg *et al.* 1968

Organic Additives

CW	coconut water	ME	malt extract
CH	casein hydrolysate	YE	yeast extract

II. GENERAL METHODOLOGY

A. Explants

Callus growth has been obtained from explants of many species of plant cells cultured *in vitro*. Callus cells are generally characterized as being unorganized, parenchyma-like, and rapidly proliferating. Most viable plant cells can be induced to undergo cell division *in vitro*. Plant regeneration has been accomplished from explants of cotyledon (Hu and Sussex 1971), hypocotyl (Kamat and Rao 1978), stem (Crocomo *et al.* 1980), leaf (Handro 1977), shoot apex (Kartha *et al.* 1974a), root (Gunckel *et al.* 1972), young inflorescences (Majumdar 1970), flower petals (Heuser and Apps 1976), petioles (Pierik 1972), ovular tissue (Kochba and Spiegel-Roy 1973), and embryos (Nag and Johri 1969). For any species or cultivar particular explants may be necessary for successful plant regeneration, e.g., embryonic tissue is required for cereals. Explants consisting of shoot tips or isolated meristems, which contain actively dividing cells, have been especially successful for callus initiation and subsequent plantlet regeneration (Murashige 1978). Tissue explants from mature organs also can be induced to proliferate and form callus on appropriate culture media.

B. Callus Proliferation

Only a small percentage of cells in an explant will form callus. Callus initiation generally occurs either at the surface of the inoculum or at the excised surface. The latter event is related to the wound response. For example, callus is initiated on the bulbil of *Dioscorea* from storage parenchyma cells in which a wound periderm is first formed followed by the development of a meristem inside the periderm (Rao 1969). Callus also may be commonly formed from parenchyma, as in *Nicotiana* (Ellis and Bornman 1971), or from the procambium of the leaf sheath nodes as in rice (Wu and Li 1971). Callus formation has been discussed in detail by Gautheret (1959) and Vasil'tsova (1967).

Cytological studies have been made on callus and suspension cells, notably with carrot (Halperin and Jensen 1967; Neumann 1969; Neumann *et al.* 1969) and with sycamore (Carceller *et al.* 1971). Cell size (in μm) ranges between 200 and 475 in length and 45 and 90 in width for elongated cells, 60 and 130 in diameter for spherical cells, and 120 and 175 in length and 60 and 95 in width for oval cells (Kant and Hildebrandt 1969). Average cell size increases with time in culture (Syono 1965). Growth regulator concentrations (Harada *et al.* 1972) and nutritional factors (Toren 1955) also regulate cell size. Friability, or the tendency for cells to separate from each other, can be increased at higher concentrations of auxin in the culture medium (Torrey and Reinert 1961).

The season of the year can affect callus initiation from explants, especially when the donor plant is field grown. Seasonal variations in the concentration of endogenous auxins have been observed (Wodzicki 1978) and also have been reported for establishment of potato meristem (Mellor and Stace-Smith 1969) and conifer (Harvey and Grasham 1969) cultures. For these species, spring and summer were found to be optimum seasons for starting cultures. Developmental stage and physiological state of the plant at the time of culture must be considered, because such factors as dormancy of the cambium or lateral buds, induction of flowering, etc., may affect the response and/or the success of initiating cultures.

Callus has been successfully induced from different types of explants (Table 6.1). Callus initiation is dependent on explant source. For exam-

TABLE 6.1. EXPLANTS SUCCESSFULLY USED FOR CALLUS INDUCTION *IN VITRO*

Organ	Example	Reference
Root sections	*Sinapis alba*	Bajaj and Bopp 1972
	Lycopersicon esculentum	Norton and Boll 1954
Storage root	*Daucus carota*	Komamine *et al.* 1969
	Solanum tuberosum	Steward and Caplin 1951
Hypocotyl	*Lycopersicon esculentum*	Ulrich and MacKinney 1970
	Sinapis alba	Kamat and Rao 1978
Pith	*Nicotiana tabacum*	Patau *et al.* 1957
	Pelargonium sp.	Chen and Galston 1967
Cortex	Conifers	Harvey and Grasham 1969
Cambium	*Theobroma cacao*	Archibald 1954
Stem	*Foeniculum*	Maheshwari and Gupta 1965
Shoot tip	*Rosa* sp.	Jacobs *et al.* 1970
Leaf	*Calystegia*	Harada *et al.* 1972
Petiole	*Brassica oleracea*	Vasil'tsova 1967
Cotyledon	*Arachis hypogea*	Verma and Van Huystee 1970
	Sinapis alba	Hu and Sussex 1971
Bud	*Citrus sinensis*	Altman and Goren 1971
Seed	*Arabidopsis thaliana*	Shen-Miller and Sharp 1966
Embryo	*Oryza sativa*	Maeda 1968
	Secale cereale	Carew and Schwartung 1958
Bulb scales	*Allium cepa*	Fridborg 1971
	Lilium speciosum	Robb 1957
Anther	*Ephedira foliata*	Konar 1963
	Brassica oleracea	Kameya and Hinata 1970
Stigma, style	*Solanum melongena*	Islam and Maruyama 1970
Fruit	*Cucurbita*	Schroeder *et al.* 1962
	Persea americana	Blumenfeld and Gazit 1971
Endosperm	*Zea mays*	LaRue 1949
Nucellus	*Citrus*	Rangaswamy 1958

ple, growth regulator requirements in the culture medium are different for mesocarp in comparison to cotyledon tissue for *Persea* (Blumenfeld and Gazit 1971). In *Persea* the cotyledonary tissue is autonomous for cytokinin synthesis, whereas the mesocarp tissue is cytokinin dependent. Different concentrations of 2,4-D in the culture medium are necessary for obtaining callus from root, scutellum, cotyledonary nodes, coleoptile, or leaf sheath nodes of rice (Wu and Li 1971). These variations in culture medium requirements reflect physiological differences of the different explants. Differences also occur in the morphology of callus from *Sinapis* root and hypocotyl (Bajaj and Bopp 1972) and in cell size in callus from adult and juvenile tissue of *Hedera* (Stoutmeyer and Britt 1965); however, Barker (1969) did not observe gross morphological differences between callus derived from various organs of *Tilia* or *Triticum*. Ploidy differences also have been reported to occur in callus obtained from different tissue explants of a taxon (Yamada *et al.* 1967). Soybean hypocotyl callus cultures are unable to support symbiotic growth of *Rhizobium*, whereas nodules develop in callus cultures from root explants of the same plant (Holsten *et al.* 1971).

Such variations *in vitro* may reflect the differences in the phenotypic physiological expression of the cells in the original explant. On the other hand, the phenotypic expression of a cell may be modified by isolation and culture. Explants of different origin may achieve the same state of cytodifferentiation on a given medium or express epigenetic differences which reflect tissue origin. The response for any cell in culture is determined by environmental and genetic interactions. The size of the explant and, in a few cases, the mode of culture or polarity of the explant in the medium can influence callus development. Smaller explants are more likely to form callus, while larger explants maintain greater morphogenetic potential (Okazawa *et al.* 1967). Small tissue explants frequently require a complex medium. Pine embryo cotyledons can absorb nutrients from the medium better than the roots can, resulting in improved growth (Brown and Gifford 1958).

C. Culture Media

The components of different culture media have been reviewed (Gamborg *et al.* 1976). Murashige and Skoog's (1962) MS medium, Schenk and Hildebrandt's (1972) SH medium, and Gamborg *et al.*'s (1968) B5 medium contain high concentrations of macronutrients, whereas White's (1963) and Gautheret's (1942) media contain low concentrations of these compounds. Other media commonly used include those of Blaydes (1966), Nitsch and Nitsch (1969), and Veliky and Martin (1970), V-47. Growth regulators are normally essential for callus initiation. Organic supplements are also required and include a carbon source, commonly sucrose, and vitamins, particularly thiamine, inositol, nicotinic acid, pyridoxine, calcium pantothenate, or biotin. Various plant extracts or undefined additives are sometimes added to the medium to increase the growth response.

D. Modes of Culture

Two modes of cell culture are generally used: (1) the cultivation of clusters of cells on a solid substrate (e.g., agar, gelatin, filter paper, Millipore® filters) and (2) the cultivation of cell suspensions in liquid medium. New cultures should be initiated on solid medium as essential nutrients may leach from small explants placed in large volumes of liquid medium.

A suspension cell culture is usually initiated by placing friable callus into liquid culture medium. The suspension usually consists of free cells and aggregates of 2 to 100 cells (Evans and Gamborg 1980). Suspension cultures should be subcultured regularly, whereas callus cultures of the same species usually can be subcultured less frequently. Subculturing should increase when cultures become established to achieve optimal growth rate and genetic stability (Evans and Gamborg 1980).

E. Plant Regeneration

The appropriate concentration of growth regulators is critical for control of growth and morphogenesis. Generally a high concentration of auxin and a low concentration of cytokinin in the medium promote callus formation. Alternatively, low auxin and high cytokinin concentrations in the medium generally result in the induction of shoot morphogenesis. Auxin alone is important in the induction of root primordia.

III. PLANT REGENERATION METHODOLOGY

A. Solanaceae

Solanaceous species have been used as model systems of *in vitro* studies. Totipotency was first demonstrated with *Nicotiana tabacum* by regeneration of mature plants from single cells (Vasil and Hildebrandt 1965). The first successful production of haploid plants by the *in vitro* culture of excised anthers was achieved by using *Datura innoxia* (Guha and Maheshwari 1964). Plant regeneration from isolated protoplasts was accomplished first with *N. tabacum* (Takebe *et al.* 1971), and the first somatic hybrid was obtained from two *Nicotiana* species, *N. glauca* and *N. langsdorfii* (Carlson *et al.* 1972).

1. Tobacco and Other *Nicotiana* Species.—Varieties of cultivated tobacco, *Nicotiana tabacum*, are easy to manipulate *in vitro*. The effects of auxins and cytokinins on explants and tissue cultures are quite specific and reproducible. The nutrient solution most often used for *in vitro* cultivation of plant species, Murashige and Skoog (1962) culture medium (MS), was formulated as a result of growth experiments with *N. tabacum*. Callus and suspension cultures have been initiated from leaf or stem explants of *N. tabacum* and many *Nicotiana* spp. using MS medium with the addition of 4.5 μM 2,4-D and 2 g/liter casein hydrolysate (Evans and Gamborg 1980). Callus can be initiated on MS medium with other hormone concentrations, e.g., 11.4 μM IAA and 2.3 μM KIN (Murashige and Skoog 1962), but an auxin is always necessary. Callus can be maintained on MS or a similar medium with 2,4-D; casein hydrolysate is unnecessary. Suspension cultures also can be readily obtained on B5 or MS media with 2.3 to 4.5 μM 2,4-D. Shoot regeneration from callus and suspension culture can be obtained for most *Nicotiana* species by the removal of 2,4-D and the addition of a cytokinin, e.g., subculture to a solid MS medium with 5 μM 6BA. For *N. tabacum*, numerous hormone combinations have been successfully utilized for shoot regeneration (c.f. Tran Thanh Van and Trinh 1978; Nitsch *et al.* 1967). It is also possible to induce shoot regeneration by replacing 2,4-D with combinations of auxins and cytokinins, e.g., 11.4 μM IAA and 9.3 μM KIN (Murashige and Nakano 1967) or 22.8 μM IAA and 46.5 to 66.5 μM KIN (Sacristan and Melchers 1969). Shoot formation from callus has been reported for 15 to 20 *Nicotiana* species (Table 6.2). Shoots can be obtained from *N. tabacum* callus in approximately 3 weeks, and can be multiplied using shoot culture on MS medium with 5 μM 6BA. Roots can be induced on MS, Hoagland, or White's medium with no hormones (Bourgin *et al.* 1979; Murashige and Nakano 1967; Nagata and Takebe 1971) or on one-half strength MS medium with 25 to 75 μM 3-aminopyridine (Phillips and Collins 1979; Evans *et al.* 1980). Roots appear rapidly, and it is likely that root primordia are already present.

Somatic embryos have been induced in *N. tabacum* 'Samsun' when callus cultures have been exposed to high light intensity, e.g., 10,000 to 15,000 lux, in MS medium with 15% coconut water and 9.3 μM KIN (Haccius and Lakshmanan 1965), or when young dark-grown petioles were placed on White's medium with 8.9 μM 6BA, 0.6 μM IAA, and 24.7 μM adenine (Prabhudesai and Narayanaswamy 1973). Addition of 1.3 to 13.0 μM GA_3 suppresses shoot formation (Murashige 1961).

Protoplasts of *Nicotiana* species have been used in many somatic hybridization experiments (Cocking 1978). Plant regeneration has been observed from protoplasts of 16 *Nicotiana* species: *N. tabacum* (Takebe *et al.* 1971); *N. sylvestris* (Nagy and Maliga 1976); *N. otophora* (Banks and Evans 1976); *N. alata* (Bourgin and Missonier 1978); *N. plumbaginifolia* (Gill *et al.* 1978); *N. debneyi* (Smith and Mastrangelo-Hough 1979); *N. acuminata, N. glauca, N. langsdorfii, N. longiflora, N. paniculata, N. suaveolens* (Bourgin *et al.* 1979); *N. rustica* (Gill *et al.* 1979); and *N. repanda, N. stocktonii, N. nesophila* (Evans 1980). Although protoplasts from suspension cultures or callus cultures have greater plating efficiency, *Nicotiana* spp. leaf mesophyll protoplasts also will divide and produce callus capable of subsequent plant regeneration. Using prescribed combinations of auxins and cytokinins, protoplasts of many *Nicotiana* spp. cultured in the medium of Nagata and Takebe (1971) or Kao and Michayluk (1975) will yield callus in 3 to 4 weeks, although some species, such as *N. sylvestris*, may require a modified Kao medium (Nagy

TABLE 6.2. SHOOT FORMATION FROM CALLUS AND EXPLANTS OF *NICOTIANA* SPECIES ON MS MEDIUM

Species	Explant	Growth Regulators for Shoot Formation	Reference
N. accuminata	Stem	10 μM 2iP, 1 μM IAA	Helgeson 1979
N. africana	Leaf	5 μM 6BA	Evans, unpublished
N. alata	Floral branches	10 μM KIN, 1 μM IBA	Tran Thanh Van and Trinh 1978
N. glauca	Leaf	5 μM 6BA	Evans, unpublished
N. glutinosa	Leaf	5 μM 6BA	Evans, unpublished
N. goodspeedii	Stem	10 μM 2iP, 1 μM IAA	Helgeson 1979
N. longiflora	Stem	4.7 μM KIN, 0.06 μM IAA	Ahuja and Hagen 1966
N. megalosiphon	Stem	10 μM 2iP, 1 μM IAA	Helgeson 1979
N. nesophila	Leaf	5 μM 6BA	Evans, unpublished
N. otophora	Floral branches	10 μM KIN, 1 μM IBA	Tran Thanh Van and Trinh 1978
	Leaf	5 μM KIN, 1 μM 6BA	Evans, unpublished
N. plumbaginifolia	Floral branches	10 μM KIN, 1 μM IBA	Tran Thanh Van and Trinh 1978
N. repanda	Leaf	5 μM 6BA	Evans, unpublished
N. rustica	Shoot	45.7 μM IAA, 11.9 μM KIN	Walkey and Woolfitt 1968
	Leaf	5 μM 6BA	Evans, unpublished
N. stocktonii	Leaf	5 μM 6BA	Evans, unpublished
N. suaveolens	Stem	10 μM 2iP, 1 μM IAA	Helgeson 1979
N. sylvestris	Leaf	5 μM 6BA	Flick, unpublished
		17.1 μM IAA, 0.9 μM KIN	Ogura and Tsuji 1977
N. tabacum	Cell culture	1 μM 6BA	Gamborg *et al.* 1979
N. tomentosiformis	Floral branches	10 μM KIN, 1 μM IBA	Tran Thanh Van and Trinh 1978
Sexual Hybrids			
N. debneyi × *N. tabacum*	Stem	4.7 μM KIN, 0.06 μM IAA	Ahuja and Hagen 1966
N. tabacum × *N. glauca*	Leaf	5 μM 6BA	Evans, unpublished
N. tabacum × *N. sylvestris*	Leaf	5 μM 6BA	Evans, unpublished
N. glauca × *N. tabacum*	Leaf	5 μM 6BA	Evans, unpublished

and Maliga 1976) with 0.45 μM 2.4-D, 0.89 μM 6BA, and 5.37 μM NAA. Protoplast-derived callus (p-callus), as with explant-derived callus, can be transferred to MS medium with 5 μM 6BA for shoot induction which will occur in 3 to 4 weeks. There is only one report of somatic embryogenesis from protoplasts of *N. tabacum* (Lorz *et al.* 1977), and this required specific growth regulator concentrations on Nagata and Takebe (1971) medium: 10 μM Dicamba and 50 μM 6BA or 6 μM CPA and 25 μM KIN.

Some sexual hybrids between *Nicotiana* spp. produce tumors (Kostoff 1943). It has been shown that tumorous cells of *N. glauca* × *N. langsdorfii* are hormone autotrophic when grown *in vitro* (Ahuja and Hagen 1966). This unique hormone autotrophy has been used as a basis to select somatic hybrids of *N. glauca* + *N. langsdorfii* (Carlson *et al.* 1972). Other sexual hybrids in *Nicotiana* result in tumor production, as predicted by Näf (1958), and such a selection scheme should be useful to isolate other somatic hybrids (Smith and Mastrangelo-Hough 1979). Auxin autotrophy is not limited to tumors, as cells of a number of *Nicotiana* species are also auxin autotrophic, including *N. debneyi* (Smith and Mastrangelo-Hough 1979) and *N. repanda* (Evans, unpublished). Two *Nicotiana* species which cannot be regenerated using the protocols above are *N. knightiana* from leaf explants or protoplasts (Maliga *et al.* 1977) and *N. glutinosa* from protoplasts (Bourgin *et al.* 1979).

2. Potato and Related Species.—Dihaploid lines of *Solanum tuberosum* are also amenable to tissue culture studies (Binding *et al.* 1978). Tuber, shoot tip, hypocotyl, leaf, and stem explants have been used to initiate callus with morphogenetic potential. Callus can be initiated from tuber explants on modified MS medium (Lam 1975) with 2.3 μM IAA, 1 μM GA_3, and 3.7 μM KIN, or from shoots or stems on MS medium with 9.1 μM 2,4-D (Wang and Huang 1975). Callus formation from wild *Solanum* spp. requires a high auxin:cytokinin ratio. NAA and 2,4-D generally have been utilized, although IAA and 6BA also have been used successfully (Table 6.3). Shoot induction has been observed in stem or shoot-derived callus when 4.7 to 46.5 μM KIN has been substituted for 2,4-D. Root formation occurs when young shoots are transferred to MS medium with 2.2 μM 6BA and 0.3 μM GA_3 or to MS medium without hormones.

A similar protocol can be followed for *S. xanthocarpum* (Rao and Narayanaswamy 1968), *S. sisymbriifolium* (Fassuliotis 1975), *S. dulcamara, S. nigrum* (Zenkteler 1972), *S. khasianum (Bhatt et al.* 1979), and chlorophyll-deficient *S. chacoense* (Gamborg, personal communication). A modified MS or White (1963) medium with 2.3 to 27.1 μM 2,4-D can be used to initiate or maintain callus of each species. For plant regeneration 2,4-D is removed or the concentration reduced from 4.5 to 0.5 μM 2,4-D for *Solanum xanthocarpum*. In most cases a cytokinin is

TABLE 6.3. PLANT REGENERATION OF *SOLANUM* SPECIES

Species	Growth Regulators		Shoot Medium	Explant	Mode of Regeneration	Reference
	Callus	Shoot Formation				
S. tuberosum (potato)	2.28 μM IAA 1.04 μM GA_3 3.73 μM KIN	1.8 μM 6BA	MS	Tuber	O	Lam 1975
S. tuberosum	9.05 μM 2,4-D	4.7–46.5 μM KIN	MS	Shoot, stem	O	Wang and Huang 1975
S. melongena (eggplant)	4.3 μM NAA 1 μM 6BA	1 μM 6BA	MS	Hypocotyl	E, O	Matsuoka and Hinata 1979
S. melongena	5 μM NOA	4.7 μM KIN 5.7 μM IAA	MS	Hypocotyl	O	Kamat and Rao 1978
S. dulcamara	Not reported	4.7 μM KIN 5.7 μM IAA	MS	Leaf	O	Zenkteler 1972
S. kahasianum	Not reported	10 μM 6BA 0–10 μM IAA	MS	Leaf	O	Bhatt *et al.* 1979
S. laciniatum	10 μM IAA 0.1 μM ZEA	10 μM 6BA 0.1–10 μM IBA	MS	Leaf	O	Davies and Dale 1979
S. nigrum	10 μM IAA 1–10 μM 6BA	10 μM IAA 10 μM 6BA	MS	Leaf	O	Bhatt *et al.* 1979
S. nigrum	Not reported	4.7 μM KIN 5.7 μM IAA	MS	Leaf	O	Zenkteler 1972
S. sisymbriifolium	2.26 μM 2,4-D 17.1 μM IAA 49.2 μM 2iP	0.03–1.7 μM IAA 73.8 μM 2iP	MS	Stem pith	O	Fassuliotis 1975
S. xanthocarpum	27.15 μM 2,4-D ± 5.37 μM NAA	0.09 μM 2,4-D	White or MS	Shoot	O	Rao and Narayanaswamy 1968

also added to shoot regeneration medium (Table 6.3). Root induction occurs for *S. sisymibriifolium* following transfer to Nitsch medium with 7.4 μM IAA. *Solanum melongena* (eggplant) hypocotyl sections do not respond to 2,4-D (Kamat and Rao 1978); however, callus can be initiated by using 5 μM NOA or 43 μM NAA (Matsuoka and Hinata 1979). Shoots were obtained when callus was subcultured onto MS medium with 2.3 μM ZEA or 4.7 μM KIN, or when NAA was replaced with 1 μM 6BA. MS medium with no hormones was used to obtain root formation.

Plants have been obtained from isolated leaf mesophyll protoplasts of *S. tuberosum* (Shepard and Totten 1977) using the culture medium for tubers (Lam 1975): MS medium with 2.28 μM ZEA and 0.57 μM IAA. Plants also have been obtained from protoplasts of *S. nigrum* (Nehls 1978) and *S. dulcamara* (Binding and Nehls 1977) by plating p-calli onto B5 medium with 5 μM 6BA and 2 μM NAA. Roots could be obtained in all species by removing hormones from either B5 or MS medium.

3. Tomato and Related Species.—MS medium generally has been used for tissue culture studies with species of *Lycopersicon*, particularly the cultivated tomato, *L. esculentum* (Padmanabhan *et al.* 1974). Callus can be induced from most tomato explants, but leaf or hypocotyl sections are most often used (Table 6.4). The auxin 2,4-D has been used only sparingly in regeneration studies. Combinations of IAA and NAA with KIN and 6BA have been used to initiate callus proliferation. Phenolic oxidation in leaf and hypocotyl sections can be overcome either by placing cultures in the dark until callus forms or by adding 500 mg/liter polyvinylpyrrolidone (PVP) to the culture medium (Flick and Evans, unpublished information). Shoot formation can occur on callus induction medium or on media with a higher cytokinin:auxin ratio or with cytokinin alone (Kartha *et al.* 1976). Roots can be obtained on MS medium with 10.7 μM NAA or 11.8 μM IAA (Tal *et al.* 1977). Coleman and Greyson (1977) have shown that GA_3 at 0.01 to 100 μM enhances root formation. Similar techniques for plant regeneration have been applied successfully to two wild *Lycopersicon* species (Table 6.4). Tomato does not seem to be as amenable to tissue culture techniques as do other solanaceous species (Herman and Haas 1978), since callus cultures lose the ability to undergo shoot morphogenesis when subcultured. Plant regeneration has not yet been obtained from long term suspension cultures (e.g., Meredith 1978). Normal plantlets could be obtained from 20% of tomato callus lines after 4 months in culture (Meredith 1979), whereas only abnormal plantlets incapable of root formation could be obtained after 17 months. Plant regeneration from protoplasts in *Lycopersicon* has been successful only with *L. peruvianum* (Zapata *et al.* 1977). Shoot regeneration from callus has been obtained by using 6BA and IAA but not KIN and NAA (Zenkteler 1972). The cytokinin:auxin ratio appears to be more impor-

TABLE 6.4. PLANT REGENERATION OF TOMATO AND RELATED SPECIES

Species	Growth Regulators		Shoot Medium	Explant	Mode of Regeneration	Reference
	Callus	Shoot Formation				
L. esculentum 'Rutgers'	11.4–22.8 μM IAA 9.3–18.6 μM KIN	22.8 μM IAA 18.6 μM KIN	MS	Leaf	O	Padmanabhan *et al.* 1974
L. esculentum 'Starfire'	1 μM 6BA 1 μM NAA	1–10 μM 6BA 0.1 μM IAA	MS	Leaf	O	Kartha *et al.* 1976
L. esculentum 'Rheinlands Rhum'	2.3–27.9 μM KIN 2.7–32.2 μM NAA	8.9 μM 6BA 1.1 μM IAA	MS	Leaf	O	Tal *et al.* 1977
L. esculentum 'Apidice', 'Porphyre'	0.5–2.5 μM IAA 1–20 μM 6BA	10 μM 6BA 0.5 μM NAA	MS	Leaf	O	Ohki *et al.* 1978
L. esculentum 'Rutgers', 'EP-7'	4.5 μM 2,4-D	5 μM 6BA	MS	Hypocotyl, leaf, shoot	O	Flick and Evans, unpublished
L. esculentum hybrid Pol × Pusa Pol. cultivar	5 μM 2iP 2 μM IAA	1 μM ZEA 1 μM IAA	MS	Leaf	O	Dhruva *et al.* 1978
L. esculentum 'Bizon'	22.8–34.3 μM IAA 9.3–18.6 μM KIN	91.3 μM IAA 18.6 μM KIN	MS	Stem, leaf, cotyledon	O	Vnuchkova 1978
L. esculentum 'VFNT Cherry'	10.7 μM NAA 4.4 μM 6BA	9.1 μM ZEA 0–2.9 μM IAA	MS	Leaf	O	Meredith 1979
L. esculentum 'Karnatak'	2.8 μM IAA	9 μM 6BA 2.8 μM IAA	MS	Hypocotyl, cotyledon	O	Gunay and Rao 1980
L. peruvianum	Not reported	198 μM ADE	White	Root	O	Norton and Boll 1954
L. peruvianum	0–27.9 μM KIN 10.7–32.2 μM NAA	8.9 μM 6BA 0–11.4 μM IAA 27.9 μM KIN	MS	Leaf	O	Tal *et al.* 1977
Solanum penellii	27.9 μM KIN 2.9 μM NAA	8.9 μM 6BA 0–0.1 μM IAA	MS	Leaf	O	Tal *et al.* 1977

tant for the control of tomato shoot formation than do the specific hormones utilized.

4. *Petunia* and *Datura* Species.—*Petunia* spp. can be manipulated easily *in vitro*. Research has been concentrated on *Petunia hybrida*, while investigations with other species have centered on somatic hybridization experiments. Consequently, most of the information about these species is concerned with the feasibility of plant regeneration from protoplasts. All *Petunia* spp. have been cultured on MS medium (Table 6.5). Stem or leaf explants have been used for plant regeneration, but roots (Colijn *et al.* 1979) also have been cultured successfully. Callus can be obtained when explants are cultured on combinations of 0 to 107.4 μM NAA and 0.4 to 35.2 μM 6BA, although 2,4-D may be used alone (Sangwan and Harada 1976) or in combination with a low concentration (0.9 μM) of 6BA (Rao *et al.* 1973a). In *Petunia hybrida* cell cultures 2,4-D has been implicated in somatic embryogenesis (Sangwan and Harada 1976). Shoot regeneration has been achieved in all species by reducing the concentration of auxin while simultaneously increasing the concentration of cytokinin. This has been accomplished in a few cases by substituting a weak auxin for a strong auxin, such as 22.8 μM IAA for 22.6 μM 2,4-D (Frearson *et al.* 1973), or by substituting a strong cytokinin for a weak cytokinin, such as 0.5 μM ZEA for 0.9 μM 6BA (Rao *et al.* 1973a). All six species of *Petunia* examined can be regenerated *in vitro* (Table 6.5) and somatic hybrids have been recovered following protoplast fusion.

The genus *Datura* has not been studied in great detail. Although *D. innoxia* represents ideal material for studies of anther culture (Guha and Maheshwari 1964), plant regeneration (Engvild 1973), and somatic hybridization (Schieder 1978), there are a few species, *D. ferox, D. stramonium,* and *D. discolor*, which cannot be regenerated *in vitro* (Schieder 1977). Callus of *D. innoxia* can be induced and maintained on MS medium with 1 μM 2,4-D and plants can be regenerated from callus with 1 μM 6BA. Roots can be obtained on growth regulator-free MS medium or on Nitsch medium with 1 μM IAA (Sopory and Maheshwari 1976). *Datura metel* and *D. meteloides* have been regenerated *in vitro* on B5 medium with 2.2 μM 6BA and 8 μM NAA, and shoot formation has occurred when NAA was eliminated and 6BA was increased to 4.4 μM. Shoots have been obtained in *D. innoxia* from cell suspension cultures (Hiraoka and Tabata 1974), stem segments (Engvild 1973), leaf sections (Evans and Gamborg, unpublished), and protoplasts derived from both suspension cultures (Furner *et al.* 1978) and leaf cells (Schieder 1975).

5. Other Solanaceous Species.—Of the remaining solanaceous species, only *Atropa belladonna* has been found to undergo somatic embryo-

TABLE 6.5. PLANT REGENERATION OF *PETUNIA* AND *DATURA* SPECIES

Species	Growth Regulators		Shoot Medium	Explant	Mode of Regeneration	Reference
	Callus	Shoot Formation				
Petunia axillaris	4.5 μM 2,4-D 2.3 μM 6BA	2.2 μM 6BA 0.3 μM NAA	MS	Leaf protoplasts	O	Power *et al.* 1976
P. hybrida 'Blue Dansy', 'Comanche', 'Gypsy'	22.6 μM 2,4-D 0.9 μM KIN	11.9 μM KIN 22.8 μM IAA	MS	Leaf	O	Frearson *et al.* 1973
'Celestial' × 'Blue Dedder'	0–107.4 μM NAA 0.4–35.2 μM 6BA	0.3–2.7 μM NAA 1.1–8.9 μM 6BA	MS	Root	O	Colijn *et al.* 1979
'Rose du Ciel', 'Cascade'	4.5 μM 2,4-D 0.9 μM 6BA	0.5 μM ZEA	MS	Stem, leaf	E	Rao *et al.* 1973a
P. inflata	4.5 μM 2,4-D	2.3 μM 6BA	MS	Leaf	O	Hayward and Power 1975
P. parodii	4.5 μM 2,4-D	2.3 μM 6BA	MS	Leaf	O	Hayward and Power 1975
P. parviflora	10.7 μM NAA 2.2 μM 6BA	11.4 μM IAA 4.4 μM 6BA	MS	Leaf protoplasts	O	Sink and Power 1977
P. pendula	Not reported	8.9 μM 6BA 5.4 μM NAA	MS	Floral peduncle	O	Pelletier and Delise 1969
P. violacea	26.9 μM NAA 4.4 μM 6BA	4.6 μM ZEA or 4.4 μM 6BA	MS	Leaf protoplasts	O	Power *et al.* 1976
Hybrid						
P. hybrida × *P. parodii*	26.9 μM NAA 2.3 μM 6BA	4.4 μM 6BA 11.4 μM IAA	MS	Leaf protoplasts	O	Power *et al.* 1976
Datura innoxia	1 μM 2,4-D or 10 μM NAA	1 μM 6BA	MS	Stem	O	Engvild 1973
D. innoxia	1 μM 2,4-D	100 μM KIN	MS	Suspension	O	Hiraoka and Tabata 1974
D. innoxia (haploid)	15% CW	9.3 μM KIN 9.9 μM ADE	Nitsch	Stem, leaf	O	Sopory and Maheshwari 1976
D. metel	2.2 μM 6BA 8 μM NAA	4.4 μM 6BA	B5	Leaf protoplasts	O	Schieder 1977
D. meteloides	2.2 μM 6BA 8 μM NAA	4.4 μM 6BA	B5	Leaf protoplasts	O	Schieder 1977

genesis in cell suspension cultures (Thomas and Street 1970). Embryogenesis was achieved by transferring callus grown in the presence of auxin to auxin-free medium containing 0.5 μM KIN (Gosch *et al.* 1975). Plants have been regenerated via organogenesis from eight other solanaceous species (Table 6.6). In each case callus could be induced with an auxin:cytokinin ratio ≤ 1 with shoots regenerated with auxin:cytokinin ratio >1. MS, White's, or Poirier-Hamon *et al.*'s (1974) basal medium was used in each case and root formation was obtained on basal medium or medium with auxin (NAA or IAA). Nearly all possible explants have been used (Table 6.6) for induction of embryogenesis among these nine solanaceous species. In addition to the species listed in Table 6.6, three species of *Scopolia, S. carniolica, S. lurida,* and *S. physaloides* (Wernicke and Kohlenbach 1975), have been regenerated from anther cultures. *Physalis alkekengi* (Zenkteler 1972) could not be regenerated *in vitro.* Plants have been regenerated from haploid protoplasts of *Hyoscyamus muticus* (Wernicke *et al.* 1979).

A total of 40 solanaceous species have been regenerated *in vitro.* Somatic embryogenesis has been observed in only three species, as organogenesis is the primary mode of plant regeneration in this family. In two of these embryogenic species, *P. hybrida* and *A. belladonna*, callus was induced and proliferated in the presence of high concentrations of auxin, with somatic embryos obtained when the auxin was reduced or eliminated. In *N. tabacum*, which normally undergoes organogenesis, embryogenesis is possible only in high light intensity (Haccius and Lakshmanam 1965) or with very specific hormone requirements (Lorz *et al.* 1977; Prabhudesai and Narayanaswamy 1973). MS medium has been used for 41 of the 44 species cultured *in vitro*, although B5 medium may be equally useful. Callus can be induced in most species by using 4.5 μM 2,4-D, and 5 μM 6BA is sufficient for plant regeneration in most species. In those species in which this treatment has not been successful, a combination of auxins and cytokinins generally has been used. Among solanaceous species, if the auxin:cytokinin ratio is >1, callus formation is enhanced; if <1, shoot formation is obtained. This general rule applies to 40 of the 44 species tested (Tables 6.2 through 6.6). The only exceptions are *S. dulcamara, S. nigrum, Scopolia parviflora,* and *P. parviflora*, and in each case IAA, known to degrade *in vitro*, and also not as effective an auxin as 2,4-D or NAA, was used as the auxin in combination with a cytokinin at a lower concentration for shoot induction.

B. Umbelliferae

Species of the Umbelliferae are quite amenable to tissue culture. Callus cultures are easily established and maintained over long periods of time.

TABLE 6.6. PLANT REGENERATION OF MISCELLANEOUS SOLANACEOUS SPECIES

Species	Growth Regulators		Shoot Medium	Explant	Mode of Regeneration	Reference
	Callus	Shoot Formation				
Atropa belladonna	10.7 μM NAA 2.3 μM KIN	2.7 μM NAA 12.4 μM NOA	White, WB	Roots	E	Thomas and Street 1970
A. belladonna (haploid)	10.7 μM NAA 2.3 μM KIN	4.4 μM 6BA	MS	Leaf	E	Eapen *et al.* 1978
A. belladonna	18.5 μM KIN 5.4 μM CPA	0.5–0.9 μM KIN	MS	Leaf protoplasts	E	Lorz and Potrykus 1979
A. belladonna	11.4 μM NAA 0.5 μM KIN	0.5 μM KIN	MS	Cell culture protoplasts	E	Gosch *et al.* 1975
Browallia viscosa	10.7 μM NAA 2.2 μM 6BA	4.6 μM ZEA	MS	Leaf, leaf protoplasts	O	Power and Berry 1979
Capsicum annuum (pepper)	4.5 μM 2,4-D	4.4–8.9 μM 6BA 0–5.7 μM IAA	MS	Cotyledon, hypocotyl	O	Gunay and Rao 1978
Hyoscyamus muticas	21.4 μM CPA 4.4 μM 6BA	1.1 μM 6BA 0.3 μM NAA or 4.6 μM KIN 0.5 μM NAA	MS	Leaf protoplasts	O	Lorz *et al.* 1979
Hyoscyamus niger (henbane)	0.5 μM 2,4-D 0.5 μM KIN	None	MS	Seedlings	O	Dhoot and Henshaw 1977
Physalis minima	4.5 μM 2,4-D	4.4 μM 6BA	White	Stem, leaf	O	Bapat and Rao 1977
Physalis peruviana	Not reported	18.6 μM KIN 11.4 μM IAA	MS	Leaf	O	Zenkteler 1972
Salpiglossis sinuata	0.5 μM KIN 0.5–5.4 μM NAA	9.3 μM KIN 0.5 μM NAA	MS	Leaf, flower petal	O	Lee *et al.* 1977
Scopolia parviflora	10 μM 2,4-D	10 μM IAA 0.1–10 μM KIN	MS	Stem	O	Tabata *et al.* 1972

The first sustained plant cell culture was carrot, *Daucus carota* (Gautheret 1939). Steward (1958) reported the first case of somatic embryogenesis in *D. carota*, and Halperin and Wetherell (1964) described this process in detail. Kato and Takeuchi (1963) claimed to obtain entire plants from single cells of a carrot cell culture, although it is likely that these plants were of multicellular origin. Sussex and Frei (1968) reported the induction of embryogenesis from cultures that had been maintained for over 10 years. Most plants obtained were morphologically normal but sexually sterile (Sussex and Frei 1968). Cultural conditions can be manipulated to induce either organogenesis (Kessell and Carr 1972) or embryogenesis (Jones 1974) from carrot cell cultures. Protoplasts can be isolated, cultured as callus, and induced to undergo somatic embryogenesis (Kameya and Uchimiya 1972). Umbelliferous species other than *D. carota* also have been regenerated successfully from a number of explants (Table 6.7).

Among umbelliferous species the effects of auxins and cytokinins are quite consistent. Callus and suspension cultures are usually cultured in either White's or MS medium. The embryogenic potential of callus cultures is lost in White's medium, but can be retained in MS medium. Callus can be initiated in the presence of 0.5 to 4.5 μM 2,4-D in the absence of cytokinin; higher concentrations of 2,4-D, i.e., 2.3 to 4.5 μM, enhance the cell growth rate and are most often used. Other auxins have been used, both alone and with cytokinins, e.g., 5.7 μM IAA (Sussex 1972) or 16 μM NAA and 0.2 μM KIN (Kessell and Carr 1972). Coconut water, casein hydrolysate, or yeast extract is unnecessary in the culture medium for callus initiation (Halperin and Wetherell 1964). The presence of a cytokinin does not inhibit callus growth, as 0.93 and 2.8 μM KIN have been added to callus cultures (Williams and Collin 1976a; Smith and Street 1974). The source of the explant for initiation of callus cultures of *D. carota* and other umbelliferous species is unimportant. Sections of the root have been used most often, but petiole (Halperin 1966), seedling root (Smith and Street 1974), hypocotyl (Fujimura and Komamine 1979a), protoplasts (Kameya and Uchimiya 1972), stem (Maheshwari and Gupta 1965), and young ovule (Sehgal 1968) explants have been successfully cultured. Suspension cultures have been initiated and maintained in MS (Jones 1974), B5 (Huber *et al.* 1978; Dudits *et al.* 1976), and Eriksson's media (Fridborg and Eriksson 1975), each with 4.5 μM 2,4-D. Long term culture of carrot cells *in vitro* is believed to result in changes in the chromosome number and rearrangements of chromosomes (Smith and Street 1974). Coconut water, 2,4-D, and casein hydrolysate have been implicated in the production of polyploidy in celery (Williams and Collin 1976b). Smith and Street (1974) have correlated increased chromosome number with the loss of embryogenic potential in carrot cell cultures.

TABLE 6.7. PLANT REGENERATION OF UMBELLIFERA SPECIES

Species	Growth Regulators		Shoot Medium	Explant	Mode of Regeneration	Reference
	Callus	Shoot Formation				
Ammi majus	28.5 μM IAA	11.4 μM IAA	MS	Hypocotyl	E	Grewal *et al.* 1976
Anethum graveolens (dill)	10.7 μM NAA	None	MS	Embryo	E	Steward *et al.* 1970
A. graveolens (dill)	2.3 μM 2,4-D 2.3 μM KIN	None	White	Inflorescence	E	Williams and Collin 1976a Zee and Wu 1979
Carum carvi (caraway)	10.7 μM NAA	None	MS	Petiole	E	Ammirato 1974
Conium maculatum (poison hemlock)	10.7 μM NAA	None	MS	Embryo	E	Steward *et al.* 1970
Coriandrum sativum (coriander)	10.7 μM NAA	None	MS	Embryo	E	Steward *et al.* 1970
Daucus carota (carrot)	4.5 μM 2,4-D	None	MS	Storage root	E	Halperin and Wetherell 1964
Foeniculum vulgare (fennel)	27.6 μM 2,4-D ±1 μM KIN	None	Nitsch	Stem	E	Maheshwari and Gupta 1965
Petroselinum hortense (parsley)	27 μM 2,4-D or 0.5 μM NAA	None	Hildebrandt Medium C	Petiole	E	Vasil and Hildebrandt 1966b
Pimpinella anisum (anise)	5 μM 2,4-D	None	B5	Hypocotyl	E	Huber *et al.* 1978
Sium suave (water parsnip)	10.7 μM NAA	None	MS	Embryo	E	Ammirato and Steward 1971

Callus cultures are dependent on the presence of an auxin for rapid cell growth, and reduction of the auxin concentration in the culture medium is necessary for embryogenesis to be expressed. At concentrations above 0.5 μM, 2,4-D strongly inhibits somatic embryogenesis in carrot cell cultures (Fujimura and Komamine 1975). Concentrations as low as 0.01 μM IAA strongly inhibit embryogenesis (Fujimura and Komamine 1979b). Addition of antiauxins also inhibits embryogenesis (Fujimura and Komamine 1979b). Presumably endogenous auxins are important in regulating embryogenesis. It is apparent that high concentrations of cytokinins, e.g., KIN at 0.5 to 47 μM, may inhibit embryogenesis (Fujimura and Komamine 1975), whereas low concentrations of ZEA, e.g., 0.1 μM, may enhance embryogenesis (Fujimura and Komamine 1975). Embryogenesis also can be induced by the removal of coconut water, a cytokinin source, from the culture medium (Sussex 1972). The effects of other plant growth regulators on embryogenesis from *D. carota* cell cultures also have been examined. Abscisic acid (ABA) decreases the frequency of embryos induced in a cell culture. GA_3 arrests embryos in the heart and torpedo stage (Fujimura and Komamine 1975). Ethephon, an ethylene-generating compound, causes a small decrease in the number of embryos produced and chlorosis of embryos (Tisserat and Murashige 1977). Bromodeoxyuridine has been used to reversibly block somatic embryogenesis (Dudits *et al.* 1979a).

Both cell density and aggregate size are important in the induction of embryogenesis. Free single cells in suspension cultures are usually highly vacuolate (Halperin 1966) and have lost their embryogenic potential. Cell clusters of 3 to 10 cells that are densely cytoplasmic are usually embryogenic (Fujimura and Komamine 1979a). Embryogenesis can be synchronized by isolating cell aggregates in this narrow size range (Fujimura and Komamine 1979a). The initial density at which cell aggregates are cultured in embryogenic culture medium is important. *Pimpinella anisum* cells have an optimum cell density of about 10^5 cells/ml for embryogenesis (Huber *et al.* 1978). Smaller embryologically competent aggregates fail to undergo embryogenesis in dilute suspensions, probably because of diffusion of critical metabolites from the cells into the culture medium (Halperin 1967). Embryos that do develop at low cell population densities are smaller than those developing at high cell densities (Halperin 1967).

Replacement of sucrose by other carbohydrates has little effect on embryogenesis, if the carbohydrate will support growth in the absence of 2,4-D (Verma and Dougall 1977). Potassium is involved in the induction of embryogenesis. While 20 mM K^+ is required for embryogenesis, only 1 mM K^+ is necessary for optimal cell growth (Brown *et al.* 1976). The potassium effects on embryogenesis can be separated from nitrogen effects (Brown *et al.* 1976). In some carrot cell lines 10 mM NaCl is inhibitory to embryogenesis (Brown *et al.* 1976). The source of nitrogen in a culture medium, i.e., reduced vs. non-reduced nitrogen, is critical in the induction of embryogenesis. A high concentration of NO_3, e.g., 60 mM (Tazawa and Reinert 1969), permits embryogenesis, but improved embryogenesis is obtained if reduced nitrogen (40 mM KNO_3 and 10 mM NH_4Cl) is also present (Wetherell and Dougall 1976). White's medium does not permit embryogenesis, presumably because it is ammonia-free and has a low concentration of NO_3^-; however, White's medium supplemented with 2.5 mM NH_4NO_3 will permit embryogenesis (Tazawa and Reinert 1969). The NH_4^+ requirement can be partially replaced with 1 mM alanine, 1 mM glutamine, or 10 mM casein hydrolysate. MS medium, which is rich in NH_4^+ and NO_3^-, permits somatic embryogenesis (Tazawa and Reinert 1969).

It is difficult to separate the nitrogen source and pH effects on the induction of embryogenesis in carrot cell cultures. Carrot cells exhibit a narrow optimum pH for embryogenesis at pH 5.4 (Wetherell and Dougall 1976). In the absence of NH_4^+ or the presence of low concentrations of NO_3^-, the medium is not buffered and pH fluctuation may inhibit embryogenesis (Wetherell and Dougall 1976; Dougall and Verma 1978).

Kessell and Carr (1972) first discussed the effects of the dissolved oxygen (DO_2) content of the culture medium on embryogenesis. Below a critical DO_2 concentration (16% saturation), organogenesis, i.e., separate root and shoot formation, is induced. The concentration of ATP in carrot cells increases if the DO_2 concentration falls below 16% (Kessell *et al.* 1977). The addition of adenine or adenosine to cultures with a low DO_2 concentration increases the concentration of ATP in carrot cells, and consequently induces embryogenesis (Kessell *et al.* 1977).

Loss of embryogenic potential by cell cultures long has been noted and related to several phenomena. Growth of suspension cultures at lower than normal temperature in non-inductive medium (partial inhibition of growth) will prolong the length of time during which embryogenesis may be induced in a culture (Meyer-Teuter and Reinert 1973). The loss of embryogenic potential may be related to the number of generations of cell growth, rather than to a period of absolute time (Meyer-Teuter and Reinert 1973). Smith and Street (1974) alleviated the loss of embryogenic potential of carrot cell cultures by less frequent subculturing. Bayliss (1977) found that tetraploid clones were progressively eliminated from a mixture with a diploid line. The elimination of tetraploids was not correlated with embryogenesis in this instance. Carrot cell suspension cultures that apparently have lost their embryogenic potential have been induced to undergo embryogenesis by the addition of 1% or more activated charcoal to the culture medium (Fridborg and Eriksson 1975; Drew 1979) through the absorption of polyphenolic growth regulator inhibitors.

Protoplasts have been isolated enzymatically from carrot root tissues (Kameya and Uchimiya 1972) and carrot cell suspension cultures grown in B5 medium containing 4.5 μM 2,4-D (Dudits *et al.* 1976). Cell wall regeneration and callus proliferation have been achieved in both cases in culture medium containing coconut water and auxin (cytokinin also was present for suspension cultured cells). Embryogenesis was induced when callus was transferred to auxin-free medium. Cells derived from carrot root underwent embryogenesis in the presence of 5% coconut water or 4.7 μM KIN (Kameya and Uchimiya 1972). Dudits *et al.* (1977) regenerated *D. capillifolius* plants from protoplasts as well as somatic hybrids of *D. carota* with *D. capillifolius*. Hybridized cells of *D. carota* and *Aegopodium podagrarea* have been regenerated, although *A. podograrea* alone could not be regenerated (Dudits *et al.* 1979b).

Several other umbelliferous species have been shown to undergo embryogenesis (Table 6.7). Celery (*Apium graveolens* L.) has been studied extensively. It behaves similar to carrot in cell culture (Williams and Collin 1976a; Al-Abta and Collin 1978a,b; Zee and Wu 1979). Additional species that undergo embryogenesis from callus culture are listed in Table 6.7. All were treated similarly to *D. carota*, i.e., 2.3 to 27 μM auxin, for callus culture while the removal of auxin induced embryogenesis. Only *Ammi majus* has been regenerated in the presence of IAA (Grewal *et al.* 1976).

C. Cruciferae

The Cruciferae include many important agricultural crops (e.g., broccoli, cabbage, brussels sprouts, rapeseed, kale, Chinese kale, cauliflower, and horseradish) that have been regenerated *in vitro*. With the exception of horseradish, all of these species belong to the genus *Brassica*. Another cruciferous species, *Arabidopsis thaliana*, has been valuable as a genetic tool (Redei 1975) and is also amenable to tissue culture and plant regeneration. The emphasis on research in the Cruciferae has been on the application of tissue culture to crop improvement and vegetative propagation.

Callus can be induced routinely and maintained from many explants. Shoots can be induced from callus, usually by increasing the cytokinin: auxin ratio. The potential for organogenesis in callus cultures of cruciferous species decreases rapidly with time, particularly after 6 to 8 months *in vitro* (Negrutiu and Jacobs 1978b). Organogenic potential of *Arabid-*

opsis cultures can be increased by manipulation of the culture medium. Protoplasts can be isolated from several cruciferous species, and callus proliferation from protoplast cultures with subsequent shoot and root regeneration has been attained with a few *Brassica* species (Thomas *et al.* 1976; Kartha *et al.* 1974b).

At least eight species of Cruciferae have been regenerated *in vitro* (Tables 6.8 and 6.9). In *Brassica oleracea* (cauliflower) and *Arabidopsis* cellular response to hormones in the culture medium varies among cultivars or geographical races. MS medium is used exclusively for regeneration of cruciferous species (Tables 6.8 and 6.9). The source of the explant used to induce callus can be important in determining the organogenic response of a culture.

1. *Brassica*.—Agriculturally important *Brassica* species include *B. oleracea* (cauliflower, brussels sprouts, cabbage, kale), *B. napus* (rape), and *B. alboglabra* (Chinese kale). Callus cultures of *B. oleracea* are usually initiated in the presence of both an auxin and cytokinin (Table 6.8). Although 2.3 μM KIN is most often used (Clare and Collin 1974; Pareek and Chandra 1978b; Lustinec and Horak 1970; Horak *et al.* 1975), concentrations as low as 0.5 μM (Bajaj and Nietsch 1975) and as high as 14 μM KIN (Baroncelli *et al.* 1973) have been used successfully. Numerous hormones can be used to satisfy the auxin requirement for callus initiation and maintenance, i.e., 2,4-D, IAA, and NAA, at concentrations ranging from 0.9 μM for 2,4-D to 11.4 μM for IAA (Table 6.8). Callus induction of kale and Chinese kale has been promoted by 10.7 μM NAA and 2.2 μM 6BA (Horak *et al.* 1975; Zee and Hui 1977). *Brassica napus* is the only *Brassica* species that requires no cytokinin for callus growth, and has been cultured successfully on 2.3 to 4.5 μM (Stringham 1979; Kartha *et al.* 1974b). Growth responses to 5 μM 6BA in *B. oleracea* (cauliflower) vary among different genotypes (Baroncelli *et al.* 1973).

Plants can be readily regenerated from callus cultures of *Brassica*. Normally a cytokinin, e.g., 2.3 to 93 μM KIN, must be included in the regeneration medium (Table 6.8). Auxin concentration in the regeneration medium can be quite variable. In some cases no auxin is required (Clare and Collin 1974), whereas other species may require that a low concentration of a strongly active auxin e.g., 0.05 to 0.5 μM NAA for cauliflower (Pareek and Chandra 1978b), or a high concentration of a less active auxin, e.g., 5.7 to 11.4 μM IAA for red cabbage (Bajaj and Nietsch 1975), be added to the cytokinin to achieve plant regeneration. *Brassica alboglabra* shoot regeneration has been obtained only in the presence of both 5.4 to 21.4 μM NAA and 2.3 to 4.7 μM KIN (Zee and Hui 1977; Zee *et al.* 1978). Shoot regeneration of *B. napus* is induced by the presence of 5 μM 6BA (Stringham 1979; Thomas *et al.* 1976; Kartha *et al.* 1974b). In

TABLE 6.8. PLANT REGENERATION OF *BRASSICA* SPECIES

Species	Growth Regulators: Callus	Growth Regulators: Shoot Formation	Shoot Medium	Explant	Mode of Regeneration	Reference
B. oleracea (cauliflower)	0.9 μM 2,4-D 14 μM KIN	23.2 μM KIN	MS	Leaf vein, petal	O	Baroncelli *et al.* 1973
	5.7 μM IAA 2.3 μM KIN	0.05–0.5 μM NAA 2.3 μM KIN	MS	Leaf	E	Pareek and Chandra 1978b
(kale)	10.7 μM NAA 2.3 μM KIN	2.3 μM KIN	MS	Stem, pith	O	Horak *et al.* 1975
	4.5 μM 2,4-D 2.3 μM KIN	2.3 μM KIN	MS	Leaf, root	O	Lustinec and Horak 1970
(brussels sprouts)	11.4 μM IAA 2.3 μM KIN 4.9 μM IBA	None	MS	Petiole, stem	O	Clare and Collin 1974
(red cabbage)	4.5 μM 2,4-D 0.5 μM KIN	11.4 μM IAA 9.3 μM KIN or	MS	Seeds, cotyledon	O	Bajaj and Nietsch 1975
		5.7 μM IAA 2.3 μM KIN		Hypocotyl		
(broccoli)	5.7–11.4 μM IAA 4.7 μM KIN	46–51 μM IAA 14–28 μM KIN	MS	Leaf	O	Johnson and Mitchell 1978
	Not reported	51–57 μM IAA 37–42 μM KIN	MS	Leaf rib	O	Johnson and Mitchell 1978
	Not reported	51–57 μM IAA 14–93 μM KIN	MS	Stem	O	Johnson and Mitchell 1978
B. alboglabra (chinese kale)	10.7 μM NAA 2.2 μM BA	21.4 μM NAA 2.3 μM KIN	MS	Cotyledon	O	Zee and Hui 1977
	10.7 μM NAA 2.2 μM 6BA	5.4 μM NAA 4.7 μM KIN	MS	Hypocotyl	O	Zee and Hui 1977
B. napus (haploid rape)	4.5 μM 2,4-D	5 μM 6BA 0.1–0.4 μM GA_3	MS	Leaf	O	Stringham 1979
(diploid rape)	1 μM BA 1 μM NAA 2.3 μM 2,4-D or	5 μM 6BA	B5 (callus) MS (shoots)	Mesophyll Protoplast	O	Kartha *et al.* 1974b
	2.3 μM 2,4-D	0.1 μM GA_3				
B. juncea (leaf-mustard cabbage)	Not reported	2.7 μM NAA 8.9 μM 6BA	MS	Hypocotyl Cotyledon	O	Hui and Zee 1978

TABLE 6.9. REGENERATION OF CRUCIFEROUS SPECIES

Species	Growth Regulators		Shoot Medium	Explant	Mode of Regeneration	Reference
	Callus	Shoot Formation				
Amoracia lapathifolia (horseradish)	Not reported	5.4 μM NAA 0.5–2.3 μM KIN	MS	Leaf	O	Meyer and Milbrath 1977
Arabidopsis thaliana	4.5 μM 2,4-D 0.2 μM KIN	0.2 μM IAA 4.7 μM KIN	B5 or PG_2	Seed, leaf, stem, anther	O	Negrutiu *et al.* 1975
Cheiranthus cheiri	4.5 μM 2,4-D	0.5 μM 2,4-D	MS	Seedling	E	Khanna and Staba 1970
Crambe maritima	11.4 μM IAA 3.7 μM KIN 10% CW	11.4 μM IAA 3.7 μM KIN 10% CW	MS	Root	O	Bowes 1976
Lobularia maritima	10 μM IAA 20 μM 2,4-D	600 μM ADE 5 μM 6BA 0.1–1 μM IAA	MS	Internodal stem	O	Khanna and Chopra 1977
Sinapis alba	4.5 μM 2,4-D	0.9 μM 2,4-D 9.3 μM KIN	MS	Cotyledon, hypocotyl, root	O	Bajaj and Bopp 1972
Sisymbrium irio	5.7 μM IAA 2.3 μM KIN	2.9 μM IAA 14–23 μM KIN	MS	Stem	O	Pareek and Chandra 1978a

contrast to regeneration from most callus cultures, 0.1 to 2.8 μM GA_3 is necessary for plant regeneration. In the absence of GA_3 green callus is induced, but no shoots are formed (Kartha *et al.* 1974b).

2. ***Arabidopsis.***—Callus growth of *Arabidopsis thaliana* is enhanced by addition of a cytokinin and an auxin to the cultured medium as in other *Brassica* species. Callus can be initiated on modified B5 medium (Negrutiu *et al.* 1975) with 10 μM 2,4-D and 0.25 μM KIN and maintained on 5 μM 2,4-D and 0.25 μM KIN (Negrutiu *et al.* 1978b). Callus also has been initiated and maintained on 43 μM NAA and 0.25 μM KIN (Negrutiu *et al.* 1975). The source of the explant has been shown to effect subsequent plant regeneration from callus (Negrutiu 1978b). Callus initiated from anthers has the highest organogenic potential, as well as retaining the capacity to regenerate for as long as 18 months. Seed-, stem-, and leaf-derived callus retain the capacity to regenerate for only 6 to 8 months. Embryoids have been detected only in anther-derived callus.

Regeneration of shoots from *Arabidopsis* callus cultures can be obtained by the removal or reduction of the auxin concentration in the medium with simultaneous increase in the cytokinin concentration. A low concentration of auxin is sometimes retained in the shoot regeneration medium, e.g., 1 μM KIN and 0.1 μM GA_3 (Negrutiu *et al.* 1978a) and 5 μM KIN and 0.2 μM IAA (Negrutiu *et al.* 1978b). GA_3 is sometimes added to shoot regeneration medium, e.g., 10 μM 6BA and 1 μM GA_3 (Negrutiu *et al.* 1978a), although no growth requirement for GA_3 has been demonstrated.

In vitro morphogenesis of *Arabidopsis thaliana* can be influenced by many factors. Shoot induction from callus can be enhanced by the following: (1) selection of a geographic race of this regenerative capacity (Negrutiu 1976; Negrutiu *et al.* 1975); (2) subculturing callus at 4-week rather than 8-week intervals (Negrutiu and Jacobs 1978b); (3) growth of callus in low light (Negrutiu and Jacobs 1978a); (4) filter sterilization of culture medium (Negrutiu and Jacobs 1978a); (5) replacement of NH_4^+ by glutamine as the nitrogen source (Negrutiu and Jacobs 1978a); (6) treatment of cultures for 3 to 6 days at 4°C (Negrutiu and Jacobs 1978b); (7) removal of all auxins from the culture medium prior to transfer to regeneration medium (Negrutiu and Jacobs 1978b); and (8) a 20-day passage of young cultures in glucose-free medium or transfer of old cultures to culture medium containing 6% glucose as high glucose concentrations inhibit morphogenesis in young cultures (Negrutiu and Jacobs 1978b). A progressive decline in shoot regeneration accompanies increasing age of callus. The origin of explants also may affect the longevity of the regenerative ability of a callus culture (Negrutiu *et al.* 1978b). Chromosome instabilities, e.g., increase in ploidy observed with age of callus (Negrutiu *et al.* 1975), may contribute to a decline in organogenesis with callus age.

3. Other Cruciferous Species.—Five other cruciferous species have been regenerated *in vitro* (Table 6.9). Generally, with the exception of *Amoracia lapathiofolia*, callus has been initiated and maintained on MS medium in the presence of an auxin:cytokinin ratio >1. Cytokinin is not always included in callus medium. Regeneration is induced by reduction of the auxin concentration in the medium with an increase in the cytokinin concentration. Shoots were regenerated directly from leaf pieces of *A. lapathiofolia* with 5.4 μM NAA and 0.5 to 2.3 μM KIN (Meyer and Milbrath 1977).

Plants have been regenerated from leaf protoplasts of haploid *Brassica napus* (Thomas *et al.* 1976). Diploid plants of *B. napus* also have been regenerated from leaf mesophyll protoplasts (Kartha *et al.* 1974b). The enzymatic isolation of protoplasts from *Arabidopsis* has been reported, but callus formation and plant regeneration did not occur (Negrutiu *et al.* 1975). Callus resulting from the fusion of *Arabidopsis* and *B. campestris* protoplasts has been maintained in culture for at least 7 months. Chromosomes as well as isoenzymes from both parents are present in this callus (Gleba and Hoffman 1978).

Although cruciferous species are quite amenable to growth and plant regeneration *in vitro*, genotype has considerable influence on the success of callus culture and plant regeneration. In addition, explant origin from which callus is initiated may influence the relative success of plant regeneration (Murashige 1974). Consequently, hormone concentration for callus and shoot culture can vary considerably. Although the potential for regeneration from callus culture can be lost within 6 months of initial culture, specific modifications of the culture technique can prolong regenerative viability. *In vitro* systems are now being utilized for improvement of many cruciferous species (Anderson *et al.* 1977).

D. Leguminosae

Plant regeneration has been quite difficult among the legumes. Forage legumes, e.g., clovers, are more amenable to *in vitro* plant regeneration than are seed legumes (Phillips and Collins 1979). Eighteen legume species have been regenerated *in vitro*, but in most cases regeneration is at low frequency or limited by the source of explants.

At least eight different media have been used successfully to obtain plant regeneration (Table 6.10). Phillips and Collins (1979) have devised a legume tissue culture medium (PC-L2) which, when tested with four different legume genera, resulted in increased callus growth in each case.

TABLE 6.10. REGENERATION OF LEGUME SPECIES

Species	Growth Regulators		Shoot Medium	Explant	Subculture of Callus (Months)	Mode of Regeneration	Reference
	Callus	Shoot					
Acacia koa (koa)	1% CW + 11.3 μM 2,4-D	22 μM 6BA	MS	Root sucker tip	ca. 2	O	Skolmen and Mapes 1976
Cajanus cajan 'T21' (pigeon pea)	Not reported	4.7 μM KIN 0.06 μM IAA 400 mg/liter CH	White	Hypocotyl	1	O	Shama Rao and Narayanaswamy 1975
Crotalaria burhia	1.1 μM 2,4-D 1.3 μM NAA 1.2 μM KIN	0.5–1.3 μM NAA 2.2 μM 6BA	MS	Stem, leaf	6	O	Raj Bhansali *et al.* 1978a
Crotalaria juncea	26.9 μM NAA	23.3–46.5 KIN	MS	Stem, leaf	1.5	O	Ramawat *et al.* 1977
Crotalaria medicagenia	2.3–14.3 μM 2,4-D	2.2–8.9 μM 6BA	MS	Stem, leaf	Not reported	O	Raj Bhansali *et al.* 1978b
Indigofera enneaphylla	2.3 μM 2,4-D 4.7 μM 6BA	2.9 μM IAA 4.7 μM 6BA	B5	Cotyledon, hypocotyl	1–2	O	Bharal and Rashid 1979a
Lathyrus sativus 'LSD-6'	2.3 μM 2,4-D 4.7 μM 6BA	2.9 μM IAA 4.7 μM 6BA	B5	Shoot apex	12	O	Mukhopadhyay and Bhojwani 1978
Medicago sativa (alfalfa)	9 μM 2,4-D 9.3 μM KIN 10.7 μM NAA	None 2 g/liter YE	Blaydes	Hypocotyl	1	O	Bingham *et al.* 1975
M. sativa (alfalfa)	50 μM 2,4-D 5 μM KIN	None	MS	Ovary	1	E	Walker *et al.* 1979

TABLE 6.10. *(Continued)*

Species	Growth Regulators		Shoot Medium	Explant	Subculture of Callus (Months)	Mode of Regeneration	Reference
	Callus	Shoot					
Phaseolus vulgaris 'Bico de Ouro' (pinto bean)	18.6 μM KIN 5.7 μM IAA ¼ bean seed/ml	11.4 μM IAA 5.4 μM NAA 0.9 μM KIN ¼ bean seed/ml	67-V	Leaf	1.5	O	Crocomo *et al.* 1976
Pisum sativum (pea)	10.7 μM NAA 4.7 μM 6BA	22.2 μM 6BA 1.1 μM IAA	MS	Epicotyl	4–6	O	Malmberg 1979
Psophocarpus tetragonolobus (winged bean)	5.4–26.9 μM NAA 0.5 μM KIN	None	Not reported	Hypocotyl, cotyledon	2–3	O	Venketeswaran and Huhtinen 1978
Stylosanthes hamata	9 μM 2,4-D 0.2 μM KIN	14 μM KIN	SH	Radicle, cotyledon	15	O	Scowcroft and Adamson 1976
Trifolium alexandrinum (Berseem clover)	7 μM KIN 5.4 μM NAA	2.3 μM KIN 2.7 μM NAA	MS	Hypocotyl, suspension culture	4.5	O	Mokhtarzadeh and Constantin 1978
Trifolium incarnatum (crimson clover)	10 μM 2,4-D 11 μM NAA 10 μM KIN	11 μM NAA 15 μM ADE	B5, SH	Hypocotyl	1	O	Horvath Beach and Smith 1979
Trifolium pratense (red clover)	0.5 μM 6BA 0.25 μM picloram	0.05–44.4 μM 6BA 0.03 μM picloram	PC-L2	Cotyledon, meristem	2–3	O	Phillips and Collins 1979
Trifolium repens (Ladino clover)	19.6 μM 2,4,5T 0.5 μM KIN	2.3 μM 2,4-D 0.5 μM KIN	BM	Seed	1	O	Oswald *et al.* 1977
Trigonella corniculata *T. foenum-graecum*	2.7 μM NAA 15% CW	2.7 μM NAA 15% CW	MS	Leaf	5	O	Sen and Gupta 1979

The PC-L2 medium, when compared to MS, SH, B5, and Miller's media, has increased concentrations of Ca^{+2}, Mg^{+2}, K^{+2}, and PO_4^{-3}, but is most similar to MS medium. MS medium is the medium most often used for the regeneration of legumes. Based on these media variations, it is likely that the media traditionally used for plant regeneration in other plant families may not be appropriate for legumes.

Unlike most plant families, for legume species few generalizations with regard to the role of plant hormones in regeneration can be made. The auxins most often used for callus induction were 2,4-D, NAA, IAA, 2,4,5-T, and picloram (Table 6.10). A cytokinin, either 6BA or KIN, has been used for callus formation in each of the species except *Acacia koa* (Skolmen and Mapes 1976) and *Trigonella* spp. (Sen and Gupta 1979), in which coconut water was used. In most legume species the auxin:cytokinin ratio is high for callus initiation, but a number of exceptions is evident (Table 6.10). Most legume species require higher concentrations of cytokinins than do other plant families.

Most seed legumes have a higher propensity for root formation than for shoot formation. For most species, the frequency of root initiation is quite high despite the concentrations of auxins and cytokinins. Only root initiation was observed in attempts to obtain plant regeneration for *Psophocarpus tetragonolobus* (winged bean) (Bottino *et al.* 1979), *Glycine max* (soybean) (Evans *et al.* 1976), and *Phaseolus vulgaris* (French bean) (Haddon and Northcote 1976). Root formation occurs prior to shoot regeneration with *Stylosanthes hamata* (Scowcroft and Adamson 1976). Forage legumes will form roots but at a lower frequency than seed legumes. Difficulty in root formation has been reported in both *Trifolium pratense* (Phillips and Collins 1979) and *Lathyrus sativus* (Mukhopadhyay and Bhojwani 1978).

Plant regeneration has been obtained for seed legumes by either reducing the concentration of auxin or increasing the concentration of cytokinin. The hormone concentrations successfully used for callus initiation in seed legumes were usually specific (Table 6.10). Bean seed extract was required for regeneration of *Phaseolus vulgaris* (Crocomo *et al.* 1976), while 5kRad of gamma irradiation was used to obtain plant regeneration in *Cajanus cajan* (Shama Rao and Narayanaswamy 1975). Soybean cannot yet be routinely regenerated *in vitro*, although shoot primordia have been obtained (Kimball and Bingham 1973; Oswald *et al.* 1977).

The hormone requirements are much less specific for forage legumes, e.g., no hormones are required for plant regeneration in alfalfa (Walker *et al.* 1979), whereas a wide range of concentrations of various hormones have resulted in plant regeneration in red clover (see Table 4 in Phillips and Collins 1979). Although meristematic tissue has been used to regenerate plants among legumes, plant regeneration also can be initiated from non-meristematic tissue of a number of species. Plants have been obtained from callus derived from hypocotyl, ovary, cotyledon, meristem, leaf, radicle, seed, shoot apex, roots, and cell suspension cultures (Table 6.10); however, explant source can greatly affect the frequency of plant regeneration. In red clover the frequency of plant regeneration was 1% from cotyledons and 30 to 80% from meristematic tissues (Phillips and Collins 1979).

Initial regeneration frequencies of 12% were obtained from hypocotyl explants of alfalfa (Bingham *et al.* 1975), but this frequency was increased by selecting for plant regeneration. After 2 cycles of selection, the frequency of regeneration was increased in one genetic line of alfalfa from 12 to 67%. In addition, intervarietal differences have been observed in plant regeneration. Of 14 genetic lines of *Pisum sativum* tested for regeneration only 6 could be regenerated after 2 months *in vitro* (Malmberg 1979). Only 2 of 6 lines could be regenerated after 6 months *in vitro*. Intervarietal differences in plant regeneration also were reported for 5 cultivars of red clover (Phillips and Collins 1979) and 9 cultivars of alfalfa (Bingham *et al.* 1975). The report of distinct phenotypic variation among lines and the relative ease with which selection can increase the regeneration frequency suggest that in legumes the ability to regenerate plants is inherited. Selection for regeneration has not yet been tested in other plant families. Malmberg (1979) has suggested that screening a large number of genetic lines may be useful in attempts to achieve plant regeneration.

Shoots can be rooted very rapidly from seed legumes. In nearly all cases roots were obtained in medium with an auxin. One μM IBA *(Acacia)*, 0.6 to 26.9 μM NAA *(Crotalaria)*, and 0.6 μM IAA *(Stylosanthes)* have been used. In most cases when a cytokinin is present in the medium (KIN or 6BA), the auxin:cytokinin ratio has been greater than 10, e.g., 5.7 μM IAA:0.4 μM 6BA (berseem clover); 0.5 μM NAA:0.05 μM KIN *(Psophocarpus)*; and 10.7 μM NAA:0.9 μM 6BA *(Indigofera)*. In *Lathyrus, Medicago, Trifolium pratense*, and other forage legumes, root formation has been more difficult to induce. Phillips and Collins (1979) developed an effective rooting medium for red clover which used reduced minerals and 3-aminopyridine.

Rapidly growing cell suspension cultures capable of plant regeneration have been established from a few legume species. Suspension cultures of alfalfa were initiated in Blaydes medium with 9 μM 2,4-D and 9.3 μM KIN (McCoy and Bingham 1977). After 3 weeks in suspension culture, cells plated onto basal medium produced plants from 96% of the colonies. Regeneration capacity was reduced to 9% after 3 months in culture (McCoy and Bingham 1977). Suspension cultures of berseem clover capable of plant regeneration could be induced in MS medium with 10.7 μM

NAA and 0.95 to 1.0 μM 2iP (Mokhtarzadeh and Constantin 1978). Plating efficiency of the suspension was 4%, with resulting colonies undergoing plant regeneration when placed on shoot medium (Table 6.10).

Plant regeneration has been obtained from protoplasts of only one legume—alfalfa (Kao and Michayluk 1980)—despite the frequent use of legume protoplasts for research in cell genetics and somatic hybridization (Constable 1978). Callus has been established from protoplasts of *Vicia hajastana* (vetch) (Kao and Michayluk 1975); *Vicia narbonensis* (Donn 1978); *Vigna sinensis* (cowpea) (Bharal and Rashid 1979b); *Pisum sativum* (pea) (Gamborg *et al.* 1975); *Phaseolus vulgaris* (bean) (Pelcher *et al.* 1974); and *Glycine max* (soybean) (Kao *et al.* 1971). Nutritional requirements of both protoplasts and single cells have been elucidated for *Vicia hajastana* (Kao and Michayluk 1975).

E. Compositae

Few species of Compositae have been regenerated from callus cultures. Only *Cichorium endiva* (endive) (Vasil and Hildebrandt 1966a), *Lactuca sativa* (lettuce) (Doerschug and Miller 1967), and *Cynara scolymus* (globe artichoke) (Devos *et al.* 1979) have been of agricultural interest.

Tissue culture propagation of *Chrysanthemum morifolium* (chrysanthemum) has been studied extensively (Table 6.11). *Brachycome dichromosomatica* and *Crepis capillaris* have been used for cytogenetic studies of callus culture because of their low chromosome number.

Eleven species of Compositae have been regenerated from *in vitro* callus cultures (Table 6.11). Callus has been initiated from shoot tips of *Chrysanthemum* (Bush *et al.* 1976; Earle and Langhans 1974a,b), mature embryos of *Cichorium endiva* (Vasil and Hildebrandt 1966a), and parts of seedlings of *Stevia rabaudiana* (Yang and Chang 1979), *Lactuca sativa* (Doerschug and Miller 1967), *Cynara scolymus* (Devos *et al.* 1979), and *Crepis capillaris* (Jayakar 1971). Regeneration from callus derived from mature tissue rarely has been attained, e.g., *Brachycome dichromosomatica* leaf sections (Gould 1978, 1979), *Gazania* leaf sections (Landova and Landa 1974), and *Chrysanthemum morifolium* petal segments (Bush *et al.* 1976). Callus has been initiated from leaves of *Crepis capillaris* (Sacristan 1971) and stem of *Centaurea cyanus* (bachelor's buttons) (Torrey 1975), but plants were not regenerated. The isolation and culture of protoplasts have not been reported from any composite species.

Callus cultures are usually initiated and maintained in culture medium with an auxin:cytokinin ratio >1 (Table 6.11). Although 0.5 to 53.7 μM NAA is most commonly used, 3 to 27 μM 2,4-D, 22.8 to 28.5 μM IAA, and 4.9 μM IBA have been used (Table 6.11). In most instances the auxin is

TABLE 6.11. REGENERATION OF COMPOSITAE SPECIES

Species	Growth Regulators		Explant	Shoot Medium	Mode of Regeneration	Reference
	Callus Initiation	Shoot Formation				
Brachycome dichromosomatica	2.7 μM NAA 2.3 μM KIN	21.5 μM NAA 9.3–14 μM KIN	Leaf	Miller B	O	Gould 1978, 1979
Centaurea cyanus (bachelor's buttons)	3 μM 2,4-D	Not reported	Stem	Bonner	O	Torrey 1975
Chrysanthemum morifolium (chrysanthemum)	46.5 μM KIN 5.4 μM NAA	9.3 μM KIN 0.11 μM NAA 29 μM GA_3	Shoot tips, petals	LS or MS	O	Bush *et al*. 1976; Earle and Langhans 1974a,b
Cichorium endiva (endive)	27 μM 2,4-D or 0.5 μM NAA	0.19 μM KIN	Embryo	MS or Hildebrandt D	O	Vasil and Hildebrandt 1966a
Crepis capillaris (hawk's beard)	4.9 μM IBA 4.7 μM KIN	18.6 μM KIN	Hypocotyl	White	O	Jayakar 1971
Cynara scolymus (globe artichoke)	1 μM ZEA 1 μM 2,4-D	1 μM 6BA 0.1 μM 2,4-D	Cotyledon, petiole	MS	O	Devos *et al*. 1979
Gazania splendens	4.5 μM 2,4-D 2.3 μM KIN	11.4 μM IAA 2.3 μM KIN	Leaf	MS	O	Landova and Landa 1974
Lactuca sativa (lettuce)	28.5 μM IAA 2.3 μM KIN	2.3–4.6 μM KIN	Seedling	O	O	Doerschug and Miller 1967
Parthenium hysterophorus	10 μM 2,4-D 1 μM KIN	5 μM 6BA 10 μM IAA	Stem	MS	O	Subramanian and Subba Rao 1980
Pferotheca falconeri	11.4 μM IAA 15% CW	11.4 μM IAA	Roots, hypocotyl, stem, leaf, petiole, cotyledon	MS	O	Mehra and Mehra 1971
Stevia rabaudiana	53.7 μM NAA 9.3 μM KIN	4.4–8.8 μM 6BA	Seed leaflet	MS	O	Yang and Chang 1979
Taraxacum officinale (dandelion)	11.4 μM IAA	1.1 μM NAA 2.8 μM KIN	Secondary roots	White	O	Bowes 1970

combined with 0.57 to 4.7 μM KIN (Table 6.11). Only in *Chrysanthemum* has callus initiation and maintenance in the presence of an auxin: cytokinin ratio <1 been reported; however, callus also may be cultured with 10.7 μM NAA alone (Bush *et al.* 1976; Earle and Langhans 1974a,b).

Plant regeneration is induced by reducing the auxin concentration and/or increasing the concentration of cytokinin. Kinetin is most effective in regeneration media at concentrations from 0.19 to 19 μM. Only in *Stevia rabaudiana* (Yang and Chang 1979) is 6BA used, at 4.4 to 8.8 μM. Regeneration of *Chrysanthemum* callus is enhanced by 29 μM GA_3. Embryogenesis occurs in cultivars of *Cichorium endiva* (Vasil and Hildebrandt 1966a) and *Brachycome dichromosomatica* (Gould 1978). Callus cultures of composite species do not readily lose the ability to regenerate. Regeneration from 3- to 5-year-old *Chrysanthemum* callus (Earle and Langhans 1974b), 4½-year-old *Crepis capillaris* callus (Husemann and Reinert 1976), and 14-month-old *B. dichromosomatica* callus (Gould 1979) has been reported.

Brachycome dichromosomatica has been maintained as a diploid in callus culture for 14 months (Gould 1979). Plants regenerated from this callus were also diploid. The chromosome stability of haploid and diploid *Crepis capillaris* was examined over a 1-year period in callus culture (Sacristan 1971). Cultures that were initially haploid produced more polyploid than diploid cultures. Chromosome rearrangements and aneuploidy were more common in polyploids as well. Regeneration from these callus cultures was not reported.

F. Gesneriaceae

Four species of the horticulturally important family Gesneriaceae have been regenerated *in vitro*. The commercially important species *Saintpaulia ionantha* (African violet) is the only gesneriad species that has been maintained as callus which subsequently regenerated shoots. Shoots have been regenerated from leaf explants of *Sinningia speciosa* (florist's gloxinia) (Johnson 1978a) and *Episcia cupreata* (flame violet) (Johnson 1978b). Inexpensive vegetative propagation has been emphasized for *in vitro* regeneration studies involving gesneriads.

Callus has been initiated from *S. ionantha* (African violet 'Blue Rhapsody') by culturing flower petals, sepals, or ovaries on MS medium with 10.7 μM NAA and 0.89 μM 6BA (Vasquez and Short 1978). Under these conditions organogenesis occurs in light, but not in dark. This callus retains the capacity to regenerate for at least 2 years (Vasquez and Short 1978). Shoots generally have been regenerated on MS medium. The choice of cultivar may not be crucial to success since as many as 15 cultivars of *S. ionantha* have been regenerated successfully under identical conditions (Start and Cumming 1976); however, Cooke (1977) reported that 4 of 17 cultivars of *S. ionantha* tested did not regenerate well *in vitro*. Usually shoots are initiated from explants of callus in the presence of a cytokinin:auxin ratio >1 (Table 6.12). Shoot initiation has been obtained with 6BA at 1.4 to 22 μM or 0.93 μM KIN (Johnson 1978b). Vasquez and Short (1978) used 4.7 μM 6BA and 5.4 μM NAA for shoot proliferation, although 0.54 to 5.4 μM NAA and 1.1 to 11.4 μM IAA also have been used as auxin concentrations (Table 6.12). Shoots are rooted on MS medium without phytohormones.

Saintpaulia ionantha plants regenerated *in vitro* show no leaf or flower abnormalities (Bilkey *et al.* 1978; Cooke 1977; Vasquez and Short 1978). No change in chromosome number (2n=28) was found in *S. ionantha* 'Blue Rhapsody' regenerated *in vitro* (Vasquez and Short 1978). Although protoplast isolation was reported by Bilkey and McCown (1978), no details of the procedure were described.

G. Rubiaceae and Rutaceae

Each of these families contains an agriculturally important genus which can be regenerated readily *in vitro*. In each case regeneration occurs via somatic embryogenesis.

Four species of the genus *Coffea* (Rubiaceae) have been regenerated *in vitro* via somatic embryogenesis (Table 6.13). In all cases, MS or modified MS medium has been used for both callus initiation and subsequent embryogenesis. Callus formation is regulated by the auxin and cytokinin concentrations in the culture medium (Sondahl and Sharp 1979). The induction of callus composed of cells capable of embryogenesis is restricted to leaf explant cultures established on a primary culture medium (callus medium) containing specific concentration ratios of NAA and KIN, NAA and KIN plus undefined growth substances, or 2,4-D and KIN. Embryos develop from cells capable of embryogenesis following subculture onto regeneration medium. A detailed study of cultural conditions and the developmental anatomy of high (HFSE) and low (LFSE) frequency somatic embryogenesis (Sondahl and Sharp 1979) in cultured leaf explants has allowed for the establishment of a protocol for induction of HFSE in *C. arabica, C. canephora, C. congensis*, and *C. dewervei* (Sondahl and Sharp 1977, 1979; Sondahl *et al.* 1979).

The HFSE pattern of development occurs for *C. arabica* 'Bourbon' (Sondahl and Sharp 1977) when leaf explant cultures are established on MS medium with 30 μM thiamine, 550 μM inositol, 117 μM sucrose, 20 μM KIN, and 5 μM 2,4-D, and thereafter subcultured onto regeneration medium during either the second or third transfer on callus medium. The

TABLE 6.12. REGENERATION OF GESNERIACEAE SPECIES

Species	Growth Regulators		Shoot Medium	Explant	Mode of Regeneration	Reference
	Callus	Shoot Formation				
Episcia cupreata (flame violet)	Not reported	1.1 μM IAA 0.9 μM KIN	MS	Leaf	O	Johnson 1978b
Saintpaulia ionantha (African violet)	Not reported	5.4 μM NAA 22 μM 6BA or 11.4 μM IAA 0.4 μM 6BA	MS	Leaf	O	Start and Cumming 1976 Cooke 1977
	Not reported	0.5 μM NAA 2.2 μM 6BA	MS	Petiole	O	Bilkey *et al.* 1978
	10.7 μM NAA 0.9 μM 6BA	2.7 μM NAA 4.4 μM 6BA	MS	Flower petal, sepal, ovary	O	Vasquez and Short 1978
Sinningia speciosa (gloxinia)	Not reported	9.1–10.3 μM IAA 1.4–3.3 μM 6BA	MS	Leaf	O	Johnson 1978a
Streptocarpus nobilis	Not reported	1.6 μM 6BA 0.6 μM IAA	Knop's	Leaf	O	Handro 1977

TABLE 6.13. REGENERATION IN RUBIACEAE AND RUTACEAE SPECIES

Species	Growth Regulators		Shoot Medium	Explant	Mode of Regeneration	Reference
	Callus	Shoot Formation				
Rubiaceae						
Coffea arabica 'Mundo Novo' (coffee)	2 μM KIN 2 μM 2,4-D	2.5 μM KIN 0.5 μM NAA	MS	Leaf	E	Sondahl and Sharp 1977
C. canephora	2 μM KIN 10 μM 2,4-D	2.5 μM KIN 0.5 μM NAA	MS	Leaf	E	Sondahl and Sharp 1979
C. canephora	0.5 μM KIN 0.5 μM 2,4-D	0.5 μM KIN 0.5 μM 2,4-D	MS	Shoot	E	Staritsky 1970
C. congensis	2 μM KIN 4 μM 2,4-D	2.5 μM KIN 0.5 μM NAA	MS	Leaf	E	Sondahl and Sharp 1979
C. dewevrei	10 μM KIN 6 μM 2,4-D	2.5 μM KIN 0.5 μM NAA	MS	Leaf	E	Sondahl and Sharp 1979
Rutaceae						
Citrus madurensis	0.4–4.4 μM 6BA 0.5 μM NAA	None	MS	Stem	O	Grinblat 1972
C. sinensis 'Shamouti' (orange)	0.5 μM KIN 5.7 μM IAA	1 g/liter ME	Murashige Tucker	Ovule	E	Kochba and Spiegel-Roy 1973
C. sinensis	None	None	MT	Protoplasts	E	Vardi *et al.* 1975

regeneration medium contains one-half concentration of MS inorganic salts with 2X KNO_3, 2.5 μM KIN, and 0.5 μM NAA. Initial callus cultures are grown in the absence of illumination at 24° to 28°C while regeneration cultures are grown under illumination at 460 foot-candles provided by incandescent and fluorescent lamps, for a 12-hour photoperiod at 24° to 28°C.

A number of difficulties have been encountered when culturing *Coffea*. Somatic embryogenesis is limited by various factors. (1) Phenolic oxidation may destroy most leaf explants unless 210 μM L-cysteine, an antioxidant, is added to the culture medium (Sondahl and Sharp 1979). (2) Few explants have been tested. Leaf tissue is the most often used explant, but shoot explants also have been cultured successfully (Staritsky 1970). (3) It has been observed that a minimum of 6 to 7 months is required to obtain plant regeneration from leaf explants in coffee species (Sondahl and Sharp 1979). As sexual embryo development from fertilization to germination requires 10 to 11 months in *Coffea*, it is as yet uncertain if this slow pattern of development is inherent to the perennial habitat of *Coffea* or is the result of suboptimal growth conditions (Sondahl and Sharp 1979).

Somatic embryogenesis also has been obtained in numerous Rutaceae species. Unlike *Coffea* species, the explant most often used in Rutaceae is embryonic nucellar tissue. Somatic embryos have been obtained from nucellar tissue of species in 7 Rutaceae genera and up to 17 *Citrus* species (Tisserat *et al.* 1979). When nucellar tissue is cultured to obtain somatic embryos, it has been observed that formation of embryos is independent of hormone concentration. A critical distinction between the experiments with *Coffea* and *Citrus* somatic embryogenesis has been emphasized (Sondahl *et al.* 1980). For *Coffea* leaf tissue, specific growth regulator concentrations must be used to *induce* somatic embryogenesis, whereas for *Citrus*, nucellar cells are predetermined as embryogenic cells and variation of the hormone concentrations is not necessary for the induction of somatic embryogenesis. Callus initiated from ovular tissue, suspension cultures derived from callus, single cells isolated from callus (Button and Botha 1975), and protoplasts isolated from suspension cultures (Vardi *et al.* 1975) can be regenerated *in vitro*. It has been suggested that in *Citrus* high internal concentrations of auxin may limit the frequency of somatic embryogenesis. Consistent with this hypothesis, embryogenesis is inhibited by exogenous application of the auxins IAA and NAA (Kochba and Spiegel-Roy 1977a) and the growth regulator GA_3 (Kochba *et al.* 1978a), and the frequency of embryogenesis is reduced by addition of the cytokinins KIN, 6BA, and 2iP (Kochba and Spiegel-Roy 1977a). On the other hand, somatic embryogenesis was stimulated by inhibitors of auxin synthesis (5-hydroxynitrobenzylbromide

[HNB] and 7 aza-indole [Kochba and Spiegel-Roy 1977a]), inhibitors of GA_3 synthesis (2-chloroethyltrimethyl-ammonium chloride [CCC] and butane dioic acid mono- [2,2-dimethylhydrazide] [Alar] [Kochba *et al.* 1978a]), and other compounds known to interfere with endogenous and exogenous auxins, such as abscisic acid (Kochba *et al.* 1978a), galactose and galactose-yielding sugars, raffinose and lactose (Kochba *et al.* 1978b), and gamma-irradiation (Kochba and Spiegel-Roy 1977b).

H. Scrophulariaceae

Six species of the family Scrophulariaceae have been regenerated successfully *in vitro*. The most extensively studied species, *Antirrhinum majus* (snapdragon), undergoes somatic embryogenesis, while the remaining species regenerate via organogenesis. Soft friable callus with globular embryos was induced from stem explants of *A. majus* on MS medium with 0.5 to 9.1 μM 2,4-D. However, when 6BA was added to this medium, the globular embryos appeared at lower frequency and callus growth was compact (Rao and Harada 1974). The 2,4-D must be removed from the medium to elicit further development of the embryos. When stem explants were cultured on MS medium with 1.2 μM NOA and 10% CW, embryos rapidly developed into complete plants (Sangwan and Harada 1975). Alternatively, shoot formation by somatic organogenesis could be obtained on MS or White medium with 9.3 to 18.6 μM KIN and 11.4 μM IAA (Sangwan and Harada 1975), and shoots could be rooted on MS medium with 5.4 to 134.3 μM NAA.

Torenia fournieri also has been studied extensively (e.g., Bajaj 1972). Under conditions sufficient for somatic embryogenesis in *A. majus* (MS with 4.5 μM 2,4-D with leaf explants), excellent callus proliferation has been obtained, but there was no plantlet regeneration. Shoots were initiated in numerous hormone treatments for both leaf (Bajaj 1972) and internode (Kamada and Harada 1979) explants. Five different cytokinins could be used to initiate shoot formation in *T. fournieri*. In most cases 4.4 μM 6BA, 4.6 μM ZEA, or 7.3 μM 4-phenylurea, was sufficient to initiate shoot formation (Kamada and Harada 1979), whereas numerous combinations of cytokinin and auxin also have been successful (Bajaj 1972). Shoot formation was induced from internode segments when cultured in 0.5 μM NAA with either 3.25 μM SD8339, a cytokinin that has been used with *N. tabacum* (Nitsch *et al.* 1967), or 4.7 μM KIN. Organogenesis of *Torenia* also could be regulated by application of amino acids; shoot formation was observed in cultures with glutamic acid and aspartic acid (Kamada and Harada 1979).

Organogenesis alone was obtained in cultures of *Verbascum thapsus, Digitalis purpurea, Limnophila chinensis*, and *Mazus pumilus*, but few hormone concentrations were investigated for any of these species (Table 6.14).

I. Ranunculaceae

Five species of Ranunculaceae undergo plant regeneration. In most cases regeneration is obtained by somatic embryogenesis. Embryogenesis has been observed in two species of *Nigella* and in *Ranunculus sceleratus*. The conditioning medium, used to initiate callus formation prior to plant regeneration, appears to be critical for *N. damascema* (Raman and Greyson 1974), as 2,4-D and CW must be present in liquid medium. Regeneration medium resulting in embryo formation, on the other hand, can contain either auxin (2,4-D) or cytokinin (KIN), as embryos can undergo normal development with 2,4-D present in the medium. As with carrot cultures, White medium could not be substituted for MS medium for embryogenesis. Similar results were obtained with *Nigella sativa*, in which only organogenesis was obtained from leaf explants in White medium (Banerjee and Gupta 1975a) while somatic embryogenesis was obtained in MS medium (Banerjee and Gupta 1975b). Up to 50% of root or leaf-derived somatic embryos developed into complete plants. A single leaf callus could yield 37 plants. Callus from leaf sections of *N. sativa* was obtained on White medium with 15% CW and 2.8 μM NAA. Embryogenesis was obtained when leaf callus was grown on MS with 100 to 500 mg/liter CH and 2.9 μM IAA. Both KIN and 2,4-D inhibited embryo formation. Numerous embryoids were obtained with 1000 mg/liter CH, but these failed to develop into complete plants. *Ranunculus sceleratus* somatic embryos could be obtained from numerous explants including mesophyll protoplasts (Dorion *et al.* 1975). *Ranunculus sceleratus* embryos also could be obtained from flower buds on modified White medium supplemented with CW and IAA (Konar *et al.* 1972). No other species of Ranunculaceae has been studied as extensively *in vitro*. Plants have been obtained from two other species, *Consolida orientalis* and *Coptis japonica* (Table 6.15), by organogenesis. In both *Consolida* and *N. sativa* only organogenesis was obtained in White medium with CW and IAA. Unfortunately, conditions which result in somatic embryogenesis in *N. sativa* have not been tested for either *C. orientalis* or *C. japonica*. *In vitro* regeneration of *Delphinium brunonianum* and *Clematis gouriana* has been unsuccessful (Nataraja 1971).

J. Miscellaneous Dicots

Plants from at least 30 additional dicotyledonous families including at least 39 additional species have been regenerated *in vitro* (Table 6.16).

TABLE 6.14. REGENERATION OF SCROPHULARIACEAE SPECIES

Species	Growth Regulators: Callus	Growth Regulators: Shoot Formation	Shoot Medium	Explant	Mode of Regeneration	Reference
Antirrhinum majus 'Kymosy blanc' (snapdragon)	4.5 μM 2,4-D	1.2 μM NOA 10% CW	MS	Leaf, stem	E	Sangwan and Harada 1975
Digitalis purpurea (foxglove)	4.5 μM 2,4-D 0.5 μM KIN	0.6 μM IAA 4.7 μM KIN	MS	Seedling	O	Hirotani and Furuya 1977
Limnophila chinensis	2.3 μM KIN	4.7–9.3 μM KIN	MS	Stem	O	Sangwan *et al.* 1976
Mazus pumilus	0.9 μM KIN 1.1 μM IAA	0.9–2.3 μM KIN	MS	Floral internodes	O	Raste and Ganapathy 1970
Torenia fournieri	4.5 μM 2,4-D	5.7 μM IAA 9.3 μM KIN	MS	Leaf	O	Bajaj 1972
Verbascum thapsus (mullein)	Not reported	None	MS	Stem	O	Caruso 1971

TABLE 6.15. PLANT REGENERATION IN RANUNCULACEAE SPECIES

Species	Growth Regulators		Shoot Medium	Explant	Mode of Regeneration	Reference
	Callus	Shoot Formation				
Consolida orientalis	9.1–22.6 μM 2,4-D 10% CW	4.5 μM 2,4-D 10% CW	White	Flower bud	O	Nataraja 1971
Coptis japonica (goldthread)	4.5 μM 2,4-D 0.5 μM KIN	None	MS	Petiole	O	Syono and Furuya 1972b
Nigella damascema (fennell flower)	9.1 μM 2,4-D 10% CW	9.1 μM 2,4-D	MS	Flower meristem, bud, pedicel	E	Raman and Greyson 1974
N. sativa	2.7 μM NAA 15% CW	2.9 μM IAA 500 mg/liter CH	MS	Leaf, root, stem	E	Banerjee and Gupta 1976
Ranunculus sceleratus (cursed crowfoot)	5.7 μM IAA 10% CW	5.7 μM IAA 10% CW	White	Stem	E	Konar *et al.* 1972
R. sceleratus	4.4 μM 6BA 16.1 μM NAA	None	MS	Mesophyll protoplasts	E	Dorion *et al.* 1975

TABLE 6.16. REGENERATION IN MISCELLANEOUS DICOTYLEDONOUS SPECIES

Family	Species	Growth Regulators: Callus	Growth Regulators: Shoot Formation	Medium	Explant	Mode of Regeneration	Reference
Aquifoliaceae	*Ilex aquifolium* (holly)	Not reported	None	LS	Cotyledon	E	Hu and Sussex 1971
Araliaceae	*Hedera helix* (English ivy)	21.8 μM NAA 9.3 μM KIN	5.4 μM NAA 2.3 μM KIN 200 mg/liter CH	MS	Stem	O	Banks 1979
Asclepiaceae	*Pergularia minor*	9.1 μM 2,4-D 10% CW	0.6 μM IAA	MS	Stem	E	Prabhudesai and Narayanaswamy 1974
	Tylophora indica	4.5 μM 2,4-D 5.2 μM BTOA	0–0.5 μM 2,4-D	White	Stem	E	Rao and Narayanaswamy 1972
Begoniaceae	*Begonia* × *chiemantha* 'Astrid'	Not reported	0.5 μM NAA 2.2–4.4 μM 6BA	White	Petiole	O	Fonnesbech 1974
Betulaceae	*Corylus avellana* (hazel)	4.5 μM 2,4-D 4.7 μM KIN	None	MS	Embryo	E	Radojevic *et al.* 1975
Cactaceae	*Mammillaria woodsii* (barrel cactus)	Not reported	9.8 μM IBA 9.3 μM KIN	MS	Stem	O	Kolar *et al.* 1976
Caricaceae	*Carica papaya* (papaya)	1 μM NAA 10 μM 2iP	0.1 μM NAA 0.01 μM 6BA	White	Petiole	E	de Bruijne *et al.* 1974
	Carica stipulata	1 μM NAA 2 μM 6BA	1 μM NAA 2 μM BA 1% charcoal	MS MS	Peduncle	E	Litz and Conover 1980
Chenopodiaceae	*Beta vulgaris* (sugar beet)	5.7 μM IAA 0.5 μM KIN	4.7 μM KIN 0.5 μM GA_3	PBO	Leaf	O	de Greef and Jacobs 1979
Convolvulaceae	*Convolvulus arvensis*	0.2 μM 2,4-D 15% CW	4.7 μM KIN 15% CW	LS	Stem	O	Hill 1967
	Ipomoea batatas (sweet potato)	5.4 μM NAA	3.8 μM ABA 0.1 μM KIN 0.2 μM 2,4-D	White	Root tuber	O	Yamaguchi and Nakajima 1974

Crassulaceae	*Sedum telephium* (stonecrop)	4.5 μM 2,4-D	44.4 μM 6BA	B5	Immature leaf	O	Brandao and Selema 1977
Cucurbitaceae	*Cucurbita pepo* (pumpkin)	4.9 μM IBA	1.4 μM 2,4-D 10% watermelon sap	MS	Cotyledon, hypocotyl	E	Jelaska 1972, 1974
Ebenaceae	*Diospyros kaki* (Japanese persimmon)	16.1 μM NAA 0.5 μM KIN	5.4 μM NAA 0.5–4.7 μM KIN	MS	Immature embryo	O	Yokoyama and Takeuchi 1976
Euphorbiaceae	*Euphorbia pulcherrima* (poinsettia)	20 μM NAA 2 μM KIN	10–50 μM 2iP 0.5 μM NAA	MS	Internode, petiole	O	de Langhe *et al.* 1974
	Manihot esculenta (cassava)	0.4 μM 6BA 1.1 μM NAA 0.9 μM GA_3	0.4 μM 6BA 1.1 μM NAA	MS	Stem	O	Tilquin 1979
	Putranjiva roxburghii	12 μM IAA 24 μM KIN	12 μM IAA 24 μM KIN	White	Endosperm	O	Srivastava 1973
Geraniaceae	*Pelargonium* spp. (geranium)	5.4 μM NAA	0.5 μM NAA	MS	Stem, petiole root	O	Skirvin and Janick 1976
Hydrophyllaceae	*Phlox drummondii*	4.5 μM 2,4-D 10% CW	4.5 μM IAA 10% CW	White	Flower bud	O	Konar and Konar 1966
Linaceae	*Linum usitatissimum* (flax)	4.5 μM 2,4-D	10 μM 6BA	MS	Hypocotyl	O	Gamborg and Shyluk 1976
Loranthaceae	*Nuytsia floribunda*	24.6 μM IBA 23.2 μM KIN 2 g/liter CH	24.6 μM IBA 23.2 μM KIN 2 g/liter CH	White	Embryo	E	Nag and Johri 1969
Malvaceae	*Gossypium klotzschianum* (wild cotton)	0.5 μM 2,4-D	11.4 μM IAA 4.7 μM KIN	MS	Hypocotyl	E	Price and Smith 1979
Moraceae	*Broussonetia kazinoki* (paper mulberry)	Not reported	0.5 μM 6BA	MS	Hypocotyl	O	Ohyama and Oka 1978

TABLE 6.16. *(Continued)*

Family	Species	Growth Regulators: Callus	Growth Regulators: Shoot Formation	Medium	Explant	Mode of Regeneration	Reference
Papaveraceae	*Macleaya cordata* (plume-poppy)	5.0 μM 2,4-D 5.0 μM KIN	None	BM	Leaf	E	Kohlenbach 1978
Passifloraceae	*Passiflora suberosa* (passion-flower)	4.4 μM 6BA 5.4 μM NAA	4.4 μM 6BA 5.4 μM NAA	MS	Leaf	O	Scorza and Janick 1976
Piperaceae	*Peperomia* 'Red Ripple'	11.6–46.5 μM KIN 26.9 μM NAA	11.6–23.2 μM KIN 2.7 μM NAA	SH	Leaf	O	Henry 1978
Polygonaceae	*Fagopyrum esculentum* (buckwheat)	22.6–45.2 μM 2,4-D	3 g/liter YE	MS	Cotyledon, hypocotyl	O	Yamane 1974
Portulacaceae	*Mesembryanthemum floribundum* (mid-day flower)	11.4 μM IAA 20% CW	None	MS	Root, stem hypocotyl	O	Mehra and Mehra 1972
Primulaceae	*Anagallis arvensis* (pimpernel)	4.5 μM 2,4-D 0.5 μM KIN	2.9 μM IAA 4.9 μM 2iP 3 g/liter CH	MS	Stem, leaf, hypocotyl	O	Bajaj and Mader 1974
	Cyclamen persicum	53.7 μM NAA 11.6 μM KIN	None	White	Corn	O	Loewenberg 1969
Rosaceae	*Rosa hybrida* (rose)	Not reported	8.9 μM 6BA 0.5 μM NAA	MS	Shoot tip	O	Skirvin and Chu 1979
	Prunus amygdalis (almond)	26.9 μM NAA 10% CW	26.9 μM NAA 2.3–4.7 μM KIN	MS	Leaf, cotyledon, embryo	O	Mehra and Mehra 1974
	Rubus sp. (blackberry)	Not reported	0.4 μM 6BA 0.3 μM GA_3 4.9 μM IBA	MS	Shoot tip	O	Broome and Zimmerman 1978
Saliacaceae	*Populus tremuloides* (2n, quaking aspen)	0.9–9.1 μM 2,4-D	0.4–1.3 μM 6BA	Wolter & Skoog	Stem	O	Wolter 1968

	Populus tremuloides (3n, quaking aspen)	0.2 μM 2,4-D 4.7 μM KIN	0.7 μM 6BA	Wolter & Skoog	Root, stem	O	Winton 1968, 1970
	Populus heterophylla (black cottonwood)	5.7 μM IAA 0.9 μM KIN	0.7 μM 6BA	Wolters	Flower	O	Bawa and Stettler 1972
Santalaceae	*Santalum album* (sandalwood)	9.1 μM 2,4-D 23.2 μM KIN	None	White	Embryo	E	Rao 1965
	S. album	4.5 μM 2,4-D 0.9–2.3 μM KIN	1.5–5.8 μM GA_3	MS, White	Stem, shoot tip	E	Lakshmi Sita *et al.* 1979
Sterculiaceae	*Theobroma cacao* (cacao)	None 10% CW	6.4 μM NAA 10% CW	MS	Immature embryo, cotyledon	E	Pence *et al.* 1979
Vitaceae	*Vitis* sp. (grapes)	4.5 μM 2,4-D 0.4 μM 6BA	10.7 μM NAA 0.4 μM 6BA	MS	Flower, leaf	E	Krul and Worley 1977
Zingiberaceae	*Zingiber officinale* (ginger)	Not reported	4.4 μM 6BA	MS	Rhizome	E	Hosoki and Sagawa 1977

Most of these species examined are economically important. Regeneration was achieved by organogenesis in most species, although at least nine species, including two species of asclepiads, undergo somatic embryogenesis. Among species capable of plant regeneration, MS medium is most often used (24 species), although White medium has been used for the culture of 8 species in 8 different families. Five additional media were used for plant regeneration. Medium variation has not been studied extensively or demonstrated to be family-specific for plant regeneration. In Convolvulaceae, Asclepiaceae, and Primulaceae, one species of each family was regenerated in White medium and one species in MS medium (Table 6.17). Eight of the eleven species capable of somatic embryogenesis were cultured on MS medium. These eleven species are difficult to compare, as at least seven different types of explants were used. Callus was initiated with auxin, in most cases 0.5 to 9.1 μM 2,4-D (6 of 9 cases, Table 6.16), or in the case of pumpkin, 4.9 μM IBA, suggesting that embryogenesis occurred via induced embryogenic determined cells (Sondahl *et al.* 1980). In most cases, auxin present in the callus medium was reduced, eliminated, or substituted with a weaker auxin in regeneration medium in order to achieve somatic embryogenesis (Table 6.16).

Shoot regeneration via organogenesis was achieved in 23 of 29 species by subculturing onto medium with a higher cytokinin:auxin ratio. In four additional species the cytokinin:auxin ratio was held constant to achieve regeneration. Up to 13 different explants have been used to achieve organogenesis. In general, the frequency of regeneration has not been published, but of the reported cases, the frequency ranged from 8% for leaf explants of *Prunus amygdalus* (Mehra and Mehra 1974) to 80% for cotyledons of *Theobroma cacao* (Pence *et al.* 1979) or 100% from embryos of *Nuytsia floribunda* (Nag and Johri 1969). A maximum of 11 to 50 shoots per leaf explant and 20 to 40 shoots per shoot tip were reported for *Peperomia* sp. and *Rubus* sp., respectively (Henry 1978; Broome and Zimmerman 1978).

Few angiosperm trees of economic importance have been regenerated *in vitro* from cell cultures. In most cases, an explant from juvenile tissue is used, e.g., embryo, hypocotyl, cotyledon, root sprout, stem segments, flower bud primordia. In only two instances has mature tissue been used. Wolter (1968) regenerated plants from 7-year-old callus initiated from the cambial region of the stem of diploid *Populus tremuloides* (quaking aspen). Mehra and Mehra (1974) regenerated diploid plants from *Prunus amygdalus* (almond) when stem and leaf sections were used as explants. In both of these instances, regeneration was by somatic organogenesis. *Santalum album* (sandalwood) has been regenerated from callus initiated from shoot stem segments and shoot tip callus of trees over 20 years old (Lakshmi Sita *et al.* 1979). Embryogenesis from shoot stem

TABLE 6.17. PLANT REGENERATION AMONG SPECIES OF THE SUBFAMILY POACOIDEAE (GRASSES)

Species	Growth Regulators		Shoot Medium	Subculture (months)	Explant	Mode of Regeneration	Reference
	Callus	Shoot					
Andropogon gerardii (big bluestem)	2.26 μM 2,4-D, 0.9 μM KIN	Less than 4.5 μM 2,4-D	MS	15	Inflorescence	O	Chen *et al.* 1977
Bromus inermus (bromegrass)	4.5 μM 2,4-D	None	B5	24–48	Mesocotyl	E	Gamborg *et al.* 1970
Dactylis glomerata (orchard grass)	6.78 μM 2,4-D 1.0 μM KIN	4.5 μM 2,4-D 1.0 μM KIN	SH	2	Caryopses (explant-down)	O	Conger and Carabia 1978
Festuca arundinacea (tall fescue)	40.7 μM 2,4-D	2.3 μM 2,4-D	MS	2–4	Embryo	O	Lowe and Conger 1979
Lolium multiflorum *Lolium perenne* (ryegrass)	37.1 μM IAA 1.0 μM KIN 6.8 μM 2,4-D	18.6 μM IAA 0.5 μM KIN 3.4 μM 2,4-D	MS	18	Plumule, embryo	O	Ahloowalia 1975
Phragmites communis (Indian grass)	4.5 μM 2,4-D	None	MS	1	Stem	O	Sangwan and Gorenflot 1975
Saccharum spp. (sugarcane)	13.6 μM 2,4-D	0 or 26.9 μM NAA	MS	2–5	Leaf, inflorescence	O	Heinz and Mee 1969 Nadar and Heinz 1977
Sorghastrum nutans (Indian grass)	22.6 μM 2,4-D 0.9 μM KIN	Less than 4.5 μM 2,4-D	MS	3–10	Inflorescence	O	Chen *et al.* 1979
Hybrids							
Lolium multiflorum × *Festuca arundinacea*	9–18 μM 2,4-D	1.1 μM 2,4-D	MS	2–5	Stem, peduncle	O	Kasperbauer *et al.* 1979

segments was induced by 1.5 to 5.8 μM GA_3. Plantlets developed when embryoids were transferred to hormone-free culture medium. Rao (1965) also reported development of embryoids from *S. album* callus when transferred to hormone-free medium. Of the remaining dicot species representing numerous plant families (Table 6.17), both organogenesis and embryogenesis have been observed. In most cases these *in vitro* processes are achieved using similar protocols as for most previously reported dicot species.

K. Graminaceae

In vitro plant regeneration can be induced in species of numerous families of dicotyledons, but monocotyledonous species have been more difficult to culture. This is unfortunate as graminaceous species represent one of the most important sources of nutrition. Because of the large number of agriculturally important species, most investigators using graminaceous species for studies on plant regeneration have restricted themselves to the cultivated crops. Comparisons of regenerative capacity between cultivated crops and wild relatives have not been possible.

Species within two subfamilies, Poacoideae (grasses) and Panicoideae (cereals), have been cultivated *in vitro* with limited success. Meristematic tissues are generally used to initiate callus cultures capable of plant regeneration. Plant regeneration from p-calli (callus regenerated from protoplasts) has not been obtained in any Graminaceae. It has been suggested that regenerated plants have been derived solely from preorganized structures (Thomas and Wernicke 1978). There has been evidence that plants regenerated from tissue culture may be useful in crop improvement in both corn (Gengenbach *et al.* 1977) and sugarcane (Heinz *et al.* 1977).

1. Poacoideae (Grasses).—*In vitro* propagation may prove to be particularly useful among the grasses. These species are often propagated vegetatively in nature, and may be both polyploid and uniquely variant in chromosome number (Brown 1972). Polyploidy and chromosome instability exist in *Saccharum* spp. ($2n=10X=80-120$) permitting genotypes to withstand changes in chromosome number. These instabilities may be useful for the initiation of genetic variation in vegetatively propagated grasses. Chromosomal variation has been observed in callus-derived plants of both sugarcane (Heinz and Mee 1969) and Italian ryegrass (Ahloowalia 1975), and those variants have been used successfully in breeding programs to select for disease resistance (Heinz *et al.* 1977) and other agronomic traits (Ahloowalia 1976).

Sugarcane is certainly the most malleable of the graminaceous species examined *in vitro*. Although the immature inflorescence is the most successful explant, various explants may be used for plant regeneration, including apical meristems, young leaves, and pith parenchyma (Liu and Chen 1976). As plants have been regenerated from long term callus cultures in sugarcane, it appears that plants may arise from completely unorganized tissue (Nadar and Heinz 1977). Some evidence exists that plants obtained *in vitro* may originate from single cells via somatic embryogenesis (Nadar *et al.* 1978). MS medium, when supplemented with 2 to 13.6 μM 2,4-D, has been used for callus initiation. Plants could be obtained when the 2,4-D was removed from the medium. Cytokinin (9.3 μM KIN) has been present in some regeneration media used for sugarcane (e.g., Heinz *et al.* 1977), but may be unnecessary (Liu *et al.* 1972). Shoots, when obtained in sugarcane, reportedly have been difficult to root (Heinz *et al.* 1977); however, young shoots, when separated from growing callus and placed on hormoneless medium or medium with IAA, form roots in a high percentage of cultures.

MS medium has been used to culture most other grasses except *Bromus inermis* (B5 medium) and *Dactylis glomerata* (SH medium). For *D. glomerata*, growth in SH medium, a medium devised specifically for monocots, was greater than on either B5 or MS medium (Conger and Carabia 1978).

Explant source may limit callus initiation and proliferation. In most cases young organized tissue such as immature inflorescences, embryos, caryopses, peduncle, or mesocotyl has been used to initiate callus. High 2,4-D or 2,4-D with KIN was used to initiate callus formation from these complex explants (Table 6.17). Removal or reduction of 2,4-D results in shoot formation in most grass species (Table 6.17). Shoot formation probably originates from unorganized tissue in at least some grass species as plants can be regenerated via somatic embryogenesis from 2-year-old suspension cultures of bromegrass (Gamborg *et al.* 1970) and callus cultures of sugarcane (Nader and Heinz 1977). In addition, plants can be obtained from callus cultures of some species that are over 1 year old (Table 6.17). The frequency of plant regeneration has not been reported for most grass species; nonetheless, plant regeneration in the reported species varied from 3.1% for *Festuca* (Lowe and Conger 1979) to 66% for *Phragmites* (Sangwan and Gorenflot 1975). Large numbers of shoots per explant have been reported for ryegrass (20 to 50 plantlets per culture, Ahloowalia 1975) and Indian grass (5 to 20 plantlets per culture, Chen *et al.* 1979).

Somatic embryos have been obtained when suspension cultures of bromegrass, normally grown in B5 medium with 2.3 μM 2,4-D, are filtered to remove large aggregates and cultured in liquid B5 medium without hormones. All plants obtained via embryogenesis from these suspension cultures were albino (Gamborg *et al.* 1970). Albinos also have been recovered in tissue cultures of other grass species, e.g., sugarcane (Evans and Crocomo, unpublished information), following plant regeneration from callus. This phenotypic variation may reflect an underlying variation in chromosome number. Root formation has been initiated when shoots were transferred to fresh medium with no hormones (Kasperbauer *et al.* 1979; Chen *et al.* 1979), with reduced mineral concentration (Conger and Carabia 1978; Ahloowalia 1975; Lowe and Conger 1979), or with a high auxin:cytokinin ratio (Sangwan and Gorenflot 1975).

2. Panicoideae (Cereals).—Emphasis has been placed exclusively on agriculturally important species in studies of plant regeneration in the cereals. Nonetheless, plant regeneration has been reported to occur in most cultivated cereal species, but probably involved existing meristematic centers (Rice *et al.* 1978; King *et al.* 1978). Extensive research has been carried out with *Zea mays* (corn), *Triticum aestivum* (wheat), and *Oryza sativa* (rice). Corn is certainly the most recalcitrant of these species, as plants have been obtained only from organized explants.

a. Zea mays.—Callus can be initiated from a number of explants but the ability to regenerate plants has been limited (Green 1977). Immature embryos have been used most often for initiation of callus capable of plant regeneration (e.g., Green and Phillips 1975). MS medium has been used in all cases resulting in a totipotent callus. While 2,4-D has been the sole hormone used for callus initiation, the concentration has been varied from 2.3 (Rice *et al.* 1978) to 67.8 μM (Harms *et al.* 1976). Callus proliferation can be enhanced by the addition of 21.5 μM NAA and 0.25 μM 2iP to 4.5 μM 2,4-D (Green and Phillips 1975). The scutellum of the immature embryo must be oriented upwards when immature embryos are cultured (Green 1977) and downwards when young seedlings are cultured (Harms *et al.* 1976) for maximum callus induction. Plant regeneration has been obtained by removing the 2,4-D (Green and Phillips 1975; Harms *et al.* 1976; Freeling *et al.* 1976; Rice *et al.* 1978). King *et al.* (1978) have suggested that reports of plant regeneration from cultured corn explants and most other cereals represent repression of shoot primordia during callus initiation followed by derepression of preexisting primordia when the 2,4-D is removed. Nonetheless the capacity for plant regeneration has been maintained following subculture of scutellar-derived callus for 19 to 20 months (Green and Phillips 1975; Freeling *et al.* 1976). At least 16 *Zea mays* cultivars have been tested for *in vitro* regenerative capacity with varied results. Only 3 of these 16 genetic lines have resulted in consistent plant regeneration: lines A188 and A188 X R-navajo (Green 1977) and cultivar 'Prior' (Harms *et al.* 1976). Green (1977) has concluded that regeneration in corn is genotype-dependent.

b. Triticum aestivum.—Cultivated wheat has been investigated in greater detail than has corn. Of 15 cultivars examined in 6 different laboratories, plant regeneration has occurred in only 5 cultivars. Callus proliferation has been classified extensively for both wild and cultivated species of *Triticum*. Gosch-Wackerle *et al.* (1979) initiated callus from the rachis and embryos of 7 species of *Triticum*, including diploid (*T. monococcum, T. longissimum, T. speltoides,* and *T. tauschii*, 2n=14), tetraploid (*T. timephevii* and *T. turgidum*, 2n=28), and hexaploid (*T. aestivum*, 2n=42) species. T-medium (after Dudits *et al.* 1975) with 4.5 μM 2,4-D was used to initiate callus from embryos. *Triticum tauschii* formed the least amount of callus on these media. Results from other laboratories collaborate the ability to initiate callus from these 7 species (Shimada *et al.* 1969; Prokhorov *et al.* 1974). In addition, callus has been induced from an aneuploid series of ditelosomics (Gosch-Wackerle *et al.* 1979). Only one of seven lines, ditelo $1A^L$, has a callus growth rate less than that of normal wheat. Plant regeneration has been obtained with only two *Triticum* species (Table 6.18). Unlike *Zea, Triticum* plant regeneration is auxin-dependent. IAA (Gosch-Wackerle *et al.* 1979; Dudits *et al.* 1975), NAA (Chin and Scott 1977), or CPA (O'Hara and Street 1978) has been used for shoot formation; reducing the concentration of 2,4-D is also effective (Shimada 1978). The cultivar 'Chinese Spring' has been used widely but regeneration also has been achieved with cultivars 'Salmon', 'Maris Ranger', 'Mengavi', and 'Tobari 66' (Table 6.18). Plant regeneration has been observed from callus derived from different explants including rachises, shoots, seeds, and embryos (Table 6.18). The greatest frequency of regeneration (45 to 68%) was obtained from embryo-derived callus that had been initiated on MS medium with 9 μM 2,4-D and was subcultured after the first or second transfer on medium with 4.6 μM ZEA and 5.7 μM IAA (Gosch-Wackerle *et al.* 1979). The regenerative callus has rapidly lost its organogenic capability.

c. Oryza sativa.—Rice is perhaps the easiest cereal species to regenerate *in vitro*. Callus can be obtained from numerous young explants of rice using MS medium with 9 to 45.2 μM 2,4-D (Davoyan and Smetanin 1979). Shoot regeneration has been obtained from callus derived from seeds (Nishi *et al.* 1968), endosperm (Nakano *et al.* 1975), 3- to 4-day-old roots and leaves (Henke *et al.* 1978), 2-week-old roots and shoots (Nishi *et al.* 1973), immature and mature embryos, root tips, scutellum, plumule, stem, and panicle (Davoyan and Smetanin 1979). In each case shoots can be obtained following removal of 2,4-D from the medium. Although a cytokinin or auxin is unnecessary for shoot formation, it can be enhanced by addition of 0.3 to 19.7 μM 2iP (Henke *et al.* 1978) or 79.9 μM IAA and 9.3 μM KIN (Davoyan and Smetanin 1979). The frequency of

TABLE 6.18. PLANT REGENERATION IN CEREALS

Species	Growth Regulators		Shoot Medium	Subculture Duration (months)	Explant	Frequency	Mode of Regeneration	Reference
	Callus	Shoots						
Avena sativa (oats) 12 cultivars	2.3–13.6 μM 2,4-D	None	B5	18	Immature embryo	Not reported	O	Cummings *et al.* 1976
A. sativa 'Tiger'	9.1 μM 2,4-D 10.7 μM CPA	30 μM NAA 10 μM IAA 1.5 μM 6BA	SH	12	Hypocotyl	57%	O	Lorz *et al.* 1976
Eleusine coracana (finger millet)	45.2 μM 2,4-D 10–15% CW	1.1 μM NAA or 1.1 μM IAA	MS	3	Mesocotyl	80%	O	Rangan 1976
Hordeum vulgare 'Himalaya' (barley)	10 μM IAA 15 μM 2,4-D 1.5 μM 2iP	None	MS	3	Apical meristem	60%	O	Cheng and Smith 1975
H. vulgare 'Akka'	4.5 μM 2,4-D	None	B5	2	Immature embryo	50%	O	Dale and Deambrogio 1979
Oryza sativa 'Kyote Ashi' (rice)	10 μM 2,4-D	None	MS	2	Root	64–91%	O	Nishi *et al.* 1968
O. sativa CI 8970-S	9–27.1 μM 2,4-D	0–9.8 μM 2iP	MS	4	Roots, leaves	61%	O	Henke *et al.* 1978
O. sativa 'Krasnodarskii'	9–18.1 μM 2,4-D	79.9 μM IAA 9.3 μM KIN	MS	2–3	Immature embryo, root, stem, plumule	68%	O	Davoyan and Smetanin 1979
Panicum miliaceum (common millet)	45.2 μM 2,4-D	None	MS	3	Mesocotyl	56%	O	Rangan 1974
Paspalum scrobiculatum (Koda millet)	45.2 μM 2,4-D	1.1 μM NAA	MS	3	Mesocotyl	65–80%	O	Rangan 1976
Pennisetum typhoideum (bulrush millet)	45.2 μM 2,4-D	1.1 μM IAA	MS	4	Mesocotyl	36%	O	Rangan 1976
Sorghum bicolor 'N. Dakota' (sorghum)	22.6–67.8 μM 2,4-D	26.9 μM NAA	MS	Not reported	Seedling shoot	Not reported	O	Masteller and Holden 1970

S. bicolor line X4004	5 μM 2,4-D 10 μM ZEA	0–5 μM IAA	MS	2	Immature embryo	20–50%	O	Gamborg *et al.* 1977
Triticum aestivum 'Tobari 66' (wheat)	10 μM Dicamba	1 μM IAA 0.1 μM 6BA	B5, T	Not reported	Rachis	Not reported	O	Dudits *et al.* 1975
T. aestivum 'Mengavi'	22.6 μM 2,4-D 5.4 μM CPA	23.3 μM KIN 5.4 μM NAA	Basal	2	Embryo	7–10%	O	Chin and Scott 1977
T. aestivum 'Chinese Spring'	4.5–9 μM 2,4-D	4.6 μM ZEA 5.7 μM IAA	T, B5, MS	5	Rachis, embryo, seed	33%	O	Gosch-Wackerle *et al.* 1979
T. longissiumum wild species	0.1 μM 2,4-D	5.7 μM IAA 4.6 μM ZEA	MS	1–2	Immature embryo	27–46%	O	Gosch-Wackerle *et al.* 1979
Zea mays 'Ai88', 'R-navajo' (corn)	9 μM 2,4-D	None	MS	19	Immature embryo	21–43%	O	Green and Phillips 1975
Z. mays 'Prior', 'Inrakorn' lines M9473, M0003	67.8 μM 2,4-D	None	MS	4	Mesocotyl	16%	O	Harms *et al.* 1976
Hybrids								
Triticale Triticum × *Secale*	27.1 μM 2,4-D	None	MS	Not reported	Immature embryo	20%	O	Sharma *et al.* 1978

plant regeneration may reach 100%, e.g., 4-month-old callus cultures derived from young root and leaf explants. Age of rice callus is inversely proportional to its regenerative capacity.

d. Oats, Barley and Other Cereals.—Plants can be regenerated from at least 7 other cereal species (Table 6.18). In each species the medium for callus proliferation includes 2,4-D. In no case has 2,4-D been included in the regeneration medium. Young explants which almost certainly have contained organized shoot apices have been used for callus initiation. In *Avena sativa* regenerative capacity can be retained in callus cultures maintained for 12 to 18 months, but in most species the ability to regenerate is lost after only a few subcultures (1 to 4 months). In *Avena* (oats), genotype represents an important factor in plant regeneration. Of 25 genotypes initiated *in vitro* (Cummings *et al.* 1976) 9 could not be regenerated, while an additional 5 had very low frequency of plant regeneration. Callus was initiated from seeds on MS medium with 22 μM 2,4-D (Carter *et al.* 1967) or SH medium with 9.1 μM 2,4-D and 10.7 μM CPA (Lorz *et al.* 1976) or from immature embryos on B5 medium with 2.3 to 13.6 μM 2,4-D. The 2,4-D must be removed for plant regeneration in each case and, although hormoneless medium is sufficient for plant regeneration from each explant, Lorz *et al.* (1976) have used a combination of three hormones to achieve regeneration (Table 6.18).

Hordeum vulgare (barley) plants have been obtained from callus derived from shoot apices (Cheng and Smith 1975; Koblitz and Saalbach 1976), mature embryos (Kartel and Maneshina 1978), and immature embryos (Dale and Deambrogio 1979). Plants have been obtained from embryo-derived callus of four out of seven barley cultivars. Greater shoot regeneration was obtained for immature embryos on B5 medium (50%) than on MS medium (10%) when callus was transferred from medium with 2,4-D to hormoneless medium (Dale and Deambrogio 1979). Plants of *Sorghum bicolor* have been regenerated from callus derived from seedling shoots (Masteller and Holden 1970), immature embryos (Gamborg *et al.* 1977), and mature embryos (Thomas, King and Potrykus 1977). Although organized tissue was used, multiple shoot formation was obtained. MS is better than B5 medium for plant regeneration of sorghum (Gamborg *et al.* 1977). Callus can be induced with 5 to 67.8 μM 2,4-D (Table 6.18). Plants can be regenerated if subcultured on regeneration medium within 1 to 2 months. The frequency of regeneration for immature embryos is 20 to 50%. Plant regeneration of *Sorghum* is auxin-dependent. Shoots could be obtained from immature embryos in the presence of 2,4-D, but only if a high concentration of cytokinin was also present, e.g., 10 to 50 μM ZEA. Maximum shoot and plantlet production was obtained when callus was subcultured on medium with IAA (Gamborg *et al.* 1977), NAA (Masteller and Holden 1970), or with a combination of 6BA, NAA, and GA_3 (Thomas, King and Potrykus 1977).

Four species of millets have been regenerated from mesocotyl tissue on MS medium *in vitro* (Rangan 1974, 1976). Callus was induced with 45.2 μM 2,4-D with or without 10 to 15% coconut water. Regeneration was obtained when the 2,4-D was eliminated *(P. miliaceum)* or replaced with IAA or NAA. The frequency of regeneration for all 4 species has ranged from 36 to 80% 3 to 4 months after callus initiation.

Secale cereale (rye) is one major cereal species that has not been tested for *in vitro* regeneration, although anther cultures of rye have been regenerated (Thomas *et al.* 1975). Nonetheless, hybrids of wheat and rye (*Triticale* AD-20) have been regenerated *in vitro* and behave similarly to wheat (Prokhorov *et al.* 1974).

Although organized tissues of cereals can be induced to form callus capable of *in vitro* plant regeneration, single differentiated cereal cells generally do not undergo rapid cell division. The routine production of rapidly dividing dedifferentiated cells may be impossible (King *et al.* 1978). Nonetheless, the production of suspension cultures in some cereals, albeit not capable of plant regeneration, represents unorganized meristematic cell populations which may prove to be useful for cell genetics. Rapidly growing cell suspension cultures have been initiated from callus obtained from root tissue of *Triticum monococcum* (Kao *et al.* 1970; Evans and Gamborg 1980), young shoots of *Zea mays* (Green 1977), hypocotyls of *Pennisetum americanum* (Vasil and Vasil 1979), and stem protoplasts of *Zea mays* (Potrykus *et al.* 1977).

Protoplasts cannot be regenerated to plants from any cereal. Cell division has been reported from cell culture protoplasts of *Hordeum* (Koblitz 1976), *Triticum* (Dudits 1976), *Zea* (Motoyashi 1971), and *Pennisetum* (Vasil and Vasil 1979) and "mesophyll" protoplasts of *Avena* (Brenneman and Galston 1975), *Zea* (Potrykus *et al.* 1977), *Secale* (Wenzel 1973), and *Oryza* (Deka and Sen 1976). Organogenesis of roots in *Oryza* (Deka and Sen 1976) has been reported. Recently, cell division was reported in heterokaryocytes of corn and sorghum produced by protoplast fusion (Brar *et al.* 1980).

L. Liliaceae

Several agriculturally important species of Liliaceae have been regenerated from callus culture, i.e., *Allium cepa* (onion), *Allium sativum* (garlic), and *Asparagus officinalis* (asparagus). Other horticulturally important Liliaceae also regenerate *in vitro* (Table 6.19).

At least 15 species of Liliaceae have been regenerated from callus (Table 6.19). A successful regeneration protocol for several *Haworthia*

TABLE 6.19. REGENERATION OF LILIACEAE SPECIES

Species	Growth Regulators		Shoot Medium	Explant	Mode of Regeneration	Reference
	Callus	Shoot Formation				
Allium cepa (onion)	5 μM 2,4-D	5 μM 2iP	B5	Bulb	O	Fridborg 1971
Allium sativum (garlic)	10 μM CPA 2 μM 2,4-D 0.5 μM KIN	10 μM IAA 10 μM KIN	AZ	Stem, bulb, leaf	O	Abo El-Nil 1977
		10 μM IAA 20 μM KIN	AZ	Stem	E	
A. sativum (garlic)	9.5 μM 2,4-D 11.7 μM IAA 9.3 μM KIN	11.4 μM IAA 93 μM KIN	MS	Leaf	O	Havranek and Novak 1973
Aloe pretoriensis	Not reported	0.91 μM 2,4-D 4.7 μM KIN	LS	Seed	O	Groenewald *et al.* 1975
Asparagus officinalis (asparagus)	5.4 μM NAA 4.7 μM KIN or 4.4 μM 6BA or 1.4 μM ZEA	0.5–5.7 μM IAA 0.44–17.7 μM 6BA	LS or MS	Shoot cladode protoplasts	E,O	Reuther 1977a,b, 1978 Bui Dang Ha and MacKenzie 1973 Bui Dang Ha *et al.* 1975
Cordyline terminalis (ti)	5.4 μM NAA	4.4 μM 6BA	MS	Stem	O	Kunisaki 1975
Haworthia spp.	0.91–9.1 μM 2,4-D 0.93–9.3 μM KIN	4.7 μM KIN	MS	Inflorescence, gynoecia	O	Kaul and Sabharwal 1972
Haworthia planifolia 'Setulifera'	4.7 μM KIN 0.9 μM 2,4-D	18.6 μM KIN	LS	Leaf	O	Wessels *et al.* 1976
Heloniopsis orientalis	1 μM NAA 1 μM 6BA	1 μM 6BA	MS	Stem, leaf	O	Kato 1975

Hemerocallis (daylily)	4.5 μM 2,4-D 4.7 μM KIN	4.7 μM KIN	MS	Petal	O	Heuser and Apps 1976
Hyacinthus hybrids	2.7–43 μM NAA 0.5–9.1 μM 2,4-D	<45.7 μM IAA <2.7 μM NAA <0.1 μM 2,4-D	MS	Bulb, leaf, inflorescence, stem, ovary	O	Hussey 1976
Lilium auratum *L. speciosum*	0.5 μM NAA	5.4 μM NAA 4.4 μM 6BA	MS	Peduncle, bulbscale, petal	O	Takayama and Misawa 1979
Lilium longiflorum (Easter lily)	11.4 μM IAA	0.2 μM NAA	LS	Bulb, stem, apex	O	Stimart and Ascher 1978 Sheridan 1968
Muscari botryoides	11.4–45.6 μM IAA 0.69–43 μM NAA 0.54–99.1 μM 2,4-D	<11.4 μM IAA <0.16 μM NAA <0.14 μM 2,4-D	MS	Bulb, leaf, inflorescence, stem, ovary	O	Hussey 1975
Ornithogalum thyrsoides	11.4–45.6 μM IAA 0.6–43 μM NAA 2.3–9.1 μM 2,4-D	None or 0.2–0.6 μM NAA	MS	Stem, leaf, sepal, bulb	O	Hussey 1976
Scilla sibirica	0.69–43 μM NAA 0.54–9.1 μM 2,4-D	<11.4 μM IAA <0.64 μM NAA	MS	Bulb, leaf, inflorescence, stem, ovary	O	Hussey 1975

species has been reported (Kaul and Sabharwal 1972). Callus cultures usually have been initiated and maintained on MS or LS medium. Explant source does not affect the success of culture as dormant tissues, e.g., bulbs, and metabolically active tissues, e.g., stem segments, leaf, seed, petal, and inflorescence segments, have been successfully cultured (Table 6.19). Liliaceae callus is usually initiated and maintained in the presence of high concentrations of auxin, i.e., 11.4 to 45.6 μM IAA, 0.64 to 43 μM NAA, or 0.54 to 9.1 μM 2,4-D. In most instances a cytokinin, usually 1 to 10 μM KIN, is also included in the culture medium. *Lilium longiflorum* callus can be initiated in the absence of phytohormones (Sheridan 1968).

Reduction of auxin concentration or increase of cytokinin concentration induces the regeneration of shoots from callus cultures of Liliaceae species (Table 6.19). Shoot regeneration is often enhanced by culturing in the dark (Kato 1978; Stimart and Ascher 1978; Havranek and Novak 1973). Chromosome instabilities have been associated with *in vitro* culture of liliaceous species, although these do not occur if plants are directly regenerated from explants (Reuther 1978; Hussey 1976). Increasing chromosome number is correlated with length of time in callus culture in *Allium sativum* (garlic) (Havranek and Novak 1973), *Ornithogalum thyrsoides* (Hussey 1976), and probably *Allium cepa* (onion) (Davey *et al.* 1974; Fridborg 1971). Some liliaceous species also lose the capacity to regenerate shoots after prolonged culture *in vitro* (Davey *et al.* 1974; Hussey 1976; Fridborg 1971). Shoot regeneration from onion callus >1 year old is rare (Fridborg 1971). Sheridan (1975), however, has maintained diploid callus of *Lilium longiflorum* for over 6 years with no change in chromosome number or regenerative ability. Plants regenerated from this callus have been diploid (Sheridan 1975).

Protoplasts have been isolated from cladodes of asparagus (Bui Dang Ha *et al.* 1975), and callus proliferation has been observed in a culture medium containing 5.4 μM NAA and 1.4 μM ZEA (Bui Dang Ha and Mackenzie 1973). Shoots are induced from p-callus in the presence of 4 μM 6BA with or without 0.5 to 2.5 μM IAA or NAA.

M. Iridaceae and Amaryllidaceae

Few species of Iridaceae and Amaryllidaceae have been regenerated from callus (Table 6.20). Iris callus can be initiated from inflorescence slices and maintained in the dark on MS medium with 13.4 μM NAA and 2.3 μM KIN (Meyer *et al.* 1975). Transfer of the callus to light will induce shoot regeneration without changing hormone concentration. Callus of *Furcraea gigantea* and *Narcissus* hybrids (Amaryllidaceae) can be maintained on 2.3 μM and 4.5 μM 2,4-D, respectively (Lakshmanan and Janardhanan 1977; Seabrook *et al.* 1976). *Hippeastrum* (amaryllis) cal-

TABLE 6.20. REGENERATION IN IRIDACEAE AND AMARYLLIDACEAE SPECIES

Species	Growth Regulators		Shoot Medium	Explant	Mode of Regeneration	Reference
	Callus	Shoot Formation				
Gladiolus hortulans 'Hit Parade' (Iridaceae)	2.3 μM KIN 0.5 μM NAA	2.3 μM KIN	MS	Stem tips	O	Simonsen and Hildebrandt 1971
Iris (Iridaceae)	13.4 μM NAA 2.3 μM KIN (dark)	None or 13.4 μM NAA 2.3 μM KIN (light)	MS	Inflorescence	O	Meyer *et al.* 1975
Iris (Iridaceae)	4.5 μM 2,4-D	0.6 μM IAA 0.4 μM 6BA	MS	Shoot apex	E	Reuther 1977a
Furcraeae gigantia (Amaryllidaceae)	2.3 μM 2,4-D	2.7–5.4 μM NAA or 2.9 μM IAA	Norstog & Rhamstine	Leaf	O	Lakshmanan and Janardhanan 1977
Hippeastrum (Amaryllidaceae, amaryllis)	11 μM NAA 18 μM 6BA	11 μM NAA 18 μM 6BA	MS	Ovary, peduncle	O	Seabrook and Cumming 1977
Narcissus (Amaryllidaceae)	80 μM NAA	44 μM 6BA	MS	Leaf, shoot, root	O	Seabrook *et al.* 1976

lus is initiated and maintained on 11 μM NAA and 18 μM 6BA (Seabrook and Cumming 1977). Callus of all of these species can be initiated from leaf, ovary, inverted scape, peduncle, shoot apex, and root apex (Table 6.20).

Shoot regeneration from *Narcissus* and *F. gigantea* is induced by simultaneously reducing the concentration of auxin in the culture medium and increasing the cytokinin concentration (Table 6.20). *Hippeastrum* callus develops shoots 8 weeks after culture initiation on the same medium (Seabrook and Cumming 1977). Although Hussey (1975) reported the regeneration of shoots from callus of *Freesia* (Iridaceae) and *Ipheion* (Amaryllidaceae), exact cultural conditions were not described.

N. Miscellaneous Monocots

Six additional monocot species, representing five plant families, have been regenerated *in vitro* (Table 6.21). All of these monocot species require auxin to initiate callus proliferation with reduction of auxin necessary to obtain shoot development. All but *D. deltoidea* require cytokinin in the callus medium, while all but *A. andraeanum* and date palm require reduced auxin concentrations in the shoot induction medium. In most species young embryonic tissue has been used, but plant regeneration also has been successful from mature leaf explants (Pierik 1976). These monocot species may be maintained as callus for extended time periods prior to shoot formation. Callus of *D. deltoidea* has been subcultured for at least 12 months and retains the capability for plant regeneration (Grewal and Atal 1976). *Agave* species and *Anthurium andraeanum* also have been maintained *in vitro* for nearly a year prior to plant regeneration.

Although MS medium has been used for each species, specific nutrient requirements also may exist for these monocot species. Growth of *A. andraeanum* is enhanced if the NH_4NO_3 concentration is reduced from 1650 to 206 mg/liter (Pierik 1976), while proliferation of *Ananas sativus* is enhanced by the addition of 170 mg/liter NaH_2PO_4 to MS medium (Mathews *et al.* 1976). These variations suggest that common media formulations often used for dicots may be insufficient for monocots.

The *in vitro* regeneration of only two monocotyledonous trees has been reported: *Elaeis guineensis* (oil palm) (Rabechault *et al.* 1970) and *Phoenix dactylifera* (date palm) (Reynolds and Murashige 1979). For *Elaeis guineensis* shoot buds were produced, without leaf development (Rabechault *et al.* 1970). *Elaeis guineensis* shoot buds were regenerated only in the presence of 200 mg/liter ascorbic acid. *Phoenix dactylifera* ovule callus that had been induced on high 2,4-D (Table 6.21) was

TABLE 6.21. PLANT REGENERATION IN MISCELLANEOUS MONOCOTYLEDONOUS SPECIES

		Growth Regulators					
Family	Species	Callus	Shoot	Medium	Explant	Mode of Regeneration	Reference
Agavaceae	*Agave*	4.5 μM 2,4-D 23.2 μM KIN	0.9 μM 2,4-D 4.7 μM KIN	LS	Seed	O	Groenewald *et al.* 1977
Araceae	*Anthurium andraeanum*	0.4 μM 2,4-D 3.2 μM PBA	3.2 μM PBA	MS	Leaf	O	Pierik 1976
Bromeliaceae	*Ananas sativus* (pineapple)	28.1 μM NAA 29.7 μM IBA 9.8 μM KIN	9.7 μM NAA 9.8 μM IBA 9.8 μM KIN	MS	Axillary bud	O	Mathews *et al.* 1976
Dioscoreaceae	*Dioscorea deltoidea* (yam)	4.5 μM 2,4-D	1.2 μM IBA 2.2 μM 6BA	MS	Hypocotyl	O	Grewal and Atal 1976
	D. deltoidea	5.4–26.9 μM NAA 10% CW	10% CW	MS	Tuber	O	Mascarenhas *et al.* 1975
Arecaceae (Palmae)	*Elaeis guineensis* (oil palm)	4.5 μM 2,4-D 2.3 μM KIN	5.7 μM IAA	Heller	Embryo	E	Rabechault *et al.* 1970
	Phoenix dactylifera (date palm)	452 μM 2,4-D 4.9 μM 2iP	None	MS	Ovule	E	Rabechault *et al.* 1970

transferred to hormoneless medium to produce somatic embryos (Reynolds and Murashige 1979).

O. Gymnosperm Tree Species

Plants have been regenerated from at least 18 gymnosperm tree species (Winton 1978). Both organogenesis and embryogenesis occur. Most research has been concerned with vegetative propagation of forest trees. The chief limiting factor for regeneration of gymnosperm trees is usually explant source, and it is necessary to use juvenile tissue, e.g., embryos or seedlings, for *in vitro* propagation (Table 6.22). The use of juvenile tissue precludes the selective propagation of a commercially desirable mature tree. Regeneration of tree species has been reviewed (Bonga 1977; Winton 1978; Sommer and Brown 1979). Although in his review Winton (1978) includes an extensive list of tree species regenerated *in vitro*, culture is also included. In other cases regeneration merely represents derepression of preexisting shoot primordia. There are few reports of isolation of protoplasts from tree species (Sommer and Brown 1979). Plants have not been regenerated from protoplasts of tree species, though callus has been obtained from cotyledon protoplasts of *Pinus pinaster* (David and David 1979).

Achievements in regeneration of plants from gymnosperms are similar to those with angiosperm trees. In at least one instance, *Pseudotsuga menziesii* (Douglas fir) regeneration has been achieved from mature needles (Winton and Verhagen 1977) and lateral branch shoot tips (Coleman and Thorpe 1977), as well as from juvenile tissues (Table 6.22). Most gymnosperms can be regenerated only from juvenile tissue such as embryos and seedlings (Table 6.22). In many instances, only regeneration directly from the explant with no intervening callus growth has been reported (*Picea glauca*, Campbell and Durzan 1975; *Pinus banksiana*, Campbell and Durzan 1975; *Pinus palustris*, Sommer *et al.* 1975; *Pinus pinaster*, David and David 1977). Regeneration is usually via organogenesis. Embryogenesis has been reported in *Biota orientalis* (Konar and Oberoi 1965) and *Pinus palustris* (Sommer *et al.* 1975).

Various culture media have been used to regenerate gymnosperms. Because regeneration has been achieved from Douglas fir using several different culture media (Table 6.22), the composition of the culture medium may not be crucial to success. Shoots usually can be regenerated in the presence of 0.5 to 50 μM 6BA in the culture medium. A low concentration of auxin is included in the regeneration medium. Cheng (1976) demonstrated that as little as 0.5 to 5 μM NAA enhances shoot regeneration in Douglas fir. Regeneration can occur when the cytokinin:auxin ratio is <1. *Biota orientalis* may be regenerated on 0.57 μM IAA in the absence of cytokinin (Konar and Oberoi 1965). Douglas fir has been re-

TABLE 6.22. REGENERATION OF PINACEAE SPECIES (GYMNOSPERM)

Species	Growth Regulators		Shoot Medium	Explant	Mode of Regeneration	Reference
	Callus	Shoot Formation				
Biota orientalis	0.6 μM IAA	0.6 μM IAA	White	Cotyledon	E	Konar and Oberoi 1965
B. orientalis	5.5 μM NAA	2.2 μM 6BA	Lin & Staba	Cotyledon	O	Thomas, Duhoux and Vazart 1977
	10 μM IAA or IBA or 5 μM 2,4-D or NAA	5 μM 2iP	LP (von Arnold + Eriksson 1977)	Embryo	O	Von Arnold and Eriksson 1978
Picea glauca (white spruce)	Not reported	0–0.1 μM NAA 10 μM 6BA	Campbell & Durzan	Hypocotyl	O	Campbell and Durzan 1975
Pinus banksiana (gray pine)	Not reported	0–0.01 μM NAA 10 μM 6BA	Campbell & Durzan	Hypocotyl	O	Campbell and Durzan 1975
Pinus palustris (long leaf pine)	Not reported	22.2 μM 6BA 10.7 μM NAA	Gresshoff & Doy	Embryo	E	Sommer *et al.* 1975
Pinus pinaster	Not reported	10 μM 6BA	Campbell & Durzan	Cotyledon	O	David and David 1977
Pseudotsuga menziesii (Douglas Fir)	4.5 μM 2,4-D	45 μM 2,4-D	Winton & Verhagen	Stem	O	Winton and Verhagen 1977
	0.4 μM 6BA (24.7 μM NOA)	0.4 μM 6BA (24.7 μM NOA)	Winton & Verhagen	Needle	O	Winton and Verhagen 1977
	5 μM IAA, IBA, 6BA, 2iP	0.5–1 μM 6BA	Modified MS	Embryo	O	Cheng 1975
	15–30 μM NAA 5 μM 6BA, IBA	5 μM 6BA 0.5–5 μM NAA	Modified MS	Cotyledon	O	Cheng 1975 Cheng and Voqui 1977
Pinus strobus (white pine)	0.05–10 μM 2,4-D, NAA or IBA	0.5–25 μM IBA 0.44–4.4 μM 6BA	MS	Embryo	O	Minocha 1980
Thuja plicata (western red cedar)	Not reported	1 μM 6BA 0.1 μM NAA	MS	Juvenile tissue: cotyledon, stem, shoot tip	O	Coleman and Thorpe 1977
	Not reported	50 μM 6BA 0.1 μM NAA	MS	Mature tissue: branch, shoot tip	O	Coleman and Thorpe 1977
Thuja occidentalis (white cedar)	Not reported	4.4 μM 6BA	Lin & Staba	Cotyledon	O	M.J. Thomas *et al.* 1977
Cupressus sempervirens	Not reported	2.2 μM 6BA	Lin & Staba	Cotyledon	O	M.J. Thomas *et al.* 1977
Cupressus macrocarpa	Not reported	2.2 μM 6BA	Lin & Staba	Cotyledon	O	M.J. Thomas *et al.* 1977
Cupressus arizonica	Not reported	2.2 μM 6BA	Lin & Staba	Cotyledon	O	M.J. Thomas *et al.* 1977

generated on medium with 4.5 μM 2,4-D or on medium with 24.7 μM NOA with 0.4 μM 6BA (Winton and Verhagen 1977). Harvey and Grasham (1969) reported callus culture of 12 conifer species, but apparently were not able to regenerate plants.

Transfer of *in vitro* regenerated shoots of gymnosperm tree species to soil is quite often unsuccessful, as sometimes there is no viable vascular connection between roots and shoots (Bonga 1977). Excised conifer shoots are very difficult to root (Winton 1978). *In vitro* regeneration of gymnosperm tree species will be economically feasible only if high frequency regeneration from tissue of mature trees and rooting of shoots can be achieved.

IV. GENETIC REGULATION OF PLANT REGENERATION

A. Phenotypic Variation

The phenotypic variation of cells in culture may represent reversible epigenetic changes or more permanent genetic modifications. Visual identification of a variant is insufficient to conclude that genetic modification has occurred. This is particularly evident in the published examples of genetic changes in plant cells, e.g., pigment variation in carrot cell cultures (Mok *et al.* 1976), and cytokinin habituation of tobacco tissue cultures (Dix 1977). In each case, the phenotypic variant is distinguishable from normal cells, but it is *not* accompanied by an underlying genetic change. Epigenetic changes may account for variations in some leaky mutations reported in the literature. A genetic change must be inherited in a stable fashion. Unless stability is demonstrated, it must be assumed that variation observed in tissue culture may represent either epigenetic or genetic variation. Unfortunately, the crucial experiments necessary to statistically distinguish between tissue culture-induced variability and existing somatic genetic differences have not been devised.

The variability recovered from somatic cells regenerated *in vitro* is often astounding. Skirvin and Janick (1976) systematically compared plants propagated *in vitro* from stem, root, and petiole cuttings to plants regenerated from callus of five cultivars of *Pelargonium* spp. Plants obtained from geranium stem cuttings *in vivo* were uniform, whereas plants from *in vivo* root and petiole cuttings and plants regenerated from callus were quite variable. Changes were observed in plant and organ size, leaf and flower morphology, essential oil constituents, fasciation, pubescence, and anthocyanin pigmentation. The existence of both genetic and tissue culture-induced variability is supported by three results. (1) Because evidence suggests that adventitious shoots of geranium are derived from single cells (Broertjes *et al.* 1968), the variability found in regenerated plants must, to some extent, reflect preexisting cell variability of leaf and root cells. (2) In each cultivar, variability in plants derived from callus was greater than in those derived from adventitious shoots, implying that at least a portion of the total variability was tissue culture-induced. (3) A decrease in variability in subsequent callus cell cycles suggests that variability existed in cells originally placed *in vitro*, but was reduced in clonal lines derived from the callus. Consequently, although it appears that variability in plants derived from callus, leaf, or root is greater than in stem-derived plants, it is impossible to ascribe this variability to only genetic or environmental factors. It also has been observed that plants regenerated *in vitro* from tomato stem sections (de Langhe and de Bruijne 1976) are quite uniform, while plants regenerated from leaf sections are variable (Sharp, unpublished information). Variability, resulting in the isolation of disease-resistant clones, also has been obtained in protoplast-derived plants of *Solanum tuberosum* (Shepard *et al.* 1980).

To distinguish between epigenetic and genetic variability, the genetic segregation must be analyzed. Syono and Furuya (1972a) recovered phenotypic variability in plants regenerated from stem-derived callus, particularly abnormal floral morphology, but only in plants regenerated from long term cultures. Long term cultures do result in tissue culture-induced variability in chromosome number in callus (Murashige and Nakano 1967) and in plants regenerated from callus (Sacristan and Melchers 1969). Phenotypic changes associated with change of chromosome number probably represent genetic changes. In addition to chromosome aberrations (summarized below), phenotypic variability manifested in plants regenerated from somatic cells could be the result of mitotic crossing over (Evans and Paddock 1978) or somatic mutation (Peterson 1974). Both processes are continuous ongoing genetic events occurring in somatic cells and are subject to environmental modification.

B. Chromosome Variation

Variation of chromosome number of plants in long term cell suspension cultures has been well documented (Kao *et al.* 1970). Chromosomal variation in cell suspension and callus cultures have been reviewed extensively by Sheridan (1975), Sunderland (1977), and D'Amato (1978). It has been suggested that the progressive increase in variation of cultured cells is proportional to a progressive loss in organ-forming capacity (Torrey 1967). Murashige and Nakano (1967) have shown that shoot-forming capacity of aneuploid *Nicotiana tabacum* is severely reduced, but Sacristan and Melchers (1969) were able to regenerate numerous aneuploids of *N. tabacum* with relative ease. Unfortunately, the chromosome instability of cells in suspension culture has been compared with cells of regenerated plants in only a few cases (D'Amato 1977). The chromosome variability of regenerated plants is always less than in the callus from which plants were derived; however, despite the occurrence of a wide range of chromosome numbers in callus cultures, only diploid plants have been regenerated from *Daucus carota* (Mitra *et al.* 1960), *Oryza sativa* (Nishi *et al.* 1968), *Prunus amygdalus* (Mehra and Mehra 1974), and *Triticum aestivum* (Shimada *et al.* 1969). These species range in somatic chromosome number from 2n=2x=16 for *P. amygdalus*, to 2n=6x=42 for *T. aestivum*. Consequently, under these plant regeneration conditions, diploids are selectively favored for these species. Orton (1980) has compared chromosome number of suspension and callus cultures of *Hordeum vulgare, H. jubatum*, and their interspecific hybrid with plants regenerated from these cultures. In each case it is evident that no polyploid and a greatly reduced number of aneuploid callus cells are capable of plant regeneration. Identical chromosome number is insufficient to conclude genetic stability. Only in long term cultures of *Lilium longiflorum* have regenerated plants been shown to have the normal karyotype (Sheridan 1975). Polyploid plants have been recovered from a number of cultures (D'Amato 1977), including *Asparagus officinalis* (Takatori *et al.* 1968), *Nicotiana tabacum* (Kasperbauer and Collins 1972), *Lilium longiflorum* (Sheridan 1975), and haploid *Pelargonium* (Bennici 1974). Polyploid changes are quite common in plants regenerated from anthers whether regenerated directly or via a callus intermediate (D'Amato 1977).

A wide range of aneuploid plants has been recovered from tissue cultures of numerous species. Reduction in chromosome number has been observed in plants regenerated from callus cultures of triploid (2n=3x=21) ryegrass hybrids (Ahloowalia 1976), while a wide range of aneuploids having additions and reductions in chromosome number were obtained in sugarcane (Liu and Chen 1976) and *N. tabacum* (Sacristan and Melchers 1969). Each of these chromosomal variants is associated with phenotypic variation, including agriculturally useful traits such as disease resistance (Krishnamurthi and Tlaskal 1974). Variability associated with aneuploidy, if not accompanied by a concomitant depression of yield, is particularly valuable in vegetatively propagated agricultural crops. Aneuploidy and morphological variability have been observed in plants regenerated from protoplasts (Matern *et al.* 1978) and in most somatic hybrid plants (e.g., Melchers and Sacristan 1977).

Chromosome number mosaicism of regenerated plants has been reported in *Nicotiana* (Ogura 1975), *Hordeum* (Orton 1980), *Triticum durum* (Bennici and D'Amato 1978), *Saccharum* (Liu and Chen 1976), and *Lycopersicon peruvianum* (Sree Ramulu *et al.* 1976). The common occurrence of chromosome number mosaicism in regenerated plants suggests that plantlets originate from two or more initial cells (Bennici and D'Amato 1978) or that new chromosome variability is generated *in vivo* after plant regeneration (Orton 1980). Somatic mosaicism also has been reported in somatic hybrid plants derived from protoplast fusion where plants are presumably derived from single cells (Maliga *et al.* 1978). There have been suggestions that the chromosome number mosaicism of regenerated plantlets is reduced during the subsequent development of regenerants to mature plants. Nonetheless, this mosaicism may be established as periclinal or mericlinal "chimeras" (Sree Ramulu *et al.* 1976) or transmitted to subsequent generations (Ogura 1976). The maintenance of chromosome mosaicism *in vivo* in regenerated plants may be under genetic control (Ogura 1978).

C. Utilization of Genetic Variability

Attempts to extend established plant regeneration protocols to closely related species often have been unsuccessful. For example, despite using growth regulator concentrations sufficient for plant regeneration in nearly all closely related species, plants could not be regenerated from *Nicotiana knightiana* (Maliga *et al.* 1977). Differences in regenerative capacity observed among donor plants may reflect underlying genetic differences. If plant regeneration *in vitro* is viewed as a phenotype, it is not surprising that differences exist among species, cultivars, or genotypes that are otherwise closely related. The value of screening diverse genetic lines for plant regeneration has been emphasized (Malmberg 1979). Experiments with reported differences in regenerative capacity between donor plants have been summarized in Table 6.23. In each example listed, at least one genetic line was incapable of plant regeneration. The value of screening genetic lines is more evident in attempts to regenerate recalcitrant species, such as peas (6 of 16 lines regenerate), oats (1 of 3 lines regenerate), corn (4 of 5 lines regenerate), and wheat (2 of 16 lines regenerate). In species presently incapable of plant regeneration, screening genotypes may prove to be extremely useful, especially when hormone variations are simultaneously tested. Similar variation between species and cultivars of wheat has been observed for growth of callus cultures (Gosch-Wackerle *et al.* 1979).

Mutant cell lines have proven to be quite useful in recent years in studies of somatic cell genetics. Chlorophyll-deficient and analogue-resistant lines have been used in protoplast fusion (Vasil *et al.* 1979). In most cases attempts to regenerate plants from induced mutations or from fusion products have proven to be unsuccessful (Widholm 1977; White and Vasil 1979). In some cases, loss of regenerative capacity is dependent

TABLE 6.23. VARIATION OBSERVED WITHIN PLANT SPECIES FOR ABILITY TO UNDERGO PLANT REGENERATION

Genus or Species Examined	Variations Observed Between . . .	Number Screened	Number Regenerating	Range of Regeneration Frequencies	Multiple Media Tested	Reference
Triticum	Species cultivars	12	1	Not reported	yes	Prokhorov *et al.* 1974
Triticum	Species cultivars	4	1	Not reported	yes	Shimada *et al.* 1969
Avena sativa	Cultivars	3	1	0–57%	yes	Lorz *et al.* 1976
Avena sativa	Genotypes	25	16	Not reported	no	Cummings *et al.* 1976
Medicago sativa	Cultivars	9	8	0–36%	no	Bingham *et al.* 1975
Pisum sativum	Genotypes	16	6	Not reported	no	Malmberg 1979
Saccharum	Cultivars	11	10	Not reported	yes	Heinz and Mee 1969
Saccharum	Cultivars	4	3	Not reported	yes	Barba and Nickell 1969
Trifolium pratense	Cultivars	5	4	0–80%	yes	Phillips and Collins 1979
Zea mays	Genotypes	5	4	Not reported	yes	Green and Phillips 1975
Arabidopsis thaliana	Genotypes (mutants)	6	4	0–100%	yes	Goto 1979
Lycopersicon esculentum	Genotypes (mutants)	15	12	Not reported	yes	Behki and Lesley 1976
Nicotiana tabacum	Genotypes (mutants)	5	3	0–40%	yes	Vyskot and Novak 1977

on the specific mutant. This is particularly true for chlorophyll-deficient mutants. Of five chlorophyll mutants of *N. tabacum* examined, plants were regenerated from three mutants (*Su, Ws,* and *Ys*), but could not be regenerated from the white plastid and spontaneous chimera lines despite using all genotypes of each mutant on three regeneration media (Vyskot and Novak 1977). Similarly, of 8 chlorophyll-deficient mutants of tomato, 2 mutants (*Xa*-2 and *yg*-4) were incapable of shoot formation despite being tested on 13 media for plant regeneration (Behki and Lesley 1976). Similarly, dwarf mutants, associated with reduced viability, do not regenerate as well as normal lines. When five dwarf mutants of *Arabidopsis thaliana* were cultured in eight combinations of NAA and 6BA sufficient for plant regeneration from normal donor plants, two mutants (*le* and *g*) were incapable of plant regeneration. Of the three dwarf mutants capable of regeneration, *t* and St_1 could be regenerated on only one or two of the eight growth regulator combinations, while the *F* mutant regenerated on all eight combinations (Goto 1979).

Mutants induced *in vitro* may have further reduced regeneration capabilities due both to pleiotropic effects of the mutants isolated and to tissue culture-induced variation. For example, the WOO1 and WOO2 5-methyl tryptophan (5MT) resistant mutants of *Daucus carota* form less than 0.01% embryos as normal *D. carota* in optimal media for each (Sung *et al.* 1979). The reduction in frequency is the result of tryptophan overproduction in the 5MT resistant lines. Tryptophan is readily converted to IAA and high auxin concentrations inhibit somatic embryogenesis. Other regulatory or auxotrophic mutants with altered metabolism may respond in a similar manner. Kanamycin-resistant cells of *N. sylvestris* are incapable of plant regeneration (Dix *et al.* 1977), yet restoration of regenerative capacity can be achieved by fusing mutant protoplasts with *N. knightiana* protoplasts (Maliga *et al.* 1977). In some cases, mutant isolation has resulted in aneuploid clonal lines. As discussed earlier, aneuploid lines have reduced regenerative capacity. Examples include analog-resistant *N. sylvestris* cells (White and Vasil 1979) and temperature-sensitive *N. tabacum* cells (Malmberg *et al.* 1980). Consequently, it is not surprising that mutants induced *in vitro* have reduced capacity for plant regeneration.

V. CONCLUDING REMARKS

Plant regeneration may occur via organogenesis or embryogenesis, yet some plant families undergo embryogenesis more rapidly than species in other plant families. The plant family with the highest frequency of somatic embryogenesis is Umbelliferae, while Graminaceae contains few, if any, embryogenic species. Such observations imply an underlying genetic basis for embryogenesis vs. organogenesis. In addition it is evident that growth regulator requirements exist for the expression of somatic organogenesis or embryogenesis. Some species that may by either embryogenic or organogenic, depending on the growth regulators used, have been discovered. These species include *Solanum melongena, Brassica oleracea, Nicotiana tabacum*, and *Petunia hybrida*. Consequently, genetic constitution and growth regulator and nutritional factors, as well as explant source, are each involved in the regulation of the developmental pathway for plant regeneration. In general, high auxin together with low cytokinin concentrations promotes callus proliferation. Alternately, low auxin in combination with high cytokinin concentrations promotes shoot or plantlet formation. Although definitive experiments have not been completed for all taxa, it appears that the commitment for embryo or shoot morphogenesis occurs in the callus or primary culture medium. For embryogenesis, the shoot or secondary culture medium requires either elimination or reduction of the auxin present in the callus or primary culture medium. The secondary culture medium of numerous embryogenic species contains no growth regulator.

Plant regeneration has numerous agricultural applications. These include mass clonal plant propagation, creation of genetic variability in regenerated plants, viral elimination, and cloning for artificial seed production. The rate of tissue culture propagation must be faster than conventional propagation to be economically viable. Usually shoot apex or axillary buds are used as explants. Mass propagation would be particularly useful for multiplication of novel variants, new recombinants, or hybrid crops. Genetic variability has been recovered in plants regenerated from tissue culture in numerous cases. Of particular interest are spontaneous mutants isolated *in vitro* (e.g., Widholm 1977) and/or variants recovered from protoplast-derived plants (Shepard *et al.* 1980). Plants derived from long term cultures may exhibit greater genetic variability following *in vitro* regeneration. In addition, plants regenerated from callus cultures or leaf explants also may be virus- or bacteria-free (Murakishi and Carlson 1976). Theoretically, somatic embryos could be used for artificial seed production, as progress with celery and carrot appears to be promising (Sharp, unpublished). It is evident that research designed to achieve more efficient plant regeneration protocols must be pursued for economically important plant species.

VI. LITERATURE CITED

ABO EL-NIL, M.M. 1977. Organogenesis and embryogenesis in callus cultures of garlic (*Allium sativum* L.) *Plant Sci. Letters* 9:259–264.

AHLOOWALIA, B.S. 1975. Regeneration of ryegrass plants in tissue culture.

Crop Sci. 15:449–452.

AHLOOWALIA, B.S. 1976. Chromosomal changes in parasexually produced ryegrass. p. 115–122. *In* K. Jones and P.E. Brandham (eds.) Current chromosome research. Elsevier/North-Holland, Amsterdam.

AHUJA, M.R. and G.L. HAGEN. 1966. Morphogenesis in *Nicotiana debneyi-tabacum, N. longiflora* and their tumor-forming hybrid derivatives *in vitro. Dev. Biol.* 13:408–423.

AL-ABTA, S. and H.A. COLLIN. 1978a. Control of embryoid development in tissue cultures of celery. *Ann. Bot.* 42:773–782.

AL-ABTA, S. and H.A. COLLIN. 1978b. Cell differentiation in embryoids and plantlets of celery tissue cultures. *New Phytol.* 80:517–521.

ALSKIEF, J. 1977. Sur le gefofage *in vitro* d'apex sur des plantules decapitees, de Pecher. *C. R. Acad. Sci., Paris* 284:2499–2502.

ALTMAN, A. and R. GOREN. 1971. Promotion of callus formation by abscisic acid in citrus bud cultures. *Plant Physiol.* 47:844–846.

AMMIRATO, P.V. 1974. The effects of abscisic acid on the development of somatic embryos from cells of caraway (*Carum carvi* L.). *Bot. Gaz.* 135:328–337.

AMMIRATO, P.V. and F.C. STEWARD. 1971. Some effects of environment on the development of embryos from cultured free cells. *Bot. Gaz.* 133:149–158.

ANDERSON, W.C., G.W. MEAGHER, and A.G. NELSON. 1977. Cost of propagating broccoli plants through tissue culture. *HortScience* 12:543–544.

ARCHIBALD, J.F. 1954. Culture *in vitro* of cambial tissues of cacao. *Nature* 173:351.

BAJAJ, Y.P.S. 1972. Effect of some growth regulators on bud formation by excised leaves of *Torenia fournieri. Z. Pflanzenphysiol.* 66:284–287.

BAJAJ, Y.P.S. and M. BOPP. 1972. Growth and organ formation in *Sinapis alba* tissue cultures. *Z. Pflanzenphysiol.* 66:378–381.

BAJAJ, Y.P.S. and M. MADER. 1974. Growth and morphogenesis in tissue cultures of *Anagallis arvensis. Physiol. Plant.* 32:43–48.

BAJAJ, Y.P.S. and P. NIETSCH. 1975. *In vitro* propagation of red cabbage (*Brassica oleracea* L. var. capitata). *J. Expt. Bot.* 26:883–890.

BANERJEE, S. and S. GUPTA. 1975a. Morphogenesis in tissue cultures of different organs of *Nigella sativa. Physiol. Plant.* 33:185–187.

BANERJEE, S. and S. GUPTA. 1975b. Embryoid and plantlet formation from stock cultures of *Nigella* tissues. *Physiol. Plant.* 34:243–245.

BANERJEE, S. and S. GUPTA. 1976. Embryogenesis and differentiation in *Nigella sativa* leaf callus *in vitro. Physiol. Plant.* 38:115–120.

BANKS, M.S. 1979. Plant regeneration from callus from two growth phases of English ivy, *Hedera helix* L. *Z. Pflanzenphysiol.* 92:349–353.

BANKS, M.S. and P.K. EVANS. 1976. A comparison of the isolation and culture of mesophyll protoplasts from several *Nicotiana* species and their hybrids. *Plant Sci. Letters* 7:409–416.

BAPAT, V.A. and P.S. RAO. 1977. Experimental control of growth and differentiation in organ cultures of *Physalis minima* Linn. *Z. Pflanzenphysiol.* 85:403–416.

BARBA, R. and L.G. NICKELL. 1969. Nutrition and organ differentiation in tissue cultures of sugarcane, a monocotyledon. *Planta* 89:299–302.

BARKER, W.G. 1969. Behavior *in vitro* of plant cells from various sources within the same organism. *Can. J. Bot.* 47:1334–1336.

BARONCELLI, S., M. BUIATTI, and A. BENNICI. 1973. Genetics of growth and differentiation *"in vitro"* of *Brassica oleracea* var. *botrytis*. *Z. Pflanzenzuchtg.* 70:99–107.

BAWA, K.S. and R.F. STETTLER. 1972. Organ culture with black cottonwood: morphogenetic response of female catkin primordia. *Can. J. Bot.* 50: 1627–1631.

BAYLISS, M.W. 1977. Factors effecting the frequency of tetraploid cells in a predominantly diploid suspension culture of *Daucus carota*. *Protoplasma* 92: 109–115.

BEHKI, R.M. and S.M. LESLEY. 1976. *In vitro* plant regeneration from leaf explants of *Lycopersicon esculentum* (tomato). *Can. J. Bot.* 54:2409–2414.

BENNICI, A. 1974. Cytological analysis of roots, shoots and plants regenerated from suspension and solid *in vitro* cultures of haploid *Pelargonium*. *Z. Pflanzenzuchtg.* 72:199–205.

BENNICI, A. and F. D'AMATO. 1978. *In vitro* regeneration of Durum wheat plants. 1. Chromosome numbers of regenerated plantlets. *Z. Pflanzenzuchtg.* 81:305–311.

BHARAL, S. and A. RASHID. 1979a. Regeneration of plants from tissue cultures of the legume, *Indigofera enneaphylla* Linn. *Z. Pflanzenphysiol.* 92: 443–447.

BHARAL, S. and A. RASHID. 1979b. Hypocotyl, stem, callus, and leaves of the legume *Vigna sinensis* as systems for isolation of protoplasts. *Z. Pflanzenphysiol.* 91:465–468.

BHATT, P.N., D.P. BHATT, and I.M. SUSSEX. 1979. Organ regeneration from leaf disks of *Solanum nigrum, S. dulcamara,* and *S. khasianum*. *Z. Pflanzenphysiol.* 95:355–362.

BILKEY, P.C. and B.H. MCCOWN. 1978. Towards true red, orange and yellow-flowering African violets. Asexual hybridization of *Saintpaulia* and *Episcia*. *African Violet Mag.* 31:64–66.

BILKEY, P.C., B.H. MCCOWN, and A.C. HILDEBRANDT. 1978. Micropropagation of African violet from petiole cross sections. *HortScience* 13:37–38.

BINDING, H. and R. NEHLS. 1977. Regeneration of isolated protoplasts to plants in *Solanum dulcamara* L. *Z. Pflanzenphysiol.* 85:279–280.

BINDING, H., R. NEHLS, O. SCHIEDER, S.K. SOPORY, and G. WENZEL. 1978. Regeneration of mesophyll protoplasts isolated from dihaploid clones of *Solanum tuberosum*. *Physiol. Plant.* 43:52–54.

BINGHAM, E.T., L.V. HURLEY, D.M. KAATZ, and J.W. SAUNDERS. 1975. Breeding alfalfa which regenerates from callus tissue in culture. *Crop Sci.* 15:719–721.

BLAYDES, D.F. 1966. Interaction of kinetin and various inhibitors in the growth of soybean tissue. *Physiol. Plant.* 19:748–753.

BLUMENFELD, A. and S. GAZIT. 1971. Growth of avocado fruit callus and its relation to exogenous and endogenous cytokinins. *Physiol. Plant.* 25:369–371.

BONGA, J.M. 1977. Applications of tissue culture in forestry. p. 93–107. *In* J. Reinert and Y.P.S. Bajaj (eds.) Plant cell, tissue, and organ culture. Springer-Verlag, Berlin.

BOTTINO, P.J., C.E. MAIRE, and L.M. GOFF. 1979. Tissue culture and organogenesis in the winged bean. *Can. J. Bot.* 57:1773–1776.

BOURGIN, J.P., Y. CHUPEAU, and C. MISSONIER. 1979. Plant regeneration from mesophyll protoplasts of several *Nicotiana* species. *Physiol. Plant.* 45:288–292.

BOURGIN, J.P. and C. MISSONIER. 1978. Culture de protoplastes de mesophylle de *Nicotiana alata* Link et Otto haploide. *Z. Pflanzenphysiol.* 87:55–64.

BOWES, B.G. 1970. Preliminary observations on organogenesis in *Taraxacum officinale* tissue cultures. *Protoplasma* 71:197–202.

BOWES, B.G. 1976. *In vitro* morphogenesis of *Crambe maritima* L. *Protoplasma* 89:185–188.

BRANDAO, I. and R. SALEMA. 1977. Callus and plantlets development from cultured leaf explants of *Sedum telephium* L. *Z. Pflanzenphysiol.* 85:1–8.

BRAR, D.S., S. RAMBOLD, F. CONSTABEL, and O.L. GAMBORG. 1980. Isolation, fusion, and culture of sorghum and corn protoplasts. *Z. Pflanzenphysiol.* 96:269–275.

BRENNEMAN, F.N. and A.W. GALSTON. 1975. Experiments on the cultivation of protoplasts and calli of agriculturally important plants. I. Oats (*Avena sativa* L.). *Biochem. Physiol. Pflanzen.* 168:453–457.

BROERTJES, C., B. HACCIUS, and S. WEIDLICH. 1968. Adventitious bud formation on isolated leaves and its significance for mutation breeding. *Euphytica* 17:321–344.

BROOME, O.C. and R.H. ZIMMERMAN. 1978. *In vitro* propagation of blackberry. *HortScience* 13:151–153.

BROWN, C.L. and E.M. GIFFORD. 1958. The relation of the cotyledons to root development of pine embryos grown *in vitro*. *Plant Physiol.* 33:57–64.

BROWN, S., D.F. WETHERELL, and D.K. DOUGALL. 1976. The potassium requirement for growth and embryogenesis in wild carrot suspension cultures. *Physiol. Plant.* 37:73–79.

BROWN, W.V. 1972. Textbook of cytogenetics. The C.V. Mosby Company, St. Louis.

BUI DANG HA, D. and I.A. MACKENZIE. 1973. The division of protoplasts from *Asparagus officinalis* L. and their growth and differentiation. *Protoplasma* 78:215–221.

BUI DANG HA, D., B. NORREEL, and A. MASSET. 1975. Regeneration of *Asparagus officinalis* L. through callus cultures derived from protoplasts. *J. Expt. Bot.* 26:263–270.

BUSH, S.R., E.D. EARLE, and R.W. LANGHANS. 1976. Plantlets from petal segments, petal epidermis, and shoot tips of the periclinal chimera, *Chrysanthemum morifolium* 'Indianapolis'. *Amer. J. Bot.* 63:729–737.

BUTTON, J. and C.E.J. BOTHA. 1975. Enzymic maceration of *Citrus* callus and the regeneration of plants from single cells. *J. Expt. Bot.* 26:723–729.

CAMPBELL, R.A. and D.J. DURZAN. 1975. Induction of multiple buds and needles in tissue cultures of *Picea glauca*. *Can. J. Bot.* 53:1652–1657.

CARCELLER, M., M.R. DAVEY, M.W. FOWLER, and H.E. STREET. 1971. The influence of sucrose, 2,4-D, and kinetin on the growth, fine structure, and lignin content of cultured sycamore cells. *Protoplasma* 73:367–385.

CAREW, D.P. and A.E. SCHWARTUNG. 1958. Production of rye embryo callus. *Bot. Gaz.* 119:237–239.

CARLSON, P.S., H.H. SMITH, and R.D. DEARING. 1972. Parasexual interspecific plant hybridization. *Proc. Natl. Acad. Sci. (USA)* 69:2292–2294.

CARTER, O., Y. YAMADA, and E. TAKAHASHI. 1967. Tissue culture of oats. *Nature* 214:1029–1030.

CARUSO, J.L. 1971. Bud formation in excised stem segments of *Verbascum thapsus*. *Amer. J. Bot.* 58:429–431.

CHEN, C.H., P.F. LO, and J.G. ROSS. 1979. Regeneration of plantlets from callus cultures of Indian grass. *Crop Sci.* 19:117–118.

CHEN, C.H., N.E. STENBERG, and J.G. ROSS. 1977. Clonal propagation of big bluestem by tissue culture. *Crop Sci.* 17:847–850.

CHEN, H.R. and A.W. GALSTON. 1967. Growth and development of *Pelargonium* pith cells *in vitro* II. Initiation of organized development. *Physiol. Plant.* 20:533–539.

CHENG, T.-Y. 1975. Adventitious bud formation in culture of Douglas-fir (*Pseudotsuga menziesii* Mirb. Franco). *Plant Sci. Letters* 5:97–102.

CHENG, T.-Y. 1976. Vegetative propagation of western hemlock *(Tsuga heterophylla)* through tissue culture. *Plant Cell Physiol.* 17:1347–1350.

CHENG, T.-Y and H.H. SMITH. 1975. Organogenesis from callus culture of *Hordeum vulgare*. *Planta* 123:307–310.

CHENG, T.-Y. and T.H. VOQUI. 1977. Regeneration of Douglas-fir plantlets through tissue culture. *Science* 198:306–307.

CHIN, J.C. and K.J. SCOTT. 1977. Studies on the formation of roots and shoots in wheat callus cultures. *Ann. Bot.* 41:473–481.

CLARE, M.V. and H.A. COLLIN. 1974. The production of plantlets from tissue cultures of brussels sprout (*Brassica oleracea* L. var. *gemmifera* D.C.). *Ann. Bot.* 38:1067–1076.

COCKING, E.C. 1978. Selection and somatic hybridization. p. 151–158. *In* T.A. Thorpe (ed.) Frontiers of plant tissue culture. Univ. of Calgary Press, Calgary.

COLEMAN, W.K. and R.I. GREYSON. 1977. Promotion of root initiation by gibberellic acid in leaf discs of tomato *(Lycopersicon esculentum)* cultured *in vitro*. *New Phytol.* 78:47–54.

COLEMAN, W.K. and T. THORPE. 1977. *In vitro* culture of western red cedar (*Thuja plicata* Donn.) I. Plantlet formation. *Bot. Gaz.* 138:298–304.

COLIJN, C.M., A.J. KOOL, and H.J.J. NIJKAMP. 1979. Induction of root and shoot formation from root meristems of *Petunia hybrida*. *Protoplasma* 99: 335–340.

CONGER, B.V. and J.V. CARABIA. 1978. Callus induction and plantlet regeneration in orchardgrass. *Crop Sci.* 18:157–159.

CONSTABEL, F.C. 1978. Development of protoplast fusion products, heterokaryocytes and hybrid cells. p. 141–150. *In* T.A. Thorpe (ed.) Frontiers of plant tissue culture. Univ. of Calgary Press, Calgary.

COOKE, R.C. 1977. Tissue culture propagation of African violets. *HortScience* 12:549.

CROCOMO, O.J., W.R. SHARP, and M.T.V. DECARVALHO. 1980. Controle da morfogênese e desenvolvimento de plantas em cultura de tecido de canadeacucar—resultados experimentais. Proc. 1st Congr. da Sociedade Brasileira de Técnicos Acucareiros (STAB), Maceio, Alagoas, Brasil, Jan. 21–22, 1979. (in press)

CROCOMO, O.J., W.R. SHARP, and J.E. PETERS. 1976. Plantlet morphogenesis and the control of callus growth and root induction of *Phaseolus vulgaris* with the addition of a bean seed extract. *Z. Pflanzenphysiol.* 78:456–460.

CUMMINGS, D.P., C.E. GREEN, and D.D. STUTHMAN. 1976. Callus induction and regeneration in oats. *Crop Sci.* 16:465–470.

DALE, P.J. and E. DEAMBROGIO. 1979. A comparison of callus induction and plant regeneration from different explants of *Hordeum vulgare*. *Z. Pflanzenphysiol.* 94:65–77.

D'AMATO, F. 1977. Cytogenetics of differentiation in tissue and cell cultures. p. 343–357. *In* J. Reinert and Y.P.S. Bajaj (eds.) Plant cell, tissue, and organ culture. Springer-Verlag, Berlin.

D'AMATO, F. 1978. Chromosome number variation in cultured cells and regenerated plants. p. 287–295. *In* T.A. Thorpe (ed.) Frontiers of plant tissue culture. Univ. of Calgary Press, Calgary.

DAVEY, M.R., I.A. MACKENZIE, G.G. FREEMAN, and K.C. SHORT. 1974. Studies of some aspects of the growth, fine structure and flavour production of onion tissue grown *in vitro*. *Plant Sci. Letters* 3:113–120.

DAVID, A. and H. DAVID. 1977. Manifestations de diverses potentialites organogenes d'organes ou de fragments d'organes de Pin maritime (*Pinus pinaster* Sol.) en cultive *in vitro*. *C. R. Acad. Sci., Paris* 284:627–630.

DAVID, A. and H. DAVID. 1979. Isolation and callus formation from cotyledon protoplasts of pine *(Pinus pinaster). Z. Pflanzenphysiol.* 94:173–177.

DAVIES, M.E. and M.M. DALE. 1979. Factors affecting *in vitro* shoot regeneration on leaf discs of *Solanum laciniatum* Ait. *Z. Pflanzenphysiol.* 92:51–60.

DAVOYAN, E.I. and A.P. SMETANIN. 1979. Callus production and regeneration of rice plants. *Fiziologiya Rastenii* 26:323–329.

DE BRUIJNE, E., E. DE LANGHE, and R. VAN RIJK. 1974. Action of hormones and embryoid formation in callus cultures of *Carica papaya. Meded. Fak. Landbouwwet., Gent.* 39:637–645.

DE GREEF, W. and M. JACOBS. 1979. *In vitro* culture of the sugarbeet: description of a cell line with high regeneration capacity. *Plant Sci. Letters* 17:55–61.

DE LANGHE, E. and E. DE BRUIJNE. 1976. Continuous propagation of tomato plants by means of callus cultures. *Scientia Hort.* 4:221–227.

DE LANGHE, E., P. DEBERGH, and R. VAN RIJK. 1974. *In vitro* culture as a method for vegetative propagation of *Euphorbia pulcherrima. Z. Pflanzenphysiol.* 71:271–274.

DEKA, P.C. and S.K. SEN. 1976. Differentiation in calli originated from isolated protoplasts of rice (*Oryza sativa* L.) through plating technique. *Molec. Gen. Genet.* 145:239–243.

DEVOS, P., E. DE LANGHE, and E. DE BRUIJNE. 1979. Influence of 2,4-D on the propagation of *Cynara scolymus* L. *in vitro. Meded. Fak. Lanbouwwet., Gent.* 27:829–836.

DHOOT, G.K. and G.G. HENSHAW. 1977. Organization and alkaloid production in tissue cultures of *Hyoscyamus niger. Ann. Bot.* 41:943–949.

DHRUVA, B., T. RAMAKRISHNAN, and C.S. VAIDYANATHAN. 1978. Regeneration of hybrid tomato plants from leaf callus. *Curr. Sci.* 47:458–460.

DIX, P.J. 1977. Chilling resistance is not transmitted sexually in plants regenerated from *Nicotiana sylvestris* cell lines. *Z. Pflanzenphysiol.* 84:223–226.

DIX, P.J., F. JOO, and P. MALIGA. 1977. A cell line of *Nicotiana sylvestris* with resistance to kanamycin and streptomycin. *Molec. Gen. Genet.* 157:285–290.

DOERSCHUG, M.R. and C.O. MILLER. 1967. Chemical control of adventitious organ formation in *Lactuca sativa* explants. *Amer. J. Bot.* 54:410–413.

DONN, G. 1978. Cell division and callus regeneration from leaf protoplasts of *Vicia narbonensis. Z. Pflanzenphysiol.* 86:65–75.

DORION, N., Y. CHUPEAU, and J.P. BOURGIN. 1975. Isolation, culture and regeneration into plants of *Ranunculus sceleratus* L. leaf protoplasts. *Plant Sci. Letters* 5:325–331.

DOUGALL, D.K. and D.C. VERMA. 1978. Growth and embryo formation in wild-carrot suspension cultures with ammonium ion as a sole nitrogen source. *In Vitro* 14:180–182.

DREW, R.L.K. 1979. Effect of activated charcoal on embryogenesis and regeneration of plantlets from suspension cultures of carrot (*Daucus carota* L.). *Ann. Bot.* 44:387–389.

DUDITS, D. 1976. The effect of selective conditions on the products of plant protoplast fusion. p. 153–162. *In* D. Dudits *et al.* (eds.) Cell genetics in higher plants. Akademiai Kiado, Budapest.

DUDITS, D., GY. HADLACZKY, G. BAJSZAR, C. KONCZ, G. LAZAR, and G. HORVATH. 1979a. Plant regeneration from intergeneric cell hybrids. *Plant Sci. Letters* 15:101–112.

DUDITS, D., GY. HADLACZKY, E. LEVI, O. FEJER, Z. HAYDU, and G. LAZAR. 1977. Somatic hybridization of *Daucus carota* and *D. capillifolius* by protoplast fusion. *Theor. Appl. Genet.* 51:127–132.

DUDITS, D., K.N. KAO, F. CONSTABEL, and O.L. GAMBORG. 1976. Embryogenesis and formation of tetraploid and hexaploid plants from carrot protoplasts. *Can. J. Bot.* 54:1063–1067.

DUDITS, D., G. LAZAR, and G. BAJSZAR. 1979b. Reversible inhibition of somatic embryo differentiation by bromodeoxyuridine in cultured cells of *Daucus carota* L. *Cell Diff.* 8:135–144.

DUDITS, D., G. NEMET, and Z. HAYDU. 1975. Study of callus growth and organ formation in wheat *(Triticum aestivum)* tissue cultures. *Can. J. Bot.* 53:957–963.

EAPEN, S., T.S. RANGAN, M.S. CHADHA, and M.R. HEBLE. 1978. Morphogenetic and biosynthetic studies on tissue cultures of *Atropa belladonna* L. *Plant Sci. Letters* 13:83–89.

EARLE, E.D. and R.W. LANGHANS. 1974a. Propagation of *Chrysanthemum in vitro:* I. Multiple plantlets from shoot tips and the establishment of tissue cultures. *J. Amer. Soc. Hort. Sci.* 99:128–132.

EARLE, E.D. and R.W. LANGHANS. 1974b. Propagation of *Chrysanthemum in vitro:* II. Production, growth and flowering of plantlets from tissue cultures. *J. Amer. Soc. Hort. Sci.* 99:352–358.

ELLIS, R.P. and C.H. BORNMAN. 1971. Anatomical aspects of growth proliferation in *Nicotiana tabacum* tissue cultured *in vitro. J. South Afr. Bot.* 37: 109–126.

ENGVILD, K.C. 1973. Shoot differentiation in callus cultures of *Datura innoxia. Physiol. Plant.* 28:155–159.

EVANS, D.A. 1980. Chromosome stability of plants regenerated from mesophyll protoplasts of *Nicotiana* species. *Z. Pflanzenphysiol.* 95:459–463.

EVANS, D.A. and O.L. GAMBORG. 1980. Chromosome stability of plant cell suspension cultures. *Plant Sci. Letters* (in press)

EVANS, D.A. and E.F. PADDOCK. 1978. Mitotic crossing-over in higher plants. p. 315–351. *In* W.R. Sharp *et al.* (eds.) Plant cell and tissue culture: principles and applications. Ohio State Univ. Press, Columbus.

EVANS, D.A., W.R. SHARP, and E.F. PADDOCK. 1976. Variation in callus proliferation and root morphogenesis in leaf tissue cultures of *Glycine max*, Strain T219. *Phytomorphology* 26:379–384.

EVANS, D.A., L.R. WETTER, and O.L. GAMBORG. 1980. Somatic hybrid plants of *Nicotiana glauca* and *Nicotiana tabacum* obtained by protoplast fusion. *Physiol. Plant.* 48:225–230.

FASSULIOTIS, G. 1975. Regeneration of whole plants from isolated stem parenchyma cells of *Solanum sisymbriifolium*. *J. Amer. Soc. Hort. Sci.* 100:636–638.

FONNESBECH, M. 1974. The influence of NAA, BA and temperature on shoot and root development from *Begonia* × *cheimantha* petiole segments grown *in vitro*. *Plant Physiol.* 32:49–54.

FREARSON, E.M., J.B. POWER, and E.C. COCKING. 1973. The isolation, culture, and regeneration of *Petunia* leaf protoplasts. *Dev. Biol.* 33:130–137.

FREELING, M., J.C. WOODMAN, and D.S.K. CHENG. 1976. Developmental potentials of maize tissue cultures. *Maydica* 21:97–112.

FRIDBORG, G. 1971. Growth and organogenesis in tissue cultures of *Allium cepa* var. *proliferum*. *Physiol. Plant.* 25:436–440.

FRIDBORG, G. and T. ERIKSSON. 1975. Effects of activated charcoal on growth and morphogenesis in cell cultures. *Physiol. Plant.* 34:306–308.

FUJIMURA, T. and A. KOMAMINE. 1975. Effects of various growth regulators on the embryogenesis in a carrot cell suspension. *Plant. Sci. Letters* 5:359–364.

FUJIMURA, T. and A. KOMAMINE. 1979a. Synchronization of somatic embryogenesis in a carrot cell suspension culture. *Plant Physiol.* 64:162–164.

FUJIMURA, T. and A. KOMAMINE. 1979b. Involvement of endogenous auxin in somatic embryogenesis in a carrot cell suspension culture. *Z. Pflanzenphysiol.* 95:13–19.

FURNER, I.J., J. KING, and O.L. GAMBORG. 1978. Plant regeneration from protoplasts isolated from a predominantly haploid suspension culture of *Datura innoxia* (Mill). *Plant Sci. Letters* 11:169–176.

GAMBORG, O.L., F. CONSTABEL, and R.A. MILLER. 1970. Embryogenesis and production of albino plants from cell cultures of *Bromus inermis*. *Planta* 95:355–358.

GAMBORG, O.L., R.A. MILLER, and K. OJIMA. 1968. Plant cell cultures. I. Nutrient requirements of suspension cultures of soybean root cells. *Expt. Cell Res.* 50:151–158.

GAMBORG, O.L., T. MURASHIGE, T.A. THORPE, and I.K. VASIL. 1976. Plant tissue culture media. *In Vitro* 12:473–478.

GAMBORG, O.L. and J.P. SHYLUK. 1976. Tissue culture, protoplasts and morphogenesis in flax. *Bot. Gaz.* 137:301–306.

GAMBORG, O.L., J.P. SHYLUK, D.S. BRAR, and F. CONSTABEL. 1977. Morphogenesis and plant regeneration from callus of immature embryos of sorghum. *Plant Sci. Letters* 10:67–74.

GAMBORG, O.L., J.P. SHYLUK, L.C. FOWKE, L.R. WETTER, and D.A. EVANS. 1979. Plant regeneration from protoplasts and cell cultures of *N. tabacum* sulfur mutant *(Su/Su)*. *Z. Pflanzenphysiol.* 95:255–264.

GAMBORG, O.L., J. SHYLUK, and K.K. KARTHA. 1975. Factors affecting the isolation and callus formation in protoplasts from the shoot apices of *Pisum sativum* L. *Plant Sci. Letters* 4:285–292.

GAUTHERET, R.J. 1939. Sur la possibilite de realiser la culture indefinie des tissus de tubercule de carotte. *C. R. Acad. Sci., Paris* 208:118.

GAUTHERET, R.J. 1942. Manuel technique de culture des tissus vegetaux. Masson et Cie, Paris.

GAUTHERET, R.J. 1959. La culture des tissus vegetaux. Masson et Cie, Paris.

GENGENBACH, B.G., C.E. GREEN, and C.M. DONOVAN. 1977. Inheritance of selected pathotoxin resistance in maize plants regenerated from cell cultures. *Proc. Natl. Acad. Sci. (USA)* 74:5113–5117.

GILL, R., A. RASHID, and S.C. MAHESHWARI. 1978. Regeneration of plants from mesophyll protoplasts of *Nicotiana plumbaginifolia* Viv. *Protoplasma* 96:375–379.

GILL, R., A. RASHID, and S.C. MAHESHWARI. 1979. Isolation of mesophyll protoplasts of *Nicotiana rustica* and their regeneration into plants flowering *in vitro*. *Physiol. Plant.* 47:7–10.

GLEBA, Y.Y. and F. HOFFMAN. 1978. Hybrid cell lines *Arabidopsis thaliana* + *Brassica campestris*. No evidence for specific chromosome elimination. *Mol. Gen. Genet.* 165:257–264.

GOSCH, G., Y.P.S. BAJAJ, and J. REINERT. 1975. Isolation, culture, and induction of embryogenesis in protoplasts from cell-suspensions of *Atropa belladonna*. *Protoplasma* 86:405–410.

GOSCH-WACKERLE, G., L. AVIV, and E. GALUN. 1979. Induction, culture, and differentiation of callus from immature rachises, seeds and embryos of *Triticum*. *Z. Pflanzenphysiol.* 91:267–278.

GOTO, N. 1979. *In vitro* organogenesis from leaf explants of some dwarf mutants of *Arabidopsis thaliana* (L.) Heynh. *Japan. J. Genet.* 54:303–306.

GOULD, A.R. 1978. Diverse pathways of morphogenesis in tissue cultures of the composite *Brachycome lineariloba* (2n=4). *Protoplasma* 97:125–135.

GOULD, A.R. 1979. Chromosomal and phenotypic stability during regeneration of whole plants from tissue cultures of *Brachycome dichromosomatica* (2n=4). *Austral. J. Bot.* 27:117–121.

GREEN, C.E. 1977. Prospects for crop improvement in the field of cell culture. *HortScience* 12:131–134.

GREEN, C.E. and R.L. PHILLIPS. 1975. Plant regeneration from tissue cultures of maize. *Crop Sci.* 15:417–421.

GREWAL, S. and C.K. ATAL. 1976. Plantlet formation in callus cultures of *Dioscorea deltoidea* Wall. *Ind. J. Expt. Biol.* 14:352–353.

GREWAL, S., U. SACHDEVA, and C.K. ATAL. 1976. Regeneration of plants by embryogenesis from hypocotyl cultures of *Ammi majus* L. *Ind. J. Expt. Biol.* 14:716–717.

GRINBLAT, U. 1972. Differentiation of *Citrus* stem *in vitro*. *J. Amer. Soc.*

Hort. Sci. 97:599–603.

GROENEWALD, E.G., A. KOELEMAN, and D.C.J. WESSELS. 1975. Callus formation and plant regeneration from seed tissue of *Aloe pretoriensis* Pole Evans. *Z. Pflanzenphysiol.* 75:270–272.

GROENEWALD, E.G., D.C.J. WESSELS, and A. KOELEMAN. 1977. Callus formation and subsequent plant regeneration from seed tissue of an *Agave* species (Agavaceae). *Z. Pflanzenphysiol.* 81:369–373.

GUHA, S. and S.C. MAHESHWARI. 1964. *In vitro* production of embryos from anthers of *Datura*. *Nature* 204:497.

GUNAY, A.L. and P.S. RAO. 1978. *In vitro* plant regeneration from hypocotyl and cotyledon explants of red pepper *(Capsicum)*. *Plant Sci. Letters* 11:365–372.

GUNAY, A.L. and P.S. RAO. 1980. *In vitro* propagation of hybrid tomato plants (*Lycopersicon esculentum* L.) using hypocotyl and cotyledon explants. *Ann. Bot.* 45:205–207.

GUNCKEL, J.E., W.R. SHARP, B.W. WILLIAMS, W.C. WEST, and W.O. DRINKWATER. 1972. Root and shoot initiation in sweet potato explants as related to polarity and nutrient media variations. *Bot. Gaz.* 133:254–262.

HACCIUS, B. and K.K. LAKSHMANAN. 1965. Adventive embryonen aus *Nicotiana*—Kallus, der bei hohen Lichtintensitaten kultiviert wurde. *Planta* 65:102–104.

HADDON, L. and D.H. NORTHCOTE. 1976. The influence of gibberellic acid and abscisic acid on cell and tissue differentiation of bean callus. *J. Cell Sci.* 20:47–55.

HALPERIN, W. 1964. Morphogenetic studies with partially synchronized cultures of carrot embryos. *Science* 146:408.

HALPERIN, W. 1966. Alternative morphogenetic events in cell suspensions. *Amer. J. Bot.* 53:443–453.

HALPERIN, W. 1967. Population density effects on embryogenesis in carrot-cell culture. *Expt. Cell Res.* 48:170–173.

HALPERIN, W. and W.A. JENSEN. 1967. Ultrastructural changes during growth and embryogenesis in carrot cell cultures. *J. Ultrastruct. Res.* 18:428–443.

HALPERIN, W. and D.F. WETHERELL. 1964. Adventive embryony in tissue cultures of the wild carrot, *Daucus carota*. *Amer. J. Bot.* 51:274–283.

HANDRO, W. 1977. Structural aspects of the neo-formation of floral buds on leaf discs of *Streptocarpus nobilis* cultured *in vitro*. *Ann. Bot.* 41:303–305.

HARADA, H., K. OHYAMA, and J. CHERRUEL. 1972. Effects of coumarin and other factors on the modification of form and growth of isolated mesophyll cells. *Z. Pflanzenphysiol.* 66:307–324.

HARMS, C.T., H. LORZ, and I. POTRYKUS. 1976. Regeneration of plantlets from callus cultures of *Zea mays* L. *Z. Pflanzenzuchtg.* 77:347–351.

HARVEY, A.E. and J.L. GRASHAM. 1969. Procedures and media for obtaining tissue cultures of 12 conifer species. *Can. J. Bot.* 47:547–549.

HAVRANEK, P. and F.J. NOVAK. 1973. The bud formation in the callus cultures of *Allium sativum* L. *Z. Pflanzenphysiol.* 68:308–318.

HAYWARD, C. and J.B. POWER. 1975. Plant production from leaf protoplasts of *Petunia parodii*. *Plant Sci. Letters* 4:407–410.

HEINZ, D.J., M. KRISHNAMURTHI, L.G. NICKELL, and A. MARETZKI. 1977. Cell, tissue and organ culture in sugarcane improvement. p. 3–17. *In* J. Reinert and Y.P.S. Bajaj (eds.) Plant cell, tissue and organ culture. Springer-Verlag, Berlin.

HEINZ, D.J. and G.W.P. MEE. 1969. Plant differentiation from callus tissue of *Saccharum* species. *Crop Sci.* 9:346–348.

HELGESON, J.P. 1979. Tissue and cell-suspension culture. p. 52–59. *In* R.D. Durbin (ed.) *Nicotiana*: Procedures for experimental use. USDA, Washington, D.C.

HENKE, R.R., M.A. MANSUR, and M.J. CONSTANTIN. 1978. Organogenesis and plantlet formation from organ- and seedling-derived calli of rice *(Oryza sativa)*. *Physiol. Plant.* 44:11–14.

HENRY, R.J. 1978. *In vitro* propagation of *Peperomia* 'Red Ripple' from leaf discs. *HortScience* 13:150–151.

HERMAN, E.B. and G.J. HAAS. 1978. Shoot formation in tissue cultures of *Lycopersicon esculentum* (Mill.) *Z. Pflanzenphysiol.* 89:467–470.

HEUSER, C.W. and D.A. APPS. 1976. *In vitro* plantlet formation from flower petal explants of *Hemerocallis* cv. Chipper Cherry. *Can. J. Bot.* 54:616–618.

HILL, G.P. 1967. Morphogenesis in stem callus of *Convolvulus arvensis*. *Ann. Bot.* 31:437–446.

HIRAOKA, N. and M. TABATA. 1974. Alkaloid production by plants regenerated from cultured cells of *Datura innoxia*. *Phytochemistry* 13:1671–1675.

HIROTANI, M. and T. FURUYA. 1977. Restoration of cardenolide synthesis in redifferentiated shoots from callus cultures of *Digitalis purpurea*. *Phytochemistry* 16:610–611.

HOLSTEN, R.D., R.C. BURNS, R.W.F. HARDY, and R.R. HERBERT. 1971. Establishment of symbiosis between *Rhizobium* and plant cells *in vitro*. *Nature* 232:173–176.

HORAK, J., J. LUSTINEC, J. MESICEK, M. KAMINEK, and D. POLACOKOVA. 1975. Regeneration of diploid and polyploid plants from the stem pith explants of diploid marrow stem kale (*Brassica oleracea* L.). *Ann. Bot.* 39:571–577.

HORVATH BEACH, K. and R.R. SMITH. 1979. Plant regeneration from callus of red and crimson clover. *Plant Sci. Letters* 16:231–237.

HOSOKI, T. and Y. SAGAWA. 1977. Clonal propagation of ginger (*Zingiber officinale* Roscoe) through tissue culture. *HortScience* 12:451–452.

HU, C.Y. and I.M. SUSSEX. 1971. *In vitro* development of embryoids on cotyledons of *Ilex aquifolium*. *Phytomorphology* 21:103–107.

HUBER, J., F. CONSTABEL, and O.L. GAMBORG. 1978. A cell counting procedure applied to embryogenesis in suspension cultures of anise (*Pimpinella anisum* L.). *Plant Sci. Letters* 12:209–215.

HUI, I.H. and S.-Y. ZEE. 1978. *In vitro* plant formation from hypocotyls and cotyledons of leaf-mustard cabbage (*Brassica juncea* Coss). *Z. Pflanzenphysiol.* 89:77–80.

HUSEMANN, W. and J. REINERT. 1976. Regulation of growth and morphogenesis in cell cultures of *Crepis capillaris* by light and phytohormones. *Protoplasma* 90:353–367.

HUSSEY, G. 1975. Totipotency in tissue explants and callus of some members of the Liliaceae, Iridaceae, and Amaryllidaceae. *J. Expt. Bot.* 26:253–262.

HUSSEY, G. 1976. Plantlet regeneration from callus and parent tissue in *Ornithogalum thrysoides*. *J. Expt. Bot.* 27:375–383.

ISLAM, A.S. and K. MARUYAMA. 1970. Response of various organs of *Solanum melongena* to different nutrients in callus formation. *Pak. J. Bot.* 2: 47–52.

JACOBS, G., P. ALLAN, and C.H. BORNMAN. 1970. Tissue culture studies on rose: use of shoot tip explants. 2. Cytokinin gibberellin effects. *Agroplantae* 2:25–28.

JAYAKAR, M. 1971. *In vitro* flowering of *Crepis capillaris*. *Phytomorphology* 20:410–412.

JELASKA, S. 1972. Embryoid formation by fragments of cotyledons and hypocotyls in *Curcurbita pepo*. *Planta* 103:278–280.

JELASKA, S. 1974. Embryogenesis and organogenesis in pumpkin explants. *Physiol. Plant.* 31:257–261.

JOHNSON, B.B. 1978a. *In vitro* propagation of *Gloxinia* from leaf explants. *HortScience* 13:149–150.

JOHNSON, B.B. 1978b. *In vitro* propagation of *Episcia cupreata*. *HortScience* 13:596.

JOHNSON, B.B. and E.D. MITCHELL. 1978. *In vitro* propagation of broccoli from stem, leaf, and leaf rib explants. *HortScience* 13:246–247.

JONES, L.H. 1974. Factors influencing embryogenesis in carrot cultures (*Daucus carota* L.). *Ann Bot.* 38:1077–1088.

KAMADA, H. and H. HARADA. 1979. Influence of several growth regulators and amino acids on *in vitro* organogenesis of *Torenia fournieri*. *J. Expt. Bot.* 30:27–36.

KAMAT, M.G. and P.S. RAO. 1978. Vegetative multiplication of eggplants (*Solanum melongena*) using tissue culture techniques. *Plant Sci. Letters* 13: 57–65.

KAMEYA, T. and K. HINATA. 1970. Induction of haploid plants from pollen grains of *Brassica*. *Japan. J. Breeding* 20:82–87.

KAMEYA, T. and H. UCHIMIYA. 1972. Embryoids derived from isolated protoplasts of carrot. *Planta* 103:356–360.

KANT, U. and A.C. HILDEBRANDT. 1969. Morphology of edible plant cells and tissues *in vitro*. *Can. J. Bot.* 47:849.

KAO, K.N., F. CONSTABEL, M.R. MICHAYLUK, and O.L. GAMBORG. 1971. Cell divisions in cells regenerated from protoplasts of soybean and *Haplopappus gracilis*. *Nature* 232:124.

KAO, K.N. and M.R. MICHAYLUK. 1975. Nutritional requirements for growth of *Vicia hajastana* cells and protoplasts at a very low population density in liquid media. *Planta* 126:105–110.

KAO, K.N. and M.R. MICHAYLUK. 1980. Plant regeneration from mesophyll protoplasts of alfalfa. *Z. Pflanzenphysiol.* 96:135–141.

KAO, K.N., R.A. MILLER, O.L. GAMBORG, and B.L. HARVEY. 1970. Variations in chromosome number and structure in plant cells grown in suspension cultures. *Can. J. Genet. Cytol.* 12:297–301.

KARTEL, N.A. and T.V. MANESHINA. 1978. Regeneration of barley plants in callus-tissue cultures. *Soviet Plant Physiol.* 25:223–226.

KARTHA, K.K., O.L. GAMBORG, F. CONSTABEL, and J.P. SHYLUK. 1974a. Regeneration of cassava plants from apical meristems. *Plant Sci. Letters* 2:107–113.

KARTHA, K.K., O.L. GAMBORG, J.P. SHYLUK, and F. CONSTABEL. 1976. Morphogenetic investigations on *in vitro* leaf culture of tomato (*Lycopersicon esculentum* (Mill.) cv. Starfire) and high frequency plant regeneration. *Z. Pflanzenphysiol.* 77:292–301.

KARTHA, K.K., M.R. MICHAYLUK, K.N. KAO, O.L. GAMBORG, and F. CONSTABEL. 1974b. Callus formation and plant regeneration from mesophyll protoplasts of rape plants (*Brassica napus* L. cv. Zephyr). *Plant Sci. Letters* 3:265–271.

KASPERBAUER, M.J., R.C. BUCKNER, and L.P. BUSH. 1979. Tissue culture of annual ryegrass × tall fescue F_1 hybrids: Callus establishment and plant regeneration. *Crop Sci.* 19:457–460.

KASPERBAUER, M.J. and G.B. COLLINS. 1972. Reconstitution of diploids from anther-derived haploids in tobacco. *Crop Sci.* 12:98–101.

KATO, H. and M. TAKEUCHI. 1963. Morphogenesis *in vitro* starting from single cells of carrot root. *Plant Cell Physiol.* 4:274–293.

KATO, Y. 1975. Adventitious bud formation in etiolated stem segments and leaf callus of *Heloniopsis orientalis* (Liliaceae). *Z. Pflanzenphysiol.* 75:211–216.

KATO, Y. 1978. Induction of adventitious buds on undetached leaves, excised leaves and leaf fragments of *Heloniopsis orientalis*. *Physiol. Plant.* 42:39–44.

KAUL, K. and P.S. SABHARWAL. 1972. Morphogenetic studies on *Haworthia*: establishment of tissue culture and control of differentiation. *Amer. J. Bot.* 59:377–385.

KESSELL, R.H.J. and A.H. CARR. 1972. The effect of dissolved oxygen concentration on growth and differentiation of carrot (*Daucus carota*) tissue. *J. Expt. Bot.* 23:996–1007.

KESSELL, R.H.J., C. GOODWIN, J. PHILIP, and M.W. FOWLER. 1977. The relationship between dissolved oxygen concentration, ATP, and embryogenesis in carrot *(Daucus carota)* tissue cultures. *Plant Sci. Letters* 10:265–274.

KHANNA, P. and J. STABA. 1970. *In vitro* physiology and morphogenesis of *Cheiranthus cheiri* var. Cloth of Gold and *C. cheiri* var. Goliath. *Bot. Gaz.* 131:1–5.

KHANNA, R. and R.N. CHOPRA. 1977. Regulation of shoot-bud and root formation from stem explants of *Lobularia maritima*. *Phytomorphology* 27: 267–274.

KIMBALL, S.L. and E.T. BINGHAM. 1973. Adventitious bud development of soybean hypocotyl sections in culture. *Crop Sci.* 13:758–760.

KING, P.J., I. POTRYKUS, and E. THOMAS. 1978. *In vitro* genetics of cereal: problems and perspectives. *Physiol. Vegetale* 16:381–399.

KOBLITZ, H. 1976. Isolation and cultivation of protoplasts from callus cultures of barley. *Biochem. Physiol. Pflanzen.* 170:287–293.

KOBLITZ, H. and G. SAALBACH. 1976. Callus cultures from apical meristems of barley *(Hordeum vulgare)*. *Biochem. Physiol. Pflanzen.* 170:97–102.

KOCHBA, H. and P. SPIEGEL-ROY. 1973. Effect of culture media on embryoid formation from ovular callus of 'Shamouti' orange *(Citrus sinensis)*. *Z. Pflanzenzuchtg.* 69:156–162.

KOCHBA, J. and P. SPIEGEL-ROY. 1977a. The effects of auxins, cytokinins and inhibitors on embryogenesis in habituated ovular callus of the 'Shamouti' orange *(Citrus sinensis)*. *Z. Pflanzenphysiol.* 81:283–288.

KOCHBA, J. and P. SPIEGEL-ROY. 1977b. Embryogenesis in gamma-irradiated habituated ovular callus of the 'Shamouti' orange as affected by auxin and by tissue age. *Envir. Expt. Bot.* 17:151–159.

KOCHBA, J., P. SPIEGEL-ROY, H. NEUMANN, and S. SAAD. 1978a. Stimulation of embryogenesis in citrus ovular callus by ABA, Ethephon, CCC and Alar and its suppression by GA_3. *Z. Pflanzenphysiol.* 89:427–432.

KOCHBA, J., P. SPIEGEL-ROY, S. SAAD, and H. NEUMANN. 1978b. Stimulation of embryogenesis in *Citrus* tissue culture by galactose. *Naturwissenschaften* 65:261–262.

KOHLENBACH, H.W. 1978. Basic aspects of differentiation and plant regeneration from cell and tissue cultures. p. 355–366. *In* W. Barz *et al.* (eds.) Plant tissue culture and its bio-technological application. Springer-Verlag, Berlin.

KOLAR, Z., J. BARTEK, and B. VYSKOT. 1976. Vegetative propagation of the cactus *Mamillaria woodsii* Craig through tissue cultures. *Experientiae* 32:668–669.

KOMAMINE, A., Y. MOROHASHI, and M. SHIMOKORIYAMA. 1969. Changes in respiratory metabolism in tissue cultures of carrot root. *Plant & Cell Physiol.* 10:411–423.

KONAR, R.N. 1963. A haploid tissue from the pollen of *Ephedra foliata* Boiss. *Phytomorphology* 13:170–174.

KONAR, R.N. and A. KONAR. 1966. Plantlet and flower formation in callus cultures from *Phlox drummondii* Hook. *Phytomorphology* 16:379–382.

KONAR, R.N. and Y.P. OBEROI. 1965. *In vitro* development of embryoids on the cotyledons of *Biota orientalis*. *Phytomorphology* 15:137–140.

KONAR, R.N., E. THOMAS, and H.E. STREET. 1972. Origin and structure of embryoids arising from epidermal cells of the stem of *Ranunculus sceleratus* L. *J. Cell Sci.* 11:77–93.

KOSTOFF, D. 1943. Cytogenetics of the genus *Nicotiana*. States Printing House, Sofia.

KRISHNAMURTHI, M. and J. TLASKAL. 1974. Fiji disease resistant *Saccharum officinarum* var. Pindar sub-clones from tissue cultures. *Proc. Intern. Soc. Sugar Cane Technol.* 15:130–137.

KRUL, W.R. and J.F. WORLEY. 1977. Formation of adventitious embryos in callus cultures of 'Seyval', a French hybrid grape. *J. Amer. Soc. Hort. Sci.* 102:360–363.

KUNISAKI, J.T. 1975. *In vitro* propagation of *Cordyline terminalis* (L.) Kunth. *HortScience* 10:601–602.

LAKSHMANAN, K.K. and K. JANARDHANAN. 1977. Morphogenesis in leaf culture of *Furcraea gigantea*. *Phytomorphology* 27:85–87.

LAKSHMI SITA, G., N.V. RAGHAVA RAM, and C.S. VAIDYANATHAN. 1979. Differentiation of embryoids and plantlets from shoot callus of Sandalwood. *Plant Sci. Letters* 15:265–270.

LAM, S.L. 1975. Shoot formation in potato tuber discs in tissue culture. *Amer. Pot. J.* 52:103–106.

LANDOVA, B. and Z. LANDA. 1974. Organogenesis in callus culture of *Gazania splendens* Moore induced on new medium. *Experientia* 30:832–834.

LARUE, C.D. 1949. Culture of the endosperm of maize. *Amer. J. Bot.* 34:585–586.

LEE, C.W., R.M. SKIRVIN, A.I. SOLTERO, and J. JANICK. 1977. Tissue culture of *Salpiglossis sinuata* L. from leaf discs. *HortScience* 12:547–549.

LITZ, R.E. and R.A. CONOVER. 1980. Somatic embryogenesis in cell cultures of *Carica stipulata*. *HortScience* 15. (in press)

LIU, M.C. and W.H. CHEN. 1976. Tissue and cell culture as aids to sugarcane breeding I. Creation of genetic variation through callus culture. *Euphytica* 25:393–403.

LIU, M.C., Y.J. HUANG, and S.C. SHIH. 1972. The *in vitro* production of plants from several tissues of *Saccharum* species. *J. Agr. Assoc. China* 77:52–58.

LOEWENBERG, J.R. 1969. Cyclamen callus culture. *Can. J. Bot.* 47:2065–2067.

LORZ, H., C.T. HARMS, and I. POTRYKUS. 1976. Regeneration of plants from callus in *Avena sativa* L. *Z. Pflanzenzuchtg.* 77:257–259.

LORZ, H. and I. POTRYKUS. 1979. Regeneration of plants from mesophyll protoplasts of *Atropa belladonna*. *Experientia* 35:313–314.

LORZ, H., I. POTRYKUS, and E. THOMAS. 1977. Somatic embryogenesis from tobacco protoplasts. *Naturwissenschaften* 64:439.

LORZ, H., W. WERNICKE, and I. POTRYKUS. 1979. Culture and plant-regeneration of *Hyoscyamus* protoplasts. *Planta Medica* 36:21–29.

LOWE, K.W. and B.V. CONGER. 1979. Root and shoot formation from callus cultures of tall fescue. *Crop Sci.* 19:397–400.

LUSTINEC, J. and J. HORAK. 1970. Induced regeneration of plants in tissue cultures of *Brassica oleracea*. *Experientia* 26:919–920.

MAEDA, E. 1968. Subculture and organ formation in the callus derived from rice embryos *in vitro*. *Proc. Crop Sci. Soc. Japan* 37:51–58.

MAHESHWARI, S.C. and G.R.P. GUPTA. 1965. Production of embryoids *in vitro* from stem callus of *Foeniculum vulgare*. *Planta* 67:384–386.

MAJUMDAR, S.K. 1970. Culture of *Haworthia* inflorescences *in vitro*. *J. S. Afr. Bot.* 36:63–68.

MALIGA, P., Z.R. KISS, A.H. NAGY, and G. LAZAR. 1978. Genetic instability in somatic hybrids of *Nicotiana tabacum* and *Nicotiana knightiana*. *Molec. Gen. Genet.* 163:145–151.

MALIGA, P., G. LAZAR, F. JOO, A.H. NAGY, and L. MENCZEL. 1977. Restoration of morphogenic potential in *Nicotiana* by somatic hybridization. *Molec. Gen. Genet.* 157:291–296.

MALMBERG, R.L. 1979. Regeneration of whole plants from callus cultures of diverse genetic lines of *Pisum sativum* L. *Planta* 146:243–244.

MALMBERG, R.L., P.J. KOIVUNIEMI, and P.S. CARLSON. 1980. Plant cell genetics—stuck between a phene and its genes. p. 15–30. *In* F. Sala, B. Parisi, R. Cella, and O. Ciferri (eds.) Plant cell culture: results and perspectives. Elsevier North Holland Press, Amsterdam.

MASCARENHAS, A.F., M. PATHAK, R.R. HENDRE, D.D. GHUGALE, and V. JAGANNATHAN. 1975. Tissue cultures of maize, wheat, rice and sorghum: Part IV—studies of organ differentiation in tissue cultures of maize, wheat and rice. *Ind. J. Expt. Biol.* 13:116–119.

MASTELLER, V.J. and D.J. HOLDEN. 1970. The growth of and organ formation from callus tissue of sorghum. *Plant Physiol.* 45:362–364.

MATERN, U., G. STROBEL, and J. SHEPARD. 1978. Reaction to phytotoxins in a potato population derived from mesophyll protoplasts. *Proc. Natl. Acad. Sci. (USA)* 75:4935–4939.

MATHEWS, V.H., T.S. RANGAN, and S. NARAYANASWAMY. 1976. Micro-propagation of *Ananas sativus in vitro*. *Z. Pflanzenphysiol.* 79:450–454.

MATSUOKA, H. and K. HINATA. 1979. NAA-induced organogenesis and embryogenesis in hypocotyl callus of *Solanum melongena* L. *J. Expt. Bot.* 30:363–370.

MCCOY, T.J. and E.T. BINGHAM. 1977. Regeneration of diploid alfalfa plants from cells grown in suspension culture. *Plant Sci. Letters* 10:59–66.

MEHRA, A. and P.N. MEHRA. 1972. Differentiation in callus cultures of *Mesembryanthemum floribundum*. *Phytomorphology* 22:171–176.

MEHRA, A. and P.N. MEHRA. 1974. Organogenesis and plantlet formation *in vitro* in almond. *Bot. Gaz.* 135:61–73.

MEHRA, P.N. and A. MEHRA. 1971. Morphogenetic studies in *Pterotheca falconeri*. *Phytomorphology* 21:174–191.

MEINS, F. 1974. Mechanisms underlying the persistence of tumour anatomy in crown-gall disease. p. 233–264. *In* H.E. Street (ed.) Tissue culture and plant science. Academic Press, New York.

MELCHERS, G. and M.D. SACRISTAN. 1977. Somatic hybridization of plants by fusion of protoplasts II. The chromosome numbers of somatic hybrid plants of four different fusion experiments. p. 169–177. *In* R.J. Gautheret (ed.) La culture des tissus et des cellules des vegetaux. Masson et Cie, Paris.

MELLOR, F.C. and R. STACE-SMITH. 1969. Development of excised potato buds in nutrient culture. *Can. J. Bot.* 47:1617–1621.

MEREDITH, C.P. 1978. Response of cultured tomato cells to aluminum. *Plant Sci. Letters* 12:17–24.

MEREDITH, C.P. 1979. Shoot development in established callus cultures of cultivated tomato (*Lycopersicon esculentum* Mill.). *Z. Pflanzenphysiol.* 95: 405–411.

MEYER, M.M., L.H. FUCHIGAMI, and A.N. ROBERTS. 1975. Propagation of tall bearded irises by tissue culture. *HortScience* 10:479–480.

MEYER, M.M. and G.M. MILBRATH. 1977. *In vitro* propagation of horseradish with leaf pieces. *HortScience* 12:544–545.

MEYER-TEUTER, H. and J. REINERT. 1973. Correlation between rate of cell division and loss of embryogenesis in longterm tissue cultures. *Protoplasma* 78:273–283.

MINOCHA, S.C. 1980. Callus and adventitious shoot formation in excised embryos of white pine *(Pinus strobus)*. *Can. J. Bot.* 58:366–370.

MITRA, J., M.O. MAPES, and F.C. STEWARD. 1960. Growth and organized development of cultured cells IV. The behavior of the nucleus. *Amer. J. Bot.* 47:357–368.

MOK, M.C., W.H. GABELMAN, and F. SKOOG. 1976. Carotenoid synthesis in tissue cultures of *Daucus carota* L. *J. Amer. Soc. Hort. Sci.* 101:442–449.

MOKHTARZADEH, A. and M.J. CONSTANTIN. 1978. Plant regeneration from hypocotyl and anther-derived callus of berseem clover. *Crop Sci.* 18: 567–572.

MOTOYASHI, F. 1971. Protoplasts isolated from callus cells of maize endosperm. *Expt. Cell Res.* 68:452–456.

MUKHOPADHYAY, A. and S.S. BHOJWANI. 1978. Shoot-bud differentiation in tissue cultures of leguminous plants. *Z. Pflanzenphysiol.* 88:263–268.

MURAKISHI, H.H. and P.S. CARLSON. 1976. Regeneration of virus-free plants from dark-green islands of tobacco mosaic virus infected tobacco leaves. *Phytopathology* 66:931–932.

MURASHIGE, T. 1961. Suppression of shoot formation in cultured tobacco cells by gibberellic acid. *Science* 134:280.

MURASHIGE, T. 1974. Plant propagation through tissue cultures. *Ann. Rev.*

Plant. Physiol. 25:135–166.

MURASHIGE, T. 1978. The impact of tissue culture on agriculture. p. 15–26. *In* T.A. Thorpe (ed.) Frontiers of plant tissue culture. Univ. of Calgary Press, Calgary.

MURASHIGE, T. and R. NAKANO. 1967. Chromosome complement as a determinant of the morphogenetic potential of tobacco cells. *Amer. J. Bot.* 54:963–970.

MURASHIGE, T. and F. SKOOG. 1962. A revised medium for rapid growth and bioassays with tobacco tissue cultures. *Physiol. Plant.* 15:473–497.

NADAR, H.M. and D.J. HEINZ. 1977. Root and shoot development from sugarcane callus tissue. *Crop Sci.* 17:814–816.

NADAR, H.M., S. SOEPRAPTOPO, D.J. HEINZ, and S.L. LADD. 1978. Fine structure of sugarcane (*Saccharum* sp.) callus and the role of auxin in embryogenesis. *Crop Sci.* 18:210–216.

NÄF, U. 1958. Studies on tumor formation in *Nicotiana* hybrids. *Growth* 22:167–180.

NAG, K.K. and B.M. JOHRI. 1969. Organogenesis and chromosomal constitution in embryo callus of *Nuytsia floribunda*. *Phytomorphology* 19:405–408.

NAGATA, T. and I. TAKEBE. 1971. Plating of isolated tobacco mesophyll protoplasts on agar medium. *Planta* 99:12–20.

NAGY, J.I. and P. MALIGA. 1976. Callus induction and plant regeneration from mesophyll protoplasts of *Nicotiana sylvestris*. *Z. Pflanzenphysiol.* 78: 453–455.

NAKANO, H., T. TASHIRO, and E. MAEDA. 1975. Plant differentiation in callus tissue induced from immature endosperm of *Oryza sativa* L. *Z. Pflanzenphysiol.* 76:444–449.

NARAYANASWAMY, S. 1977. Regeneration of plants from tissue cultures. p. 179–206. *In* J. Reinert and Y.P.S. Bajaj (eds.) Plant cell, tissue, and organ culture. Springer-Verlag, Berlin.

NATARAJA, K. 1971. Morphogenic variations in callus cultures derived from floral buds and anthers of some members of *Ranunculaceae*. *Phytomorphology* 21:290–296.

NEGRUTIU, I. 1976. *In vitro* morphogenesis in *Arabidopsis thaliana*. *Arabidopsis Inform. Serv.* 13:181–187.

NEGRUTIU, I., F. BEEFTINK, and M. JACOBS. 1975. *Arabidopsis thaliana* as a model system in somatic cell genetics I. Cell and tissue culture. *Plant Sci. Letters* 5:293–304.

NEGRUTIU, I. and M. JACOBS. 1978a. Factors which enhance *in vitro* morphogenesis of *Arabidopsis thaliana*. *Z. Pflanzenphysiol.* 90:423–430.

NEGRUTIU, I. and M. JACOBS. 1978b. Restoration of morphogenetic capacity in long-term callus culture of *Arabidopsis thaliana*. *Z. Pflanzenphysiol.* 90:431–441.

NEGRUTIU, I., M. JACOBS, and D. CACHITA. 1978a. Some factors controlling *in vitro* morphogenesis of *Arabidopsis thaliana*. *Z. Pflanzenphysiol.* 86:

113–124.

NEGRUTIU, I., M. JACOBS, and W. DE GREEF. 1978b. *In vitro* morphogenesis of *Arabidopsis thaliana*: The origin of the explant. *Z. Pflanzenphysiol.* 90:363–372.

NEHLS, R. 1978. Isolation and regeneration of protoplasts from *Solanum nigrum* L. *Plant Sci. Letters* 12:183–187.

NEUMANN, K.H. 1969. Mikroskopische Untersuchungen zum Wachstumsverlauf von Explantaten verscheidener Gewebepartien der Karottenwurzel in Gewebekultur. *Mikroskopie* 25:261–271.

NEUMANN, K.H., E. CIRELLI, and B. CIRELLI. 1969. Untersuchungen uber Beziehungen zwischen Zellteilung und Morphogenese bei Gewebekulturn von *Daucus carota*. III. Der Einfluss des Kinetins auf die Ultrastruktur der Zellen. *Physiol. Plant.* 22:787–800.

NISHI, T., Y. YAMADA, and E. TAKAHASHI. 1968. Organ redifferentiation and plant restoration in rice callus. *Nature* 219:508–509.

NISHI, T., Y. YAMADA, and E. TAKAHASHI. 1973. The role of auxins in differentiation of rice tissues cultured *in vitro*. *Bot. Mag. (Tokyo)* 86:183–188.

NITSCH, J.P. and C. NITSCH. 1969. Haploid plants from pollen grains. *Science* 163:85–87.

NITSCH, J.P., C. NITSCH, L.M.E. ROSSINI, and D. BUI DANG HA. 1967. The role of adenine in bud differentiation. *Phytomorphology* 17:446–453.

NORTON, J.P. and W.G. BOLL. 1954. Callus and shoot formation from tomato roots *in vitro*. *Science* 119:220–221.

OGURA, H. 1975. The effects of a morphactin, chloroflurenol, on organ redifferentiation from calluses cultured *in vitro*. *Bot. Mag. (Tokyo)* 88:1–8.

OGURA, H. 1976. The cytological chimeras in original regenerates from tobacco tissue cultures and in their offsprings. *Japan. J. Genet.* 51:161–174.

OGURA, H. 1978. Genetic control of chromosomal chimerism found in a regenerate from tobacco callus. *Japan. J. Genet.* 53:77–90.

OGURA, H. and S. TSUJI. 1977. Differential responses of *Nicotiana tabacum* L. and its punative progenitors to de- and redifferentiation. *Z. Pflanzenphysiol.* 83:419–426.

O'HARA, J.F. and H.E. STREET. 1978. Wheat callus culture: The initiation, growth and organogenesis of callus derived from various explant sources. *Ann. Bot.* 42:1029–1038.

OHKI, S., C. BIGOT, and J. MOUSSEAU. 1978. Analysis of shoot-forming capacity *in vitro* in two lines of tomato (*Lycopersicon esculentum* Mill.) and their hybrids. *Plant Cell Physiol.* 19:27–42.

OHYAMA, K. and S. OKA. 1978. Bud and root formation in hypocotyl segments of *Broussonetia kazinoki* Sieb. *in vitro*. p. 33. Proc. 4th Intern. Congr. Plant Tissue and Cell Culture. Univ. of Calgary Press, Calgary. Aug. 20–25, 1978. Univ. Calgary, Alberta, Canada.

OKAZAWA, Y., N. KATSURA, and T. TAGAWA. 1967. Effect of auxin and kinetin on the development and differentiation of potato tissue cultured *in vitro. Physiol. Plant.* 20:862–869.

ORTON, T.J. 1980. Chromosomal variation in tissue cultures and regenerated plants of *Hordeum. Theor. Appl. Genet.* 56:101–112.

OSWALD, T.H., A.E. SMITH, and D.V. PHILLIPS. 1977. Callus and plant regeneration from cell cultures of Ladino clover and soybean. *Physiol. Plant.* 38:129–134.

PADMANABHAN, V., E.F. PADDOCK, and W.R. SHARP. 1974. Plantlet formation from *Lycopersicon esculentum* leaf callus. *Can. J. Bot.* 52:1429–1432.

PAREEK, L.K. and N. CHANDRA. 1978a. Differentiation of shoot buds *in vitro* in tissue cultures of *Sisymbrium irio* L. *J. Expt. Bot.* 29:239–244.

PAREEK, L.K. and N. CHANDRA. 1978b. Somatic embryogenesis in leaf callus from cauliflower (*Brassica oleracea* var. *Botrytis*). *Plant Sci. Letters* 11:311–316.

PATAU, K., N.K. DAS, and F. SKOOG. 1957. Induction of DNA synthesis by kinetin and indoleacetic acid in excised tobacco pith tissue. *Physiol. Plant.* 10:949–966.

PELCHER, L.E., O.L. GAMBORG, and K.N. KAO. 1974. Bean mesophyll protoplasts: production, culture and callus formation. *Plant Sci. Letters* 3: 107–111.

PELLETIER, G. and B. DELISE. 1969. Sur la faculté de regeneration des plantes entieres par culture *in vitro* du pedoncule floral de *Petunia pendula. Ann. Amelior. Plantes* 19:353–355.

PENCE, V.C., P.M. HASEGAWA, and J. JANICK. 1979. Asexual embryogenesis in *Theobroma cacao* L. *J. Amer. Soc. Hort. Sci.* 104:145–148.

PETERSON, P.A. 1974. Unstable genetic loci as a probe in morphogenesis. Brookhaven Symp. Biol. 25:244–261.

PHILLIPS, G.C. and G.B. COLLINS. 1979. *In vitro* tissue culture of selected legumes and plant regeneration from callus cultures of red clover. *Crop Sci.* 19:59–64.

PIERIK, R.L.M. 1972. Adventitious root formation in isolated petiole fragments of *Lunaria annua* L. *Z. Pflanzenphysiol.* 60:343–351.

PIERIK, R.L.M. 1976. *Anthurium andraeanum* plantlets produced from callus tissues cultivated *in vitro. Physiol. Plant.* 37:80–82.

PIERIK, R.L.M. 1979. *In vitro* culture of higher plants. Posen en Looijen, Wageningen.

POIRIER-HAMON, S., P.S. RAO, and H. HARADA. 1974. Culture of mesophyll protoplasts and stem segments of *Antirrhinum majus* (snapdragon): growth and organization of embryoids. *J. Expt. Bot.* 25:752–758.

POTRYKUS, I., C.T. HARMS, H. LORZ, and E. THOMAS. 1977. Callus formation from stem protoplasts of corn (*Zea mays* L.). *Molec. Gen. Genet.* 156:347–350.

POWER, J.B. and S.F. BERRY. 1979. Plant regeneration from protoplasts of *Browallia viscosa. Z. Pflanzenphysiol.* 94:469–471.

POWER, J.B., E.M. FREARSON, D. GEORGE, P.K. EVANS, S.F. BERRY, C. HAYWARD, and E.C. COCKING. 1976. The isolation, culture and regeneration of leaf protoplasts in the genus *Petunia*. *Plant Sci. Letters* 7:51–55.

PRABHUDESAI, V.R. and S. NARAYANASWAMY. 1973. Differentiation of cytokinin-induced shoot buds and embryoids on excised petioles of *Nicotiana tabacum*. *Phytomorphology* 23:133–137.

PRABHUDESAI, V.R. and S. NARAYANASWAMY. 1974. Organogenesis in tissue cultures of certain Asclepiads. *Z. Pflanzenphysiol.* 71:181–185.

PRICE, H.J. and R.H. SMITH. 1979. Somatic embryogenesis in suspension cultures of *Gossypium klotzschianum* Anderss. *Planta* 145:305–307.

PROKHOROV, M.N., L.K. CHERNOVA, and B.V. FILIN-KOLDAKOV. 1974. Growing wheat tissues in culture and the regeneration of an entire plant. *Doklady Akad. Nauk* 214:472–475.

RABECHAULT, H., J. AHEE, and G. GUENIN. 1970. Colonies cellulaires et formes embryoides de Palmier à huile (*Elaeis guineenses* Jacq. var. dara BECC.). *C. R. Acad. Sci., Paris* 270:3067–3070.

RADOJEVIC, L., R. VUJICIC, and M. NESKOVIC. 1975. Embryogenesis in tissue culture of *Corylus avellana* L. *Z. Pflanzenphysiol.* 77:33–41.

RAJ BHANSALI, R., A. KUMAR, and H.C. ARYA. 1978a. Morphogenesis in somatic callus culture of *Crotalaria* spp. p. 34. Proc. 4th Intern. Congr. Plant Tissue and Cell Culture. Aug. 20–25, 1978. Univ. of Calgary Press, Calgary, Canada.

RAJ BHANSALI, R., K.G. RAMAWAT, A. KUMAR, and H.C. ARYA. 1978b. Callus initiation and organogenesis in *in vitro* cultures of *Crotalaria burhia*. *Phytomorphology* 28:98–102.

RAMAN, K. and R.I. GREYSON. 1974. *In vitro* induction of embryoids in tissue cultures of *Nigella damascena*. *Can. J. Bot.* 52:1988–1989.

RAMAWAT, K.G., R.R. BHANSALI, and H.C. ARYA. 1977. Differentiation in *Crotalaria* callus cultures. *Phytomorphology* 27:303–307.

RANGAN, T.S. 1974. Morphogenic investigations on tissue cultures of *Panicum miliaceum*. *Z. Pflanzenphysiol.* 72:456–459.

RANGAN, T.S. 1976. Growth and plantlet regeneration in tissue cultures of some Indian millets: *Paspalum scrobienlatum* L., *Eleusine coracana* Gaertn and *Pennisetum typhoideum* Pers. *Z. Pflanzenphysiol.* 78:208–216.

RANGASWAMY, N.S. 1958. Culture of nucellar tissue of *Citrus in vitro*. *Experientia* 14:111–112.

RAO, A.N. 1969. Tissue culture from bulbil of *Dioscorea sansiberensis*. *Can. J. Bot.* 47:565–566.

RAO, P.S. 1965. *In vitro* induction of embryonal proliferation in *Santalum album* L. *Phytomorphology* 15:175–179.

RAO, P.S., W. HANDRO, and H. HARADA. 1973a. Hormonal control of differentiation of shoots, roots and embryos in leaf and stem cultures of *Petunia inflata* and *Petunia hybrida*. *Physiol. Plant.* 28:458–463.

RAO, P.S., W. HANDRO, and H. HARADA. 1973b. Bud formation and embryo differentiation in *in vitro* cultures of *Petunia. Z. Pflanzenphysiol.* 69: 87–90.

RAO, P.S. and H. HARADA. 1974. Hormonal regulation of morphogenesis in organ cultures of *Petunia inflata, Antirrhinum majus* and *Pharbitis nil.* p. 1113–1120. *In* S. Tamura (ed.) Plant growth substances. Hirokawa-Shoten, Tokyo.

RAO, P.S. and S. NARAYANASWAMY. 1968. Induced morphogenesis in tissue cultures of *Solanum xanthocarpum. Planta* 81:372–375.

RAO, P.S. and S. NARAYANASWAMY. 1972. Morphogenetic investigations in callus cultures of *Tylophora indica. Physiol. Plant.* 27:271–276.

RASTE, A.P. and P.S. GANAPATHY. 1970. *In vitro* behavior of inflorescence segments of *Mazus pumilus. Phytomorphology* 20:367–374.

REDEI, G.P. 1975. *Arabidopsis* as a genetic tool. *Annu. Rev. Genet.* 9:111–127.

REYNOLDS, J.F. and T. MURASHIGE. 1979. Asexual embryogenesis in callus cultures of palms. *In Vitro* 15:383–387.

REUTHER, G. 1977a. Embryoide differenzierungsmuster im Kallus der Gattungen *Iris* und *Asparagus. Ber. Deutsch. Bot. Ges.* 90:417–437.

REUTHER, G. 1977b. Adventitious organ formation and somatic embryogenesis in callus of asparagus and iris and its possible application. *Acta Hort.* 78: 217–224.

REUTHER, G. 1978. Cloning of female and male asparagus strains by tissue culture. *Gartenbauwissenschaft* 43:1–10.

RICE, T.B., R.K. REID, and P.N. GORDON. 1978. Morphogenesis in field crops. p. 262–277. *In* K. Hughes, R. Henke and M. Constantin (eds.) Propagation of higher plants through tissue culture. Tech. Info. Center, U.S. Dept. of Energy, GPO, Washington, D.C.

ROBB, S.H. 1957. The culture of excised tissue from bulb scales of *Lilium speciosum* Hun. *J. Expt. Bot.* 8:348–352.

SACRISTAN, M.D. 1971. Karyotypic changes in callus cultures from haploid and diploid plants of *Crepis capillaris* (L.) Wallr. *Chromosoma* 33:273–283.

SACRISTAN, M.D. and G. MELCHERS. 1969. The karyological analysis of plants regenerated from tumorous and other callus cultures of tobacco. *Molec. Gen. Genet.* 105:317–333.

SANGWAN, R.S. and R. GORENFLOT. 1975. *In vitro* culture of *Phragmites* tissues. Callus formation, organ differentiation and cell suspension culture. *Z. Pflanzenphysiol.* 75:256–259.

SANGWAN, R.S. and H. HARADA. 1975. Chemical regulation of callus growth, organogenesis, plant regeneration, and somatic embryogenesis in *Antirrhinum majus* tissue and cell cultures. *J. Expt. Bot.* 26:868–881.

SANGWAN, R.S. and H. HARADA. 1976. Chemical factors controlling morphogenesis of *Petunia* cells cultured *in vitro. Biochem. Physiol. Pflanzen* 170:77–84.

SANGWAN, R.S., B. NORREEL, and H. HARADA. 1976. Effects of kinetin and gibberellin A_3 on callus growth and organ formation in *Limnophila chinensis* tissue culture. *Biol. Plant.* 18:126–131.

SCHENK, R.U. and A.C. HILDEBRANDT. 1972. Medium and techniques for induction and growth of monocotyledonous and dicotyledonous plant cell cultures. *Can. J. Bot.* 50:199–204.

SCHIEDER, O. 1975. Regeneration of haploid and diploid *Datura innoxia* Mill. mesophyll-protoplasts to plants. *Z. Pflanzenphysiol.* 76:462–466.

SCHIEDER, O. 1977. Attempts in regeneration of mesophyll protoplasts of haploid and diploid wild type lines, and those of chlorophyll-deficient strains from different *Solanaceae. Z. Pflanzenphysiol.* 84:275–281.

SCHIEDER, O. 1978. Somatic hybrids of *Datura innoxia* Mill. + *Datura discolor* Bernh. and of *Datura innoxia* Mill + *Datura stramonium* L. var. tatula L. *Molec. Gen. Genet.* 162:113–119.

SCHROEDER, C.A., E. KAY, and L.H. DAVIS. 1962. Totipotency of cells from fruit pericarp tissue *in vitro. Science* 138:595–596.

SCORZA, R. and J. JANICK. 1976. Tissue culture in *Passiflora.* p. 179–183. *In* Proc. 24th Congr. Amer. Soc. Hort. Sci., Tropical Region. Mayaguez, Puerto Rico. Amer. Soc. Hort. Sci., Alexandria, Va.

SCOWCROFT, W.R. and J.A. ADAMSON. 1976. Organogenesis from callus cultures of the legume, *Stylosanthes hamata. Plant Sci. Letters* 7:39–42.

SEABROOK, J. and B.G. CUMMING. 1977. The *in vitro* propagation of amaryllis (*Hippeastrum* spp. hybrids). *In Vitro* 13:831–836.

SEABROOK, J., B.G. CUMMING, and L.A. DIONNE. 1976. The *in vitro* induction of adventitious shoot and root apices on *Narcissus* (daffodil and narcissus) cultivar tissue. *Can. J. Bot.* 54:814–819.

SEHGAL, C.B. 1968. *In vitro* development of neomorphs in *Anethum graveolens* L. *Phytomorphology* 18:509–514.

SEHGAL, C.B. 1978. Differentiation of shoot-buds and embryoids from inflorescence of *Anethum graveolens* in cultures. *Phytomorphology* 28:291–297.

SEN, B. and S. GUPTA. 1979. Differentiation in callus cultures of leaf of two species of *Trigonella. Physiol. Plant.* 45:425–428.

SHAMA RAO, H.K. and S. NARAYANASWAMY. 1975. Effect of gamma irradiation on cell proliferation and regeneration in explanted tissues of pigeon pea, *Cajanus cajan* (L.) Mills P. *Radiat. Bot.* 15:301–305.

SHARMA, G.C., L.L. BELLO, and V.T. SAPRA. 1978. Callus induction and regeneration of plantlets from immature embryos in hexaploid triticale (× *Tritcosecale* Wittmack). p. 258. *In* K. Hughes, R. Henke and M. Constantin (eds.) Propagation of higher plants through tissue culture. Tech. Info. Center, U.S. Dept. of Energy, GPO, Washington, D.C.

SHARP, W.R., M.R. SONDAHL, L.S. CALDAS, and S.B. MARAFFA. 1980. The physiology of asexual embryogenesis. p. 268–310. *In* J. Janick (ed.) Horticultural reviews, Vol. 2. AVI Publishing, Westport, Conn.

SHEN-MILLER, J. and W.R. SHARP. 1966. An improved medium for rapid initiation of *Arabidopsis* tissue culture from seed. *Bul. Torr. Bot. Club* 93:68–69.

SHEPARD, J.F., D. BIDNEY, and E. SHAHIN. 1980. Potato protoplasts in crop improvement. *Science* 208:17–24.

SHEPARD, J.P. and R.E. TOTTEN. 1977. Mesophyll cell protoplasts of potato. *Plant Physiol.* 60:313–316.

SHERIDAN, W.F. 1968. Tissue culture of the monocot *Lilium. Planta* 82: 189–192.

SHERIDAN, W.F. 1975. Plant regeneration and chromosome stability in tissue cultures. p. 263–295. *In* L. Ledoux (ed.) Genetic manipulations with plant materials. Plenum Press, London.

SHIMADA, T. 1978. Plant regeneration from the callus induced from wheat embryo. *Japan. J. Genet.* 53:371–374.

SHIMADA, T., T. SASAKUMA, and K. TSUNEWALEI. 1969. *In vitro* culture of wheat tissues. I. Callus formation, organ redifferentiation and single cell culture. *Can. J. Genet. Cytol.* 11:294–304.

SIMONSEN, J. and A.C. HILDEBRANDT. 1971. *In vitro* growth and differentiation of *Gladiolus* plants from callus cultures. *Can. J. Bot.* 49:1817–1819.

SINK, K.C. and J.B. POWER. 1977. The isolation, culture and regeneration of leaf protoplasts of *Petunia parviflora* Juss. *Plant Sci. Letters* 10:335–340.

SKIRVIN, R.M. and M.C. CHU. 1979. *In vitro* propagation of 'Forever yours' rose. *HortScience* 14:608–610.

SKIRVIN, R.M. and J. JANICK. 1976. Tissue culture-induced variation in scented *Pelargonium* spp. *J. Amer. Soc. Hort. Sci.* 101:281–290.

SKOLMEN, R.G. and M.O. MAPES. 1976. *Acacia koa* Gray plantlets from somatic callus tissue. *J. Hered.* 67:114–115.

SMITH, H.H. and I.A. MASTRANGELO-HOUGH. 1979. Genetic variability available through cell fusion. p. 265–285. *In* W.R. Sharp *et al.* (eds.) Plant cell and tissue culture: principles and applications. Ohio State Univ. Press, Columbus.

SMITH, S.M. and H.E. STREET. 1974. The decline of embryogenic potential as callus and suspension cultures of carrot are serially subcultured. *Ann. Bot.* 38:223–241.

SOMMER, H.E. and C.L. BROWN. 1979. Applications of tissue cultures to forest tree improvement. p. 461–492. *In* W.R. Sharp *et al.* (eds.) Plant cell and tissue culture: principles and applications. Ohio State Univ. Press, Columbus.

SOMMER, H.E., C.L. BROWN, and P.P. KORMANIK. 1975. Differentiation of plantlets in long-leaf pine (*Pinus palustris* Mill.) tissue cultured *in vitro. Bot. Gaz.* 135:196–200.

SONDAHL, M.R. and W.R. SHARP. 1977. High frequency induction of somatic embryos in cultured leaf explants of *Coffea arabica* L. *Z. Pflanzenphysiol.* 81:395–408.

SONDAHL, M.R. and W.R. SHARP. 1979. Research in *Coffea* and applications of tissue culture methods. p. 527–584. *In* W.R. Sharp *et al.* (eds.) Plant cell and tissue culture: principles and applications. Ohio State Univ. Press, Columbus.

SONDAHL, M.R., D.A. SPAHLINGER, and W.R. SHARP. 1979. A histological study of high frequency and low frequency induction of somatic embryos in cultured leaf explants of *Coffea arabica* L. *Z. Pflanzenphysiol.* 94:101–108.

SOPORY, S.K. and S.C. MAHESHWARI. 1976. Morphogenetic potentialities of haploid and diploid vegetative parts of *Datura innoxia*. *Z. Pflanzenphysiol.* 77:274–277.

SREE RAMULU, K., M. DEVREUX, G. ANCORA, and U. LANERI. 1976. Chimerism in *Lycopersicon peruvianum* plants regenerated from *in vitro* cultures of anthers and stem internodes. *Z. Pflanzenzuchtg.* 76:299–319.

SRIVASTAVA, P.S. 1973. Formation of triploid "plantlets" in endosperm cultures of *Putranjiva* × *roxburghii*. *Z. Pflanzenphysiol.* 69:270–273.

STARITSKY, G. 1970. Embryoid formation in callus tissue of coffee. *Acta Bot. Neerl.* 19:509–514.

START, N.D. and B.G. CUMMING. 1976. *In vitro* propagation of *Saintpaulia ionantha* Wendl. *HortScience* 11:204–206.

STEWARD, F.C. 1958. Interpretations of growth from free cells to carrot plants. *Amer. J. Bot.* 45:709–713.

STEWARD, F.C., P.V. AMMIRATO, and M.O. MAPES. 1970. Growth and development of totipotent cells: some problems, procedures, and perspectives. *Ann. Bot.* 34:761–787.

STEWARD, F.C. and S.M. CAPLIN. 1951. A tissue culture from potato tuber: the synergistic action of 2,4-D and of coconut milk. *Science* 113:518–520.

STIMART, D.P. and P.D. ASCHER. 1978. Tissue culture of bulb scale sections for asexual propagation of *Lilium longiflorum* Thunb. *J. Amer. Soc. Hort. Sci.* 103:182–184.

STOUTMEYER, V.T. and O.K. BRITT. 1965. The behavior of tissue cultures from English and Algerian ivy in different growth phases. *Amer. J. Bot.* 52:805–810.

STRINGHAM, G.R. 1979. Regeneration in leaf-callus of haploid rapeseed (*Brassica napus* L.). *Z. Pflanzenphysiol.* 92:459–462.

SUBRAMANIAN, V. and P.V. SUBBA RAO. 1980. *In vitro* culture of the allergenic weed *Parthenium lysterophorus* L. *Plant Sci. Letters* 17:269–277.

SUNDERLAND, N. 1977. Nuclear cytology. p. 177–206. *In* H.E. Street (ed.) Plant tissue and cell culture, 2nd edition. Univ. of California Press, Berkeley.

SUNG, Z.R., R. SMITH, and J. HOROWITZ. 1979. Quantitative studies of embryogenesis in normal and 5-methyltryptophan-resistant cell lines of wild carrot. *Planta* 147:236–240.

SUSSEX, I.M. 1972. Somatic embryos in long term carrot tissue cultures: histology, cytology, and development. *Phytomorphology* 22:50–59.

SUSSEX, I.M. and K.A. FREI. 1968. Embryoid development in long term tissue cultures of carrot. *Phytomorphology* 18:339–349.

SYONO, K. 1965. Physiological and biochemical changes of carrot root callus during successive cultures. *Plant Cell Physiol.* 6:371–392.

SYONO, K. and T. FURUYA. 1972a. Studies on plant-tissue culture 18. Abnormal flower formation of tobacco plants regenerated from callus cultures. *Bot. Mag. (Tokyo)* 85:273–284.

SYONO, K. and T. FURUYA. 1972b. The differentiation of *Coptis* plants *in vitro* from callus cultures. *Experientia* 28:236.

TABATA, M., H. YAMAMOTO, N. HIRAOKA, and M. KONOSHIMA. 1972. Organization and alkaloid production in tissue cultures of *Scopolia parviflora. Phytochemistry* 11:949–955.

TAKATORI, F.H., T. MURASHIGE, and J.I. STILLMAN. 1968. Vegetative propagation of asparagus through tissue culture. *HortScience* 3:20–22.

TAKAYAMA, S. and M. MISAWA. 1979. Differentiation in *Lilium* bulb scales grown *in vitro*. Effect of various cultural conditions. *Physiol. Plant.* 46:184–190.

TAKEBE, I., G. LABIB, and G. MELCHERS. 1971. Regeneration of whole plants from isolated mesophyll protoplasts of tobacco. *Naturwissenschaften* 58:318–320.

TAL, M., K. DEHAN, and H. HEIKIN. 1977. Morphogenetic potential of cultured leaf sections of cultivated and wild species of tomato. *Ann. Bot.* 41:937–941.

TAZAWA, M. and J. REINERT. 1969. Extracellular and intracellular chemical environments in relation to embryogenesis *in vitro. Protoplasma* 68:157–173.

THOMAS, E., F. HOFFMAN, I. POTRYKUS, and G. WENZEL. 1976. Protoplast regeneration and stem embryogenesis of haploid androgenetic rape. *Mol. Gen. Genet.* 145:245–247.

THOMAS, E., F. HOFFMANN, and G. WENZEL. 1975. Haploid plantlets from microspores of rye. *Z. Pflanzenzuchtg.* 75:106–113.

THOMAS, E., P.J. KING, and I. POTRYKUS. 1977. Shoot and embryo-like structure formation from cultured tissues of *Sorghum bicolor. Naturwissenschaften* 64:587.

THOMAS, E. and H.E. STREET. 1970. Organogenesis in cell suspension cultures of *Atropa belladonna* L. and *Atropa belladonna* cultivar *lutea* Doll. *Ann. Bot.* 34:657–669.

THOMAS, E. and G. WENZEL. 1975. Embryogenesis from microspores of rye. *Naturwissenschaften* 62:40–41.

THOMAS, E. and W. WERNICKE. 1978. Morphogenesis in herbaceous crop plants. p. 403–410. *In* T.A. Thorpe (ed.) Frontiers of plant tissue culture. Univ. of Calgary Press, Calgary.

THOMAS, M.J., E. DUHOUX, and J. VAZART. 1977. *In vitro* organ initiation in tissue cultures of *Biota orientalis* and other species of the Cupressaceae. *Plant Sci. Letters* 8:395–400.

TILQUIN, J.P. 1979. Plant regeneration from stem callus of cassava. *Can. J. Bot.* 57:1761–1763.

TISSERAT, B., E.B. ESAN, and T. MURASHIGE. 1979. Somatic embryogenesis in angiosperms. p. 1–78. *In* J. Janick (ed.) Horticultural reviews, Vol. 1. AVI Publishing, Westport, Conn.

TISSERAT, B. and T. MURASHIGE. 1977. Repression of asexual embryogenesis *in vitro* by some plant growth regulators. *In Vitro* 13:799–805.

TOREN, J. 1955. Effects des carences minérales sur l'anatomie des tissus végétaux cultivés *in vitro*. *Rev. Gén. Bot.* 62:392–422.

TORREY, J.G. 1967. Morphogenesis in relation to chromosomal constitution in long-term plant tissue cultures. *Physiol. Plant.* 20:265–275.

TORREY, J.G. 1975. Tracheary element formation from single isolated cells in culture. *Physiol. Plant.* 35:158–165.

TORREY, J.G. and J. REINERT. 1961. Suspension cultures of higher plant cells in synthetic media. *Plant. Physiol.* 36:483–491.

TRAN THANH VAN, K. and H. TRINH. 1978. Morphogenesis in thin cell layers: concept, methodology and results. p. 37–48. *In* T.A. Thorpe (ed.) Frontiers of plant tissue culture. Univ. of Calgary Press, Calgary.

ULRICH, J.M. and G. MACKINNEY. 1970. Callus cultures of tomato mutants. II. Carotenoid formation. *Physiol. Plant.* 23:88–92.

VARDI, A., P. SPIEGEL-ROY, and E. GALUN. 1975. Citrus cell culture: isolation of protoplasts, plating densities, effect of mutagens and regeneration of embryos. *Plant Sci. Letters* 4:231–236.

VASIL, I.K., M.R. AHUJA, and V. VASIL. 1979. Plant tissue cultures in genetics and plant breeding. *Adv. Genet.* 20:127–215.

VASIL, I.K. and A.C. HILDEBRANDT. 1965. Differentiation of tobacco plants from single, isolated cells in micro cultures. *Science* 150:889–892.

VASIL, I.K. and A.C. HILDEBRANDT. 1966a. Variation of morphogenetic behavior in plant tissue cultures I. *Cichorium endiva*. *Amer. J. Bot.* 53:860–869.

VASIL, I.K. and A.C. HILDEBRANDT. 1966b. Variations of morphogenetic behavior in plant tissue cultures II. *Petroselinum hortense*. *Amer. J. Bot.* 53:869–874.

VASIL, V. and I.K. VASIL. 1979. Isolation and culture of cereal protoplasts I. Callus formation from pearl millet *(Pennisetum americanum)* protoplasts. *Z. Pflanzenphysiol.* 92:379–383.

VASIL'TSOVA, T.M. 1967. Studies in callusogenesis and somatic embryogenesis in leaf and stem explants of cabbage. *Sel'skokhozyaystvennaya Biologiya* 2:233–248.

VASQUEZ, A.M. and K.C. SHORT. 1978. Morphogenesis in cultured floral parts of African violet. *J. Expt. Bot.* 29:1265–1271.

VELIKY, I.A. and S.M. MARTIN. 1970. A fermenter for plant cell suspension cultures. *Can. J. Microbiol.* 16:223–226.

VENKETESWARAN, S. and O. HUHTINEN. 1978. *In vitro* root and shoot differentiation from callus cultures of a legume, winged bean, *Psophocarpus tetragonolobus*. *In Vitro* 14:355.

VERMA, D.C. and D.K. DOUGALL. 1977. Influence of carbohydrates on quantitative aspects of growth and embryo formation in wild carrot suspension cultures. *Plant. Physiol.* 59:81–85.

VERMA, D.P.S. and R.B. VAN HUYSTEE. 1970. Cellular differentiation and peroxidase isozymes in cell culture of peanut cotyledons. *Can. J. Bot.* 48:429–431.

VNUCHKOVA, V.A. 1978. Development of a method for obtaining regenerate tomato plants under tissue-culture conditions. *Soviet Plant Physiol.* 24:884–889.

VON ARNOLD, S. and T. ERIKKSON. 1978. Induction of adventitious buds on embryos of Norway spruce grown *in vitro*. *Physiol. Plant.* 44:283–287.

VYSKOT, B. and F.J. NOVAK. 1977. Habituation and organogenesis in callus cultures of chlorophyll mutants of *Nicotiana tabacum* L. *Z. Pflanzenphysiol.* 81:34–42.

WALKER, K.A., M.L. WENDELN, and E.G. JAWORSKI. 1979. Organogenesis in callus tissue of *Medicago sativa*. The temporal separation of induction processes from differentiation processes. *Plant Sci. Letters* 16:23–30.

WALKEY, D.G.A. and J.M.G. WOOLFITT. 1968. Clonal multiplication of *Nicotiana rustica* L. from shoot meristems in culture. *Nature* 220:1346–1347.

WANG, P.-J. and L.-C. HUANG. 1975. Callus cultures from potato tissues and the exclusion of potato virus X from plants regenerated from stem tips. *Can. J. Bot.* 53:2565–2567.

WENZEL, G. 1973. Isolation of leaf protoplasts from haploid plants of *Petunia,* rape, and rye. *Z. Pflanzenzuchtg.* 69:58–61.

WERNICKE, W. and H.W. KOHLENBACH. 1975. Antherenkulturen bei *Scopolia*. *Z. Pflanzenphysiol.* 77:89–93.

WERNICKE, W., H. LORZ, and E. THOMAS. 1979. Plant regeneration from leaf protoplasts of haploid *Hyoscyamus muticus* L. produced via anther culture. *Plant Sci. Letters* 15:239–249.

WESSELS, D.C.J., E.G. GROENEWALD, and A. KOELEMAN. 1976. Callus formation and subsequent shoot and root development from leaf tissue of *Haworthia planifolia* var. *setulifera* v. Poelln. *Z. Pflanzenphysiol.* 78:141–145.

WETHERELL, D.F. and D.K. DOUGALL. 1976. Sources of nitrogen supporting growth and embryogenesis in cultured wild carrot tissue. *Physiol. Plant.* 37:97–103.

WHITE, D.W.R. and I.K. VASIL. 1979. Use of amino acid analogue-resistant cell lines for selection of *Nicotiana sylvestris* somatic cell hybrids. *Theor. Appl. Genet.* 55:107–112.

WHITE, P.R. 1963. A handbook of plant tissue culture. Jacques Cottell Press, Lancaster, Pa.

WIDHOLM, J. 1977. Selection and characterization of amino acid analog resistant plant cell cultures. *Crop Sci.* 17:597–600.

WILLIAMS, L. and H.A. COLLIN. 1976a. Embryogenesis and plantlet formation in tissue cultures of celery. *Ann. Bot.* 40:325–332.

WILLIAMS, L. and H.A. COLLIN. 1976b. Growth and cytology of celery plants derived from tissue cultures. *Ann. Bot.* 40:333–338.

WINTON, L. 1968. Plantlets from aspen tissue cultures. *Science* 160:1234–1235.

WINTON, L. 1970. Shoot and tree production from aspen tissue cultures. *Amer. J. Bot.* 57:904–909.

WINTON, L. 1978. Morphogenesis in clonal propagation of woody plants. p. 419–426. *In* T.A. Thorpe (ed.) Frontiers of plant tissue culture. Univ. of Calgary Press, Calgary.

WINTON, L. and S.A. VERHAGEN. 1977. Shoots from Douglas-fir cultures. *Can. J. Bot.* 55:1246–1250.

WODZICKI, T.J. 1978. Seasonal variation of auxin in stem cambial region of *Pinus silvestris. Acta Soc. Bot. Polon.* 47:225–231.

WOLTER, K.E. 1968. Root and shoot initiation in aspen callus cultures. *Nature* 219:509–510.

WU, L. and H.W. LI. 1971. Induction of callus tissue initiation from different somatic organs of rice plant by various concentrations of 2,4-dichlorophenoxyacetic acid. *Cytologia* 36:411–416.

YAMADA, Y., T. NISHI, T. YASUDA, and E. TAKAHASHI. 1967. Sterile culture of rice cells, *Oryza sativa*, and its application. p. 377–386. *In* M. Miyakawa and T.D. Luckey (eds.) Advances in germfree research gnotobiology. CRC Press, Cleveland, Ohio.

YAMAGUCHI, T. and T. NAKAJIMA. 1974. Hormonal regulation of organ formation in cultured tissue derived from root tuber of sweet potato. p. 1121–1127. *In* S. Tamura (ed.) Plant growth substances. Hirokawa-Shoten, Tokyo.

YAMANE, Y. 1974. Induced differentiation of buckwheat plants from subcultured calluses *in vitro. Japan. J. Genet.* 49:139–146.

YANG, Y.W. and W.C. CHANG. 1979. *In vitro* plant regeneration from leaf explants of *Stevia rebaudiana* Bertoni. *Z. Pflanzenphysiol.* 93:337–343.

YOKOYAMA, T. and M. TAKEUCHI. 1976. Organ and plantlet formation from callus in Japanese persimmon *(Diospyros kaki). Phytomorphology* 26: 273–275.

ZAPATA, F.J., P.K. EVANS, J.P. POWER, and E.C. COCKING. 1977. The effect of temperature on the division of leaf protoplasts of *Lycopersicon esculentum* and *Lycopersicon peruvianum. Plant Sci. Letters* 8:119–124.

ZEE, S.Y. and L.H. HUI. 1977. *In vitro* plant regeneration from hypocotyl and cotyledons of Chinese kale (*Brassica alboglabra* Bailey). *Z. Pflanzenphysiol.* 82:440–445.

ZEE, S.Y. and S.C. WU. 1979. Embryogenesis in the petiole explants of Chinese celery. *Z. Pflanzenphysiol.* 93:325–335.

ZEE, S.Y., S.C. WU, and S.B. YUE. 1978. Morphogenesis of the hypocotyl explants of chinese kale. *Z. Pflanzenphysiol.* 90:155–163.

ZENKTELER, M. 1972. *In vitro* formation of plants from leaves of several species of the *Solanaceae* family. *Biochem. Physiol. Pflanzen.* 163:509–512.

7

Stock and Scion Growth Relationships and the Dwarfing Mechanism in Apple[1]

R.G. Lockard and G.W. Schneider
Department of Horticulture, University of Kentucky,
Lexington, Kentucky 40546

[1] The investigation reported in this paper (No. 80-10-60) is in connection with a project of the Kentucky Agricultural Experiment Station and is published with approval of the Director.

I. INTRODUCTION

The practical, economic, and scientific importance of stock effects on apple tree growth and fruit production has been recognized for many years. This paper is concerned with the possible mechanism whereby a dwarfing apple rootstock restricts growth of a scion cultivar and thereby limits size and/or weight of the entire tree. An acceptable explanation of the mechanism involved in rootstock and interstock dwarfing effects has not been developed even though a number of articles and reviews have been published on the subject (Roberts 1949; Sax 1954; Beakbane 1956; Dickinson and Samuels 1956; Rogers and Beakbane 1957; Scholz 1957; Robitaille 1970; Tubbs 1973; Jones 1976a). This report contains primarily recent findings that provide new information and concepts which enable us to refine and advance ideas concerning possible mechanisms for stock control of top growth.

The early explanations of rootstock dwarfing assumed that a reduced water or nutrient supply reached the scion from the roots. This concept is inconsistent with the findings that scion leaf nutrients differ very little among different rootstocks (Awad and Kenworthy 1963; Lockard 1976; Schneider *et al.* 1978a; Chaplin and Westwood 1980).

Since these concepts were inadequate, others that may be applicable to both rootstocks and interstocks were developed. Tubbs (1967) suggested that there must be a feedback mechanism from the scion to the rootstock. He also proposed a theory based on destruction or inhibition of auxin by substances in or coming from the stock rather than on the production of auxin in the scion. This would fit many of the observed rootstock and interstock effects. Sax (1953) suggested that the dwarfing effect of an inverted ring of bark on the stem may be due to altering auxin flow to the roots.

There must be more than one type of plant dwarfing mechanism. Several dwarf vegetables do not transmit their dwarfing characteristics from the root to the stem, as shown by the work of Denna (1963) on squash, Lockard and Grunwald (1970) on peas, and Rick (1952) on a tomato mutant. In these cases, final scion growth control must reside in the scion, and is not greatly affected by the type of root that supports it.

Tree size can be limited by factors other than rootstocks. Such factors include length of internodes, precocity, yield, incompatible graft unions, nutrient supply, soil type, rate of water use, and climatic conditions. This review will not consider any of these growth-limiting factors, but will discuss possible mechanisms in apple trees by which the size of the tree is controlled by the dwarfing characteristics of the stock.

There is limited data that deals directly with the mechanism of apple tree dwarfing by rootstocks or interstocks. Thus, this review includes many studies not directly concerned with apples, but which provide information potentially important to an understanding of growth control or regulation. The sections on the various growth substances are not intended to be exhaustive, but are selective of concepts that may elucidate the possible mechanism of growth control as exercised by apple rootstocks or interstocks on scion growth.

II. TOP-ROOT RELATIONSHIPS

The interdependence of plant shoots and roots is apparent since roots provide water and nutrients for most vascular plants whereas shoots and their foliage provide most metabolites required for root growth.

A study of the effect of 5 root and 1 leaf temperature on 12 grass species led Davidson (1969) to the conclusion that partitioning of photosynthate is controlled by relative rates of photosynthesis and root absorption by inverse proportion according to the following formula: root mass × rate of absorption/leaf mass × rate of photosynthesis. Later a similar relationship was reported (Hunt *et al.* 1975; Hunt 1975) in cranberry and perennial ryegrass. The top-root ratio of a given cultivar is not affected by vigor of the stock (Beakbane 1956; Rogers and Booth 1959; Tubbs 1973). However, the top-root ratio of one combination changed from 1:1 when grown on a sandy soil to 2:1 when grown on a fertile loam soil (Rogers and Booth 1959).

The constant stem increment/root increment ratio implies that each grows at the same rate and that even if the two parts have different growth rates before combination, a new and uniform rate of growth is established after union (Barlow 1959). This includes adjustment to treatments that affect the growth of either part, such as defoliation or root pruning (Wareing 1970).

The modifying effect of soil conditions on apple top-root relationship was demonstrated by Gur *et al.* (1976a) who showed that increased root temperature decreased photosynthetic activity and increased the shoot-root ratio. The same phenomenon was found in *Nicotiana rustica* and *Phaseolus vulgaris* (Itai *et al.* 1973). In addition, clonal apple rootstocks had different levels of tolerance to high soil temperatures (Carlson 1965), and this was also affected by the scion cultivar (Gur *et al.* 1976b).

Interdependence of root and shoot growth in peach seedlings was shown by physically restricting root volume (Richards and Rowe 1977a). Growth control factors from root and/or top must have been responsible for the 34% reduction in top dry weight associated with a 39% reduction in root dry weight, for apparently it was not due to reduced root capacity to absorb and translocate water. Buttrose and Mullins (1968) demonstrated that shoot elongation rate of young grape vines could be regulated by size of root system when controlled by root pruning, and this was not due to reduced water or nutrients. Lateral soybean roots play a role in leaf senescence (Hsia and Kao 1978), which may also be related to root control of top growth. Likewise, root pruning was followed by rapid root growth and slowed top growth until the former top-root ratio was restored. Richards and Rowe (1977b) also showed that, irrespective of nutrient solution treatment, total water and nutrient uptake by roots was directly related to total plant dry weight.

Leaf or stem cuttings usually do not grow until roots are formed (Carr 1966). Tying a stem into a complete loop inhibited lateral shoot growth until roots formed at the base of the loop even though the roots were in deionized water (Dickinson and Samuels 1956; Smith and Wareing 1964; Wareing 1970). This effect was evident even when the stem was girdled. Translocation of leaf assimilates to tea roots appeared to be dependent on a feeder root stimulus and nitrogen assimilation by the root was dependent on root reserves from leaf activity, yet both were interrupted by ringing the trunk (Kandiah and Wimaladharma 1978). They reported a reciprocal relationship between feeder roots and leaves and showed that the organs were functionally interdependent. Tea feeder root activity was greater when the terminal bud was active and root growth was greater under active plants than under dormant ones (Kulasegaram and Kathiravetpillai 1972b). Similarly, Cripps (1970) found that apple root growth was concurrent with shoot growth. Richardson (1959) found that terminal buds on *Acer* seedlings seemed to control root production and leaf surface controlled root elongation. Selvendran (1970) reported that cambial activity is initiated over the whole plant at bud break and this in turn activates hydrolysis of reserves to sugars and their translocation to actively growing areas. Earlier, Vaadia and Itai (1968) concluded that plant hormones exercise control of root-top relationships even though direct evidence was limited. Sachs (1972) proposed that growing points form signals, the strength of which depends on growth rate, which attracts nutrients and hormones from the rest of the plant. Earlier, Carr (1966) suggested the same relationship and in addition concluded that roots import organic compounds from shoots and then convert or alter and re-export at least some of them to the top. Thus, he was proposing the presence of a circulatory system in plants. He also reported that the root can regulate shoot metabolic activity by exporting reduced or unreduced nitrogeneous compounds to the shoot.

The rootstock can affect the scion in the following ways: time of leaf senescence (Kandiah 1979); the ratios of soluble sugar to protein, soluble sugar to starch, and free amino acids to protein in grape (Martin *et al.* 1974); photosynthetic rate (Carr 1966; Ferree and Barden 1971); growth rate and duration (Roberts 1949; Tubbs 1973); precocity (Tubbs 1973); thickness of bark, especially near the stock (Lockard 1976; Yadava and Doud 1978); amount of parenchymatous tissue in xylem and phloem (Rogers and Beakbane 1957); size of wood ray tissue and xylem vessels (McKenzie 1961); apical dominance (Rogers and Beakbane 1957; Tubbs 1973); leaf size (Swarbrick and Naic 1932); metabolic activity of the shoot (Carr 1966); mineral composition of foliage (Schneider *et al.* 1978a; Poling and Oberly 1979; Chaplin and Westwood 1980). The available data on stock effect on foliar mineral composition do not provide good information on composition during the critical period of active growth when nutrient composition might influence rate and/or duration of growth.

The shoot portion of the tree has been shown to affect root structure (McKenzie 1961), tree size (Vyvyan 1955; Beakbane and Rogers 1956), root growth (Lambers 1979), character of the rootstock (Tukey and Brase 1933), and precocity (Tubbs 1973). An intermediate stem piece (interstock) in double worked trees can affect scion vigor (Rogers and Beakbane 1957; Roberts and Blaney 1967), precocity (Roberts and Blaney 1967), total nitrogen concentration (Jones and Pate 1976), nutrient composition of xylem sap (Jones 1976a), and rootstock morphology and bark thickness (Scholz 1957; Roberts and Blaney 1967).

The effects of rootstock and scion on tree size are frequently, though not always, additive (Tubbs 1976). Tubbs (1977, 1980) pointed out that the influence of clonal tissue on tree size may differ when used as stock or scion. Growth of a given stock-scion combination is not necessarily indicative of performance of another scion cultivar on the same stock or of the same cultivar on another stock (Tubbs 1973). Perhaps this can be explained by what Tubbs (1967) referred to as the effects of the characteristic metabolism of each cultivar. This suggests that growth control mechanisms must differ in different stock-scion combinations.

Clearly, a balance must exist between root and shoot systems (Troughton 1974) and growth of root or of shoot is influenced, if not controlled, by the other. The mechanism involved in this influence of one plant part on growth of another must, at least partially, be what Chailakhian (1961) referred to as a coordinating role of the hormonal system, or what Wareing (1977) called a growth substance interrelationship between root and shoot that has an important role in coordinating activities in the plant. Since plants apparently generate chemical messages, such as growth substances, that are translocated through the plant, one must conclude that they are responsible for growth regulation by stock or scion. Thus, their effects on growth control are discussed below.

III. ENDOGENOUS FACTORS AFFECTING TOP-ROOT RELATIONSHIPS AND GROWTH

A. Auxins

The exact role of auxin in regulating plant growth is not fully understood (Goldsmith 1977). However, it presumably has an important role in controlling growth and size of shoots and perhaps roots. It thus may play a significant role in tree growth regulation of dwarfing apple rootstocks and scions.

Heyn (1970) reported that dextran, structurally a highly branched compound, is present throughout all cell walls and is broken down by dextranase, an enzyme sensitive to auxin. The dextran breakdown increases wall elasticity necessary for cell elongation. Later, Cleland (1977) concluded that auxin's role in growth was due to a cell wall loosening process related to hydrogen ion release which regulates cell extensibility and response to turgor pressure. However, Green *et al.* (1977) believe the cell wall is a self-regulating system with enlargement being due to enzymatic "delivery" of more wall polymers, a process that is speeded up by indoleacetic acid (IAA). The latter concept is supported by Gilkes and Hall (1977), who concluded that IAA promoted turnover of certain hemicelluloses involved in cell wall development. For additional discussion, see Masuda (1978) and/or Cleland and Rayle (1978).

1. Auxin Translocation.—A key aspect of the role of auxin is related to its translocation and site of synthesis in the plant. Auxin is generally considered to move primarily in a basipetal direction (Goldsmith 1977), even against a concentration gradient. In support of that principle, Eliezer and Morris (1979) concluded that long distance IAA transport through stems of intact plants is driven by the transporting cells themselves and is independent of the activity of sinks for the transport of auxin. However, Jacobs and Aloni (1978) have suggested that auxin does not move against a gradient.

There has also been a difference of opinion on the plant tissues that most actively transport auxin. Auxin applied to or synthesized in mature leaves appears to be translocated in phloem sieve tubes along with metabolites in either the basipetal or acropetal direction (Morris and Thomas 1978), although Bonnemain and Bourbouloux (1973) reported that auxin may move faster than the assimilates. Eliasson (1972) found that IAA easily escapes from sieve tubes during translocation. Nix and Woodzicki (1974) reported a steep gradient of IAA from the cambial region to differentiating xylem. These views are supported by Hoad's (1973) finding that the auxin concentration is higher in xylem tissue than in the cambium or phloem. However, Sheldrake (1973a) found little auxin in xylem of trees, perhaps due to presence of an auxin-destroying enzyme in the xylem. Apparently, auxin applied or synthesized in terminals moves in or near the cambial region (Sheldrake 1973b). The auxin translocated through the parenchymatous or cambial cells is quite polar in movement, i.e., basipetal (away from tip) in shoots and acropetal (toward tip) in roots (Goldsmith 1977; Goodwin 1978). Bonnemain and Bourbouloux (1973) also concluded that the cambial zone and protoxylem and protophloem parenchyma are primary paths of auxin translocation. Polarity of auxin movement may be a principal factor in auxin being of importance in correlative growth control (King 1976; Jacobs 1977; Wareing 1977), since it could be a mechanism that controls availability of auxin to the various plant parts and processes.

Once polarity of auxin translocation is established in tissue during a cell's meristematic state, it is retained indefinitely regardless of the geotropic orientation of the tissue (Sheldrake 1974; Antoszewski *et al.* 1978) because an inverted bark graft or intermediate stem piece may restrict auxin transport to the roots (Sax 1953; Antoszewski *et al.* 1978). Sachs (1975) concluded that an auxin stream was essential for cell differentiation and maintenance of phloem integrity. Since auxin supply affects many plant processes, an understanding of factors that influence its concentration in various plant structures may help explain growth control of apple stocks and scions.

Stem vascular differentiation is also affected by auxin levels. High auxin levels favor differentiation of xylem (Roberts 1969). Beakbane (1952) and Scholz (1957) reported more living tissue in the xylem area of dwarfing stocks than in vigorous ones. This is consistent with other reports of a greater amount of parenchymatous cells in stem xylem of dwarfing clones (McKenzie 1961). Low IAA favors differentiation of phloem (Digby and Wareing 1966). The latter suggests the presence of low auxin in stems or roots of apple dwarfing stocks for the more dwarfing clones have thicker bark than the vigorous ones (Beakbane and Thompson 1939; Rogers and Beakbane 1957; McKenzie 1961). However, Poniedzialek *et al.* (1979) suggest that there actually may not be lower auxin but higher carbohydrates in proportion to the auxin, resulting in the thicker bark (Lockard 1976). Dwarfing citrus (Mendel and Cohen 1967) and apple stocks (Colby 1935) contain more starch than vigorous ones. Messer and Lavee (1969) suggested that the starch accumulation may be due to a "block" in the utilization of sugar in the rootstock. If this conversion of sugars to starch in dwarfing stocks is so rapid that "functional" metabolites become low, it could reduce root growth and also NO_3^- reduction in the roots (Radin *et al.* 1978). Another possible auxin supply control is the root system, because roots may act as a sink for excess auxin synthesized in the top (Phillips 1964b). It also has been suggested (Phillips 1964a; Skene 1975) that roots may be the center for oxidative inactivation of excess shoot-synthesized auxin as well as a source of compounds that regulate top growth.

2. Auxin and Shoot Growth.—The effect of available auxin levels in regulating plant growth is also indicated by the higher auxin content of standard cultivars of barley and pea vs. dwarf cultivars (Kuraishi 1974; Kefeli 1978) and of standard 'Cortland' and 'Golden Delicious' shoots when compared to spur type (dwarf) mutants of the same cultivars (Jindal *et al.* 1974). Bark of the more dwarf apple stocks caused a higher rate of auxin destruction than bark of the less dwarf clones (Gur and Samish 1968), further suggesting a probable role for auxin in growth control in grafted apple trees. Cytokinin movement in the plant is influenced by auxin (Pilet 1968; Morris and Winfield 1972; Morris 1977), apparently by the establishment of sinks rather than by a direct effect on the transport system itself (Morris and Winfield 1972). Suppressed growth of lateral shoots of dwarf pea may be due to the "polarization" of cytokinin movement by IAA (Morris 1977). This is also one explanation of inhibition of lateral bud growth by IAA (Phillips 1975). However, Jablanovic and Neskovic (1977) found that inhibited shoots of dwarf pea had low auxin and high tryptophan (an auxin precursor) and cytokinin. Removal of the terminal of the dominant shoot increased auxin and decreased tryptophan and cytokinin in the inhibited shoot. They suggested that the high cytokinin and low growth may be due to the cytokinin being in an inactive form when auxin was low. Thus, a reduction of auxin level by the stock could influence scion growth by its effect on form of cytokinin translocated to the scion.

3. Auxin and Metabolite Availability.—Indoleacetic acid also plays a role in metabolite mobilization (Lepp and Peel 1971; Hatch and Powell 1972; Luckwill 1973; Altman and Wareing 1975; Patrick and Wareing 1978), a key factor involved in the effect of auxin in rooting of cuttings (Altman and Wareing 1975). It also apparently has a role in cell wall synthesis by increasing glucose uptake (Abdul-Baki and Ray 1971). Patrick and Wareing (1970) have suggested that the difference in sucrose movement in IAA-treated and untreated plants is due to the fact that IAA prevents senescence of the transporting tissue. This does not explain the short-term results cited above, but the long range effect could be important in affecting supply of metabolites to other plant parts and their growth and indirectly that of the rest of the plant.

Cleland's (1977) suggestion, that a plant's lack of ability to produce soluble sugars is a possible cause for reduced growth of dwarfs, is supported by Abdul-Baki and Ray (1971) who proposed that IAA affects the hexokinase reaction rate. Even though NAA suppressed starch hydrolysis in the specialized guard cells of *Stachytarpheta* (Pemadasa 1979), IAA apparently is involved in conversion of starch to a more active carbohydrate (Borthwick *et al.* 1937; Bausor 1942; Nanda and Anand 1970) in other cells and a decrease in auxin increased starch synthesis in potato (Obata-Sasamato and Suzuki 1979). The role of IAA in carbohydrate metabolism may be important in the dwarfing caused by certain apple rootstocks, because the more dwarfing stocks have higher starch content (Colby 1935).

Accumulation of ^{32}P at points of IAA application (Davies and Wareing 1964; Seth and Wareing 1967; Hatch and Powell 1971; Hall and Baker 1973) demonstrates the role of auxin in nutrient translocation. Auxin may accumulate just above and below the graft unions of a dwarfing intermediate stock and just above and at the lower end of a knotted interstock (Dickinson and Samuels 1956), just as Sax (1953) proposed auxin accumulation due to translocation inhibition as the explanation of the swelling just above an inverted bark graft. Stoltz and Hess (1966) did not find more auxin above a stem girdle on *Hibiscus*, but proposed that the basipetally moving auxin impeded by the girdle may have been metabolized in growth of the enlargement above the girdle.

4. Auxin and Root Growth.—Since plants maintain a rather constant top-root ratio, it seems probable that one or more factors, such as auxin, produced in the shoot and translocated to the root might be the mechanism for control of root growth by the shoot (Torrey 1976; Elliott 1977; Kramer and Kozlowski 1979). Shoot-produced auxin apparently plays a role in development of the root system, because auxin supplied to the root via the shoot promotes root growth (Goodwin *et al.* 1978); an inadequate supply of auxin to the root decreased *Perilla* root growth (Beever and Woolhouse 1975). The effect of auxin on metabolite translocation and carbohydrate metabolism may be involved in the role of auxin on root growth. In addition, exogenous or shoot-produced auxin may have a more direct role in regulating root growth (McDavid *et al.* 1972, 1973; Beever and Woolhouse 1975; Pilet 1977b). McDavid *et al.* (1973) concluded that cessation of root growth of etiolated pea seedlings was due to reduction of auxin supply from shoot to root rather than due to a lack of sucrose.

Transport of abscisic acid, another growth substance synthesized in root tips and apparently involved in growth control of roots, appears to be controlled by IAA (Pilet 1977a; Pilet *et al.* 1979). Pilet (1977b) also concluded that IAA from shoots moved acropetally toward root tips and that these two compounds controlled growth of corn roots. Elliott (1977) concluded that the IAA accumulated in the root cap was shoot-produced auxin that was in excess of that needed for root growth. This concept is supported by the finding that ^{14}C-IAA applied to the plant was found in high concentration in the root tip and root cap (Martin *et al.* 1978) within a short time.

Auxin concentrations in excess of 10^{-8}M inhibit root elongation and may cause swelling of root tips (Svensson 1972) apparently due to reorientation of cortex growth from longitudinal to lateral (Burstrom and Svensson 1974). Burstrom and Svensson (1974) concluded that auxin reduction of root cell elongation is due to shortening of the elongation period rather than a reduced rate of cell elongation. Indoleacetic acid inhibition of root growth is thought to be due to auxin-dependent ethylene production (Galston and Davies 1969; Chadwick and Burg 1970) which also may cause reorientation of the mitotic spindle. However, Elliott (1977) and Pilet (1977b) do not consider endogenous IAA to be the inhibitor of root growth of intact plants. Apparently, cambial activity involved in root growth is stimulated by auxin from the shoot (Digby and Wangerman 1965; Wilson 1975) but not by IAA from roots (Digby and Wangerman 1965). The role of auxin in control of root growth is further emphasized by the fact that a shoot application of 2,3,5-tri-iodobenzoic acid, which blocks IAA movement, alters root development (King 1976). Beever and Woolhouse (1974) have suggested that a hormone produced in *Perilla* shoot apices passes to the roots and there regulates production of kinetin, which affects shoot growth; this is discussed elsewhere in this review. Auxin influence on cytokinin metabolism may result in formation of less active cytokinins (Phillips 1975). The importance of this concept is emphasized by Charles-Edwards' (1979) conclusion that growth and development are less controlled by intrinsic photosynthetic activity than by photosynthate partitioning, a process Gersani *et al.* (1980) attributed to auxin and cytokinin.

Synthesis of phenols, which affect plant growth and are discussed in the following section, is increased by increasing auxin, in some cases *in vitro* (Sargent and Skoog 1960; Leonova *et al.* 1970; Shah *et al.* 1976). But Westcott and Henshaw (1976), working with *Acer* and the auxin 2,4-dichlorophenoxyacetic acid, reported a suppression of the phenol biosynthetic pathway. In addition, Sarapuu (1965) reported that IAA stimulated decomposition of phloridzin in apple trees with the formation of growth accelerating substances. Thus, polarity of auxin movement plus the effect of auxin on cell wall plasticity, metabolite metabolism, translocation of nutrients and metabolites, and root development potentially makes it a prime factor in regulation of growth of one part of the plant by another.

B. Cytokinins

1. Effect of Roots on Shoot Growth.—The effect of roots on shoots is not limited to supplying water and nutrients to the plant top. For example, the inhibition of lateral shoot growth after tying stems of *Salix viminalis* in a loop was relieved by inducing roots to grow at the base of the loop, whether the roots at the loop base were supplied with nutrients or deionized water (Smith and Wareing 1964). Sax (1957) also noted a reduction in growth when he tied the stems of apple rootstocks or interstocks into a loop, but he did not induce roots on the loop. The rootless cut shoot of a tomato stem did not grow until it formed roots (Went and Bonner 1942), nor did the excised shoot apex of a rye embryo produce additional leaves until a root was formed (De Ropp 1945). In addition, roots that developed on the petiole of excised leaves retarded senescence (Kulaeva 1962; Goodwin *et al.* 1978; Nooden and Lindoo 1978). Loo (1945) found that asparagus shoots grew much faster in tissue culture when they possessed attached roots.

Roots provide essential growth substances for shoot growth beyond water and nutrients. Went (1938) showed that roots were essential for the growth of the stem and lateral buds in pea seedlings, and also observed (1943) that aeration of tomato roots was essential for stem growth, an effect beyond the supply of water or salts by the roots. This effect was attributed to a root-synthesized hormone, "caulocaline" (Went 1938, 1943). Later it was shown that the root effects could be replaced by coconut milk and pea diffusate (Went and Bonner 1942). Both are now known to be high in cytokinins (Goodwin *et al.* 1978).

The influence of roots on shoot growth has been emphasized by Chailakhian (1961), Richards and Rowe (1977a,b), Holm and Key (1969), Kulasegaram and Kathiravetpillai (1972a,b), Kulasegaram (1969), Carr (1966), Humphries (1958), Buttrose and Mullins (1968), Jackson (1955, 1956), Smith and Wareing (1964), and Miginiac (1974). A direct effect of roots on leaf growth has also been reported (Luckwill 1959; Richards and Rowe 1977a; Kandiah and Wimaladharma 1978), as has an effect on inflorescence initiation (Mullins 1967, 1968; Smith 1969) and branching (Smith 1969). Bruinsma (1979) demonstrated that lateral buds on decapitated pea plants do not grow if roots are excised but they will grow if cytokinin is applied. He concluded that apical dominance was regulated by cytokinin from roots and auxin from shoots.

2. Effect of Roots on Shoot Metabolism.—Roots greatly extend the life of a leaf cutting (Vaadia and Itai 1968; Nooden and Lindoo 1978 and references therein), an effect explained by cytokinin translocation from the roots (Richmond and Lang 1957; Kulaeva 1962). Pruning the roots of light-grown pea seedlings every four days accelerated senescence of older leaves and reduced their rate of $^{14}CO_2$ fixation (McDavid *et al.* 1973). Removing the roots from soybean plants accelerated leaf senescence (Hsia and Kao 1978). There is evidence that roots that develop on the petioles of excised leaves supply cytokinins to the leaf blade (Engelbrecht 1972) and retard senescence (Goodwin *et al.* 1978). Substances from roots will delay leaf senescence (Kende 1964) and are required for the maintenance of chlorophyll, protein, and RNA levels in leaves. In detached leaves these factors can be replaced by cytokinins (Seth and Wareing 1965; Jordan and Skoog 1971 and references therein). This effect apparently is not duplicated by mineral elements or purine or pyrimidine bases of RNA or IAA at concentrations of 100 to 0.0001 mg/liter (Kulaeva 1962). Roots have a considerable influence on assimilation in leaves (Carr 1966), and the net assimilation rate of detached dwarf bean leaves is controlled by the size of the root system regenerated on the petiole (Humphries 1963; Humphries and Thorne 1964). Holm and Key (1969) reported that DNA synthesis in soybean seedlings was inhibited when the plants were grown without roots.

3. Cytokinins and Root Effects.—Many root effects on shoot growth can be replaced by cytokinins. For example, the application of benzyladenine (BA) relieved most of the symptoms of flooding injury on tomatoes (Railton and Reid 1973) and water stress symptoms on sugar beets (Shah and Loomis 1965). Cytokinins simulate the effects of roots in promoting normal branching of inflorescences on cultured explants of *Carex flacca* (Smith 1969) and promote the retention and development of inflorescences on unrooted single node cuttings of grape (Mullins 1967, 1968). The effect of cytokinins and growth has been covered by many reviews (Letham 1967; Srivastava 1967; Skoog and Armstrong 1970; Kende 1971; Skoog and Schmitz 1972; Hall 1973; Letham 1978a).

Subsequent cytokinin sections present evidence that cytokinins are synthesized primarily in roots and are translocated through the xylem to the shoot tip where they influence shoot growth. Cytokinins synthesized in immature fruits or seeds are probably not important in apple tree growth and will not be included in this discussion.

4. Presence of Cytokinins in Roots.—Vaadia and Itai (1968), Goodwin *et al.* (1978), and Letham (1978a) argue that cytokinins are synthesized in the plant roots. Feldman (1975 and references therein) presented evidence that cytokinins are synthesized in the root apical meristem, and Kende (1965) provided evidence for the presence of cytokinins in sunflower roots. Beever and Woolhouse (1974) and Henson and Wareing (1977) postulated that some hormone from the growing shoot apex passes to the roots and there regulates production of kinetin. Following decapitation of sunflower plants the level of cytokinins in the sap remained constant for four days, indicating that the cytokinins were synthesized in the roots (Kende 1965; Kende and Sitton 1967). In his review, Kende (1971) stated that "root systems are the primary source of cytokinin." Henson and Wareing (1977) found that ringing the stem of *Xanthium strumarium* L. below the leaves prior to short day treatments prevented the usual decrease in cytokinin level in the sap, indicating that the bark ring had prevented a signal from passing from leaves to roots, counteracting the short day effect on cytokinins. Skene (1972) did not detect an increase in cytokinin level above a ring in the bark of grapevines.

Weiss and Vaadia (1965) found high levels of kinetin-like activity in the root apex of sunflower plants, but little, if any, was found in the physiologically older root tissue. Similarly, Short and Torrey (1972) reported that the terminal 0 to 1 mm segment of the pea root tip contained 40 times more free cytokinin than the 1 to 5 mm section. They were unable to detect cytokinin activity in the root cap of maize. Cytokinins have been found in the root sap of many plant species (Vaadia and Itai 1968; Goodwin 1978 and references therein), including apple (Greene 1975).

The cytokinin level in the root varies with stage of growth and development. In lateral roots of the sugar maple, the cytokinin level was maximal just before bud burst (Dumbroff and Brown 1976). Sitton *et al.* (1967a) found that the cytokinin concentration of sunflower root exudate increased during the exponential growth phase and dropped by a factor of ten when the plants had reached their final size. The cytokinin level in radish roots increased with the initiation of cambial activity (Radin and Loomis 1971), whereas the level in rice roots declined markedly at flowering (Oritani and Yoshida 1971).

Root stress affects cytokinin levels in roots, which indicates that roots may be the source of cytokinin synthesis. Roots subjected to water stress contained lower levels of cytokinin activity than did the controls (Itai and Vaadia 1965, 1971; Vaadia and Itai 1968). Roots under osmotic stress (Itai *et al.* 1968) or salinity stress (Torrey 1976) also exported less cytokinin than did the controls.

Nutritional stress can also lower the cytokinin concentration in the roots. The level of cytokinin in the root exudate was lower in tomato plants grown with low phosphate than in the controls (Menary and Van Staden 1976). A low plant nitrogen status reduced the level of cytokinin activity in roots and exudate of sunflower (Wagner and Michael 1971) and in roots of *Solarium* (Woolley and Wareing 1972b). Conversely, the application of nutrients to the soil increased cytokinin activity in the root sap of sunflower (Wagner and Michael 1971) and rice (Yoshida and Oritani 1974).

5. Translocation of Cytokinins.—Other reviewers have argued that cytokinins are exported from the plant roots and are translocated to the shoot tip through the xylem (Carr 1966; Pate 1975 and references therein; Wareing 1977 and references therein). Much of the evidence for the presence of cytokinin in the xylem sap has been determined using plants under stress. Vaadia and Itai (1968) considered it a general rule that roots under stress provide reduced levels of cytokinin to the shoots. Supra-optimal root temperatures on ungrafted apple rootstocks caused a reduction in cytokinins in the leaves (Gur *et al.* 1972). In maize the export of total cytokinin was greatest at a room temperature of 28°C and declined as the root temperature was lowered to 8°C (Atkin *et al.* 1973). A 2-minute treatment of *Nicotiana rustica* and *Phaseolus vulgaris* roots at 46°C reduced the cytokinin levels in the xylem sap (Itai *et al.* 1973), and flooding of the plants' roots produced a similar effect (Burrows and Carr 1969; Goodwin *et al.* 1978). Water stress applied to shoots of *Nicotiana rustica* by increased evaporation markedly reduced the cytokinin activity in the xylem sap (Itai and Vaadia 1971).

There is other evidence for the presence of cytokinin in the xylem sap. The cytokinin level in the xylem sap of trees increased markedly prior to or at bud burst in apple (Luckwill and Whyte 1968) and *Populus robusta* (Hewett and Wareing 1973, 1974). The cytokinin level in the xylem sap also decreased following exposure to floral induction (Beever and Woolhouse 1973, 1974), but may be high during flowering (Phillips and Cleland 1972). It declined markedly prior to growth cessation (Luckwill and Whyte 1968) and senescence (Sitton *et al.* 1967a). Even short days decreased the level of cytokinin in the sap (Henson and Wareing 1974; Skene 1975). Yoshida and Oritani (1974) found that application of potassium sulphate or ammonium sulphate to the soil markedly increased the level of cytokinins in the xylem sap of rice plants. Cytokinin levels in the bleeding sap of grape vines were affected by rootstock (Skene and Antcliff 1972).

Carr and Burrows (1966) detected cytokinin in the whole xylem sap and in ethanolic extracts of sap from balsam, *Xanthium*, field pea, and blue lupin. There seems to be little doubt that cytokinin-like compounds are translocated in the xylem from the roots to the shoot tip of most and, perhaps, all plants.

6. Effect of Cytokinins on Shoot Metabolism.—The synthesis of cytokinins in plant roots and their translocation to the shoot tips explains the influence of roots on shoot growth (Sachs and Thimann 1967; Audus 1972; Railton and Reid 1973; Goodwin *et al.* 1978). The importance of cytokinin on the growth of apple shoots cultured *in vitro* was demonstrated by Jones (1967, 1973). The cytokinins found in xylem sap were usually zeatin and its nucleoside or nucleotide (Van Staden and Smith 1978). Goodwin *et al.* (1978) consider zeatin riboside to be the most common cytokinin in xylem sap. It is translocated to the shoot tip where it is utilized or re-exported to the older leaves (Van Staden 1976; Davey and Van Staden 1976, 1978; Beever and Woolhouse 1973) where, if it is in the free form, it is converted to the glucoside and re-exported to other parts of the plant (Van Staden 1976; Wareing 1977) via the phloem tissue (Vonk 1978, 1979).

When cytokinin reaches the shoot tip it probably functions in a number of different ways. One of its properties may be to enhance the "sink" effect of the tissues (Carr 1966; Morris and Winfield 1972), and it probably makes the shoot tip a more efficient competitor for carbohydrates and amino acids. It may also affect cell membrane integrity (Shaw and Manocha 1965) and facilitate the movement of compounds in the tip region (Turvey and Patrick 1979). Other effects are on promotion of cell division (Miller 1961a,b; Fox 1968; Wilkins 1969) and its influence on lateral cell expansion (Shibaoka *et al.* 1974). A function of cytokinins in plant tissues is the increase in DNA, RNA, and proteins (Fox 1968; Srivastava 1968; Audus 1972; Leshem 1973) which is probably its main contribution to shoot tip growth.

Cytokinins undoubtedly interact with other plant growth substances at the shoot tip, especially with auxin. There is evidence that IAA and kinetin have a synergistic effect on growth at low concentrations in bioassay systems (Katsumi and Kazama 1978) including the Avena straight growth test (Hemberg and Larsson 1972). There is other evidence of an interrelationship between auxin and cytokinin which could affect shoot growth. Auxin may regulate both distribution and metabolism of cytokinins (Legerstedt and Langston 1967; Woolley and Wareing 1972a; Phillips 1975; Letham 1978a). Letham (1978a) concluded that growth and auxin production in the coleoptile tip depend on a supply of root-produced cytokinins. Cytokinins are potentially capable of regulating auxin status of the shoot (Letham 1978a). Lee (1971a,b, 1972, 1974) has demonstrated an interrelationship between IAA and kinetin in the synthesis of IAA oxidase isoenzymes in tobacco callus cells. Auxin and cytokinin also interact on the release of buds from apical dominance (Wickson and Thimann 1958; Sachs and Thimann 1964; Carr 1966), indicating that the endogenous cytokinin supply may be involved in this process. The cytokinin supply to the lateral buds may be controlled by auxin from the apical bud (Phillips 1975). Auxin and kinetin interact to regulate vascular differentiation in the lateral buds of *Pisum sativum*. Auxin induced vascular strand formation, but kinetin was needed to convert these to functional xylem units (Sorokin and Thimann 1966; Morgan and Morgan 1974). Sachs (1969, 1972) observed that kinetin applied to a dormant bud led to the linking of the vascular strand tissue with that of the main stem, but auxin supplied from the tip suppressed the linkup of the vascular strands.

Cytokinins also appear to interact with gibberellin in the control of shoot growth. Endogenous cytokinins may normally have some controlling influence over endogenous gibberellin levels and stem growth (Reid and Railton 1974a). BA applied as a foliar spray to flooded tomato plants restored the overall gibberellin levels in the shoots to that in nonflooded plants (Burrows and Carr 1969; Reid and Railton 1974b). These authors and Goodwin *et al.* (1978) suggested that cytokinins exert some control over gibberellin metabolism in the shoot and influence stem growth.

C. Gibberellins

The importance of gibberellins in the growth and metabolism of plants is well established (MacMillan 1968, 1971a,b; Lang 1970; Paleg and West 1972; R.L. Jones 1973; Krishnamoorthy 1975; Graebe and Roper 1978; Hedden *et al.* 1978). Gibberellins are synthesized in shoots (Lockhart 1957; Jones and Phillips 1966; Jones and Lang 1968; Frydman and Wareing 1973; Harris and Atherton 1976; Goodwin *et al.* 1978) and root tips (Butcher 1963; Phillips 1964a; Jones and Phillips 1966; Kende and Sitton 1967; Sitton *et al.* 1967b; Carr and Reid 1968; Crozier and Reid 1971; Goodwin *et al.* 1978). They have also been detected in the phloem (Hoad and Forrest 1972; Hoad 1973; Graebe and Roper 1978 and references therein) as well as in the xylem of numerous plants (Carr *et al.* 1964; Phillips and Jones 1964; Reid and Carr 1967; Jones and Lacey 1968; Burrows and Carr 1969; Reid *et al.* 1969; Selvendran and Sabaratnam 1971; Reid and Railton 1974a; Ziegler 1975; Goodwin *et al.* 1978; Graebe and Roper 1978). Gibberellin-like substances occur in apple trees (Kato and Ito 1962; Jones and Lacey 1968; Luckwill and Whyte 1968; Ibrahim and Dana 1971; Hoad and Forrest 1972), although most identification has been carried out on the seed (Dennis and Nitsch 1966; MacMillan 1968; Luckwill *et al.* 1969; Dennis 1976) or fruit flesh (Hayashi *et al.* 1968). However, it is not very clear how gibberellin may contribute to dwarfing in apple trees. No consistent relationship between dwarfism and gibberellin content has been found in higher plants (Graebe and Roper 1978), and there is as yet no evidence that shoot growth is dependent on root-synthesized gibberellins (Kende and Sitton 1967; Carr and Reid 1968).

Rootstocks may be able to convert gibberellins into a more active or less active form. Plants can metabolize gibberellins when they are applied exogenously (Durley *et al.* 1973, 1974; Railton *et al.* 1974a,b; Davies and Rappaport 1975a,b; Frydman and MacMillan 1975; Yamane *et al.* 1979). Crozier and Reid (1971) and Hoad (1973) consider that shoot-synthesized gibberellin may be translocated to the roots where it is converted to another gibberellin and then recirculated to the shoot tip. The gibberellin may then be active in shoot metabolism, but the root conversion may alter the effectiveness of the hormone. However, Goodwin *et al.* (1978) suggest that the gibberellins found in xylem sap are largely produced by the roots and do not merely circulate through them. Sankhla and Huber (1974) reported that exogenous GA not only reduced the rate of $^{14}CO_2$ fixation, but affected the compound into which it was transformed.

McComb (1964) considered the possibility that dwarf pea plants metabolize gibberellins more rapidly than do tall plants, but suggested that the observed stability of gibberellin makes this theory very unlikely. There is an extensive literature on the interactions between gibberellins and other plant hormones, but this would not account for differences in growth of apples on different stocks unless, as noted above, the root-synthesized gibberellin has a specific function or the activity of the shoot-synthesized gibberellin is altered in the root system. We agree with Robitaille (1970) that there is little evidence to support a role for gibberellin in the rootstock effect.

D. Abscisic Acid

Abscisic acid (ABA) has long been known to increase markedly in leaves of water stressed plants (Milborrow 1978), which usually results in stomatal closing. Likewise, nutritionally stressed tobacco showed a continuing increase in ABA for seven days (Mizrahi and Richmond 1972). Dorffling (1972) concluded that the ABA increase was due to synthesis and not release from a bound form in the leaf. Thus, stressing a plant appears to increase ABA content. Waterlogging, cold temperatures, and some pathological conditions cause a similar response (Milborrow 1978), although the latter factors may cause a slight water loss and thus trigger the usual water stress response of increasing ABA. Milborrow and Robinson (1973) concluded that water stressing roots had no effect on root ABA synthesis although it did affect its formation in leaves. However, Hoad (1975) found that increasing the osmotic concentration of rooting media increased ABA in xylem exudate of cut shoots.

1. Effects of Abscisic Acid.—Abscisic acid has been reported to inhibit root growth (Pilet 1977b; Thimann 1977); decrease potassium and phosphorus accumulation by corn roots, perhaps due to an effect on the transport mechanism of intact plants (Shaner *et al.* 1975), although it does not affect energy availability; increase cell permeability to water (Glinka and Reinhold 1971; Dorffling 1972); inhibit stem elongation (El-Antably *et al.* 1967; Robitaille and Carlson 1971; Rehm and Cline 1973; Thimann 1977), perhaps by inhibiting cell elongation (Riseuno *et al.* 1971); accelerate leaf senescence (El-Antably *et al.* 1967; Galston and Davies 1969); increase leaf abscission (Addicott and Lyon 1969; Abeles 1973); increase ethylene production (Dorffling 1972; Abeles 1973), though Lieberman and Kunishi (1971) reported that it decreased ethylene evolution and acts independently of it; induce formation of resting buds (Galston and Davies 1969; Robitaille and Carlson 1971) and thereby stop growth; inhibit response to some growth promoters, especially gibberellin (Kefeli and Kadyrov 1971), perhaps by inhibiting the assimilation of the amylase involvement in starch metabolism (Chrispeels and Varner 1966); and accumulate in wood plants during short days (Kefeli and Kadyrov 1971). Thus, ABA can, under specified conditions, have varied and extensive effects on a plant depending on species and environmental conditions. Thus, any factor, such as stock or scion, that influences ABA content of either part could be a factor in growth control such as dwarfing.

2. Abscisic Acid Translocation.—Movement of ABA in young roots is apparently primarily basipetal with lateral translocation confined to the apex and perhaps controlled by IAA (Pilet 1977a). Juniper (1976) had concluded earlier that geotropic-stimulated lateral movement of ABA occurred in the root cap with its subsequent basipetal translocation to the cells of the extension zone where it inhibits cell extension, resulting in the typical geotropic response of roots. Hartung's (1976) report of uniform distribution of exogenous ^{14}C-ABA in horizontal bean roots does not support this concept.

Milborrow (1978) concluded that ABA is moved by more than one transport system or mechanism and that it occurs in both xylem and phloem of stems, with perhaps the greatest amount in the phloem (Hoad 1978). Ingersall and Smith (1971) consider ABA translocation to be an active process because low temperature or low O_2 reduced it in cotton by

80 to 98%. Apparently, much of the ABA synthesis occurs in the plastids (Milborrow 1978), although Loveys (1977) concluded that extra-chloroplastic synthesis cannot be excluded. Both agree that ABA is quite mobile as it is found in various tissues in stems, leaves, fruit buds, and roots. Apparently, movement in the stems may be in either a basipetal or acropetal direction, depending on site of synthesis and/or exogenous application (Torrey 1976). Girdling below a leaf receiving ABA prevented its movement below the girdle (Hocking *et al.* 1972). Understanding how ABA moves in the plant and the effect of various factors on its movement can help determine its possible role in dwarfing by rootstocks.

Although ABA is synthesized in leaves (Zeevaart 1974), its ready translocatability may account for high ABA levels in young shoots and leaves, because Hoad (1978) concluded that it was not synthesized in those tissues. Thimann (1977) concurs that the ABA in leaves was not synthesized so rapidly during wilting, but thought that its origin was due to hydrolysis of a complex carbohydrate residue rather than translocation, because there is evidence that hydrolysis starts very rapidly on wilting. When fed to apple seedlings ABA is rapidly converted to a water soluble complex of ABA and glucose, suggesting another mechanism that affects ABA level in plants (Powell and Seeley 1974). There is higher ABA in some apple dwarfing stocks (Yadava and Dayton 1972), and higher ABA levels in stems of dwarfed apple trees than in more vigorous ones from full bloom till summer dormancy (Robitaille and Carlson 1976). Since ABA influences plant growth, it could have a significant effect on tree growth and size. Milborrow (1974) reported that ABA-treated guard cells accumulate starch. In contrast to the above findings with apple, Feucht *et al.* (1974) found no correlation between ABA level in *Prunus* species and dwarfing potential. They suggest that the amount or concentration of ABA present may not be as important in dwarfing as is sensitivity of a species or cultivar to ABA. The concept of different sensitivity of cultivars to ABA may also be relevant for other plant growth substances, and could be an explanation and/or mechanism involved in specificity of response of various stock-scion combinations.

Other possible mechanisms for ABA influence on growth could be its effect on translocation of other growth substances. Although Bellandi and Dorffling (1974) reported that ABA did not affect transport of IAA, Basler and McBride (1977) reported an inhibition of auxin translocation by ABA. Exogenous ABA increased sugar transport from shoot to root in bean (Karmaker and Van Stevenind 1979). It is not known if this factor is involved in the higher starch levels found in dwarfing apple stocks (Colby 1935).

E. Nitrogen Compounds

Nitrogen metabolism in apple, as in many species, is an important process in controlling tree vigor and size. Colby (1935) referred to the moderately vigorous 'Whitney' as a low nitrogen tree and the very vigorous 'Snow' and 'York Imperial' as high nitrogen trees. Apparently, nitrogen can be readily taken up by apple roots in either the NO_3^- or NH_4^+ form (Frith and Nichols 1975). In young plants, nitrogen increased the total and relative intensity of assimilate movement from the leaf to roots (Anisimov *et al.* 1973). Sunflower plants supplied with nitrate nitrogen had higher levels of cytokinin than those supplied with ammonium sulfate or ammonium nitrate, and reducing the nitrogen level in nutrient solution also reduced cytokinin in roots, root exudate, and leaves (Salama and Wareing 1979). To be used by the tree the NO_3^- must be reduced for incorporation into amino acids and protein. Jones and Pate (1976) found that the interstock M 9 (between 'Cox' scion and M 2 roots) affected total nitrogen of xylem sap, but such effects were non-selective since range and proportion of amino compounds were not affected. In apple sap, the relative amounts of the various nitrogenous compounds change during a growing season, i.e., arginine is the principal form in March whereas asparagine is the principal form in late May (Cooper *et al.* 1972).

1. Nitrogen Utilization.—Luckwill (1959) indicated that nitrate reduction by nitrate reductase (NR) in apple and peach occurred only in roots, and the ammonia produced combined with organic acids from carbohydrate oxidation to form amino acids. Grasmanis and Nicholas (1967) first isolated NR in apple roots. Later Klepper and Hageman (1969) reported NR in all parts of young apple trees including roots, stems, and leaves of mature trees, which could mean that the NO_3^- ion is distributed throughout the tree. Even though Jones and Sheard (1979) concluded that the leaf is the major site of NO_3^- assimilation, others (Cooper *et al.* 1972; Hill-Cottingham and Lloyd-Jones 1979; Tromp and Ovaa 1979) have concluded that incorporation of absorbed nitrogen in apple took place in young roots, and that nitrogen metabolism of above-ground parts is not directly affected by the form of nitrogen applied to the soil. One might suspect that differences in root capacity to reduce NO_3^- could be important in rootstock control of scion growth except that interstocks have an effect similar to rootstocks on scion growth. Jackson (1978) found that NO_3^- and reduced NO_3^- deposition into root xylem are very sensitive to substances translocated from other tissues into the roots. Since both interstock and rootstock could affect substances translocated to the xylem loading area, it is possible that nitrogen metabolism has at least an indirect role in apple dwarfing. Frith (1972) reported that effective use of NO_3^- by apple depends on its reduction in roots and that NR was highest when NO_3^- was the sole nitrogen source. The difference in the two views may exist because (1) Tromp and Ovaa (1979) reported on the form of nitrogen in above-ground portions and (2) Klepper and Hageman (1969) reported on NR levels. In any case, NR has a high rate of turnover (half-life, 2 to 6 hours), so anything affecting NR synthesis would soon be reflected in the nitrogen metabolism of the plant (Knypl 1979). Exogenous sucrose (Schag *et al.* 1979) and cytokinin have been shown to increase NR level by *de novo* synthesis of the enzyme (Hirschberg *et al.* 1972; Knypl 1979), although Sahulka (1972, 1974) reported a decrease in NR level due to exogenous kinetin. Nitrate and benzyladenine increase NR level, with the latter acting directly and not through NO_3^- (Kende 1971; Kende *et al.* 1971, 1974; Knypl 1979). Nitrate induces NR synthesis in some species, but not all, and thus cannot be considered to be a universal inductor of NR (Kulaeva *et al.* 1976). Gibberellin also increases NR, but increasing kinetin decreases the amount of gibberellin required to increase NR (Roth-Bejerano and Lips 1970); hence, Roth-Bejerano and Lips (1970) believe that the effect of kinetin and gibberellin on NR is relatively specific. Photosynthesis increased NR, the main effect being due to supplying carbohydrates for respiration to drive the induction process (Aslam *et al.* 1973; Lips 1979). However, Schag *et al.* (1979) reported that its effect on NR activity is determined by blue light and phytochrome. Knypl (1979) also suggested that the effect of IAA, gibberellin, and ABA on NR synthesis is indirect via their effects on plant cellular metabolism.

The NR activity in corn seedlings was not affected by NH_4^+ ions (Schrader and Hageman 1967); however, Schag *et al.* (1979) reported an increase in NR in grain crops and in roses, but felt that NH_4^+ was without effect or inhibited NR synthesis in most species. It thus seems possible to affect and perhaps control NR conversion of nitrates by plants and thus affect growth. However, Huck *et al.* (personal communication) concluded that rate of protein synthesis and corresponding shoot growth could be controlled better by controlling the final step in peptide polymerization than by controlling NR production or activity. Westcott and Henshaw (1976) reported that nitrogen level affected phenol synthesis in *Acer* tissue *in vitro*. Because phenols can markedly affect plant growth, this may be still another potential pathway for nitrogen supply to influence growth.

F. Phenols

Phenols may be involved in the dwarfing mechanism in apple (see p. 342). In this section, the possible role of phenols in plant growth will be reviewed.

1. Effect on Growth.—A number of phenolic compounds inhibit plant growth (Thimann 1977; Kefeli and Kutacek 1977). There are several books and reviews on the subject (Kefeli and Kadyrov 1971; Ribereau-Gayon 1972; Harborne 1973; Hess 1975; Kefeli 1978; Letham 1978b), as well as papers on experimental methods (Swain 1953; Stenlid 1968; Engelsma 1979). The level of phenol required to induce growth inhibition varies with the type of phenol and the plant tissue. Kefeli (1978) induced 50% inhibition of growth of wheat coleoptile section and of wheat seedlings with 10 mg/liter and 62.5 mg/liter (6.84×10^{-5}M and 4.28×10^{-4}M), respectively, of coumarin whereas 250 mg/liter (1.71×10^{-3}M) were required to inhibit root formation. Hara *et al.* (1973) found that 10 to 100 ppm (6.84×10^{-5} to 6.84×10^{-4}M) coumarin inhibited growth of rice, mung bean, lettuce, and clover seedling. Borner (1959) inhibited root growth of rye and wheat seedlings with several phenols at concentrations as low as 10 ppm.

Phenols inhibit growth of plant stems (Van Overbeek *et al.* 1951; Hemberg 1951; Sarapuu 1965), hypocotyls (Neumann 1959; Engelsma and Meijer 1965), and roots (San Antonio 1952) and inhibit mitosis (Avers and Goodwin 1956; Rice 1974), cell division (Croak 1972), and cell elongation (Avers and Goodwin 1956; Svensson 1971; Hess 1975). However, phenols also stimulate rooting (Poapst *et al.* 1970; Jones and Hatfield 1976) and at low concentrations may stimulate growth (Thimann *et al.* 1962; Pilet 1966). Jones (1976b) concluded that phloridzin was involved in growth of apple shoots possibly through the production of phloroglucinol. Vendrig and Buffel (1961) reported that *trans*- caffeic acid at a concentration of 1 μg/ml (5.55×10^{-9}M) or less enhanced the elongation of coleoptile sections of coleus, and Neumann (1959) found that coumarin at 10^{-5} to 10^{-3}M stimulated the elongation of *Helianthus* hypocotyls. In both cases, synergism with IAA was evident. Marigo and Boudet (1975) suggested that polyphenol levels and growth can be negatively related in tomato. Likewise, Mendel and Cohen (1962) suggested that phenols may be an important factor in growth control by citrus rootstock.

2. Effect on Auxin.—A primary effect of phenols on plant or tissue growth is through its effect on the level of auxin in plant tissues (Kefeli and Kutacek 1977; Letham 1978b). Phenolic acids that inhibit growth enhance oxidative decarboxylation of IAA, whereas phenolic acids that promote growth suppress decarboxylation of IAA (Zenk and Muller 1963; Tomaszewski and Thimann 1966). The monophenols contain a single hydroxyl group and act as cofactors of IAA oxidase and inhibit growth, whereas the polyphenols (with two or more hydroxyl groups) inhibit IAA oxidase and tend to enhance growth (Nitsch and Nitsch 1962; Tomaszewski 1964; Grochowska 1967; Schneider and Wightman

1974; Stafford 1974). Their action with IAA oxidase seems to be related to the number and position of hydroxyl groups on the ring. Lee and Skoog (1965) found that monohydroxybenzoic acids enhance IAA inactivation. The increasing order of inactivation is 2-, 3-, and 4-hydroxybenzoic acid. 2,4-dihydroxybenzoic acid also enhances IAA inactivation, but 3-, 4-dihydroxybenzoic acid does not promote inactivation of IAA. As noted previously, many phenols promote growth at very low concentrations, an effect that may be related to ready interconversion of the mono- to the polyphenols (Audus 1972).

Apple bark phenols that have been reported to be synergistic or to antagonize IAA are listed in Table 7.1. Ferulic acid appears to be both an IAA synergist and an antagonist. Gortner *et al.* (1958) report that at low levels ferulic acid can stimulate IAA oxidase activity, but at higher levels it is a powerful enzyme inhibitor. Zenk and Muller (1963) reported ferulic acid to be a growth-inhibiting phenol, whereas others (Table 7.1) reported that it is a growth-promoting phenol. These differences may be due to the use of different concentrations or possibly to the different types of plant or tissue utilized in the experiment. Phloretin reportedly stimulates wheat root growth, especially in the presence of auxin; however, its glycoside, phloridzin, is a potent stimulator of IAA oxidase (Stenlid 1968). This could be an important fact with regard to growth of apple trees, because phloridzin is present in large amounts in the bark. A test

TABLE 7.1. SOME PHENOLIC COMPOUNDS FOUND IN APPLE TREE BARK WHICH EXHIBIT A SPECIFIC REGULATORY EFFECT ON IAA CONTENT IN PLANT TISSUE

Participation in Auxin Metabolism	Compound	Reference
IAA synergist or auxin-oxidase inhibitors	Protocatechuic acid	Kefeli 1978
	Chlorogenic acid	Sondheimer and Griffin 1960; Zenk and Muller 1963; Rekoslavskaya *et al.* 1974; Kefeli 1978
	Ferulic acid	Tomaszewski and Thimann 1966; Schaeffe *et al.* 1967; Rekoslavskaya *et al.* 1974
	Caffeic acid	Zenk and Muller 1963; Konings 1964; Tafuri *et al.* 1972; Kefeli 1978
	Sieboldin	Stenlid 1968
	Sinapic acid	Zenk and Muller 1963; Tomaszewski and Thimann 1966
IAA antagonists or auxin-oxidase cofactors	*p*-Hydroxybenzoic acid	Borner 1960; Zenk and Muller 1963; Pilet 1966; Kefeli 1978
	p-Coumaric acid	Gortner *et al.* 1958; Borner 1960; Engelsma and Meijer 1965; Schaeffe *et al.* 1967; Tafuri *et al.* 1972; Kefeli 1978
	Ferulic acid	Gortner *et al.* 1958; Zenk and Muller 1963
	Phloretic acid	Zenk and Muller 1963; Grochowska 1966
	Phloridzin	Grochowska 1966; Stenlid 1963, 1968

for free phenols in MM 111 rootstock bark (Lockard, Schneider and Kemp, unpublished) revealed the presence of only phloretin. Sarapuu (1965) reports that under the influence of low temperatures, phloridzin is partially converted to phenolic growth stimulators. This process is of great significance in the emergence of the plant from dormancy.

Sieboldin is not present in the Malling or Malling Merton rootstocks, but replaces phloridzin in the Sieboldianae series (Williams 1966).

Phenols may also influence the synthesis of IAA. Kefeli (1978) reported that *p*-coumaric acid retards the synthesis of tryptophan from ^{14}C-anthranilic acid. This is considered to be a nonspecific effect, but may result in a reduction in IAA synthesis as tryptophan is a precursor for auxin (Audus 1972). This is supported by reports that IAA biosynthesis may be modified by phenols (Gordon and Paleg 1961; Kefeli and Kutacek 1977). There is a suggestion that regulation of the auxin/phenol balance may occur during the biosynthesis of both types of growth regulators (Kefeli and Kutacek 1979).

3. Site of Synthesis.—There are reports that phenol synthesis requires light (Kefeli and Turetskaya 1965; Kefeli 1978). It may be true of leaves but it is probably not true of other plant organs for there are large amounts of phenols in apple tree roots (Williams 1966; Miller 1973). There is no evidence of transport of phenolics in the phloem (Bate-Smith 1962; Towers 1964). There is, however, evidence that light enhances phenol synthesis (Siegelman 1964; Engelsma and Meijer 1965; Engelsma 1968, 1979; Kefeli and Kadyrov 1971; Davies 1972). Phenols in *Prunus domestica* were lower in older leaves collected from the shady side of trees compared with those collected from the sunny side (Hillis and Swain 1959). The evidence indicates that the precursors for phenolic synthesis are translocated, but the phenols are synthesized in the cells in which they occur (Bate-Smith 1962; Hillis 1962). This is also true for leaf-synthesized phenols which are not translocated out of the leaves (Bate-Smith 1962).

4. Effect on Phloem Translocation.—Phloridzin is apparently very effective in inhibiting sugar transport in animals, but initially was considered to have no effect in plants (Stenlid 1968). Later Van Sumere *et al.* (1975) reported that phloridzin apparently inhibits transport of sugars in plants and that phloretin is more effective than phloridzin.

Ferulic acid increased the polarity of transport of IAA and stimulated basipetal transport (Stenlid 1976), whereas coumarin enhanced acropetal translocation of auxin (Basler and McBride 1977). Adding coumarin to the nutrient medium caused epinasty of petioles and upper stems of bean seedlings, indicating enhancement of endogenous auxin movement to these areas or inhibition of its movement to lower parts of the plant.

5. Effect on Metabolism.—Coumarin is reported to decrease cell wall permeability (Burstrom 1954; Svensson 1971 and references therein; Uhrstrom and Svensson 1979), but most phenolic acids increase cell wall permeability (Glass and Dunlop 1974; Kefeli 1978). In leaves, phenols are precursors for anthocyanins and flavanoid pigments, but in the phloem of plants, including apple, they serve as precursors for lignins and tannins (Harborne 1973; Walker 1975). Phenols not utilized in the synthesis of these compounds are stored in the cell in a relatively inactive form in the cell vacuole (Tomaszewski 1964; Kefeli and Kutacek 1977), or in the cell wall (Hillis 1962) bound to proteins (Van Sumere *et al.* 1975) as polysaccharide esters (Levand and Heinicke 1968; Lamport 1970; Fry 1979), or esterified to lignin (Harris and Hartley 1976).

Datta and Nanda (1978) found that some phenols increase amylase activity (presumably starch synthesis). This could account for accumulation of starch in the more dwarfing apple rootstocks.

6. Phenols in Apple Trees.—There are considerable amounts of phenols in vegetative tissue of apple trees (Table 7.2). Table 7.2 does not include phenols found in the fruit (Williams 1960; Macheix 1974), nor does it include sieboldin or trilobatin found in certain species of apple (Williams 1960, 1966; Harborne and Simmonds 1964; Miller 1973) but not in *M. domestica* Borkh. Phenols are concentrated mainly in the bark of the apple tree where phloridzin is reported to be as high as 12% of the dry weight, and in the leaves, where it can be 1% of the fresh weight (Williams 1966; Miller 1973). Martin and Williams (1967) reported 9.5% phloridzin in the stem bark of M 16 before it started to grow. In our laboratory phloridzin in 1-year-old apple stem bark was as high as 9.2% and 7.7% of dry weight for MM 111 and M 26, respectively. In the new roots, it was as high as 6.6% and 5.1% of dry weight for MM 111 and M 26, respectively. Levels of other phenols co-chromatographed were much lower. Total phenol in the stem bark of 1-year-old apple rootstocks was 23.8% and 20.4% of the dry weight in MM 111 and M 26, respectively (Lockard, Schneider and Kemp, unpublished).

Priestly (1960) reported phloridzin in the wood of apple trees to be highest in June and near zero by August. Harborne and Simmonds (1964) do not list apple wood as a source of phenols although they did find them in the wood of other species, as did Hillis and Swain (1959). We have not found phenols in the wood of two Malling rootstocks, but we have not searched through a full season or measured the total phenols.

Phloridzin in the bark of apple trees is highest just before growth commences (Priestly 1960; Sarapuu 1964; Martin and Williams 1967). It tends to decrease during the period of rapid growth and to increase again as dormancy approaches. Sarapuu (1964) reported that the maximum

TABLE 7.2. PHENOLIC COMPOUNDS FOUND IN APPLE TREES

Phenol	Plant Part	Reference
Caffeic acid	shoots	Nachit and Feucht 1977
Chlorogenic acid	leaves	Avadhani and Towers 1961
p-Coumaric acid	shoots	Nachit and Feucht 1977
	xylem sap	Grochowska 1966, 1967
	leaves	Avadhani and Towers 1961
p-Coumaryl-quinic acid	leaves	Avadhani and Towers 1961
p-Hydroxybenzoic acid	shoots	Nachit and Feucht 1977
Kampferol	leaves	Williams 1960
	stem bark	Williams 1960
Phloretic acid	shoots	Sarapuu 1964
	xylem sap	Grochowska 1966, 1967
Phloretin[1]	shoots	Sarapuu 1964
	stem bark	Williams 1960
Phloridzin	shoots	Hutchinson *et al.* 1959; Hancock *et al.* 1961; Sarapuu 1964
	xylem sap	Grochowska 1966, 1967
	leaves	Williams 1960; Avadhani and Towers 1961; Sarapuu 1964, 1965
	stem bark	Priestly 1960; Williams 1960; Sarapuu 1964, 1965; Martin and Williams 1967
	buds	Sarapuu 1965
	flowers	Sarapuu 1965
	wood	Priestly 1960; Miidla *et al.* 1970
	roots	Borner 1959, 1960
Phloroglucinol	xylem sap	Grochowska 1966, 1967
Phlorol	shoots	Sarapuu 1964
Quercitin	leaves	Williams 1960
	stem bark	Williams 1960; Sarapuu 1965
	buds	Sarapuu 1965
Sinapic acid	shoots	Nachit and Feucht 1977
Syringic aldehyde	wood	Miidla *et al.* 1970
Vanillin	wood	Miidla *et al.* 1970

[1]Found attached to a glycoside.

levels in the bark occur during the period of deep dormancy. The phenol level in grape vines follows a similar pattern (Darne 1977). Phenol levels are higher in immature than in mature leaves (Hillis and Swain 1959; Williams 1966). The chloroplasts of spring willow leaves contain more phenolic compounds and in greater amounts than the chloroplasts of autumn leaves (Kefeli 1978).

Free phenols are seldom found in plants in more than trace amounts (Towers 1964), as most phenols are present as glycosides or esters (Towers 1964; Walker 1975). When simple phenols are administered to plant tissues, the corresponding β-glucosides are usually formed (Towers 1964 and references therein). In the bark of apple rootstocks, we (Lockard, Schneider and Kemp, unpublished) have found only phloretin (the aglycone of phloridzin) in the free form and this was estimated at 618 μg/g dry weight. Grochowska (1967) has found phloretic acid, phlorogucinol, *p*-coumaric acid, and *o*-coumaric acid in the xylem sap of apple spurs. The glycosides are usually less active in the growth processes than the aglycone (Stenlid 1968). The glucoside may provide a form of carbohydrate reserve. Phloridzin is present in such large amounts that the glucose on the molecule must be a significant carbohydrate source. Barlow (1959) considered phloridzin to be a storage material in apples. Priestly (1960) listed phloridzin as a component of the carbohydrate fraction in apples, and Kandiah (1979) did not separate phloridzin from total sugars in estimating carbohydrate reserves in apple trees.

Phenols have been regarded as possible growth controlling compounds in the control of tree size by dwarfing rootstocks. Mendel and Cohen (1962) suggested that phenols may be important inhibitors in citrus rootstocks. Scholz (1957) collected extracts from lyophilized apple interstock bark and found that the relative inhibiting effect of the extracts was correlated with the growth of the plants sampled. Malling (M) 8 and Clark were most inhibitory, M 9 intermediate, and M 2 and M 3 were least inhibitory. This order was correlated with the size of the trees in the nursery as well as their interstem dwarfing capabilities. Similarly, extracts of leaves from size-controlling rootstocks inhibited the growth of stratified apple seeds. The vigor of the rootstocks and the respective growth of apple and wheat seedlings (Miller 1965) were negatively correlated. However, Hutchinson *et al.* (1959) reported that the phloridzin content of dormant terminal twigs of *Malus* rootstock clones was not related to the vigor imparted to the scions. Martin and Williams (1967) found that phloridzin was actually higher in the bark of the more vigorous M 16 than in M 9. Similarly, Nachit and Feucht (1977) found higher levels of caffeic and sinapic acids in the more vigorous than in the less vigorous rootstocks. The less vigorous apple rootstocks contained more *p*-coumaric and *p*-hydroxybenzoic acids. We (Lockard, Schneider and Kemp, unpublished) have found total phenols in the bark of 1-year-old apple rootstocks to be higher in MM 111 than in M 26, primarily because of higher levels of phloridzin. Other phenols measured were at much lower concentrations (most were in the range of 50 to 300 $\mu g/g$ compared with phloridzin at 80 mg/g dry weight), and the differences were not great. Some were higher in MM 111, and others higher in M 26 rootstock. Extracts of M 9 apple callus tissue contained a greater amount of phenolic substances than did those of M 13 (Messer and Lavee 1969).

The types and levels of phenols in different species or strains are obviously genetically controlled. Beakbane (1956) found that the phenolic substances present in the stem bark of trees of pear on quince differed above and below the union. In such grafts, the phenols apparently are not transported across the union (Williams 1953). Carlson (1974) and Yu and Carlson (1975a,b) consider that phenols may be involved in graft incompatibility. They found 32 phenols in Mazzard and only 22 in Mahaleb, and reported that this may contribute to graft incompatibilities of these 2 sweet cherry rootstocks. Noggle (1979) found that when freshly cut surfaces of stock and scion are brought together, cell walls of contiguous cells undergo dissolution and the plasma membrane of the two come into contact. This dissolution of the cell wall may release the attached phenols into the surrounding medium. The cells of the adjacent graft may be subjected to "foreign" phenolic compounds and may not have the appropriate enzymes to break down the phenols or to convert them into a nontoxic form. The free phenols then may inhibit cell division, which results in non-union of the graft, or incompatibility.

IV. PROPOSED DWARFING MECHANISM

A. Hypothesis

Ample evidence has been presented to show that there is an interrelationship between shoot and root growth in plants. Such an interrelationship requires communication between the shoot and root. This is probably a continuing process in all plants and constitutes normal growth control. Sachs (1972) and Gersani *et al.* (1980) suggested—and we concur—that the messenger from the shoot and young leaves to the root is the auxin synthesized in these organs. The auxin flows basipetally through the phloem and cambial cells of the tree to the roots. Some of the auxin is degraded in the bark and the concentration decreases as it proceeds down through the bark (Leopold and Kriedemann 1975). The amount of degradation depends on the amount of IAA oxidase, peroxidase, and phenols and perhaps other compounds present in the phloem and cambial cells.

The levels of these compounds are genetically controlled and may vary among species and even cultivars of plants, such as apple. This variation could account for the different rates of growth among cultivars, as well as the effect of dwarfing rootstocks. The thicker bark and much higher starch levels in dwarf rootstocks indicate a low level of auxin in these tissues.

The level of auxin in an active form that reaches the roots of the plant undoubtedly influences root growth and metabolism, including the synthesis of other hormones, such as cytokinins. The amount and/or kind of cytokinin translocated up through the xylem would, therefore, reflect the amount of shoot-synthesized auxin reaching the roots. The cytokinins arriving at the shoot tip would then influence shoot growth, which would influence the synthesis of auxin, and the amount translocated to the roots. This integrated system would keep the shoot and root growth in balance under various cultural and climatic conditions. Other hormones such as gibberellin and ABA obviously play a role in plant growth, as do other compounds such as nitrogren which affects the cytokinin level. However, there is no indication at present that these compounds are utilized as signals to maintain root and shoot balance.

We propose that the manner in which a dwarfing apple rootstock or interstock may affect this growth control system is by control of the auxin passing through the bark of the rootstock or interstock. Bark of different genetic composition will permit different amounts of unchanged auxin to pass through it. This will affect root growth, which affects cytokinin synthesis and shoot growth, resulting in a small or large tree depending on the type of rootstock.

In the mechanism proposed, dwarfing theoretically should be counteracted by applying auxin to the roots or cytokinin to the shoot tip. However, neither of these compounds induces extension growth in an intact plant, dwarfed or not dwarfed.

B. Supporting Evidence

1. Bark Grafts.—The strongest evidence that the bark is the key to the dwarfing mechanism is that insertion of M 26 bark grafts in the scion of 'Gravenstein'/MM 111 trees results in a dwarfed tree (Lockard and Schneider, unpublished). In this case only the bark was transferred and a 20-cm graft gave more dwarfing effect than a 10-cm graft. There were two controls: (1) an ungrafted tree and (2) a tree with an equivalent length of 'Gravenstein' bark removed, rotated horizontally, and replaced on the same tree.

The trees bark-grafted with 20 cm M 26 bark responded similarly to interstock trees: they were smaller than the controls (Fig. 7.1), the stems swelled above the grafts (Fig. 7.2 and 7.3), and new shoot growth was inhibited.

Another treatment consisted of 10 cm of 'Gravenstein' or M 26 bark inverted and grafted onto the trees. These treatments restricted growth more than M 26 bark grafts with normal orientation. The trees with inverted bark grafts were significantly smaller than those with non-inverted bark grafts, with even greater difference between them and the controls. An unexpected result was that the inverted 'Gravenstein' bark restricted growth of the 'Gravenstein' scion significantly more than the inverted M 26 bark. In fact, several trees with inverted 'Gravenstein' bark died before the end of the first growing season, even though the grafts had healed normally (Fig. 7.4).

In our opinion, the graft of dwarfing bark reduces the translocation of auxin and possibly sugars and other compounds across it, leading to the swelling above the bark graft. Sax (1953) and Tubbs (1967) came to the

FIG. 7.1. BARK GRAFTED TREES ('GRAVENSTEIN'/MM 111) AFTER SECOND LEAF

A. No bark graft, control. B. 20 cm M 26 bark graft. C. 10 cm 'Gravenstein' bark removed, inverted and replaced.

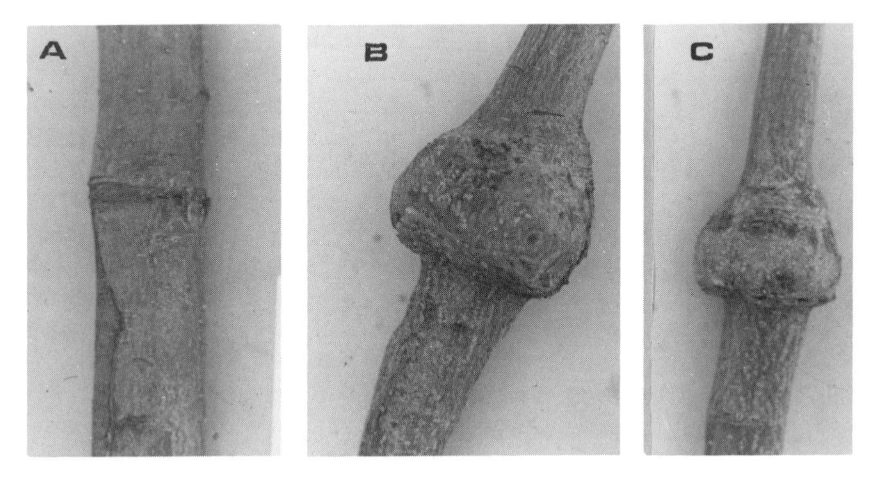

FIG. 7.2. TRUNK OF BARK GRAFTED TREES AFTER SECOND LEAF

A. 20 cm 'Gravenstein' bark, removed, rotated and replaced. B. 20 cm M 26 bark graft on 'Gravenstein' stem. Note swelling above graft. C. 10 cm M 26 graft on 'Gravenstein' stem. Note swelling above graft.

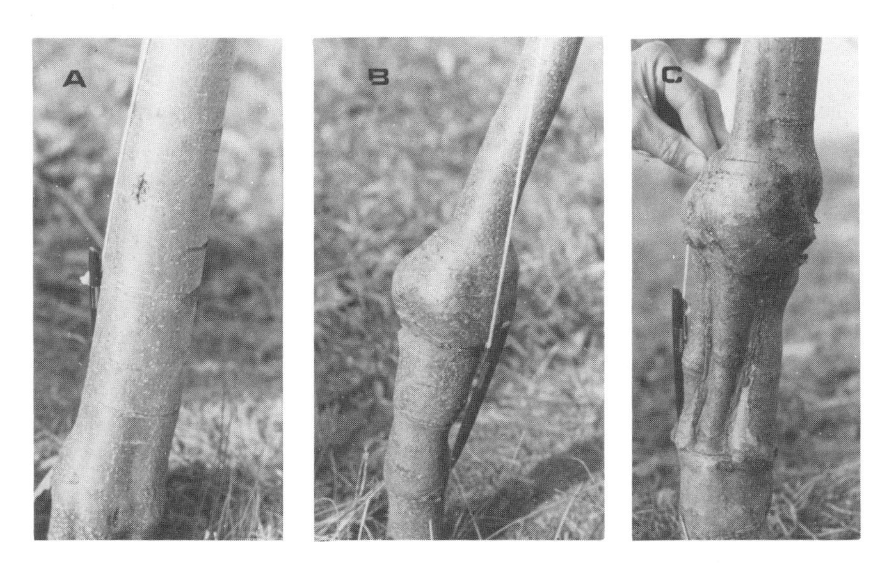

FIG. 7.3. TRUNK OF TREES AFTER FOURTH LEAF

A. No graft, control. B. 10 cm M 26 bark graft on 'Gravenstein' stem. Note swelling above graft. C. 20 cm M 26 bark graft on 'Gravenstein' stem. Note 'Gravenstein' bark tissue developed at seam of M 26 bark graft in C.

FIG. 7.4. TRUNK OF BARK GRAFTED TREES AFTER SECOND LEAF

A. 10 cm 'Gravenstein' bark, removed, rotated and replaced. B. 10 cm inverted M 26 grafted on 'Gravenstein' stem. Note extensive growth of 'Gravenstein' tissue in grafted zone. C. 10 cm 'Gravenstein' bark removed, and inverted.

same conclusion regarding the translocation of auxin across dwarfing interstocks and inverted bark rings. Such a conclusion was supported by Dickinson and Sammuels (1956) when they found that ^{32}P accumulated in the bark of an interstock graft of M 9 and above the graft when the interstock was inverted. In addition, Bukovac *et al.* (1958), using 'McIntosh' on several dwarfing rootstock, found that the proportion of ^{32}P applied to the leaves and transported to the roots was related to the vigor of the stock.

When a strip of bark is inverted and grafted back onto the stem of an apple tree the phloem cells retain their original polarity and reduce basipetal translocation of auxin (Sax and Dickson 1956; Antoszewski *et al.* 1974, 1978; Poniedzialek *et al.* 1979). The reversed polarity may persist for more than 60 weeks (Antoszewski *et al.* 1978), but the new cells formed by the cambium under the inverted bark will have the usual polarity, and normal translocation will resume after 1 or 2 years. The new wood formed under the bark graft, whether inverted or not, is of the same genetic origin as the bark, whereas the wood regenerated at the vertical seam of the bark graft is derived from the underlying wood (Sax and Dickson 1956). A double bark inversion with vertical seams on opposite sides of the trunk increased the duration of the dwarfing effect, but the lateral movement of phloem sap soon resulted in a lateral orientation of the new phloem and reversed the polarity by regeneration of new tissue (Sax and Dickson 1956).

We believe that there is sufficient evidence to conclude that an inverted bark graft will reduce auxin and other metabolites translocated through the phloem to the roots. As a result, the tree will be dwarfed and there will be a swelling on the stem above the inverted graft, similar to a dwarfing interstock. Both an inverted and a non-inverted dwarf bark graft appear to inhibit translocation.

Phenols are one type of compound that may be contributing to growth control and dwarfing in apple trees. Their precise role is unknown, but they have many properties that would enable them to perform this function. Certain phenols enhance the degradation of IAA by functioning as a synergist of IAA oxidase (Table 7.1). Some phenols may also inhibit the translocation of sugars and perhaps auxin, or they may function by regulating the polar auxin transport (Marigo and Boudet 1977). The difficulty in prescribing a specific role for these compounds is that phenols occur in high concentrations in the bark of all apple trees, dwarfing and non-dwarfing. The concentration of total phenols is usually higher in the bark from non-dwarfing trees. This can be attributed primarily to phloridzin, which comprises nearly half of all phenols in the bark and is higher in the non-dwarfing bark. However, bark from dwarfing rootstocks is thicker than from non-dwarfing stocks and there is more living tissue. This may result in a greater total amount of phenols in the bark and a greater area for interaction between IAA and phenols. Other phenols vary—some are higher in dwarfing bark and others higher in non-dwarfing bark. However, the actual function of these phenols is obscure. Most seem to be held in an inactive form, found in the cell wall or dissolved in the metabolically inactive vacuole. The levels of free phenols in apple bark are very low, yet this is the most active form. Further research is needed to clarify the role of phenols in the bark of apple trees. It should be noted that Gur and Samish (1968) found that the amount of IAA destroyed by root and shoot bark of various apple rootstocks was negatively correlated with the scion vigor induced by those rootstocks.

Phenols may also be growth promoters and at certain times during the growing season may function in this way. When growth commences in the spring phloridzin in the bark is broken down to sucrose and phloretin. The sucrose is utilized as a food reserve and the free phloretin or its degradation products may function as growth promoters (Sarapuu 1964). There is, however, no evidence to show that phenols are translocated out of the cells in which they are formed. Any effect they have must be in the cells of the bark. The larger amount of phloridzin in the bark of non-dwarf trees could result in more free phloretin in the cells of that bark.

2. Tissue Culture Investigations.—Another area of research that lends support to the dwarfing mechanism postulated in this paper is the effect of apple calli from different sources on the growth of adjacent calli in tissue culture (Schneider *et al.* 1978b). The calli were placed adjacent in a vertical position or in a horizontal position and separated by filter paper. Compounds from one callus that influenced the growth of the other had to pass through the filter paper or the medium. Calli obtained from dwarfing apple rootstocks, such as M 9 and MM 106, when grown adjacent to calli from faster growing clones, such as MM 111 or 'Golden Delicious', inhibited the growth of the latter. The fact that growth inhibition did occur is evidence for the synthesis of growth-inhibiting compounds in these calli, with larger amounts being synthesized by the calli from dwarfing rootstocks than by those from non-dwarfing clones.

One unexpected result of this work was that calli from 'Golden Delicious', when grown next to each other, grew less than when they were grown next to calli from the more dwarf M 9. At present we do not understand this effect, but it may be related to the severe dwarfing effect of an inverted 'Gravenstein' bark graft noted in the previous section.

To identify the growth-inhibiting compounds synthesized by the apple calli, the agar on which they were grown was extracted with methanol and fractionated (Chiang, Lockard, Schneider and Kemp, unpublished data). The inhibiting fraction proved to contain three compounds when it was gas chromatographed. One of these compounds co-chromatographed with ferulic acid. The other two did not co-chromatograph with any of 20 phenol standards. However, because they were recovered after hydrolysis and other procedures for phenol purification, they probably are phenolic compounds. This would lend support to the suggestion that phenols are the inhibiting compounds produced by apple bark.

C. Evidence Conflicting with the Dwarfing Theory

The series of bark grafts in the field included some made 65 cm above the original scion graft of the tree (1-year-old 'Gravenstein'/MM 111) (Lockard and Schneider, unpublished). One treatment was 20 cm of 'Gravenstein' bark removed, rotated horizontally, and replaced on the tree; the other was 20 cm of M 26 bark. One limb was allowed to grow out below each graft. On the plant with an M 26 bark graft, the limb below the graft overgrew the rest of the tree, even though it was lightly pruned. However, this limb was still heavier (258 g) than the one below a 'Gravenstein' graft (210 g). The growth above the M 26 graft was inhibited (400 g) compared to that above the 'Gravenstein' graft (826 g) so that the limb below the M 26 graft was a much larger proportion of the tree (24%) compared to the limb below the 'Gravenstein' graft (13%). We expected the tree growth, including the limb below the 20 cm M 26 bark graft, to be intermediate between the dwarfed and non-dwarfed tree. The limb below the M 26 graft was not restricted by the dwarfing M 26 bark, and could contribute more auxin to the roots. The auxin supply arriving in the roots would be the sum of that contributed by the plant top and by the limb below the graft. This level of auxin thus would regulate root metabolism and influence the level of cytokinin going up the stem in the xylem to both the top of the tree and to the lower limb. This should have resulted in roughly equal growth of the limbs above and below the graft. This response is more easily explained in terms of an inhibitor being restricted in its downward movement by the bark graft. The inhibitor would then concentrate above the graft, resulting in growth inhibition. This is more plausible in a small plant than it would be in an apple tree.

The high M 26 bark graft treatment without a branch below the graft restricted the size of the tree to the same size as a tree with a similar low graft (5 cm above the original tree graft).

V. SUMMARY

Previous concepts of the dwarfing mechanism in apples have not explained many of the known facts about dwarfing, especially the effect of an interstock. Recent advances in the knowledge of growth control in plants—especially the role of plant hormones—have enabled development of more advanced concepts of the dwarfing mechanism. Dwarfing is, probably, an alteration in the normal growth pattern.

One of the important aspects of plant growth is the constant top/root ratio. This holds for grafted apple trees as well as for plants on their own roots. Any scion/rootstock combination has a specific top/root ratio and any attempt to alter this ratio results in the plant changing its growth pattern until the ratio is re-established. This growth response requires communication between scion and rootstock. Evidence indicates that the main signal from the shoot to the root is auxin, which translocates primarily through the phloem, and the signal from the root to the shoot is cytokinin, which translocates primarily through the xylem. The literature concerning this concept is reviewed. The discussion of other growth controlling compounds, i.e., GA, ABA, and nitrogen, is more limited, not because they are less important in plant growth, but because, at this time, we see no specific role for them in the dwarfing mechanism and many reviews of their function already exist.

The dwarfing mechanism postulated is that auxin produced by the shoot tip is translocated down the phloem and the amount arriving at the root influences root metabolism and affects the amount and kind of cytokinins synthesized and translocated to the shoot through the xylem.

The auxin in the phloem is oxidized or otherwise degraded by compounds in the bark and the amount of active auxins reaching the roots will vary in different cultivars. This will account for the fast-growing and slow-growing cultivars, and for the dwarfing effect of rootstocks and interstock. Phenols may be important growth controlling compounds in apple bark as they interact with auxin in many different ways, i.e., promote oxidation, act synergistically, and influence synthesis and possibly translocation. These aspects are reviewed.

The main support for this concept of dwarfing is that when a complete ring of M 26 bark was grafted on the scion of a 'Gravenstein'/MM 111 tree it resulted in a dwarfed tree similar to one with a dwarfing interstock. A 20-cm length of bark resulted in more dwarfing than did a 10-cm length of bark. One graft combination gave a response that could not be explained by this dwarfing concept.

VI. LITERATURE CITED

ABDUL-BAKI, A.A. and P.M. RAY. 1971. Regulation by auxin of carbohydrate metabolism involved in cell wall synthesis by pea stem tissue. *Plant Physiol.* 47:537–544.

ABELES, F.B. 1973. Ethylene in plant biology. Academic Press, New York.

ADDICOTT, F.T. and J.L. LYON. 1969. Physiology of abscisic acid and related substances. *Annu. Rev. Plant Physiol.* 20:139–164.

ALTMAN, A. and P.F. WAREING. 1975. The effect of IAA on sugar accumulation and basipetal transport of ^{14}C-labelled assimilates in relation to root formation in *Phaseolus vulgaris* cuttings. *Physiol. Plant.* 33:28–32.

ANISIMOV, A., L. OBININA, and T. BULATOVA. 1973. Influence of mineral nutrition on the translocation of assimilates and auxins in plant organisms. p. 227–231. *In* R. Antoszewski (ed.) Proc. 3rd Symp. on Accumulation and Translocation of Nutrients and Regulators in Plant Organisms. Skierniewiece, Warsaw, Poland.

ANTOSZEWSKI, R., U. DZIECIOL, A. MIKA, and A. CZYNCZYK. 1974. The polarity of IAA translocation in the bark of apple tree. p. 81. *In* R. Antoszewski, L. Harrison and C.C. Zych (eds.) Proc. XIX Intern. Hort. Congr. Vol. IA. Warszawa.

ANTOSZEWSKI, R., U. DZIECIOL, A. MIKA, and A. CZYNCZYK. 1978. The polarity of IAA translocation in the bark of apple trees. *Acta Physiol. Plant.* 1:35–44.

ASLAM, M., R.C. HUFFAKER, and R.L. TRAVIS. 1973. The interaction of respiration and photosynthesis in induction of nitrate reductase activity. *Plant Physiol.* 52:137–141.

ATKIN, R.K., G.E. BARTON, and D.K. ROBINSON. 1973. Effect of root-growing temperature on growth substances in xylem exudate of *Zea mays*. *J. Expt. Bot.* 24:475–487.

AUDUS, L.J. 1972. Plant growth substances, Vol. 1. Chemistry and physiology. Leonard Hill, London.

AVADHANI, P.N. and G.H.N. TOWERS. 1961. Fate of phenylalanine-C^{14} and cinnamic acid-C^{14} in *Malus* in relation to phloridzin synthesis. *Can. J. Biochem. Physiol.* 39:1605–1617.

AVERS, C.J. and R.H. GOODWIN. 1956. Studies on roots. IV. Effects of coumarin and scopaletin on the standard root growth pattern of *Phleum pratense*. *Amer. J. Bot.* 43:612–620.

AWAD, M.M. and A.L. KENWORTHY. 1963. Clonal rootstock, scion variety and time of sampling influences in apple leaf composition. *Proc. Amer. Soc. Hort. Sci.* 83:68–73.

BARLOW, H.W.B. 1959. Root/shoot relationships in fruit trees. *Scientia Hort.* 14:35–41.

BASLER, E. and R. MCBRIDE. 1977. Interaction of coumarin, gibberellic acid and abscisic acid in the translocation of auxin in bean seedlings. *Plant & Cell Physiol.* 18:939–947.

BATE-SMITH, E.C. 1962. The simple polyphenolic constituents of plants. p. 133–158. *In* W.E. Hillis (ed.) Wood extractives and their significance to the pulp and paper industry. Academic Press, New York.

BAUSOR, S.C. 1942. Effects of growth substances on reserve starch. *Bot. Gaz.* 104:115–121.

BEAKBANE, A.B. 1952. Anatomical structure in relation to rootstock behavior. p. 152–158. *In* P.M. Synge (ed.) Rpt. 13th Intern. Hort. Congr., Vol. I. Sept. 8–15, 1952, London. Roy. Hort Soc., London.

BEAKBANE, A.B. 1956. Possible mechanism of rootstock effect. *Ann. Appl. Biol.* 44:517–521.

BEAKBANE, A.B. and W.S. ROGERS. 1956. The relative importance of stem and root in determining rootstock influence in apples. *J. Hort. Sci.* 31:99–110.

BEAKBANE, A.B. and E.C. THOMPSON. 1939. Anatomical studies of stems and roots of hardy fruit trees. II. The internal structure of the roots of some vigorous and some dwarfing apple rootstocks, and the correlation of structure with vigour. *J. Pomol. Hort. Sci.* 17:141–149.

BEEVER, J.E. and H.W. WOOLHOUSE. 1973. Increased cytokinin from root system of *Perilla frutescens* and flower and fruit development. *Nature New Biol.* 246:31–32.

BEEVER, J.E. and H.W. WOOLHOUSE. 1974. Increased cytokinin export from the roots of *Perilla frutescens* following disbudding or floral induction. p. 681–686. *In* R.L. Bielski, A.R. Ferguson and M.M. Cresswell (eds.) Mechanisms of regulation of plant growth. Royal Soc. of N.Z. Bul. 12.

BEEVER, J.E. and H.W. WOOLHOUSE. 1975. Changes in the growth of roots and shoots when *Perilla frutescens* is induced to flower. *J. Expt. Bot.* 26:451–463.

BELLANDI, D.M. and K. DORFFLING. 1974. Transport of abscisic acid in intact pea seedlings. *Physiol. Plant.* 32:365–368.

BONNEMAIN, J. and A. BOURBOULOUX. 1973. The transport and metabolism of ^{14}C-indoleacetic acid in intact plants. p. 207–214. *In* R. Antoszewski (ed.) Proc. 3rd Symp. on Accumulation and Translocation of Nutrients and Regulators in Plant Organisms. Skierniewiece, Warsaw, Poland.

BORNER, H. 1959. The apple replant problem. I. The excretion of phloridzin from apple root residues. *Contrib. Boyce Thomp. Inst.* 20:39–56.

BORNER, H. 1960. Liberation of organic substances from higher plants and their role in the soil sickness problem. *Bot. Rev.* 26:393–424.

BORTHWICK, H.A., K.C. HAMNER, and M.W. PARKER. 1937. Historical and microchemical studies of the reactions of tomato plants to IAA. *Bot. Gaz.* 98:491–519.

BRUINSMA, J. 1979. Root hormones and overground development. p. 35–48. *In* T.K. Seat (ed.) Plant regulation and world agriculture. Plenum Press, New York.

BUKOVAC, M.J., S.H. WITTWER, and H.B. TUKEY. 1958. Effect of stock-scion interrelationships on the transport of ^{32}P, ^{45}Ca in the apple. *J. Hort. Sci.* 33:145–152.

BURROWS, W.J. and D.J. CARR. 1969. Effects of flooding the root system of sunflower plants on the cytokinin content in the xylem sap. *Physiol. Plant.* 22:1105–1112.

BURSTROM, H.G. 1954. Studies on growth and metabolism of roots. XI. The influence of auxin and coumarin derivatives on the cell wall. *Physiol. Plant.* 7:548–558.

BURSTROM, H.G. and S. SVENSSON. 1974. Hormonal regulation of root development. p. 121–135. *In* J. Kolek (ed.) Structure and function of primary root tissues. Bratislava, Veda.

BUTCHER, D.N. 1963. The presence of gibberellins in excised tomato roots. *J. Expt. Bot.* 14:272–280.

BUTTROSE, M.S. and M.G. MULLINS. 1968. Proportional reduction in shoot growth of grapevines with root systems maintained at constant relative volumes by repeated pruning. *Austral. J. Biol. Sci.* 21:1095–1101.

CARLSON, R.F. 1965. Responses of Malling Merton clones and Delicious seedlings to different root temperatures. *Proc. Amer. Soc. Hort. Sci.* 86:41–45.

CARLSON, R.F. 1974. Some physiological aspects of scion/rootstocks. p. 294–302. Proc. XIXth Intern. Hort. Congr., Warszawa.

CARR, D.J. 1966. Metabolic and hormonal regulation of growth and development. p. 253–283. *In* E.G. Cutter (ed.) Trends in plant morphogenesis. Longmans Green, London.

CARR, D.J. and W.J. BURROWS. 1966. Evidence of the presence in xylem sap of substances with kinetin-like activity. *Life Sci.* 5:2061–2077.

CARR, D.J. and D.M. REID. 1968. The physiological significance of the synthesis of hormones in roots and of their export to the shoot system. p. 1169–1185. *In* F. Wightman and C. Setterfield (eds.) Proc. 6th Intern. Conf. on Plant Growth Substances, July 24–29, 1967, Carlton Univ., Ottawa. The Runge Press, Ottawa.

CARR, D.J., D.M. REID, and K.G.M. SKENE. 1964. The supply of gibberellins from the root to the shoot. *Planta* 63:382–392.

CHADWICK, A.V. and S.P. BURG. 1970. Regulation of root growth by auxin ethylene interaction. *Plant Physiol.* 45:192–200.

CHAILAKHIAN, M. KH. 1961. Principles of ontogenesis and physiology of flowering in higher plants. *Can. J. Bot.* 39:1817–1841.

CHAPLIN, M.H. and M.N. WESTWOOD. 1980. Nutritional status of 'Bartlett' pear on *Cydonia* and *Pyrus* species rootstocks. *J. Amer. Soc. Hort. Sci.* 105:60–63.

CHARLES-EDWARDS, D.A. 1979. Photosynthesis and crop growth. p. 111–124. *In* R. Marcelle, H. Clijsters and M. Van Poucke (eds.) Photosynthesis and plant development. W. Junk, The Hague.

CHRISPEELS, M.J. and J.E. VARNER. 1966. Inhibition of gibberellic acid induced formation of amylase by abscission II. *Nature* 212:1066–1067.

CLELAND, R.E. 1977. The control of cell enlargement. p. 101–115. *In* D.H. Jennings (ed.) XXXI Symp. Soc. Expt. Biol. Integration of Activity in Higher

Plants. Sept. 7–9, 1976, Univ. of Durham. Cambridge University Press, Cambridge.

CLELAND, R.E. and D.L. RAYLE. 1978. Auxin, H-excretion and cell elongation. p. 125–140. *In* H. Shibaoka, M. Furuya, M. Katsumi and A. Takimoto (eds.) Controlling factors in plant development. *The Bot. Mag.*, Special Issue 1, Tokyo.

COLBY, H.L. 1935. Stock-scion chemistry and the fruiting relationship in apple trees. *Plant Physiol.* 10:483–498.

COOPER, D.R., D.G. HILL-COTTINGHAM, and M.J. SHORTHILL. 1972. Gradients in the nitrogenous constituents of the sap extracted from apple shoots of different ages. *J. Expt. Bot.* 23:247–254.

CRIPPS, J.E.L. 1970. A seasonal pattern of apple root growth in western Australia. *J. Hort. Sci.* 45:153–161.

CROAK, M.L. 1972. Effects of phenolic inhibitors on growth, metabolism, mineral depletion, and ion uptake in Paul's scarlet rose cell suspension cultures. Ph.D. Dissertation, University of Oklahoma, Norman.

CROZIER, A. and D.M. REID. 1971. Do roots synthesize gibberellins? *Can. J. Bot.* 49:967–975.

DARNE, G. 1977. Characterization and interpretation of different phases during the development of phenolic compounds and leucoanthocyanins in fruiting canes of *Vitis vinifera.* Comptes Rendus Hebdomadaires des Sciences de l'Academie des Sciences D. 284:441–444. (*Hort. Abstr.* 47:9293.)

DATTA, K.S. and K.K. NANDA. 1978. Effect of some phenols and gibberellic acid on the growth and development of T22 *Triticale. Indian J. Agr. Sci.* 48:89–93.

DAVEY, J.E. and J. VAN STADEN. 1976. Cytokinin translocation: changes in zeatin and zeatin-riboside levels in the root exudate of tomato plants during their development. *Planta* 130:69–72.

DAVEY, J.E. and J. VAN STADEN. 1978. Cytokinin activity in *Lupinus albus. Physiol. Plant.* 43:77–81.

DAVIDSON, R.L. 1969. Effect of root/leaf temperature differentials on root/shoot ratios in some pasture grasses and clover. *Ann. Bot.* 33:561–569.

DAVIES, C.R. and P.F. WAREING. 1964. Auxin-directed transport of radiophosphorus in stems. *Planta* 65:139–156.

DAVIES, L.J. and L. RAPPAPORT. 1975a. Metabolism of tritiated gibberellins in d-5 dwarf maize. I. In excised tissues and intact dwarf and normal plants. *Plant Physiol.* 55:620–625.

DAVIES, L.J. and L. RAPPAPORT. 1975b. Metabolism of tritiated gibberellins in d-5 dwarf maize. II. (^{3}H) gibberellin A_1, (^{3}H) gibberellin A_3, and related compounds. *Plant Physiol.* 46:60–66.

DAVIES, M.E. 1972. Polyphenol synthesis in cell suspension, cultures of 'Paul's Scarlet Rose'. *Planta* 104:50–65.

DE ROPP, R.S. 1945. Studies in the physiology of leaf growth. I. The effect of various accessory growth factors on the growth of the first leaf of isolated stem tips of rye. *Ann. Bot. N.S.* 9:369–381.

DENNA, O.W. 1963. The physiological genetics of the bush and vine habit in *Cucurbita pepo* L. squash. *Proc. Amer. Soc. Hort. Sci.* 83:657–666.

DENNIS, F.G., JR. 1976. Gibberellin-like substances in apple seeds and fruit flesh. *J. Amer. Soc. Hort. Sci.* 101:629–633.

DENNIS, F.G., JR. and J.P. NITSCH. 1966. Identification of gibberellins A_4 and A_7 in immature apple seeds. *Nature* 211:781–782.

DICKINSON, A.G. and E.W. SAMUELS. 1956. The mechanism of controlled growth of dwarf apple trees. *J. Arnold Arboretum* 37:307–313.

DIGBY, J. and E. WANGERMAN. 1965. A note on the effect of the shoot and root apex on secondary thickening in pea radicles. *New Phytol.* 64:168–170.

DIGBY, J. and P.F. WAREING. 1966. The effect of applied growth hormones on cambial division and the differentiation of the cambial derivatives. *Ann. Bot.* 30:539–548.

DORFFLING, K. 1972. Recent advances in abscisic acid research. p. 281–295. *In* H. Kaldeway and G. Vardar (eds.) Hormonal regulation in plant growth and development. Verlag Chemie, Weinheim.

DUMBROFF, E.B. and D.C.W. BROWN. 1976. Cytokinin and inhibitor activity in roots and stems of sugar maple seedlings through the dormant season. *Can. J. Bot.* 54:191–197.

DURLEY, R.C., I.D. RAILTON, and R.P. PHARIS. 1973. Interconversion of gibberellin A_5 to gibberellin A_3 in seedlings of dwarf *Pisum sativum. Phytochemistry* 12:1609–1612.

DURLEY, R.C., I.D. RAILTON, and R.P. PHARIS. 1974. The metabolism of gibberellin A_1 and gibberellin A_{14} in seedlings of dwarf *Pisum sativum.* p. 285–293. *In* Plant growth substances. Proc. 8th Intern. Conf. on Plant Growth Substances. Aug. 26–Sept. 1, 1973, Tokyo. Hirokawa Publ. Co., Tokyo.

EL-ANTABLY, H.M.M., P.F. WAREING, and J. HILLMAN. 1967. Some physiological responses to DL-abscisin. *Planta* 73:74–90.

ELIASSON, L. 1972. Translocation of shoot-applied indoleacetic acid into the roots of *Populus tremula. Physiol. Plant.* 27:412–416.

ELIEZER, J. and D.A. MORRIS. 1979. Effects of temperature and sink activity on the transport of ^{14}C-labelled indoleacetic acid in the intact pea plant. *Planta* 147:216–224.

ELLIOTT, M.C. 1977. Auxins and the regulation of root growth. p. 100–108. *In* P.E. Pilet (ed.) Plant growth regulation. Proc. Intern. Conf. on Plant Growth Substances, Aug. 20–Sept. 4, 1976, Lausanne. Springer-Verlag, New York.

ENGELBRECHT, L. 1972. Cytokinins in leaf-cuttings of *Phaseolus vulgaris* during their development. *Biochem. Physiol. Pflanz.* 163:335–343.

ENGELSMA, G. 1968. Photoinduction of phenylalanine deaminase in gherkin seedlings. III. Effects of excision and irradiation on enzyme development in hypocotyl segments. *Planta* 82:355–368.

ENGELSMA, G. 1979. Effect of day length on phenol metabolism in the leaves of *Salvia occidentalis. Plant Physiol.* 63:765–768.

ENGELSMA, G. and G. MEIJER. 1965. The influence of light of different spectral regions on the synthesis of phenolic compounds in gherkin seedlings in relation to photomorphogenesis. I. Biosynthesis of phenolic compounds. *Acta. Bot. Neerl.* 14:54–72.

FELDMAN, L.J. 1975. Cytokinins and quiescent center activity in roots of *Zea*. p. 55–71. *In* J.G. Torrey and D.T. Clarkson (eds.) The development and function of roots. Academic Press, New York.

FERREE, M.E. and J.H. BARDEN. 1971. The influence of strains and rootstocks on photosynthesis, respiration and morphology of 'Delicious' apple trees. *J. Amer. Soc. Hort. Sci.* 96:453–457.

FEUCHT, W., M.M. KHAN, and J.P. DANIEL. 1974. Abscisic acid in *Prunus* trees: isolation and the effect on growth of excised shoot tissue. *Physiol. Plant.* 32:247–252.

FOX, J.E. 1968. Molecular control of plant growth. Dickenson Publishers, Calif.

FRITH, G.J.T. 1972. Effect of ammonium nutrition on the activity of nitrate reductase in the roots of apple seedlings. *Plant & Cell Physiol.* 13:1085–1090.

FRITH, G.J.T. and D.G. NICHOLS. 1975. Preferential assimilation of ammonium ions from ammonium nitrate solutions by apple seedlings. *Physiol. Plant.* 33:247–250.

FRY, S.C. 1979. Phenolic components of the primary cell wall and their possible role in the hormonal regulation of growth. *Planta* 146:343–351.

FRYDMAN, V.M. and J. MACMILLAN. 1975. The metabolism of gibberellins A_9, A_{20}, A_{29} in immature seeds of *Pisum sativum* cv. Progress No. 9. *Planta* 125:181–195.

FRYDMAN, V.M. and P.F. WAREING. 1973. Phase change in *Hedera helix* L. I. Gibberellin-like substances in the two growth phases. *J. Expt. Bot.* 24:1131–1138.

GALSTON, A.W. and P.J. DAVIES. 1969. Hormonal regulation in higher plants. *Science* 163:1288–1296.

GERSANI, M., S.H. LIPS, and T. SACHS. 1980. The influence of shoots, roots and hormones on sucrose distribution. *J. Expt. Bot.* 31:177–184.

GILKES, N.R. and M.A. HALL. 1977. The hormonal control of cell wall turnover in *Pisum sativum* L. *New Phytol.* 78:1–15.

GLASS, A.D.M. and J. DUNLOP. 1974. Influence of phenolic acids on ion uptake. *Plant Physiol.* 54:855–858.

GLINKA, Z. and L. REINHOLD. 1971. Abscisic acid raises the permeability of plant cells to water. *Plant Physiol.* 48:103–105.

GOLDSMITH, M.H.M. 1977. The polar transport of auxin. *Annu. Rev. Plant Physiol.* 28:439–478.

GOODWIN, P.B. 1978. Phytohormones and growth and development of organs of the vegetative plant. p. 31–173. *In* D.S. Letham, P.B. Goodwin and T.V. Higgins (eds.) The biochemistry of phytohormones and related compounds: a comprehensive treatise, Vol. II. Elsevier/North Holland, Amsterdam.

GOODWIN, P.B., B.I. GALLNOW, and D.S. LETHAM. 1978. Phytohormones and growth correlations. p. 215–250. *In* D.S. Letham, P.B. Goodwin and T.J.V. Higgins (eds.) The biochemistry of phytohormones and related compounds: a comprehensive treatise, Vol. II. Elsevier/North Holland, Amsterdam.

GOODWIN, R.H. and C. TAVES. 1956. The effect of coumarin derivatives on the growth of avena roots. *Amer. J. Bot.* 37:224–231.

GORDON, S.A. and L.G. PALEG. 1961. Formation of auxin from tryptophan through action of polyphenols. *Plant Physiol.* 36:838–845.

GORTNER, W.A., M.J. KENT, and G.K. SUTHERLAND. 1958. Ferulic and *p*-coumaric acids in pineapple tissue as modifiers of pineapple indoleacetic acid oxidase. *Nature* 181:630–631.

GRAEBE, J.E. and H.J. ROPER. 1978. Gibberellins. p. 107–204. *In* D.S. Letham, P.B. Goodwin and T.J.V. Higgins (eds.) The biochemistry of phytohormones and related compounds: a comprehensive treatise, Vol. I. Elsevier/ North Holland, Amsterdam.

GRASMANIS, V.O. and D.J.D. NICHOLAS. 1967. A nitrate reductase from apple roots. *Phytochemistry* 6:217–218.

GREEN, P.B., K. BAUER, and W.R. CUMMINS. 1977. Biophysical model for plant cell growth: auxin effects. p. 30–45. *In* A.M. Jungreis, T.K. Hodges, A. Kleinzeller and S.G. Schultz (eds.) Water relations in membrane transport in plants and animals. Academic Press, New York.

GREENE, D.W. 1975. Cytokinin activity in the xylem sap and extracts of MM 106 apple rootstocks. *HortScience* 10:73–74.

GROCHOWSKA, M.J. 1966. Chromatographic degradation of phloridzin. *Plant Physiol.* 41:432–436.

GROCHOWSKA, M.J. 1967. Occurrence of free phloretic acid (*p*-hydroxy-dihydrocinnamic acid) in xylem sap of the apple tree. *Bul. de L'Academie Polonaise Des Sciences* XV:455–459.

GUR, A., B. BRAVDO, and J. HEPNER. 1976a. The influence of root temperature on apple trees. III. The effect on photosynthesis and water balance. *J. Hort. Sci.* 51:203–210.

GUR, A., B. BRAVDO, and Y. MIZRAHI. 1972. Physiological responses of apple trees to supraoptimal root temperature. *Physiol. Plant.* 27:130–138.

GUR, A., Y. MIZRAHI, and R.M. SAMISH. 1976b. The influence of root temperature on apple trees. *J. Hort. Sci.* 51:195–202.

GUR, A. and R.M. SAMISH. 1968. The role of auxins and auxin destruction in the vigor effect induced by various apple rootstocks. *Beitr. Biol. Pflanzen.* 45:91–111.

HALL, R.H. 1973. Cytokinins as a probe of developmental processes. *Annu. Rev. Plant Physiol.* 24:415–444.

HALL, S. and D.A. BAKER. 1973. Aspects of the hormonal control of phloem transport in *Ricinus communis*. p. 9–16. *In* R. Antoszewski (ed.) Proc. 3rd Symp. on Accumulation and Translocation of Nutrients and Regulators in Plant Organisms, Warsaw, 1972. Skierniewiece, Warsaw, Poland.

HANCOCK, C.R., H.W.B. BARLOW, and H.J. LACEY. 1961. The behavior of phloridzin in the coleoptile straight-growth test. *J. Expt. Bot.* 12:401–408.

HARA, M., N. UMETSU, C. MIYAMOTO, and K. TAMARI. 1973. Inhibition of the biosynthesis of plant cell wall materials, especially cellulose biosynthesis, by coumarin. *Plant & Cell Physiol.* 14:11–28.

HARBORNE, J.B. 1973. Phytochemical methods. A guide to modern techniques of plant analysis. Chapman & Hall, New York.

HARBORNE, J.B. and N.W. SIMMONDS. 1964. The natural distribution of the phenolic aglycones. p. 77–127. *In* J.B. Harborne (ed.) Biochemistry of phenolic compounds. Academic Press, New York.

HARRIS, G.P. and J.G. ATHERTON. 1976. Gibberellin-like substances obtained from chilled plants of *Dianthus barbatus* L. by an agar diffusion technique. *Ann. Bot.* 40:531–536.

HARRIS, P.J. and R.D. HARTLEY. 1976. Detection of bound ferulic acid in cell walls of the *Gramineae* by ultraviolet fluorescence microscopy. *Nature* 259:508–510.

HARTUNG, W. 1976. Der basipetale (2-^{14}C) abscisinsauetransport in wurzeln intakter bohnenkeimlinge und sline bedeutung fur den wurzel-geatropismus. *Planta* 128:59–62.

HATCH, A.H. and L.E. POWELL. 1971. Hormone-directed transport of ^{32}P in *Malus sylvestris* seedlings. *J. Amer. Soc. Hort. Sci.* 96:230–234.

HATCH, A.H. and L.E. POWELL. 1972. Hormone-directed transport of certain organic compounds in *Malus sylvestris* seedlings. *J. Amer. Soc. Hort. Sci.* 96:399–400.

HAYASHI, F., R. NAITO, M.J. BUKOVAC, and H.M. SELL. 1968. Occurrence of gibberellins A_3 in parthenocarpic apple fruit. *Plant Physiol.* 43:448–450.

HEDDEN, P., J. MACMILLAN, and B.O. PHINNEY. 1978. The metabolism of the gibberellins. *Annu. Rev. Plant Physiol.* 29:149–192.

HEMBERG, T. 1951. Establishment of acid growth-inhibiting substances in plant extracts containing auxins by means of the avena test. *Physiol. Plant.* 4:437–445.

HEMBERG, T. and U. LARSSON. 1972. Interaction of kinetin and indoleacetic acid in the avena straight-growth test. *Physiol. Plant.* 26:104–107.

HENSON, I.E. and P.F. WAREING. 1974. Cytokinins in *Xanthium strumarium:* a rapid response to short day treatment. *Physiol. Plant.* 32:185–187.

HENSON, I.E. and P.F. WAREING. 1977. Cytokinins in *Xanthium strumarium* L.: some aspects of the photoperiodic control of endogenous levels. *New Phytol.* 78:35–45.

HESS, D. 1975. Plant physiology. Springer-Verlag, New York.

HEWETT, E.W. and P.F. WAREING. 1973. Cytokinins in *Populus* × *robusta*: qualitative changes during development. *Physiol. Plant.* 29:386–389.

HEWETT, E.W. and P.F. WAREING. 1974. Cytokinin changes during chilling and bud burst in woody plants. p. 693–701. *In* R.L. Bieleski, A.R. Ferguson and M.M. Cresswell (eds.) Mechanisms of regulation of plant growth. The Royal Society of New Zealand, Wellington.

HEYN, A.N.J. 1970. Dextranase activity and auxin-induced cell elongation in coleoptiles of *Avena*. *Biochem. Biophys. Res. Comm.* 38:831–837.

HILL-COTTINGHAM, D.G. and C.P. LLOYD-JONES. 1979. Translocation of nitrogenous compounds in plants. p. 397–405. *In* E.J. Hewitt and C. Cutting (eds.) Nitrogen assimilation of plants. Academic Press, New York.

HILLIS, W.E. 1962. The distribution and formation of polyphenols within the tree. p. 59–131. *In* W.E. Hillis (ed.) Wood extractives and their significance to the pulp and paper industry. Academic Press, New York.

HILLIS, W.E. and T. SWAIN. 1959. The phenolic constituents of *Prunus domestica*. II. The analysis of tissues of the Victoria plum tree. *J. Sci. Food & Agr.* 10:135–144.

HIRSCHBERG, K., G. HUBNER, and H. BORRISS. 1972. Cytokinin-induzierte-*de novo*—synthese der nitrareductase in embrymen von *Agrostemma githago*. *Planta* 108:333–337.

HOAD, G.V. 1973. Hormones in the phloem of higher plants. p. 17–30. *In* R. Antoszewski (ed.) Proc. 3rd Symp. Accumulation and Translocation of Nutrients and Regulators in Plant Organisms, Warsaw, 1972. Skierniewiece, Warsaw, Poland.

HOAD, G.V. 1975. Effect of osmotic stress on abscisic acid levels in xylem sap of sunflower. *Planta* 124:25–29.

HOAD, G.V. 1978. Effect of water stress on abscisic acid levels in white lupine (*Lupenus albus* L.) fruit, leaves and phloem exudate. *Planta* 142:287–290.

HOAD, G.V. and M.R. BOWEN. 1968. Evidence for gibberellin-like substances in phloem exudate of higher plants. *Planta* 82:22–32.

HOAD, G.V. and J.M.S. FORREST. 1972. Phloem translocation of endogenous hormones. Annu. Rpt. Bristol Univ. Res. Sta. Long Ashton 1971. p. 40–41.

HOCKING, T.J., J.R. HILLMAN, and M.B. WILKINS. 1972. Movement of abscisic acid in *Phaseolus vulgaris* plants. *Nature New Biol.* 235:124–125.

HOLM, R.E. and J.L. KEY. 1969. Hormonal regulation of cell elongation in the hypocotyl of rootless soybean: an evaluation of the role of DNA synthesis. *Plant Physiol.* 44:1295–1302.

HSIA, C.P. and C.H. KAO. 1978. The importance of roots in regulating the senescence of soybean primary leaves. *Physiol. Plant.* 43:385–389.

HUCK, M.G., N. SEONIT, J.M. DAVIS, R.E. YOUNG, and V.D. BROWNING. Control of shoot growth and water relations in cotton plants with varying root size. (personal communication)

HUMPHRIES, E.C. 1958. Effect of removal of a part of the root system on the subsequent growth of the root and shoot. *Ann. Bot. N.S.* 22:251–257.

HUMPHRIES, E.C. 1963. Dependence of net assimilation rate on root growth in isolated leaves. *Ann. Bot. N.S.* 27:175–183.

HUMPHRIES, E.C. and G.N. THORNE. 1964. The effect of root formation on photosynthesis of detached leaves. *Ann. Bot. N.S.* 28:391–400.

HUNT, R. 1975. Further observations on root-shoot equilibria in perennial ryegrass (*Lolium perenne* L.). *Ann. Bot.* 39:745–755.

HUNT, R., D.P. STRIBLEY, and D.J. READ. 1975. Root/shoot equilibria in cranberry (*Vaccinium macrocarpon* Ait). *Ann. Bot.* 39:807–810.

HUTCHINSON, A., C.D. TAPER, and G.H.N. TOWERS. 1959. Studies of phloridzin in *Malus*. *Can. J. Biochem. Physiol.* 37:901–910.

IBRAHIM, I.M. and M.N. DANA. 1971. Gibberellin-like activity in apple rootstocks. *HortScience* 6:541–542.

INGERSALL, R.B. and O.E. SMITH. 1971. Transport of abscisic acid. *Plant & Cell Physiol.* 12:301–309.

ITAI, C., A. BEN-ZIONI, and L. ORDIN. 1973. Correlative changes in endogenous hormone levels and shoot growth induced by short heat treatments to the root. *Physiol. Plant.* 29:355–360.

ITAI, C., A. RICHMOND, and Y. VAADIA. 1968. The role of root cytokinins during water and salinity stress. *Israel J. Bot.* 17:187–195.

ITAI, C. and Y. VAADIA. 1965. Kinetin-like activity in root exudate of water-stressed sunflower plants. *Physiol. Plant.* 18:941–944.

ITAI, C. and Y. VAADIA. 1971. Cytokinin activity in water-stressed shoots. *Plant Physiol.* 47:87–90.

JABLANOVIC, M. and M. NESKOVIC. 1977. Changes in endogenous level of auxins and cytokinins in axillary buds of *Pisum sativum* in relation to apical dominance. *Biol. Plant.* 19:34–39.

JACKSON, W.A. 1978. Factors influencing nitrate acquisition by plants: assimilation and fate of reduced nitrogen. p. 45–88. *In* D.R. Nielsen and J.G. Macdonald (eds.) Nitrogen in the environment, Vol. 2. Academic Press, New York.

JACKSON, W.T. 1955. The role of adventitious roots in recovery of shoots following flooding of the original root systems. *Amer. J. Bot.* 42:816–819.

JACKSON, W.T. 1956. Flooding injury studied by approach-graft and split root system techniques. *Amer. J. Bot.* 43:496–502.

JACOBS, W.P. 1977. Regulation of development by the differential polarity of various hormones as well as by effects of one hormone on the polarity of another. p. 361–380. *In* H.R. Schutte and D. Gross (eds.) Regulation of developmental processes in plants. Fisher-Verlag, Jena.

JACOBS, W.P. and R. ALONI. 1978. Evidence that IAA does not move basipetally against a concentration gradient in *Coleus blumei* internodes. *Ann. Bot.* 42:989–991.

JINDAL, K.K., S. DALBRO, A.S. ANDERSON, and L. POLL. 1974. Endogenous growth substances in normal and dwarf mutants of 'Cortland' and 'Golden Delicious' apple shoots. *Physiol. Plant.* 52:71–77.

JONES, O.P. 1967. Effect of benzyl adenine on isolated apple shoots. *Nature* 215:1514–1515.

JONES, O.P. 1973. Effects of cytokinins in xylem sap from apple trees on apple shoot growth. *J. Hort. Sci.* 48:181–188.

JONES, O.P. 1976a. Effect of dwarfing interstock on xylem sap composition in apple trees: effect on nitrogen, potassium, P, Ca, and Mg content. *Ann. Bot.* 40:1231–1235.

JONES, O.P. 1976b. Effect of phloridzin and phloroglucinol on apple shoots. *Nature* 262:392–393.

JONES, O.P. and S.G.S. HATFIELD. 1976. Root initiation in apple shoots cultured *in vitro* with auxins and phenolic compounds. *J. Hort. Sci.* 51:495–499.

JONES, O.P. and H.J. LACEY. 1968. Gibberellin-like substances in the transpiration stream of apple and pear trees. *J. Expt. Bot.* 19:526–531.

JONES, O.P. and J.S. PATE. 1976. Effect of M 9 dwarfing interstocks on the amino compounds of apple xylem sap. *Ann. Bot.* 40:1237.

JONES, R.L. 1973. Gibberellins: their physiologic role. *Annu. Rev. Plant Physiol.* 24:571–598.

JONES, R.L. and A. LANG. 1968. Extractable and diffusible gibberellins from light- and dark-grown pea seedlings. *Plant Physiol.* 43:629–634.

JONES, R.L. and I.D.J. PHILLIPS. 1966. Organs of gibberellin synthesis in light-grown sunflower plants. *Plant Physiol.* 41:1381–1386.

JONES, R.W. and R.W. SHEARD. 1979. Light factors in nitrogen assimilation. p. 521–539. *In* E.J. Hewitt and C. Cutting (eds.) Nitrogen assimilation in plants. Academic Press, New York.

JORDAN, W.R. and F. SKOOG. 1971. Effects of cytokinins on growth and auxin in coleoptiles of derooted *Avena* seedlings. *Plant Physiol.* 48:97–99.

JUNIPER, B.E. 1976. Geotropism. *Annu. Rev. Plant Physiol.* 27:385–406.

KANDIAH, S. 1979. Turnover of carbohydrates in relation to growth in apple trees. I. Seasonal variation of growth and carbohydrate reserves. *Ann. Bot.* 44:175–183.

KANDIAH, S. and S. WIMALADHARMA. 1978. Root-shoot interaction in the turnover of reserves in tea (*Camellia sinensis* L.) roots. *Ann. Bot.* 42: 931–935.

KARMAKER, J.L. and R.F.M. VAN STEVENIND. 1979. The effect of abscisic acid on sugar levels in seedlings of *Phaseolus vulgaris* 'Redland Pioneer'. *Planta* 146:25–30.

KATO, T. and H. ITO. 1962. Physiological factors associated with the shoot growth of apple trees. *Tohoku J. Agr. Res.* 13:1–21.

KATSUMI, M. and H. KAZAMA. 1978. Auxin-gibberellin relationships in their effects on hypocotyl elongation of light-grown cucumber seedlings. VI. Promotive and inhibitory effects of kinetin. *Plant & Cell Physiol.* 19:107–115.

KEFELI, V.I. 1978. Natural plant growth inhibitors and phytohormones. W. Junk, The Hague.

KEFELI, V.I. and C.S. KADYROV. 1971. Natural growth inhibitors, their chemical and physiological properties. *Annu. Rev. Plant Physiol.* 22:185–196.

KEFELI, V.I. and M. KUTACEK. 1977. Phenolic substances and their possible role in plant growth regulation. p. 181–188. *In* P.E. Pilet (ed.) Plant growth regulation. Proc. Intern. Conf. on Plant Growth Substances, Aug. 20–Sept. 4, 1976, Lausanne. Springer-Verlag, New York.

KEFELI, V. and M. KUTACEK. 1979. Effects of phenolic compounds on auxin biosynthesis and vice versa. p. 13–23. *In* M. Luckner and K. Schreiber (eds.) Regulation of secondary product and plant hormone metabolism. FEB. Soc. 12th Mtg. Symp. S8, Dresden, 1978. Pergamon Press, New York.

KEFELI, V.I. and R. TURETSKAYA. 1965. Participation of phenolic compounds in the inhibition of auxin activity and in the growth of willow shoot. *Fiziologiya Rastenii.* 12:638–645.

KENDE, H. 1964. Preservation of chlorophyll in leaf sections by substances obtained from root exudate. *Science* 145:1066–1067.

KENDE, H. 1965. Kinetin-like factors in the root exudate of sunflowers. *Proc. Nat. Acad. Sci.* 53:1302–1307.

KENDE, H. 1971. The cytokinins. *Intern. Rev. Cytol.* 31:301–338.

KENDE, H., M. FUKUYAMA-DILWORTH, and R. DEZACKS. 1974. On the control of nitrate reductase by nitrate and benzyladenine in *Agrostemma githago* embryos. p. 675–682. *In* Plant growth substances. Hirokawa Publ. Co., Tokyo.

KENDE, H., H. HAHN, and S.E. KAYS. 1971. Enhancement of nitrate reductase activity by benzyladenine in *Agrostemma githago. Plant Physiol.* 48: 702–706.

KENDE, H. and D. SITTON. 1967. The physiological significance of kinetin- and gibberellin-like root hormones. *Ann. N.Y. Acad. Sci.* 144:235–243.

KING, R.W. 1976. Implications for plant growth of the transport of regulatory compounds in phloem and xylem. p. 415–431. *In* I.F. Wardlaw and B.J. Passioura (eds.) Transport and transfer processes in plants. Academic Press, New York.

KLEPPER, L. and R.H. HAGEMAN. 1969. The occurrence of nitrate reductase in apple leaves. *Plant Physiol.* 44:110–114.

KNYPL, J.S. 1979. Hormonal control of nitrate assimilation. p. 541–546. *In* E.J. Hewett and C. Cutting (eds.) Nitrogen assimilation of plants. Academic Press, New York.

KONINGS, H. 1964. On the indoleacetic acid converting enzyme of pea roots and its relation to geotropism, straight growth and cell wall properties. *Acta. Bot. Neerl.* 13:566–622.

KRAMER, P.J. and T.T. KOZLOWSKI. 1979. Physiology of woody plants. Academic Press, New York.

KRISHNAMOORTHY, H.N. 1975. Gibberellins and plant growth. Wiley Eastern Limited, New Delhi.

KULAEVA, O.N. 1962. The effect of roots on leaf metabolism in relation to the action of kinetin on leaves. *Soviet Plant Physiol.* 9:182–189.

KULAEVA, O.N., V.V. KUZNETSOV, and V.V. KUZNETSOV. 1976. Cytokinin induction of nitrate reductase activity in isolated embryos of *Agrostemma githago. Soviet Plant Physiol.* 23:1061–1068.

KULASEGARAM, S. 1969. Studies on the dormancy of tea shoots. 2. Roots as the source of a stimulus associated with the growth of dormant buds. *Tea Quart.* 40:84–92.

KULASEGARAM, S. and A. KATHIRAVETPILLAI. 1972a. Effect of nutrition and hormones on growth and apical dominance in tea (*Camellia sinensis* L.). *J. Hort. Sci.* 47:11–24.

KULASEGARAM, S. and A. KATHIRAVETPILLAI. 1972b. Observations on roots of tea plants with active and dormant shoots. *Tea Quart.* 43:53–57.

KURAISHI, S. 1974. Biogenesis of auxin in barley. p. 209–216. *In* Plant growth substances. Proc. 8th Intern. Conf. on Plant Growth Substances, Aug. 26–Sept. 1, 1973, Tokyo. Hirokawa Publ. Co., Tokyo.

LAMBERS, H. 1979. Efficiency of root respiration in relation to growth rate, morphology, and soil composition. *Physiol. Plant.* 46:194–202.

LAMPORT, T.A. 1970. Cell wall metabolism. *Annu. Rev. Plant Physiol.* 21: 235–270.

LANG, A. 1970. Gibberellins: structure and metabolism. *Annu. Rev. Plant Physiol.* 21:537–570.

LEE, T.T. 1971a. Promotion of indoleacetic acid oxidase isoenzymes in tobacco callus cultures by indoleacetic acid. *Plant Physiol.* 48:56–59.

LEE, T.T. 1971b. Cytokinin-controlled indoleacetic acid oxidase isoenzymes in tobacco callus cultures. *Plant Physiol.* 47:181–185.

LEE, T.T. 1972. Interaction of cytokinin, auxin, and gibberellin on peroxidase isoenzymes in tobacco tissues cultured *in vitro. Can. J. Bot.* 53:2471–2477.

LEE, T.T. 1974. Cytokinin control in subcellular localization of indoleacetic acid oxidase and peroxidase. *Phytochemistry* 13:2445–2453.

LEE, T.T. and F. SKOOG. 1965. Effects of substituted phenols on bud formation and growth of tobacco tissue cultures. *Physiol. Plant.* 18:386–401.

LEGERSTEDT, H.B. and R.G. LANGSTON. 1967. The mobilization force of kinetin. *Life Sci.* 6:145–149.

LEONOVA, L.A., L.V. GAMANETS, and K.Z. GAMBURG. 1970. Effect of auxin on polyphenol content in callus tissue of tobacco grown in suspended culture. *Soviet Plant Physiol.* 17:611–616.

LEOPOLD, A.C. and P.E. KRIEDEMANN. 1975. Plant growth and development. 2nd ed. McGraw-Hill Book Co., New York.

LEPP, N.W. and A.J. PEEL. 1971. Influence of IAA upon the longitudinal and tangential movement of labeled sugar in the phloem of willow. *Planta* 97:50–61.

LESHEM, Y. 1973. The molecular and hormonal basis of plant growth regulation. Pergamon Press, New York.

LETHAM, D.S. 1967. Chemistry and physiology of kinetin-like compounds. *Annu. Rev. Plant Physiol.* 18:349–364.

LETHAM, D.S. 1978a. Cytokinins. p. 205–263. *In* D.S. Letham, P.B. Goodwin and T.J.V. Higgins (eds.) The biochemistry of phytohormones and related compounds: a comprehensive treatise, Vol. I. Elsevier/North Holland, Amsterdam.

LETHAM, D.S. 1978b. Natural-occurring plant growth regulators other than the principal hormones of higher plants. p. 349–465. *In* D.S. Letham, P.B. Goodwin and T.J.V. Higgins (eds.) The biochemistry of phytohormones and related compounds: a comprehensive treatise, Vol. I. Elsevier/North Holland, Amsterdam.

LEVAND, O. and R.M. HEINICKE. 1968. Ferulic acid as a component of a complex carbohydrate polymer of bromelain. *Phytochemistry* 7:1659–1662.

LIEBERMAN, M. and A.T. KUNISHI. 1971. Abscisic acid and ethylene production. *Plant Physiol.* 47:5–22.

LIPS, S.H. 1979. Photosynthesis and photorespiration in nitrate metabolism. p. 445–450. *In* E.J. Hewitt and C. Cutting (eds.) Nitrogen assimilation in plants. Academic Press, New York.

LOCKARD, R.G. 1976. The effect of dwarfing rootstocks and interstocks on the proportion of bark on the tree. *Hort. Res.* 15:83–94.

LOCKARD, R.G. and C. GRUNWALD. 1970. Grafting and gibberellin effects on the growth of tall and dwarf peas. *Plant Physiol.* 45:160–162.

LOCKHART, J.A. 1957. Studies on the organ of production of the natural gibberellin factor in higher plants. *Plant Physiol.* 32:204–207.

LOO, S.W. 1945. Cultivation of excised stem tips of asparagus *in vitro. Amer. J. Bot.* 32:13–17.

LOVEYS, B.R. 1977. The intracellular location of ABA in stressed and nonstressed leaf tissue. *Physiol. Plant.* 40:6–10.

LUCKWILL, L.C. 1959. The physiological relationships of root and shoot. *Scientia Hort.* 14:22–26.

LUCKWILL, L.C. 1973. Growth regulators—their potential and limitations. *Scientia Hort.* 24:153–157.

LUCKWILL, L.C., P. WEAVER, and J. MACMILLAN. 1969. Gibberellins and other growth hormones in apple seeds. *J. Hort. Sci.* 44:413–424.

LUCKWILL, L.C. and P. WHYTE. 1968. Hormones on the xylem sap of apple trees. Plant growth regulators. *Sci. Monograph* 31:87–101.

MACHEIX, J. 1974. Evolution de la teneur du fruit de *Pirus malus* L. en esters des acides *p*-coumarique et cafeique au cours de la croissance. *Physiol. Veg.* 12:25–33.

MACMILLAN, J. 1968. Direct identification of gibberellins in plant extracts by gas chromatography-mass spectrometry. p. 101–107. *In* F. Wightman and G. Setterfield (eds.) Biochemistry and physiology of plant growth substances. The Runge Press, Ottawa.

MACMILLAN, J. 1971a. Current aspects of gibberellins and plant growth regulation. p. 175–187. *In* H. Kaldeway and Y. Vardar (eds.) Hormonal regulation in plant growth and development. Izmir, Turkey.

MACMILLAN, J. 1971b. Diterpenes—the gibberellins. p. 153–180. *In* T.W. Goodwin (ed.) Aspects of terpenoid chemistry and biochemistry. Academic Press, New York.

MARIGO, G. and A.M. BOUDET. 1975. The role of polyphenols in growth. The definition of an experimental model in *Lycopersicon esculentum*. *Physiol. Plant.* 34:51–55.

MARIGO, G. and A.M. BOUDET. 1977. Relations polyphenols-croissance: Mise en evidence d'un effet inhibiteur des composes phenoliques sur le transport polarise de l'auxine. *Physiol. Plant.* 41:197–202.

MARTIN, G.C. and M.W. WILLIAMS. 1967. Comparison of some biochemical constituents of EM IX and XVI bark. *HortScience* 2:154.

MARTIN, H.V., M.C. ELLIOTT, E. WANGERMANN, and P.E. PILET. 1978. Auxin gradient along the root of the maize seedling. *Planta* 141:179–181.

MARTIN, T., A. CIOFU, M. GEORGESCU, and M. PAUNET. 1974. The influence of the rootstock on the biochemical composition of *V. vinifera* cultivars. Lucr. Stint Inst. Agron. Nicolae Balcescu Bucar Ser. B. (*Biol. Abstr.* 60: 66545, 1975).

MASUDA, Y. 1978. Auxin induced cell wall loosening. p. 103–124. *In* H. Shibaoka, M. Furuya, M. Katsumi and A. Takimoto (eds.) Controlling factors in plant development. The Botanical Magazine, Spec. Issue 1, Tokyo.

MCCOMB, A.J. 1964. The stability and movement of gibberellic acid in pea seedlings. *Ann. Bot. N.S.* 28:669–686.

MCDAVID, C.R., C. MARSHALL, and G.R. SAGAR. 1973. Factors influencing the growth of the root system of etiolated seedlings of *Pisum sativum* L. *New Phytol.* 72:269–275.

MCDAVID, C.R., G.R. SAGAR, and C. MARSHALL. 1972. The effect of auxin from the shoot on root development in *Pisum sativum* L. *New Phytol.* 71:1027–1032.

MCKENZIE, D.W. 1961. Rootstock-scion interaction in apples with special reference to root anatomy. *J. Hort. Sci.* 36:40–47.

MENARY, R.C. and J. VAN STADEN. 1976. Effect of phosphorus nutrition and cytokinins on flowering in the tomato, *Lycopersicon esculentum* Mill. *Austral. J. Plant Physiol.* 3:201–205.

MENDEL, K. and A. COHEN. 1962. Methods for rapid evaluation of rootstocks for citrus. Hebrew University, Jerusalem Division Citriculture. Spec. Bul. 46. (S.R. Miller, 1965. Growth inhibition produced by leaf extracts from size-controlling apple rootstocks. *Can. J. Plant Sci.* 45:519–524).

MENDEL, K. and A. COHEN. 1967. Starch level in the trunk as a measure of compatability between stock and scion in citrus. *J. Hort. Sci.* 42:231–234.

MESSER, G.S. and S. LAVEE. 1969. Studies on vigour and dwarfism of apple trees in an *in vitro* tissue culture system. *J. Hort. Sci.* 44:219–233.

MIGINIAC, E. 1974. Flowering and correlations between organs in *Scrofularia arguta* Sol. p. 539–545. *In* R.L. Bieleski, A.R. Ferguson and M.M. Cresswell (eds.) Mechanisms of regulation of plant growth. The Royal Society of New Zealand, Bul. 12.

MIIDLA, H., A. MILIUS, and T. VAINJARU. 1970. Content of phenolic compounds and lignification of the woody parts of apple shoots depending on mineral nutrition. *Biol. Plant (Praha)* 12:11–18.

MILBORROW, B.V. 1974. Chemistry and physiology of abscisic acid. *Annu. Rev. Plant Physiol.* 25:259–307.

MILBORROW, B.V. 1978. Abscisic acid. p. 295–348. *In* D.S. Letham, P.B. Goodwin and T. Higgins (eds.) The biochemistry of phytohormones and related compounds: a comprehensive treatise, Vol. I. Elsevier/North Holland, Amsterdam.

MILBORROW, B.V. and D.R. ROBINSON. 1973. Factors affecting the biosynthesis of abscisic acid. *J. Expt. Bot.* 24:537–548.

MILLER, C.O. 1961a. A kinetin-like compound in maize. *Proc. Natl. Acad. Sci. (U.S.A.)* 47:170–174.

MILLER, C.O. 1961b. Kinetin and related compounds in plant growth. *Annu. Rev. Plant Physiol.* 14:395–408.

MILLER, L.P. 1973. Glycosides. p. 297–375. *In* L.P. Miller (ed.) Phytochemistry. The process and products of photosynthesis, Vol. I. Van Nostrand Reinhold, New York.

MILLER, S.R. 1965. Growth inhibition produced by leaf extracts from size controlling apple rootstocks. *Can. J. Plant Sci.* 45:519–524.

MIZRAHI, Y. and A.E. RICHMOND. 1972. Abscisic acid in relation to mineral deprivation. *Plant Physiol.* 50:667–670.

MORGAN, D.G. and C.B. MORGAN. 1974. Plant growth substances. p. 129–158. *In* D. Northcote (ed.) MTP international review of science: biochemistry, Series I, Vol. 11. University Park Press, Baltimore.

MORRIS, D.A. 1977. Transport of exogenous auxin in two-branched dwarf pea seedlings. *Planta* 136:91–96.

MORRIS, D.A. and A.G. THOMAS. 1978. A microautographic study of auxin transport in the stem of intact pea seedlings. *J. Expt. Bot.* 29:147–157.

MORRIS, D.A. and P.J. WINFIELD. 1972. Kinetin transport to axillary buds of dwarf pea (*Pisum sativum* L.). *J. Expt. Bot.* 23:346–355.

MULLINS, M.G. 1967. Morphogenetic effects of roots and of some synthetic cytokinins in *Vitis vinifera* L. *J. Expt. Bot.* 18:206–214.

MULLINS, M.G. 1968. Regulation of inflorescence growth in cuttings of the grape vine (*Vitis vinifera* L.). *J. Expt. Bot.* 19:532–543.

NACHIT, M. and W. FEUCHT. 1977. Suitability of phenolic and amino acids as selection criteria for the vigor of *Malus* rootstocks. Mitteilungen Rebe und Wein Obstfaw und Fruchteverwertung 26:199–204. (*Hort. Abstr.* 47:4250.)

NANDA, K.K. and V.K. ANAND. 1970. Seasonal changes in auxin effects on rooting of stem cuttings of *Populus nigra* and its relationship with mobilization of starch. *Physiol. Plant.* 23:99–107.

NEUMANN, J. 1959. An auxin-like action of coumarin. *Science* 129:1675–1676.

NITSCH, M.P. and C. NITSCH. 1962. Composes phenoliques et croissance vegetale. *Ann. Physiol. Veg.* 4:211–225.

NIX, L.E. and T.J. WOODZICKI. 1974. The radial distribution and metabolism of IAA-^{14}C in *Pinus echinata* stems in relation to wood formation. *Can. J. Bot.* 52:1349–1355.

NOGGLE, G.R. 1979. Recognition systems in plants. *What's New in Plant Physiol.* 10:5–8.

NOODEN, L.D. and S.J. LINDOO. 1978. Monocarpic senescence. *What's New in Plant Physiol.* 9:25–28.

OBATA-SASAMATO, H. and H. SUZUKI. 1979. Activities of enzymes relating to starch synthesis and endogenous levels of growth regulators in potato stolon tips during tuberization. *Physiol. Plant.* 45:320–324.

ORITANI, T. and R. YOSHIDA. 1971. Studies on nitrogen metabolism in crop plants. XI. The changes of abscisic acid and cytokinin-like activity accompanying growth and senescence in the crop plants. *Proc. Crop Sci. Soc. (Japan)* 40:325–331.

PALEG, L.G. and G.A. WEST. 1972. The gibberellins. p. 146–181. *In* F.C. Steward (ed.) Plant physiology: a treatise, Vol. VIB. Academic Press, New York.

PATE, J.S. 1975. Exchange of solutes between phloem and xylem and circulation in the whole plant. p. 451–473. *In* M.H. Zimmerman and J.A. Milburn (eds.) Transport in plants. I. Phloem transport. Encyclopedia of plant physiology, new series, Vol. I. Springer-Verlag, New York.

PATRICK, J.W. and P.F. WAREING. 1970. Experiments on the mechanism of hormone-directed transport. p. 695–700. *In* D.J. Carr (ed.) Plant growth substances. Springer-Verlag, New York.

PATRICK, J.W. and P.F. WAREING. 1978. Auxin promoted transport of metabolism of stems of *Phaseolus vulgaris* L. *J. Expt. Bot.* 29:359–366.

PEMADASA, M.A. 1979. Stomatal response to two herbicidal auxins. *J. Expt. Bot.* 30:267–274.

PHILLIPS, D.A. and C.F. CLELAND. 1972. Cytokinin activity from the phloem sap of *Xanthium strumarium* L. *Planta* 102:173–178.

PHILLIPS, I.D.J. 1964a. Root-shoot hormone relations. I. The importance of an aerated root system in the regulation of growth hormone levels in the shoot of *Helianthus annuus*. *Ann. Bot.* 28:17–35.

PHILLIPS, I.D.J. 1964b. Root-shoot hormone relations. II. Changes in endogenous auxin concentration produced by flooding of the root system of *Helianthus annuus*. *Ann. Bot. N. S.* 28:37–45.

PHILLIPS, I.D.J. 1975. Apical dominance. *Annu. Rev. Plant Physiol.* 26: 341–367.

PHILLIPS, I.D.J. and R.L. JONES. 1964. Gibberellin-like activity in bleeding-sap of root systems of *Helianthus annuus* detected by a new dwarf pea epicotyl assay and other methods. *Planta* 63:269–278.

PILET, P.E. 1966. Effect of *p*-hydroxybenzoic acid on growth, auxin content and auxin catabolism. *Phytochemistry* 5:77–82.

PILET, P.E. 1968. *In vitro* and *in vivo* auxin and cytokinin translocation. p. 993–1004. *In* F. Wightman and G. Setterfield (eds.) Proc. 6th Intern. Conf. on Plant Growth Substances, July 24–29, 1967, Carlton Univ., Ottawa. The Runge Press, Ottawa.

PILET, P.E. 1977a. Hormone balance and endogenous interactions in root growth and georeaction. p. 331–342. *In* H.R. Schutte and D. Gross (eds.) Regulation of developmental processes in plants. Fisher-Verlag, Jena.

PILET, P.E. 1977b. Growth inhibitors in growing and geostimulated maize roots. p. 115–128. *In* P.E. Pilet (ed.) Plant growth regulation. Proc. 9th Intern. Conf. on Plant Growth Substances, Aug. 20–Sept, 4, 1976, Lausanne. Springer-Verlag, New York.

PILET, P.E., M.E. ELLIOTT, and M.M. MOLONEY. 1979. Endogenous and exogenous auxin in the control of root growth. *Planta* 146:405–408.

POAPST, P.A., A.B. DURKEE, and F.B. JOHNSTON. 1970. Root-differentiating properties of some glycosides and polycyclic phenolic compounds in apple and pear fruits. *J. Hort. Sci.* 45:69–74.

POLING, E.B. and G.H. OBERLY. 1979. Effect of rootstock on mineral composition of apple leaves. *J. Amer. Soc. Hort. Sci.* 104:799–801.

PONIEDZIALEK, W., W. LECH, and M. URBANEK. 1979. Anatomical changes in the grafted apple bark rings. *Fruit Sci. Rpt.* 6:115–134.

POWELL, L.E. and S.D. SEELEY. 1974. Metabolism of abscisic acid to a water soluble complex in apple. *J. Amer. Soc. Hort. Sci.* 99:439–441.

PRIESTLY, C.A. 1960. Seasonal changes in the carbohydrate resources of some six-year-old apple trees. p. 70–77. East Malling Res. Sta. Annu. Rpt. 1959.

RADIN, J.W. and R.S. LOOMIS. 1971. Changes in the cytokinins of radish roots during maturation. *Physiol. Plant.* 25:240–244.

RADIN, J.W., L.L. PARKER, and C.R. SELL. 1978. Partitioning of sugar between growth and nitrate reduction in cotton roots. *Plant Physiol.* 62:550–553.

RAILTON, I.D., R.C. DURLEY, and R.P. PHARIS. 1974a. Studies on gibberellin biosynthesis in etiolated shoots of dwarf pea, 'Meteor'. p. 294–304. *In* Plant growth substances. Proc. 8th Intern. Conf. on Plant Growth Substances, Aug. 26–Sept. 1, 1973, Tokyo. Hirokawa Publ. Co., Tokyo.

RAILTON, I.D., N. MUROFUSHI, R.C. DURLEY, and R.P. PHARIS. 1974b. Interconversion of gibberellin A_{20} to gibberellin A_{29} by etiolated seedlings and germinating seeds of dwarf *Pisum sativum*. *Phytochemistry* 13:793–796.

RAILTON, I.D. and D.M. REID. 1973. Effects of benzyladenine on the growth of waterlogged tomato plants. *Planta* 111:261–266.

REHM, M.M. and M.G. CLINE. 1973. Rapid growth inhibition of *Avena* coleoptile segments by abscisic acid. *Plant Physiol.* 51:93–96.

REID, D.M. and D.J. CARR. 1967. Effects of a dwarfing compound, CCC, on the production and export of gibberellin-like substances by root systems. *Planta* 73:1–11.

REID, D.M., A. CROZIER, and B.M.R. HARVEY. 1969. The effects of flooding on the export of gibberellins from the root to the shoot. *Planta* 89:376–379.

REID, D.M. and I.D. RAILTON. 1974a. The influence of benzyladenine on the growth and gibberellin content of shoots of waterlogged tomato plants. *Plant Sci. Letters* 2:151–156.

REID, D.M. and I.D. RAILTON. 1974b. Effect of flooding on the growth of tomato plants: involvement of cytokinins and gibberellins. p. 789–792. *In* R.L. Bieleski, A.R. Ferguson and M.M. Cresswell (eds.) Mechanisms of regulation of plant growth. The Royal Society of New Zealand, Bul. 12.

REKOSLAVSKAYA, N.I., K.Z. GAMBURG, and L.V. GAMANETS. 1974. Effects of endogenous and exogenous polyphenols on metabolism and activity of IAA in suspension cultures of tobacco tissue. *Soviet Plant Physiol.* 21:591–596.

RIBEREAU-GAYON, P. (Translated) 1972. Plant phenolics. Oliver & Boyd, Edinburgh.

RICE, E.L. 1974. Allelopathy. Academic Press, New York.

RICHARDS, D. and R.N. ROWE. 1977a. Effects of root restriction, root pruning and 6-benzylaminopurine on the growth of peach seedlings. *Ann. Bot.* 41:729–740.

RICHARDS, D. and R.N. ROWE. 1977b. Root-shoot interactions in peach: the function of the root. *Ann. Bot.* 41:1211–1216.

RICHARDSON, S.D. 1959. Studies of root growth in *Acer saccharinum.* VI. Further effects of the shoot system on root growth. *Proc. Kon. Ned. Akad. V. Wetenach.* C60:624–629.

RICHMOND, A.E. and A. LANG. 1957. Effect of kinetin on protein content and survival of detached *Xanthium* leaves. *Science* 125:650–651.

RICK, C.M. 1952. The grafting relations of wilty dwarf, a new tomato mutant. *Amer. Natur.* 86:173–184.

RISEUNO, M.C., J.L. DIEZ, G. GIMENEZ-MARTIN, and C. DE LA TORRE. 1971. Ultrastructural study of effect of abscisic acid on cell elongation in plant cells. *Protoplasma* 73:323–328.

ROBERTS, A.N. and L.T. BLANEY. 1967. Qualitative, quantitative, and positional aspects of interstock influence on growth and flowering of the apple. *Proc. Amer. Soc. Hort. Sci.* 91:39–49.

ROBERTS, L.W. 1969. The initiation of xylem differentiation. *Bot. Rev.* 35: 201–250.

ROBERTS, R.H. 1949. Theoretical aspects of graftage. *Bot. Rev.* 15:423–463.

ROBITAILLE, H.A. 1970. The relationship of endogenous hormones to growth characteristics and dwarfing in *Malus.* Ph.D. Dissertation, Michigan State University, East Lansing.

ROBITAILLE, H. and R.F. CARLSON. 1971. Response of dwarfed apple trees to stem injections of gibberellin and abscisic acid. *HortScience* 6:539–540.

ROBITAILLE, H.A. and R.F. CARLSON. 1976. Gibberellic and abscisic acid-like substances and the regulation of apple shoot extension. *J. Amer. Soc. Hort. Sci.* 101:388–392.

ROGERS, W.S. and A.B. BEAKBANE. 1957. Stock and scion relations. *Annu. Rev. Plant Physiol.* 8:217–236.

ROGERS, W.S. and G.A. BOOTH. 1959. The roots of fruit trees. *Scientia Hort.* 14:27–34.

ROTH-BEJERANO, N. and S.H. LIPS. 1970. Hormonal regulation of nitrate reductase activity in leaves. *New Phytol.* 69:165–169.

SACHS, T. 1969. Polarity and the induction of organized vascular tissues. *Ann. Bot.* 33:263–275.

SACHS, T. 1972. A possible basis for apical organization in plants. *J. Theor. Biol.* 37:353–361.

SACHS, T. 1975. The induction of transport channels by auxin. *Planta* 127: 201–206.

SACHS, T. and K.V. THIMANN. 1964. Release of lateral buds from apical dominance. *Nature* 201:939–940.

SACHS, T. and K.V. THIMANN. 1967. The role of auxins and cytokinins in the release of buds from dominance. *Amer. J. Bot.* 54:136–144.

SAHULKA, J. 1972. The effect of exogenous IAA and kinetin on nitrate reductase, nitrite reductase and glutamate dehydrogenase activities in excised pea roots. *Biol. Plant.* 14:330–336.

SAHULKA, J. 1974. The effect of exogenous IAA and kinetin on nitrate reductase, nitrite reductase and glutamate dehydrogenase activities in excised pea roots. p. 265–270. *In* J. Kolek (ed.) Structure and function of primary root tissues. Bratislava, Veda.

SALAMA, A.M.E.A. and P.F. WAREING. 1979. Effects of mineral nutrition on endogenous cytokinins in plants of sunflower. *J. Expt. Bot.* 30:971–981.

SAN ANTONIO, J.P. 1952. The role of coumarin in the growth of roots of *Melilotus alba. Bot. Gaz.* 114:79–95.

SANKHLA, N. and W. HUBER. 1974. Effects of abscisic acid on the activities of photosynthetic enzymes and $^{14}CO_2$ fixation products in leaves of *Pennisetum typhoides* seedlings. *Physiol. Plant.* 30:291–294.

SARAPUU, L. 1964. Phloridzin as a β-inhibitor and the seasonal dynamics of its metabolic products in apple shoots. *Soviet Plant Physiol.* 11:520–525.

SARAPUU, L. 1965. Physiological effect of phloridzin as a Beta inhibitor during growth and dormancy in the apple trees. *Soviet Plant Physiol.* 12:110–119.

SARGENT, J.A. and F. SKOOG. 1960. Effects of indoleacetic acid and kinetin on scopoletin-scopolin levels in relation to growth of tobacco tissue *in vitro. Plant Physiol.* 35:934–941.

SAX, K. 1953. Interstock effects in dwarfing fruit trees. *Proc. Amer. Soc. Hort. Sci.* 62:201–204.

SAX, K. 1954. The control of tree growth by phloem blocks. *J. Arnold Arboretum* 35:251–258.

SAX, K. 1957. The control of vegetative growth and the induction of early fruiting of apple trees. *Proc. Amer. Soc. Hort. Sci.* 69:68–74.

SAX, K. and A.Q. DICKSON. 1956. Phloem polarity in bark regeneration. *J. Arnold Arboretum* 37:173–179.

SCHAEFFE, G.W., J.G. BUTA, and F. SHARPE. 1967. Scopoletin and polyphenol-induced lag in peroxidase catalyzed oxidation of indole-3-acetic acid. *Physiol. Plant.* 20:342–347.

SCHAG, R.K., S.G. MUKHERJEE, and S.K. SOPORY. 1979. Effect of ammonium, sucrose and light on the regulation of nitrate reductase level in *Pisum sativum*. *Physiol. Plant.* 45:281–287.

SCHNEIDER, E.A. and F. WIGHTMAN. 1974. Metabolism of auxin in higher plants. *Annu. Rev. Plant Physiol.* 25:487–513.

SCHNEIDER, G.W., C.E. CHAPLIN, and D.C. MARTIN. 1978a. Effects of apple rootstock, tree spacing, and cultivar on fruit and tree size, yield and foliar mineral composition. *J. Amer. Soc. Hort. Sci.* 103:230–232.

SCHNEIDER, G.W., R.G. LOCKARD, and P.L. CORNELIUS. 1978b. Growth controlling properties of apple stem callus, *in vitro*. *J. Amer. Soc. Hort. Sci.* 103:634–638.

SCHOLZ, E.W. 1957. Physiology of interstem dwarfing in apple. *Diss. Abstr.* 18:751. No. 58-01058.

SCHRADER, L.E. and R.H. HAGEMAN. 1967. Regulation of nitrate reductase activity in corn seedlings by endogenous metabolites. *Plant Physiol.* 42:1750–1756.

SELVENDRAN, R.R. 1970. Changes in the composition of xylem exudate of tea plants *(Camellia sinensis)* during recovery from pruning. *Ann. Bot.* 34: 825–833.

SELVENDRAN, R.R. and S. SABARATNAM. 1971. Composition of the xylem sap of tea plants (*Camellia sinensis* L.). *Ann. Bot.* 35:679–682.

SETH, A.K. and P.F. WAREING. 1965. Isolation of a kinetin-like root-factor in *Phaseolus vulgaris*. *Life Sci.* 4:2275–2280.

SETH, A.K. and P.F. WAREING. 1967. Hormone-directed transport of metabolites and its possible role in plant senescence. *J. Expt. Bot.* 18:65–77.

SHAH, C.B. and R.S. LOOMIS. 1965. Ribonucleic acid and protein metabolism in sugar beet during drought. *Physiol. Plant.* 18:240–254.

SHAH, R.R., K.V. SUBBAIAH, and A.R. MEHTA. 1976. Hormonal effect on polyphenol accumulation in *Cassia* tissues cultured *in vitro*. *Can. J. Bot.* 54: 1240–1245.

SHANER, D.L., S.M. MERTZ, and C.J. ARNTZEN. 1975. Inhibition of ion accumulation in maize roots by abscisic acid. *Planta* 122:79–90.

SHAW, M. and M.S. MANOCHA. 1965. Fine structure in detached senescing wheat leaves. *Can. J. Bot.* 43:747–755.

SHELDRAKE, A.R. 1973a. The production of hormone in higher plants. *Biol. Rev.* 48:509–559.

SHELDRAKE, A.R. 1973b. Auxin transport in secondary tissue. *J. Expt. Bot.* 24:87–96.

SHELDRAKE, A.R. 1974. The polarity of auxin transport in inverted cuttings. *New Phytol.* 73:637–642.

SHIBAOKA, H., T. HOGETSU, and M. SHIMOKORIYAMA. 1974. Involvement of cellulose synthesis and colchicine-sensitive mechanism in gibberellin promotion and kinetin inhibition of stem elongation. p. 821–827. *In* Plant growth substances. Hirokawa, Tokyo.

SHORT, K.C. and J.G. TORREY. 1972. Cytokinins in seedling roots of peas. *Plant Physiol.* 49:155–160.

SIEGELMAN, H.W. 1964. Physiological studies on phenolic biosynthesis. p. 437–456. *In* J.B. Harborne (ed.) Biochemistry of phenolic compounds. Academic Press, New York.

SITTON, D., C. ITAI, and H. KENDE. 1967a. Decreased cytokinin production in the roots as a factor in shoot senescence. *Planta* 73:296–300.

SITTON, D., A. RICHMOND, and Y. VAADIA. 1967b. On the synthesis of gibberellins in roots. *Phytochemistry* 6:1101–1105.

SKENE, K.G.M. 1967. Gibberellin-like substances in root exudate of *Vitis vinefera*. *Planta* 74:250–262.

SKENE, K.G.M. 1972. The effect of ringing on cytokinin activity in shoots of the grapevine. *J. Expt. Bot.* 23:768–774.

SKENE, K.G.M. 1975. Cytokinin production by roots as a factor in the control of plant growth. p. 365–396. *In* J.G. Torrey and D.T. Clarkson (eds.) The development and function of roots. Academic Press, New York.

SKENE, K.G.M. and A.J. ANTCLIFF. 1972. A comparative study of cytokinin levels in bleeding sap of *Vitis vinefera* L. and the two grapevine rootstocks, Saltcreek and 1613. *J. Expt. Bot.* 23:283–293.

SKOOG, F. and D.J. ARMSTRONG. 1970. Cytokinins. *Annu. Rev. Plant Physiol.* 21:359–384.

SKOOG, F. and R.Y. SCHMITZ. 1972. Cytokinins. p. 181–213. *In* F.C. Steward (ed.) Plant physiology: a treatise, Vol. VIB. Academic Press, New York.

SMITH, D.L. 1969. The role of leaves and roots in the control of inflorescence development in *Carex*. *Ann. Bot.* 33:505–514.

SMITH, H. and P.F. WAREING. 1964. Gravimorphism in trees. *Ann. Bot.* 28:297–309.

SONDHEIMER, E. and D.H. GRIFFIN. 1960. Activation and inhibition of indoleacetic acid oxidase activity from peas. *Science* 131:672.

SOROKIN, H.P. and K.V. THIMANN. 1966. The histological basis for inhibition of axillary buds in *Pisum sativum* and the effects of auxins and kinetin on xylem developments. *Protoplasma* 59:326–350.

SRIVASTAVA, B.I.S. 1967. Cytokinins in plants. *Intern. Rev. Cytol.* 22:349–387.

SRIVASTAVA, B.I.S. 1968. Mechanism of action of kinetin in the retardation of senescence in excised leaves. p. 1479–1494. *In* P. Wightman and G. Setterfield (eds.) Proc. 6th Intern. Conf. on Plant Growth Substances, July 24–29, 1967, Carlton Univ., Ottawa. The Runge Press, Ottawa.

STAFFORD, H.A. 1974. The metabolism of aromatic compounds. *Annu. Rev. Plant Physiol.* 25:459–486.

STENLID, G. 1963. The effects of flavanoid compounds of oxidase phosphorylation and on the enzymatic destruction of indoleacetic acid. *Physiol. Plant.* 16:110–120.

STENLID, G. 1968. On the physiological effects of phloridzin, phloretin and some related substances upon higher plants. *Physiol. Plant.* 21:882–894.

STENLID, G. 1976. Effects of flavanoids on the polar transport of auxins. *Physiol. Plant.* 38:262–266.

STOLTZ, L.P. and C.P. HESS. 1966. The effect of girdling upon root initiation: auxin and rooting cofactors. *Proc. Amer. Soc. Hort. Sci.* 89:744–751.

SVENSSON, S. 1971. The effect of coumarin on root growth and root histology. *Physiol. Plant.* 24:446–470.

SVENSSON, S. 1972. A comparative study of the changes in root growth induced by coumarin, auxin, ethylene, kinetin and gibberellic acid. *Physiol. Plant.* 26:115–135.

SWAIN, T. 1953. The identification of coumarins and related compounds by filter-paper chromatography. *Biochem. J.* 53:200–208.

SWARBRICK, T. and K.C. NAIC. 1932. Factors governing fruitbud formation. IX. A study of the relation between leaf area and internode length in the shoots of Worchester Permain apple, as affected by six different vegetative rootstocks. *J. Pomol. Hort. Sci.* 10:42–63.

TAFURI, F., M. BUSINELLI, and L. SCARPONI. 1972. Effect of caffeic and *p*-coumaric acids on indole-3-acetic acid catabolism. *J. Sci. Food & Agr.* 23: 1417–1423.

THIMANN, K.V. 1977. Hormone action in the whole life of plants. Univ. of Massachusetts Press, Amherst.

THIMANN, K.V., M. TOMASZEWSKI, and W.L. PORTER. 1962. Growth-promoting activity of caffeic acid. *Nature* 193:1203.

TOMASZEWSKI, M. 1964. The mechanism of synergistic effects between auxin and some natural phenolic substances. p. 335–351. *In* J.P. Nitsch (ed.) Proc. 5th Intern. Conf. on Plant Growth Substances, July 15–20, 1963, Gifsur-Yvette, France.

TOMASZEWSKI, M. and K.V. THIMANN. 1966. Interactions of phenolic acids, metallic ions and chelating agents on auxin-induced growth. *Plant Physiol.* 41:1443–1454.

TORREY, J.G. 1976. Root hormones and plant growth. *Annu. Rev. Plant Physiol.* 27:435–459.

TOWERS, G.H.N. 1964. Metabolism of phenolics in higher plants and microorganisms. p. 249–294. *In* J.B. Harborne (ed.) Biochemistry of phenolic compounds. Academic Press, New York.

TROMP, J. and J.C. OVAA. 1979. Uptake and distribution of nitrogen in young apple trees after application of nitrate or ammonium with special reference to asparagine and arginine. *Physiol. Plant.* 45:23–28.

TROUGHTON, A. 1974. The growth and function of the root in relation to the shoot. p. 153–164. *In* J. Kolek (ed.) Structure and function of primary root tissues. Bratislava, Veda.

TUBBS, F.R. 1967. Tree size control through dwarfing rootstocks. *Proc. XVII Intern. Hort. Congr.* III:43–56.

TUBBS, F.R. 1973. Research fields in the interaction of rootstocks and scions in woody perennials. I and II. *Hort. Abstr.* 43:247–253 and *Hort. Abstr.* 43:325–335.

TUBBS, F.R. 1976. The largely additive relationships of the contributions by scions and by rootstock to the growth of deblossomed compound trees. *J. Hort. Sci.* 51:435–439.

TUBBS, F.R. 1977. The relative influences of fruit clones when present as rootstock or as scion. *J. Hort. Sci.* 52:37–48.

TUBBS, F.R. 1980. Growth relations of rootstock and scion in apples. *J. Hort. Sci.* 55:181–189.

TUKEY, H.B. and K.D. BRASE. 1933. Influence of the scion and of an intermediate stem-piece upon the character and development of roots of young apple trees. N.Y. Agr. Expt. Sta. Tech. Bul. 218.

TURVEY, P.M. and J.W. PATRICK. 1979. Kinetin-promoted transport of assimilates in stem of *Phaseolus vulgaris* L. Localized versus remote site(s) of action. *Planta* 147:151–155.

UHRSTROM, I. and S. SVENSSON. 1979. The effect of coumarin on Young's modulus in sunflower stems and maize roots and on water permeability in potato parenchyma. *Physiol. Plant.* 45:41–44.

VAADIA, Y. and C. ITAI. 1968. Interrelationships of growth with reference to the distribution of growth substances. p. 65–79. *In* W.J. Whittington (ed.) Root growth. Plenum Publishing Corp., New York.

VAN OVERBEEK, J., R. BLONDEAU, and V. HORNE. 1951. Trans-cinnamic acid as an anti-auxin. *Amer. J. Bot.* 38:589–595.

VAN STADEN, J. 1976. Occurrence of a cytokinin glucoside in the leaves and honeydew of *Salix babylonica*. *Physiol. Plant.* 36:225–228.

VAN STADEN, J. and A.R. SMITH. 1978. The synthesis of cytokinins in excised roots of maize and tomato under aseptic conditions. *Ann. Bot.* 43: 251–253.

VAN SUMERE, C.F., J. ALBRECHT, A. DEDONDER, H. DEPOOTER, and I. PE. 1975. Plant proteins and phenolics. p. 211–264. *In* J. B. Harborne and C.F. Van Sumere (eds.) The chemistry and biochemistry of plant proteins. Academic Press, New York.

VENDRIG, J.C. and K. BUFFEL. 1961. Growth-stimulating activity of trans-caffeic acid isolated from *Coleus rhenaltianus*. *Nature* 192:276.

VONK, C.R. 1978. Formation of cytokinin nucleotides in a detached inflorescence stalk and the occurrence of nucleotides in phloem exudate from attached yucca plants. *Physiol. Plant.* 44:161–166.

VONK, C.R. 1979. Origin of cytokinins transported in the phloem. *Physiol. Plant.* 46:235–240.

VYVYAN, M.C. 1955. Interrelation of scion and rootstalk in fruit trees. *Ann. Bot.* 19:401–423.

WAGNER, H. and G. MICHAEL. 1971. The influence of varied nitrogen supply on the production of cytokinins in sunflower roots. *Biochem. Physiol. Pflanz.* 162:147–158.

WALKER, R.L. 1975. The biology of plant phenolics. The Institute of Biology's Studies in Biology, No. 54. Edward Arnold, London.

WAREING, P.F. 1970. Growth and its coordination in trees. p. 1–21. *In* L.C. Luckwill and C. Cutting (eds.) Physiology of tree crops. Academic Press, New York.

WAREING, P.F. 1977. Growth substances and integration in the whole plant. p. 337–366. *In* D.H. Jennings (ed.) Integration of activity in the higher plant. Cambridge University Press, New York.

WEISS, C. and Y. VAADIA. 1965. Kinetin-like activity in root apices of sunflower plants. *Life Sci.* 4:1323–1326.

WENT, F.W. 1938. Specific factors other than auxin affecting growth and root formation. *Plant Physiol.* 13:55–80.

WENT, F.W. 1943. Effect of the root system on tomato stem growth. *Plant Physiol.* 18:51–65.

WENT, F.W. and D.M. BONNER. 1942. Growth factors controlling tomato stem growth in darkness. *Arch. Biochem. Biophys.* 1:439–452.

WESTCOTT, R.J. and G.G. HENSHAW. 1976. Phenolic synthesis and phenylalanine ammonia-lyase activity in suspension cultures of *Acer pseudoplatanus. Planta* 131:67–73.

WICKSON, M. and K.V. THIMANN. 1958. The antagonism of auxin and kinetin in apical dominance. *Physiol. Plant.* 11:62–74.

WILKINS, M.B. 1969. The physiology of plant growth and development. McGraw-Hill Book Co., New York.

WILLIAMS, A.H. 1953. The application of chromatographic methods in the study of the biochemistry and nutrition of plants. *Science and Fruit.* p. 205–212. (B.B. Beakbane, 1956. Possible mechanism of rootstock effect. *Ann. Appl. Biol.* 44:517–521).

WILLIAMS, A.H. 1960. The distribution of phenolic compounds in apple and pear trees. p. 3–7. *In* J.B. Pridham (ed.) Phenolics in plants in health and disease. Pergamon Press, Oxford.

WILLIAMS, A.H. 1966. Dihydrochalcones. p. 297–307. *In* T. Swain (ed.) Comparative phytochemistry. Academic Press, New York.

WILSON, B.F. 1975. Distribution of secondary thickening in tree root systems. p. 197–220. *In* J.D. Torrey and D.T. Clarkson (eds.) The development and function of roots. Academic Press, New York.

WOOLLEY, D.J. and P.F. WAREING. 1972a. The interaction between growth promoters in apical dominance. I. Hormonal interaction, movement and metabolism of a cytokinin in rootless cuttings. *New Phytol.* 71:781–793.

WOOLLEY, D.J. and P.F. WAREING. 1972b. The interaction between growth promoters in apical dominance. II. Environmental effects on endogenous cytokinin and gibberellin levels in *Solanum andigena*. *New Phytol.* 71:1015–1025.

YADAVA, U.L. and D.F. DAYTON. 1972. The relation of endogenous abscisic acid to the dwarfing capability of East Malling apple rootstocks. *J. Amer. Soc. Hort. Sci.* 97:701–705.

YADAVA, U.L. and S.L. DOUD. 1978. Effect of rootstock on the bark thickness of peach scions. *HortScience* 13:538–539.

YAMANE, H., N. TAKAHASHI, K. TAKENO, and M. FURUYA. 1979. Identification of gibberellin A_9 methyl ester as a natural substance regulating formation of reproductive organs in *Lygodium japonicum*. *Planta* 147:251–256.

YOSHIDA, R. and T. ORITANI. 1974. Studies on nitrogen metabolism in crop plants. XIII. Effects of nitrogen top-dressing on cytokinin content in the root exudate of rice plant. *Proc. Crop. Sci. Soc. (Japan)* 43:47–51.

YU, K.S. and R.F. CARLSON. 1975a. Paper chromatographic determination of phenolic compounds occurring in the leaf, bark and root of *Prunus avium* and *P. mahaleb*. *J. Amer. Soc. Hort. Sci.* 100:536–541.

YU, K.S. and R.F. CARLSON. 1975b. Gas liquid chromatography determinations of phenolic acids and coumarins in mazzard and mahaleb cherry seedlings. *HortScience* 10:401–403.

ZEEVAART, J.A.D. 1974. Levels of abscisic acid and xanthoxin in spinach under different environmental conditions. *Plant Physiol.* 53:644–648.

ZENK, M.H. and G. MULLER. 1963. *In vivo* destruction of exogenously applied indolyl-3-acetic acid as influenced by naturally occurring phenolic acids. *Nature* 200:761–763.

ZIEGLER, H. 1975. Nature of transported substances. p. 59–100. *In* M.H. Zimmerman and J.A. Melburn (eds.) Transport in plants. I. Phloem transport. Encyclopedia of plant physiology. New series Vol. I. Springer-Verlag, New York.

8

Morphology and Reproduction of Pistachio

Julian C. Crane and Ben T. Iwakiri
Department of Pomology, University of California, Davis, California 95616

I. INTRODUCTION

Within the past decade California has added the pistachio to the long list of horticultural crops that it produces commercially. From slightly more than 100 ha planted to pistachios in 1968, rapid expansion took place during the 1970s to about 12,000 ha, with over 95% being planted in the San Joaquin Valley. The first commercial crop of any size—about 2 million kilograms—was harvested in 1977, while in 1979 over 7.7 million kilograms were produced. By the mid 1980s, production is expected to surpass the annual U.S. consumption of about 11½ million kilograms. Then there will be no need for importing the product from Iran and Turkey, the two largest exporting countries.

A review of the worldwide literature in 1969, when rapid expansion of pistachio acreage began in California, revealed very little scientific information regarding the tree or its fruit. Only 4 articles dealing primarily with general cultural requirements and processing of the nuts had been published in the North American literature during the previous 15 years. This review will not encompass the commercial and technical aspects of pistachio production and processing, but will focus primarily on several characteristic morphological and physiological phenomena of the tree and its fruit that set it apart from other deciduous fruit and nut trees, the problems brought about by these differences, and the research conducted to alleviate them. The reader is referred to Whitehouse (1957), Joley (1979), and Woodroof (1979) for general information dealing with environmental and cultural requirements of pistachio.

II. HISTORY

Pistachio, *Pistacia vera* L., probably originated in central Asia (Zohary 1952; Whitehouse 1957) where large stands of wild trees are found in areas known today as Iran, Turkestan, and Afghanistan. There is little doubt, according to Whitehouse (1957), that the first commercial plantings in Iran, Turkey, Syria, and other countries adjacent to the wild pistachio stands were started with seedlings grown from the best of the wild nuts. The tree was introduced into Mediterranean Europe at approximately the beginning of the Christian era.

Pistachio nuts first became available on U.S. markets at about 1880. Various cultivars from the Mediterranean area were introduced into California by the U.S. Department of Agriculture beginning about 1904, and evaluated at the Plant Introduction Station, Chico, California, as a potential new crop. Relatively little interest was generated in producing the nut commercially because each of the cultivars had one or more faults that precluded its production under existing conditions. Also, mechanical harvesting and hulling techniques and equipment were not to be developed until years later.

As a result of a program initiated in 1929 by the U.S. Department of Agriculture to develop better cultivars, a seedling produced from seeds imported from Iran was selected by W.E. Whitehouse and Lloyd E. Joley and in 1952 was named 'Kerman', the center of pistachio production in Iran. 'Kerman' was introduced for trial in 1957, and today it is the only pistillate cultivar produced in California. 'Kerman' is a relatively upright-growing tree that may reach a height and width of 5 to 7 m. It produces relatively high yields of exceptionally large nuts having excellent kernel quality. 'Peters' was selected and named in about 1930 by A.B. Peters in Fresno, California and is the main pollenizer (staminate) cultivar. One tree is generally planted for every eight 'Kerman' trees. Both 'Peters' and 'Kerman' are grown on *P. atlantica* Desf. or *P. terebinthus* L. rootstocks because they have greater resistance to nematodes and other soil organisms than *P. vera* (Joley and Whitehouse 1953). *Pistacia integ-*

errima Stewart has been found recently to be resistant to verticillium wilt (*Verticillium albo-atrum* Reinke & Berth), and it is being used as a rootstock in new plantings and as replants for trees killed by the fungus.

III. BOTANY

Pistacia is a genus of the Anacardiaceae which comprises some widely known trees and shrubs such as cashew, mango, poison ivy, poison oak, and sumac. The latest monographic study of the genus was made by Zohary (1952) who recognized eleven species based primarily upon leaf characteristics. He pointed out that floral characters among the species are so simplified and uniform as to be of little value for diagnostic purposes. On the other hand, hybridization, which occurs readily among species (Grundwag 1975; Peebles and Hope 1937; Whitehouse and Stone 1941; Crane and Iwakiri 1980), may render leaf characters unreliable. This led to an attempt by Grundwag and Werker (1976) to characterize *Pistacia* species according to their wood anatomy. Zohary (1952) reported diploid chromosome numbers of 24 for *P. lentiscus* L., 28 for *P. atlantica*, and 30 for *P. vera*. Jones (1950), however, reported *P. vera* to have 32. Many of the species, when used as rootstocks for *P. vera*, have proven to be compatible.

The pistachio is deciduous and is characterized by imparipinnate leaves, most often two-paired (Zohary 1952). Simple leaves, however, occur during the juvenile stage of growth and also on mature growth made subsequent to insufficient chilling to completely break the rest period (Crane and Takeda 1979). Shoot extension begins the last of March and terminates the last of April to the middle of May. A leaf at each node subtends a single axillary bud (Fig. 8.1). Most of these differentiate into inflorescence buds during April and grow to their ultimate size for the season by late June (Takeda *et al.* 1979). Generally, one or two axillary buds located distally on the new growth are vegetative. They are considerably smaller than inflorescence buds, and may give rise to lateral branches the following year or they may remain dormant. Inflorescence buds begin expansion the last of the following March and anthesis occurs generally during the first half of April. Thus, pistachio bears its fruits laterally on wood produced the previous season (Fig. 8.1) (Crane and Nelson 1971).

Pistachio is dioecious and both staminate and pistillate inflorescences are panicles that may consist of one to several hundred individual flowers (Fig. 8.2). Both types of flowers are apetalous and wind is the pollinating agent. The fruit produced is a semi-dry drupe. Fruit maturity is manifested by a change in the epicarp (skin) from translucent to opaque, and a softening and loosening of the epicarp and mesocarp (hull) from the endocarp (shell) which encloses the embryo (kernel).

FIG. 8.1. RELATIONSHIP BETWEEN NUT PRODUCTION AND ABSCISSION OF INFLORESCENCE BUDS

The nut-bearing branch (left) abscised most of its inflorescence buds on the new growth, whereas the nonbearing branch (right) retained all of its inflorescence buds.

IV. BUD DEVELOPMENT AND PRUNING

All axillary buds on current growth of the pistachio are vegetative the first 4 or 5 years. The response to pruning for establishing properly spaced scaffold branches is like that of other deciduous fruit and nut trees, i.e., removal of terminal portions of branches stimulates lateral buds to grow. When the tree begins to form flower instead of vegetative buds in the leaf axils the fourth or fifth year, each year thereafter for several years more and more flower buds and fewer and fewer vegetative

FIG. 8.2. INFLORESCENCES OF TWO PISTACHIO CULTIVARS

Left, male inflorescences of 'Peters'; right, female inflorescences of 'Kerman'. Photographed on April 15 a few days past full bloom and before shoot growth and leaf expansion were complete.

buds are produced. Eventually, all axillary buds on some shoots are flower buds, while on others all are flower buds except one or two near the distal ends of the shoots. Many of the latter subsequently fail to grow because of strong apical dominance exerted by the terminal vegetative bud. It is not uncommon for a branch to continue growth in length for 8 to 10 years with no lateral branching (Fig. 8.3). Fruiting occurs progressively further and further from the center of the tree. Eventually, the branches are bent out of position from the weight of foliage and nuts and are then subject to sunburn. They also crowd and bring about shading of branches below them.

Experimental pruning has been carried out at the University of California to prevent the situation described above, as well as to keep the trees in bounds and to promote formation of renewal fruiting wood on scaffold branches. Our studies have shown that the pistachio does not respond to conventional pruning procedures. Two factors are responsible. Unlike other deciduous fruit and nut trees which generally produce at least one vegetative bud per node, a bearing pistachio tree characteristically produces relatively few lateral vegetative buds; practically all are flower buds. Consequently, when a branch is headed-back in pruning to force lateral growth, there are no lateral vegetative buds present and the portion remaining dies back to a lateral branch (Fig. 8.3). The other factor responsible for unconventional behavior of pistachio is apical dominance. Reducing the number of growing points 50% by the thinning-out type of pruning cut results in practically no new shoots arising laterally from the remaining branches. Removal of all terminal buds by the heading-back type of cut and consequent elimination of apical dominance is necessary to stimulate renewal growth from the scaffold branches (Fig. 8.3). The extent to which heading-back is done depends upon the tree size reduction desired. Pruning of this type should be done in winter prior to the "off" year of production in the alternate bearing cycle to minimize loss in yield, as will be discussed later.

V. INFLORESCENCE BUD DIFFERENTIATION AND DEVELOPMENT

Axillary bud primordia are found within the overwintering terminal buds on both staminate and pistillate trees. As these buds elongate during the last of March, the bud primordia develop rapidly in the axils

FIG. 8.3. PRUNED (TOP) AND UNPRUNED (BOTTOM) 10-YEAR-OLD 'KERMAN' PISTACHIO BRANCHES

Four years' growth was removed from the top branch by the heading-back pruning cut that destroyed apical dominance and forced lateral branching. Note that the pruned branch died back to the point where two vegetative buds were opposite each other.

of the young leaves on the new shoots (Jones 1950; Takeda *et al.* 1979). Each axillary meristem elongates and forms the inflorescence rachis with its primary lateral branches. Formation of secondary and tertiary lateral branches (each subtended by a bract) follows. Floral primordia (staminate or pistillate) subsequently develop both laterally and terminally on the branches. The first sepals may be seen the latter part of May on some branchlets. The branched inflorescence structure which eventually supports numerous flowers is fully differentiated in June. Pistillate and staminate inflorescence buds then undergo no further development until October and the following March, respectively. After a 3-month quiescent period, pistillate flower growth resumes in October in the form of pistil initiation, but does not go beyond that stage during the winter months. Near the end of March each primordium develops rapidly into a three-carpellate pistil, one large functional and two small sterile carpels. A 'Kerman' inflorescence generally contains 100 to 150 individual flowers. After a 9-month quiescent period, the stamen primordia (5 per flower), initiated the previous May, resume rapid growth, differentiation, and maturation the last of March. Time of bloom among staminate and pistillate cultivars may differ as much as 2 to 3 weeks. Staminate and pistillate cultivars having similar bloom periods must be provided to ensure adequate pollination of the latter. Before artificial pollination may become practical, conditions for satisfactory pollen storage must be determined (Crane *et al.* 1974).

VI. FRUIT GROWTH AND DEVELOPMENT

Certain tissues or organs of the pistachio fruit develop in time and space like those of drupes in general, while others develop quite differently (Crane *et al.* 1971; Crane and Al-Shalan 1974). Growth in diameter of the pericarp of fleshy drupaceous fruits such as apricot (Lilleland 1930), cherry (Lilleland and Newsome 1934), olive (Hartmann 1949), peach (Connors 1920), and raspberry (Hill 1958) occurs in three distinct periods—two cycles of rapid growth separated by one of slow growth (Fig. 8.4). Growth of the almond (Brooks 1939), a dry drupe, occurs in two periods only, i.e., a rapid growth phase followed by one of practically no growth (Fig. 8.4). Thus, almond has the equivalent of Periods I and II of fleshy drupes, but lacks a final phase of rapid growth (Period III). Pistachio, a semi-dry drupe, exhibits growth similar to that of almond, but has a second cycle of relatively accelerated growth (Period III) which adds little to the final fruit diameter (Fig. 8.4).

Period I in fleshy drupaceous fruits, as well as the almond, encompasses complete and simultaneous growth of the endocarp and the nucellus and

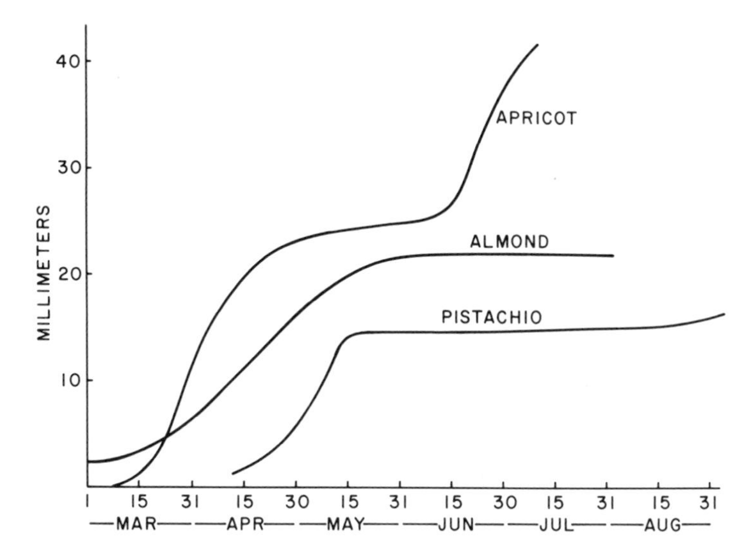

FIG. 8.4. CURVES OF GROWTH IN DIAMETER OF FRUITS OF 'NE PLUS' ALMOND, 'ROYAL' APRICOT, AND 'KERMAN' PISTACHIO

integuments of the seed. While ultimate size of the pistachio endocarp also is attained by the end of Period I, the nucellus and integuments do not undergo extensive growth until more than a month of Period II has elapsed (Fig. 8.5). Development of the embryo in fleshy drupes and in almond is delayed, and is not macroscopically evident until the end of Period I. Embryonic development in the pistachio is delayed even longer, not being macroscopically visible until Period II has been underway for about 6 weeks. Once the embryo becomes evident, it grows simultaneously with the integuments to ultimate size during a 6-week period about midway in Period II of pericarp growth. Decline in total sugars (sucrose predominantly) in the kernels from a maximum of about 35% (dry weight basis) in early July is accompanied by a continual increase in crude fat to a maximum of about 40% (dry weight basis) at fruit maturity (Crane 1978). Lignification of the endocarp begins at the initiation of growth Period II and continues for 4 to 6 weeks. Endocarp dehiscence is first noted along the ventral suture the last week in July, about the time ultimate kernel size is attained, and progresses along both sutures until physiological maturity, about the middle of September. Physiological maturity is signaled by easy separation of the hull from the shell (Crane 1978), the equivalent of flesh separation from the pit in a freestone peach.

A. Xenia and Metaxenia

Pistachio is one of few plants in which xenia and metaxenia have been observed (Peebles and Hope 1937; Whitehouse *et al.* 1964; Pontikis 1977; Crane and Iwakiri 1980). Effects such as delayed or hastened fruit maturity, increased fruit length, increased shell dehiscence, and increases and decreases in size and in weight of kernel have been reported, depending upon the pollen applied. The data indicate that the use of pollen from a source other than *P. vera* is generally undesirable.

B. Endocarp Dehiscence

Of the eleven species of the genus *Pistacia*, dehiscence of the endocarp occurs only in fruits of *P. vera* (Fig. 8.6). The bulk of the pistachio crop is marketed in-shell, and endocarp dehiscence (shell splitting) is a desirable

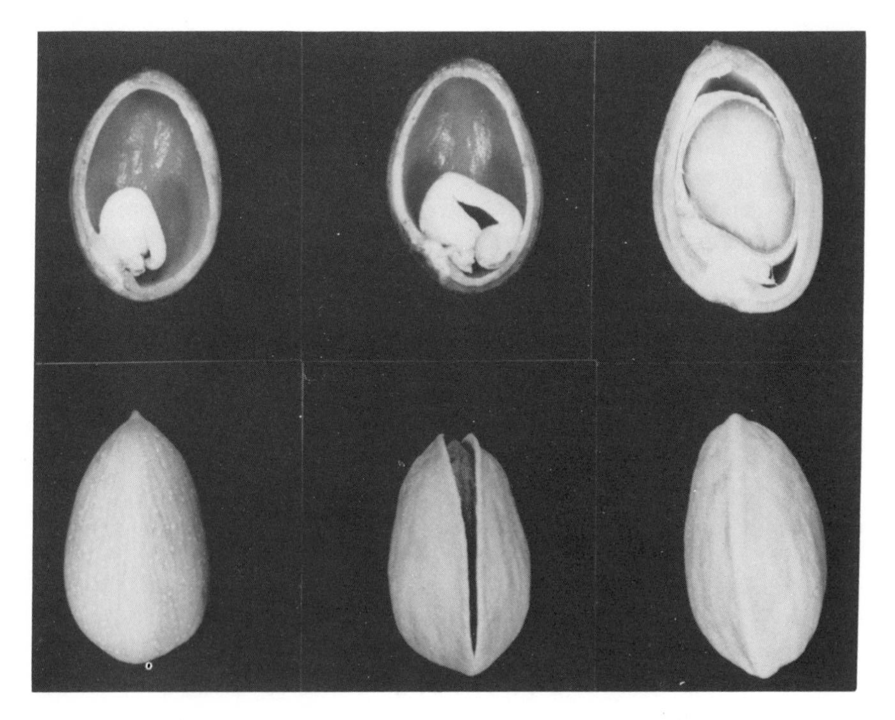

FIG. 8.5. GROWTH AND DEVELOPMENT OF 'KERMAN' PISTACHIO NUT

TOP, left, pericarp attains about ultimate size by mid-May and encloses a curved funiculus supporting a single ovule at its apex; center, elongated funiculus supporting an expanding seed (June 26); right, seed that has almost filled the locule (July 22). BOTTOM, left, mature fruit before removal of the hull; center, after hull removal and drying, showing dehisced shell enclosing seed; right, undehisced shell which may or may not contain a seed.

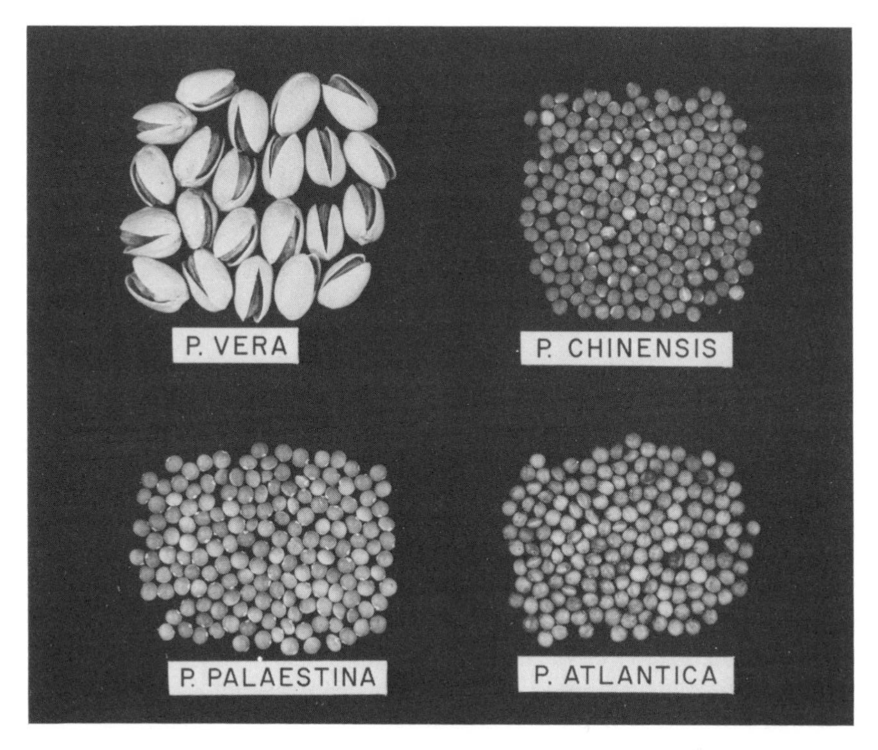

FIG. 8.6. THE FRUITS OF *P. VERA* ARE SEVERAL TIMES LARGER THAN THOSE OF OTHER *PISTACIA* SPECIES AND ARE THE ONLY ONES IN WHICH THE ENDOCARP DEHISCES

trait as it facilitates shell removal by the consumer. The shells of unsplit nuts must be removed in cracking machines prior to marketing. The extent to which shell splitting occurs varies from tree to tree and from year to year. Based on yields of 22 'Kerman' trees over a 7-year period, the greatest variation among trees in a given year was a low of 44 to a high of 80% of the total nuts having split shells, the remainder being unsplit and with or without kernels (blanks). The data in Table 8.1 show that the percentage of total nuts having split shells may have a yearly variation from 55.1 (1978) to 71.3% (1976). This variability results from variation in both percentage of blank nuts (in which shell splitting never occurs) produced from year to year and percentage of nuts with kernels in which shell splitting does not take place (Table 8.1). The exceptionally high percentage of unsplit nuts in 1978 was undoubtedly a consequence of the unusually mild winter of 1977–1978 which brought about various abnormal leaf and flower symptoms due to lack of chilling (Crane and Takeda 1979).

TABLE 8.1. YEARLY VARIATION IN 'KERMAN' PISTACHIO NUT CHARACTERS

Data are averages based on at least 2 100-nut samples from each of 22 trees.

	Type of Nut				
		Filled			
Year	Blank (%)	Unsplit (%)	Split (%)	Weight/Filled Nut (g)	Yield/Tree (kg)
1973	24.9	7.2	67.9	1.46	17.7
1974	27.3	5.1	67.6	1.48	15.4
1975	38.2	3.0	58.8	1.38	7.4
1976	19.7	9.0	71.3	1.29	28.1
1977	14.2	16.5	69.3	1.44	12.7
1978	24.7	20.2	55.1	1.41	20.9
1979	16.4	15.3	68.3	1.37	30.7

Nevo *et al.* (1974), in a detailed anatomical and morphological study of the developing pistachio fruit, observed an abscission zone extending along both the ventral and dorsal sutures of the endocarp. They found no anatomical difference in this zone between dehiscent and indehiscent fruits and concluded ". . . that it is the size of the seed which determines dehiscence." In some fruits, they stated, seed volume exceeds that of the endocarp, and dehiscence occurs. In others, a greater shrinkage of the endocarp than that of the seed brings about dehiscence upon drying.

Our observations and data indicate that pistachio endocarp dehiscence is not a physical reaction but is a biochemical phenomenon associated with seed growth and development. The endocarp in fruits in which seeds fail to develop does not dehisce nor does separation of the hull from the shell occur. Thus, both processes are associated with seed development, but it would seem unlikely that one is a physical phenomenon while the other is biochemical. Studies of growth and development of the fruit (Crane *et al.* 1971; Crane and Al-Shalan 1974) have shown that the seed reaches its maximum size at the end of July. Less than 30% of the fruits were found to have a dehisced endocarp at that time. Subsequently, however, a progressive increase to 85% occurred in endocarp dehiscence on Sept. 12, 1977 when the fruits were physiologically mature (Crane 1978). Thus, endocarp dehiscence took place primarily after the seeds had reached maximum size. If dehiscence were the result of seed volume exceeding that of endocarp, as suggested by Nevo *et al.* (1974), then the epicarp and mesocarp would be expected to rupture also. This, however, generally does not occur.

C. Blank Production

Production of seedless fruits is a characteristic common to several, if not all, *Pistacia* species (Crane 1973; Grundwag 1975). Crane (1973), for example, reported variation in blank production of 23 to 100% and 1 to

72% in seedling populations of *P. atlantica* and *P. chinensis*, respectively. Blank production in *P. vera* is a serious problem wherever pistachios are grown. The extent to which it occurs varies from cultivar to cultivar (Crane 1973), from year to year (Table 8.1), and from rootstock to rootstock (Crane 1975). 'Kerman' produces an excessive amount of blanks (Table 8.1) when compared with other cultivars.

Production of blanks in 'Kerman' has been shown to result primarily from seed abortion (Grundwag and Fahn 1969; Bradley and Crane 1975), but also from vegetative parthenocarpy (Crane 1973). Aberrations in seed also occur at various stages and result in subnormal size of some seeds at maturity. Apparently, because of the parthenocarpic tendency, fruits with aborted seeds do not abscise but grow to practically normal size (Crane *et al.* 1971). This is unusual as growth and development in most fruits, particularly single-seeded fruits, is dependent upon the presence of seeds (Crane 1964). Growth of blank fruits is identical to that of seeded fruits during Periods I and II, but there is no Period III.

The degree of blank production is not associated with yield of the individual tree (Table 8.1) nor to its position with respect to the prevailing wind and proximity to staminate trees (Crane 1973). It is apparently governed somewhat by the seedling rootstock, for some trees consistently produce high percentages of blanks and other trees consistently produce low percentages. Whether the effect of rootstock is mediated through a nutritional or a hormonal mechanism remains an unanswered question.

VII. ALTERNATE BEARING

Production of a heavy crop one year followed by little or no crop the next is characteristic of such fruits as apple, mango, orange, pear, pecan, and prune. The specific cause of alternate bearing has not been elucidated, but it has been demonstrated that the presence of a heavy crop of fruit limits the formation of flower buds for the year following. It is not known if this limitation is the result of a critical depletion of assimilates by the developing fruit or the action of an inhibitor originating in the fruit. The pistachio is also an alternate bearing species, but the mechanism involved is unusual. It produces abundant inflorescence buds every year, but they abscise in such numbers during the summer of a heavy crop that few remain to produce a light crop the next year (Fig. 8.1). Thus, alternate bearing in the pistachio is the result of abscission of premature inflorescence buds during a heavy crop year rather than lack of bud formation (Crane and Nelson 1971). The period of maximum bud abscission coincides with that of rapid seed growth and development during July and August. This, as well as adjustments in crop load and/or in leaf area, indicates that assimilate depletion is responsible for bud abscission (Crane and Nelson 1972; Crane *et al.* 1973). Carbohydrate and nitrogen levels in bark and wood of nut-bearing branches, however, were found to be similar to those in identical tissues of nonbearing branches during the period of bud abscission (Porlingis 1974; Crane *et al.* 1976; Crane and Al-Shalan 1977).

Since the data indicated that bud abscission might not be the result of carbohydrate depletion, it was speculated that the phenomenon might be hormonally controlled (Crane *et al.* 1973, 1976). An abscission-inhibiting hormone produced in the leaves may be directed away from the inflorescence buds to developing nuts with large demand for carbohydrates. Nut removal would change the source-sink translocation relationship and make the proposed inhibitor immediately available to the buds. An abscission-promoting compound originating in the developing nuts also might be a control mechanism of bud abscission. The results of a recent study of abscisic acid levels in developing seeds and inflorescence buds, however, indicated that this potential abscission-promoting compound is not involved (Takeda and Crane 1980).

In spite of similarity in carbohydrate levels in bearing and nonbearing branches during the main period of inflorescence bud drop, the presence of a nut crop depresses inflorescence bud growth on current wood during the summer (Takeda *et al.* 1979), starch storage in branches during fall and winter (Crane and Al-Shalan 1977), and vegetative growth in length the following spring (Crane and Nelson 1972). Thus, production of a crop of nuts brings about direct and long lasting effects in the pistachio tree. A study of the effect of developing nuts on translocation and distribution of photosynthates from leaves revealed that most of the ^{14}C-photosynthate transported from leaves accumulated in developing nuts. Inflorescence buds competed poorly against the developing nuts for photosynthate as those on bearing branches had about half as much as those on nonbearing branches (Takeda *et al.* 1980). Thus, carbohydrate deficiency in the buds themselves may be responsible for the bud drop phenomenon and subsequent alternate bearing.

VIII. CHILLING REQUIREMENT

The pistachio, like most deciduous fruit and nut trees, has a chilling requirement that must be satisfied during the winter months in order for the buds to open and grow normally. The requirement in terms of number of hours at or below 7°C (45°F) has not been precisely determined, but it appears to be about 1000 hours for 'Kerman'. Cultivars that originated in the eastern Mediterranean area seem to require less.

The mild winter of 1977–1978 in California, during which there were only 670 hours of temperatures below 7°C at Davis, brought about unusual responses of 'Kerman' pistachio (Crane and Takeda 1979). Delayed and irregular blooming and foliation—symptoms common to other deciduous fruit and nut trees—were manifest, but the most conspicuous and prevalent abnormality was change in leaf form. Leaves predominantly had three leaflets rather than the normal five, and simple leaves were not uncommon (Fig. 8.7). Temperature has been shown to have marked effects on leaf form in various herbaceous perennial plants (Bostrack and Millington 1962; Pound 1949), but the leaf changes in pistachio in this respect following the mild winter of 1977–1978 were the first to be described for a deciduous tree.

Another notable response of pistachio to insufficient chilling was a drastic modification of its bearing habit. Not only did fruiting occur normally on 1-year-old wood, but it also took place abnormally both laterally and terminally on current wood (Fig. 8.7). Thus, crops that normally would require 2 years to be produced were telescoped into 1 year. Procopiou (1973) reported that application of a mineral oil and dinitro-ortho-cresol spray was effective in alleviating delayed blooming due to lack of chilling.

IX. CONCLUSIONS

The research conducted on the pistachio to date has provided little more than an indication of some of the diverse problems that are associated with the culture of this unique tree. There are other problems that have received little or no attention. For example, critical comparison of the various rootstock species, as well as their hybrids, has not been made with regard to relative compatibility with *P. vera*, disease resistance, blank production, endocarp dehiscence, alternate bearing, etc. Similarly, very limited study has been made with regard to moisture (Spiegel-Roy *et al.* 1977) and nutritional (Parsa and Karimian 1975; Uriu and Crane 1977) requirements and their modification according to rootstock. Once the ideal rootstock is decided upon for a particular situation, it must be propagated vegetatively. Propagation by stem cuttings has proven to be unsuccessful. Experimentation hopefully leading to propagation by meristem or tissue culture has just been initiated.

Pollination is another area that urgently needs research. The most effective ratio of staminate to pistillate trees and their relative placement one to another have never been determined experimentally. Windless and rainy weather is not conducive to pollen dispersal and pollination and may lead to crop failure. Temperature and humidity conditions under which pollen should be stored to maintain viability must be

FIG. 8.7. BRANCH OF 'KERMAN' PISTACHIO WITH PANICLES OF YOUNG NUTS BORNE LATERALLY ON 1-YEAR-OLD WOOD (NORMAL) AND TERMINALLY ON CURRENT WOOD (ABNORMAL) AS A RESULT OF INSUFFICIENT WINTER CHILLING. NOTE ALSO THE SIMPLE LEAVES SUBTENDING SOLITARY NUTLETS (ABNORMAL)

determined before supplemental or complete artificial pollination may be accomplished effectively. Information on recognizing the period of receptivity of the female flowers and the time required for pollen tube growth and fertilization is also needed. Pollen carriers or dilutants, as well as methods of application, must be tested.

X. LITERATURE CITED

BOSTRAK, J.M. and W.F. MILLINGTON. 1962. On the determination of leaf form in an aquatic heterophyllous species of *Ranunculus*. *Bul. Torrey Bot. Club* 89:1–20.

BRADLEY, M.V. and J.C. CRANE. 1975. Abnormalities in seed development in *Pistacia vera* L. *J. Amer. Soc. Hort. Sci.* 100:461–464.

BROOKS, R.M. 1939. A growth study of the almond fruit. *Proc. Amer. Soc. Hort. Sci.* 37:193–197.

CONNORS, C.H. 1920. Growth of fruits of peach. N. J. Agr. Expt. Sta. Annu. Rpt. 40:82–88. New Brunswick, N.J.

CRANE, J.C. 1964. Growth substances in fruit setting and development. *Annu. Rev. Plant Physiol.* 15:303–326.

CRANE, J.C. 1973. Parthenocarpy—a factor contributing to the production of blank pistachios. *HortScience* 8:388–390.

CRANE, J.C. 1975. The role of seed abortion and parthenocarpy in the production of blank pistachio nuts as affected by rootstock. *J. Amer. Soc. Hort. Sci.* 100:267–270.

CRANE, J.C. 1978. Quality of pistachio nuts as affected by time of harvest. *J. Amer. Soc. Hort. Sci.* 103:332–333.

CRANE, J.C. and I. AL-SHALAN. 1974. Physical and chemical changes associated with growth of the pistachio nut. *J. Amer. Soc. Hort. Sci.* 99:87–89.

CRANE, J.C. and I. AL-SHALAN. 1977. Carbohydrate and nitrogen levels in pistachio branches as related to shoot extension and yield. *J. Amer. Soc. Hort. Sci.* 102:396–399.

CRANE, J.C., I. AL-SHALAN, and R.M. CARLSON. 1973. Abscission of pistachio inflorescence buds as affected by leaf area and number of nuts. *J. Amer. Soc. Hort. Sci.* 98:591–592.

CRANE, J.C., M.V. BRADLEY, and M.M. NELSON. 1971. Growth of seeded and seedless pistachio nuts. *J. Amer. Soc. Hort. Sci.* 96:78–80.

CRANE, J.C., P.B. CATLIN, and I. AL-SHALAN. 1976. Carbohydrate levels in the pistachio as related to alternate bearing. *J. Amer. Soc. Hort. Sci.* 101: 371–374.

CRANE, J.C., H.I. FORDE, and C. DANIEL. 1974. Pollen longevity in *Pistacia*. *Calif. Agr.* 28(11):8–9.

CRANE, J.C. and B.T. IWAKIRI. 1980. Xenia and metaxenia in pistachio. *HortScience* 15(2):184–185.

CRANE, J.C. and M.M. NELSON. 1971. The unusual mechanism of alternate bearing in the pistachio. *HortScience* 6(5):489–490.

CRANE, J.C. and M.M. NELSON. 1972. Effects of crop load, girdling, and auxin application on alternate bearing of the pistachio. *J. Amer. Soc. Hort. Sci.* 97:337–339.

CRANE, J.C. and F. TAKEDA. 1979. The unique response of the pistachio tree to inadequate winter chilling. *HortScience* 14(2):135–137.

GRUNDWAG, M. 1975. Seed set in some *Pistacia* L. (Anacardiaceae) species after inter- and intraspecific pollination. *Israel J. Bot.* 24:205–211.

GRUNDWAG, M. and A. FAHN. 1969. The relation of embryology to the low seed set in *Pistacia vera* (Anacardiaceae). *Phytomorphology* 19:225–235.

GRUNDWAG, M. and E. WERKER. 1976. Comparative wood anatomy as an aid to identification of *Pistacia* L. species. *Israel J. Bot.* 25:152–167.

HARTMANN, H.T. 1949. Growth of the olive fruit. *Proc. Amer. Soc. Hort. Sci.* 54:86–94.

HILL, R.G., JR. 1958. Fruit development of the raspberry and its relation to nitrogen treatment. *Ohio Agr. Expt. Sta. Res. Bul.* 803.

JOLEY, L.E. 1979. Pistachios. p. 163–174. *In* R.A. Jaynes (ed.) Nut tree culture in North America. The Northern Nut Growers Association, Hamden, Ct.

JOLEY, L.E. and W.E. WHITEHOUSE. 1953. Root knot nematode susceptibility—a factor in the selection of pistachio nut rootstocks. *Proc. Amer. Soc. Hort. Sci.* 61:99–102.

JONES, L.E. 1950. Fruit bud development and gametogenesis in *Pistacia vera* L. Ph.D. Thesis, Oregon State College, Corvallis.

LILLELAND, O. 1930. Growth study of the apricot fruit. *Proc. Amer. Soc. Hort. Sci.* 27:237–245.

LILLELAND, O. and L. NEWSOME. 1934. A growth study of the cherry fruit. *Proc. Amer. Soc. Hort. Sci.* 32:291–294.

NEVO, A., E. WERKER, and R. BEN-SASSON. 1974. The problem of indehiscence of pistachio (*Pistacia vera* L.) fruit. *Israel J. Bot.* 23:1–13.

PARSA, A.A. and N. KARIMIAN. 1975. Effect of sodium chloride on seedling growth of two major varieties of Iranian pistachio (*Pistacia vera* L.). *J. Hort. Sci.* 50:41–46.

PEEBLES, R.H. and C. HOPE. 1937. The influence of different pollens on the development of the pistache nut. *Proc. Amer. Soc. Hort. Sci.* 34:29–32.

PONTIKIS, K.A. 1977. Contribution to studies on the effect of pollen of different species and cultivars of the genus *Pistacia* on nut development and quality. Anōtatē Geōponikē Sholē Athhēnōn, 1975. (*Hort. Abstr.* 47: 10270.)

PORLINGIS, I.C. 1974. Flower bud abscission in pistachio (*Pistacia vera* L.) as related to fruit development and other factors. *J. Amer. Soc. Hort. Sci.* 99: 121–125.

POUND, G.S. 1949. Effect of air temperature on virus concentration and leaf morphology of mosaic-infested horse radish. *J. Agr. Res.* 78:161–170.

PROCOPIOU, J. 1973. The induction of earlier blooming in female pistachio trees by mineral oil-DNOC winter sprays. *J. Hort. Sci.* 48:393–395.

SPIEGEL-ROY, P., D. MAZIGH, and M. EVENARI. 1977. Response of pistachio to low soil moisture conditions. *J. Amer. Soc. Hort. Sci.* 102:470–473.

TAKEDA, F. and J.C. CRANE. 1980. Abscisic acid in pistachio as related to inflorescence bud abscission. *J. Amer. Soc. Hort. Sci.* 105:573–576.

TAKEDA, F., J.C. CRANE, and J. LIN. 1979. Pistillate flower bud development in pistachio. *J. Amer. Soc. Hort. Sci.* 104:229–232.

TAKEDA, F., K. RYUGO, and J.C. CRANE. 1980. Translocation and distribution of ^{14}C-photosynthates in bearing and nonbearing pistachio branches. *J. Amer. Soc. Hort. Sci.* 105:642–644.

URIU, K. and J.C. CRANE. 1977. Mineral element changes in pistachio leaves. *J. Amer. Soc. Hort. Sci.* 102:155–158.

WHITEHOUSE, W.E. 1957. The pistachio nut—a new crop for the western United States. *Econ. Bot.* 11:281–321.

WHITEHOUSE, W.E., E.J. KOCH, L.E. JONES, J.C. LONG, and C.L. STONE. 1964. Influence of pollens from diverse *Pistacia* species on development of pistachio nuts. *Proc. Amer. Soc. Hort. Sci.* 84:224–229.

WHITEHOUSE, W.E. and C.L. STONE. 1941. Some aspects of dichogamy and pollination in pistache. *Proc. Amer. Soc. Hort. Sci.* 39:95–100.

WOODROOF, J.G. 1979. Tree nuts: production, processing, products. 2nd ed. AVI Publishing, Westport, Ct.

ZOHARY, M. 1952. A monographical study of the genus *Pistacia*. *Palestine J. Bot., Jerusalem Ser.* 5:187–228.

9

Beneficial Effects of Viruses for Horticultural Plants[1]

Mortimer Cohen
Agricultural Research Center, Institute of Food and Agricultural Sciences, The University of Florida, P.O. Box 248, Fort Pierce, Florida 33450

I. INTRODUCTION

Virus infection, whether of plants or of animals, usually signifies trouble. Tobacco mosaic, tristeza disease of citrus, rabies in animals and humans, yellow fever and polio in humans, and certain forms of cancer are all dreaded diseases caused by different types of viruses. But many virus infections are latent or even benign in their effects. Maramorosch (1960) has suggested that, "It may be that viruses serve constructive functions in nature as varied and universal as those played by bacteria, but as difficult to see as the viruses themselves." This review is an examination of various ways in which viruses benefit or add value to plants in which humans are interested.

[1]Florida Agricultural Experiment Station Journal Series *2556*.

II. HISTORY

Viruses as submicroscopic infective entities which can not live apart from their hosts have been known for less than 100 years (Bawden 1964), but, looking back, virus-affected plants have influenced man's life for many centuries. One of the more spectacular effects of viruses was the color break seen in tulip flowers as a result of tulip mosaic virus infection. In the 1630s interest in unusual flower color combinations of the tulip, a plant lately introduced into Holland, Austria, and Germany from Turkey, was so great that it produced a phenomenon known as "tulipomania." A single bulb of a special cultivar brought as much as 5500 florins, and less unusual bulbs were sold for 1000 florins, at a time when an ox might cost 120 florins and a suit of clothes 80 florins (Berger 1977). Some purchasers sought to possess the bulbs as they might jewels, to admire the fantastic flowers, but many others bought tulips because they were anxious to resell them at tremendous profits. Tulipomania was as important to the history of mob psychology and finance as to the history of horticulture. Special markets and laws were established to regulate trade in tulips and speculation in the plants reached astronomical heights (Anon. 1971; Bawden 1964; Blunt 1950). Eventually many growers learned that flower breaking could be induced in any tulip plant by rubbing it with the juice of a plant showing flower break symptoms (Maramorosch 1960). The boom inevitably was succeeded by a bust in which fortunes were lost and financial panic prevailed, especially in Holland. Complete disaster was avoided only by the most drastic governmental measures, but many speculators lost tremendous sums of money.

Although the essential horticultural facts were available, the world was not ready for an understanding of the scientific principles involved. Plants with flower breaks and unique flower color continued to be highly prized for many years but on a less hysterical basis (Anon. 1971). The virus nature of tulip mosaic disease, which produced the flower breaks, was not demonstrated until 1928 (Cayley 1928).

An ornamental plant introduced into European gardens around 1870 was the variegated *Abutilon* or flowering maple (actually unrelated to maple). These shrubs were prized for their colorful mottled patterns and were classified as a horticultural variety distinct from the green wild type. Early virologists found that the variegation could be produced in the wild type by grafting into it a piece of the variegated plant, but they were puzzled as to how the variegation could persist in nature in the absence of a vector. Only recently was it discovered that variegation is due to a virus which is not transmitted in temperate climates but is freely moved from plant to plant in subtropical areas by a white fly, *Bemisia tabaci* (Gennadius) (Bawden 1964; Maramorosch 1960).

III. VIRUSES FOR CONTROL OF ANIMAL PESTS

A. Arthropods

Many insects harmful to horticultural plants are known to be affected by virus diseases (Mazzone and Tignor 1976; Stoltz and Vinson 1979), and effective natural control of harmful insects by insect viruses has been observed (Balch and Bird 1944; Entwistle 1977). With the increased interest in biological control of plant pests and diseases there has been much consideration of the possibility of using insect viruses for insect control. Three recent reviews have been those of David (1975), Falcon (1976), and Tinsley (1979).

A number of different morphologic types of insect viruses are known: nucleopolyhedrosis viruses (NPV), cytoplasmic polyhedrosis viruses, granulosis viruses, entomopoxviruses, and non-occluded viruses (Mazzone and Tignor 1976). All types mentioned have been used experimentally for insect or mite control.

Viruses are usually introduced by dispersing the virus material in water and spraying it on the plants being protected, but Ignoffo (1978) points out that a number of other approaches to the distribution of pathogens have been and should be used. These include autodissemination (use of insects to introduce and spread entomopathogens in the ecosystem); induction of an epizootic using a level of inoculum high enough to produce economically effective killing of the target insects; colonization (this leads in time to the development of an epizootic); and manipulation of the environment to provide ideal conditions for the spread of insect pathogens.

As an indication of the scientific community's interest in the whole question of insect control by microbiological agents including viruses, a workshop sponsored by the National Science Foundation, the U.S. Department of Agriculture, and the University of Florida was held in January 1978 at Gainesville, Florida (Allen *et al.* 1978). Participants discussed the role use of entomopathogens in pest management systems and various techniques for increasing the effectiveness of control.

One of the most damaging agricultural insects is *Heliothis zea* (Boddie) which, depending on the crop, is called the corn earworm, the bollworm, the tomato budworm, the tobacco budworm, or the bollworm on soybean. Numerous reports on the use of a NPV to control this insect on different crops have been published (Smith *et al.* 1977; Bell and Kanavel 1978; Ignoffo *et al.* 1978; Ignoffo 1978). NPVs have also been used in the control of the gypsy moth (*Lymantria dispar* L.) (Magnoler 1974), the European spruce saw fly (*Gilpinia hercyniae* (Hartig)) (Entwistle 1977), the cabbage looper (*Trichoplusia ni* (Hübner)) (Martignoni and Milstead 1962), the Douglas-fir tussock moth (*Orgyia pseudotsugata* (Mc-

Dunnough)) (Morris 1963), and the coconut palm rhinoceros beetle (*Oryctes rhinoceros* L.) (Young 1974; Bedford 1976). A granulosis virus epizootic was studied in the imported cabbage worm (*Pieris rapae* L.) by Wilson (1963). Both an entomopox virus and a NPV were used experimentally for the control of the spruce budworm (*Choristoneura fumiferana* (Clemens)) by Bird *et al.* (1973) with "encouraging" results. The citrus red mite (*Paratetranychus citri* Mcg.) was controlled by J.G. Shaw and J.B. Beavers (1970) using a noninclusion virus.

The effectiveness of at least some insect viruses is beyond question. In a natural occurrence of a virus epizootic, the spruce saw fly was dramatically controlled following the accidental introduction of a virus into the Canadian South in 1938. Within 4 years control moved from southern to northern Canada (Balch and Bird 1944). A similar control incident was reported for the European spruce saw fly in Wales (Entwistle 1977).

Because of the danger of mutation and side effects such as possible toxic, allergic, or teratogenic influences on the other organisms, great care must be taken before such materials can be released for general use (Tinsley 1979). Absolute assurances can never be given, but with intensive study the risk can be reduced to a reasonable level. The World Health Organization prepared a report (1973) describing procedures that must be followed before viruses can be considered safe for commercial distribution. Many problems must be faced in developing a virus insecticide for the open market (Falcon 1976; Tinsley 1979). In addition to satisfying safety and environmental requirements there are questions as to the profitability of such products. Such materials may not be patentable since they are natural products, and commercial companies may have difficulty in making a profit on their investment if others are free to copy any effective material.

Despite the difficulties involved in meeting all requirements for commercial development of virus insecticides, one material is now fully registered and is commercially available. This is a NPV material, "Elcar," designed to control *Heliothis* insects. It became available in June 1979 with the opening of a new Sandoz Inc. manufacturing facility in California. The development of this type of pesticide is encouraged by the U.S. Environmental Protection Agency (EPA). At the official opening of the facility, the Assistant Administrator of the EPA outlined measures designed to speed approval of biorational pesticides such as viruses, bacteria, protozoa, fungi, and some naturally occurring biochemicals (Anon. 1979).

Governmental agencies and universities all over the world are involved in the production of new virus insecticides. Such institutions are concerned only with developing effective materials and are not limited to materials likely to provide a good income to the developer (Falcon 1976).

Virus insecticides utilize natural principles for plant protection. Their potential is great and the next few years will show how many environmentally safe materials scientists can develop in this area.

B. Other Animals

The growth of horticultural plants is limited by animals other than insects. A classic instance of plant destruction was seen in Australia after the introduction of the European rabbit, *Oryctolagus cuniculus* L., in the 19th Century. According to Fenner and Meyers (1978), it was common in 1950–1951 to see "moving swarms of rabbits on many properties" in the valuable agricultural region of New South Wales. Although Australia contained many enemies of the rabbits including "foxes, feral cats, diurnal and nocturnal birds of prey, the domestic dog and man" (Fenner and Meyers 1978), they had a negligible effect on rabbit numbers because of the tremendous reproductive potential of the rabbits. The rabbits' principal environmental effect was injury to pastures and woodlands and consequent destructive competition with sheep, Australia's major agricultural asset.

The seriousness of the rabbit problem in Australia was well recognized, and in 1918 the Brazilian bacteriologist Aragao suggested the use of the myxoma virus to control this pest (Fenner 1959). This virus was endemic in the Brazilian wild rabbit, *Sylvilagus brasiliensis* (L.), in a relatively benign form, but was virulent in its effect on the European rabbit (Fenner and Meyers 1978). Extensive studies were made of the host range and of techniques for introducing the virus. Finally in 1950 the myxoma virus was used in Australia, with spectacular results. Initial kill of rabbits was as great as 99.8% of the infected individuals, and within 3 years the disease covered the entire southern half of the Australian continent.

The vector of the myxoma virus is the mosquito which acts as a "flying pin" in moving the disease from rabbit to rabbit. Subsequent control has been limited by the appearance of milder virus strains which have permitted a substantial rabbit population to reappear in Australia but not in the massive numbers which created the original problem.

In Europe a similar situation existed. Rabbits were a most serious agricultural pest which destroyed plants needed for the sustenance of conventional agricultural animals. In contrast to the carefully planned, scientifically controlled introduction of myxomatosis in Australia was the do-it-yourself method used in France, where a landowner inoculated two rabbits near Paris in 1952. Amazingly, results again were dramatic with an initiation of a large epizootic which drastically reduced the number of rabbits in the wild in France and elsewhere on the Continent. England followed even more informally in 1953. The disease was introduced secretly and without official sanction, again followed by a great reduction in the rabbit population. In these areas the most significant consequence of the reduced rabbit population was a strong regrowth of woody plants, especially pines and grasses.

An interesting example of a virus which protects a plant from injury by animals was recently described in Australia (Gibbs 1980). Gibbs demonstrated that the dusky coral pea (*Kennedya rubicunda* Schneev.) was partially protected from herbivore grazing when infected with *Kennedya* yellow mosiac virus. In Gibb's experiments virus-free plants set out in the field disappeared twice as fast as plants inoculated with the virus. A similar preference for virus-free leaf material was shown by caged young rabbits in palatability trials comparing powdered virus-free and virus-infected leaves of coral pea mixed with grated carrot.

IV. VIRUSES USED TO CONTROL PLANT DISEASES

A. Fungus Diseases

Viruses have been found in all major groups of fungi including Phycomycetes, Basidiomycetes, Ascomycetes, and Fungi Imperfecti. Recent reviews of fungal viruses are those of Hollings (1978), Lemke (1977), and Lemke and Nash (1974).

Although many fungi are known to contain viruses or virus-like particles (VLP), Lemke and Nash (1974) have suggested that viruses in fungi are usually latent. Most fungus viruses do not produce symptoms in the affected hosts (Hollings 1978). Our interest here is not in viruses of fungi generally but in viruses and VLP that are found in plant pathogenic fungi and the possibility of controlling these fungi with viruses. More than 35 plant pathogenic fungi are known to carry viruses or VLP, but in most cases the infected plant pathogens do not deteriorate (Lemke 1977). In a few instances, however, the pathogens are clearly injured by the infection. Castanho and Butler (1978a) described a disease of the pathogen *Rhizoctonia solani* Kühn, which produced poorer growth of the organism and greatly reduced pathogenicity to hosts like cabbage and sugar beets. The disease was associated with certain segments of double-stranded ribonucleic acid in the cytoplasm although it was not possible to obtain a purified virus or VLP from infected cultures (Castanho *et al.* 1978c). It was demonstrated that elements within the fungus cell were connected to the disease because infected isolates were freed of disease by hyphal tip subculturing, and the disease could be introduced into certain disease-free isolates by hyphal anastomosis. The possibility of using a diseased isolate of *Rhizoctonia solani* to control *Rhizoctonia* damping-off of sugar beet seedlings in the greenhouse was explored. Sugar beet seedlings were inoculated with a known virulent strain of *R. solani.* When the diseased (hypovirulent) counterpart of the virulent isolate was added to the infected flat, the number of seedlings killed was greatly reduced (Castanho and Butler 1978b). This demonstration of disease control was limited to a few selected isolates and did not apply to a wide spectrum of *R. solani* strains.

A very important fungus pathogen that might be brought under control by a virus or virus-like agent is the chestnut blight organism, *Endothia parasitica* (Murr.) And. (Anagnostakis 1978). Chestnut blight disease was first seen in the United States in 1906. Within 10 years chestnut blight had done devastating damage, and within 40 years there were no mature strands of the American chestnut (*Castanea dentata* Borkh.) to be found within the original range of this beautiful and useful tree. All trees were not killed, however, since new sprouts continued to grow from stumps of diseased trees, only to be killed back by *E. parasitica* infection within a few years.

In 1938 chestnut blight appeared in Italy, attacking the European chestnut (*C. sativa* Mill.), and a much feared epidemic was underway in Europe. Surprisingly, the epidemic did not go to completion and many trees that had been infected began to recover (Biraghi 1953). Study revealed that the destructive cankers originally formed by virulent *E. parasitica* were being replaced naturally by a hypovirulent (non-virulent) form of the fungus with a consequent recovery of chestnut trees (Grente and Sauret 1969). Chestnut blight has now ceased to be a problem in Italy (Mittempergher 1978; Turchetti 1978). Research suggested that when hypovirulent strains were introduced into blight cankers, the hyphae anastomosed with the hyphae of the virulent strain and introduced cytoplasmic material which converted the virulent strain into a relatively harmless one (Grente and Berthelay-Sauret 1978).

The dramatic recovery of chestnut tree in Europe led to a demonstration that European hypovirulent strains were promising in the United States (Anagnostakis and Jaynes 1973) and to extensive work designed to replace virulent *E. parasitica* in North American trees with hypovirulent strains (Anagnostakis 1978). Day *et al.* (1977) reported that hypovirulent strains were characterized by the presence of particles of double-stranded RNA in their cytoplasm, the material characteristic of most fungal viruses. Dodds (1978) found club-shaped virus-like structures in one hypovirulent strain, but it has not yet been established that these are generally associated with hypovirulent strains.

Ustilago maydis (DC.) Cda., the corn smut fungus, has been shown to have three categories of strains, one of which exhibits an interesting "killer" feature (Hankin and Puhalla 1971). A somewhat similar "killer" condition is also known for *Saccharomyces cervisiae* Meyen ex Hansen, brewers yeast (Hollings 1978). In *Ustilago maydis* the categories are (1) P_1—strains that release a protein toxin which quickly kills sensitive cells but that are themselves unaffected by the toxin; (2) P_2—cells killed by the toxin; and (3) P_3—strains containing cells not affected by toxin from P_1 but that themselves produce no toxin. Killer production and resistance to the toxin are transmitted in both cytoplasm and nucleus (Day and Anagnostakis 1973). Resistance to the toxin is associated with virus particles found in the cytoplasm of cells in categories P_1 and P_3. Koltin and Day (1975) tested the possibility of using the toxin for biological control of a range of fungus diseases, especially smut diseases, but observed only a limited lethal range of the toxin on fungi closely related to corn smut. This is a complex situation which needs additional study, but may be of value for biological control.

B. Bacterial Diseases

Bacterial viruses, called bacteriophages, are known for all types of bacteria (Luria *et al.* 1978). Under proper conditions bacteriophages kill large numbers of bacterial cells and potentially could be of great value in combating bacterial diseases of plants. Unfortunately, thus far, there are no reports of the practical use of bacteriophages against pathogens (Baker and Cook 1974).

Bacteriophages have been valuable for some other purposes. Koizumi (1969) has used one of the bacteriophages of the citrus canker bacterium *Xanthomonas citri* (Hasse) Dowson as an indicator of the presence of the pathogen in ecological studies.

C. Virus Diseases

Mild, attenuated, or killed strains of viruses which attack mammals have been utilized most successfully in combating diseases such as smallpox and polio. Antibodies, produced in the mammalian bloodstream following the introduction of mild forms of pathogenic viruses, effectively attack any later arrival of severe forms of those viruses. Plants do not possess an antibody mechanism, but McKinney (1929) was the first to discover that the inoculation of one strain of a virus into a plant often prevented the subsequent development of the symptoms of a second strain. This phenomenon, known as cross protection or interference, has interested virologists for many years (Bennett 1953; Matthews 1970). Fernow (1967) and Niblett *et al.* (1978) have also demonstrated cross protection among some viroids, although it has not yet been possible to show that mild strains of citrus exocortis viroid can protect against more severe strains (Kapur *et al.* 1974).

One outstanding example of the value of the cross protection method has been its use in growing citrus trees in the presence of tristeza virus in Brazil. Tristeza virus disease, attacking trees on sour orange rootstock, killed more than 20 million trees in Brazil and Argentina in the 1930s and 1940s. The citrus industry in Brazil and Argentina was rebuilt by planting trees of tristeza-tolerant combinations, but certain cultivars such as 'Galego' lime (West Indian lime), grapefruit, and the 'Pera' cultivar of sweet orange did not grow well because their own tissue was affected by the virus regardless of rootstock. Muller and Costa (1968, 1972) and Costa and Muller (1980) used cross protection techniques to reduce the damage that was done by tristeza to trees with such tristeza-intolerant tops. For mild tristeza budwood sources they selected 45 individual trees of 'Galego' lime, 'Pera' orange, and grapefruit in the field which were making outstanding growth in areas where most trees showed severe tristeza injury. New trees carrying the mild strains were then propagated, and replicated field trials were set out to study the performance of the trees. Clear-cut results were obtained most quickly with the 'Galego' limes but only 5 of the tested strains were considered to be outstanding, all originating from 'Galego' lime sources. All the mild strains that came from 'Pera' orange or grapefruit were too severe for 'Galego' limes, but some were satisfactory for grapefruit and sweet orange. The success of these cross protection trials is demonstrated by the fact that about ten million trees carrying mild strain protection—mainly 'Pera' orange, but also many trees of 'Galego' lime—have been produced in Brazil.

Selected mild strains of tristeza have also been reported to be successful in India (Balaraman and Ramakrishnan 1980) in protecting acid limes against severe tristeza disease. Mild strain protection of grapefruit trees against tristeza has also been demonstrated in Australia in trees on tristeza-tolerant rootstocks (Fraser *et al.* 1968; Cox *et al.* 1976) and even, although less successfully, in trees on sour orange rootstock (Thornton *et al.* 1980).

Sour orange has been abandoned as a rootstock for sweet orange and grapefruit in Brazil and in the tristeza areas of Argentina because even the mildest strains there are too severe to permit satisfactory growth of trees on this stock. In contrast, sour orange rootstock continues to be used widely in many districts in Florida despite the almost universal presence of mild strains of tristeza virus in Florida citrus trees (Bridges and Youtsey 1972) and the occasional outbreak of severe tristeza decline (Garnsey and Jackson 1976). What seems to be operating, although as yet unproven, is a natural system of cross protection which limits the extent of tree damage by tristeza.

Tristeza cross protection research in Florida and California has concentrated on prevention of decline in trees on sour orange rootstock. The benefits to be obtained from the introduction of certain mild tristeza strains are easily demonstrated for the short term, but long time benefits must await continued observation (Wallace and Drake 1974; Cohen 1976).

Another practical use of protective mild virus strains resulted from the work of Rast (1972) who developed a symptomless strain of tobacco mosaic virus which has been used in a number of countries to increase yield of greenhouse tomatoes. Rast's symptomless strain, M II-16, was produced following treatment of common tomato strains with nitrous acid. Tomato seedlings are inoculated in the cotyledon stage with a spray containing both virus and fine carborundum particles to produce minute injuries. More recently Rast (1979) also suggested that strain M II-16 could be used to help produce tomato seed free of tobacco mosaic virus.

V. VIRUSES USED FOR CONTROL OF WEEDS

Examples of biological control of weeds with viruses are not common, but work in this area is beginning. Charudattan and co-workers (1980) have made an interesting and complete study of factors which must be considered if a newly described virus is to be utilized in the battle against one of the more serious pests of Florida citrus, the milkweed vine. Virus-infected milkweed vine plants were found in Argentina and sent under quarantine to Florida where they were subjected to intensive study of virus characteristics, host range, and aphid transmissibility. Plants in 6 genera of the Asclepiadaceae, related to the milkweed vine, were successfully inoculated either manually or by aphids. Plants in 108 other species, representing 75 genera and 25 families, were also tested and found not to be affected. It is believed that if quarantine authorities permit the virus to be released it will be spread rapidly by vector aphids to new plants within its restricted host range. A U.S. patent on the composition and process for controlling milkweed vine with this virus has been issued.

Research on the use of a virus to control a weed pest has also been conducted in Russia. Izhevsky (1980) reported on a study of the control of *Solanum carolenense* L. (horse nettle), a weed of American origin, which is a pest of tea plantations in Soviet Georgia. The virus studied is a European strain of tobacco mosaic, highly pathogenic to *Solanaceae* but with no natural vectors to move the virus to solanaceous plants of economic importance.

VI. VIRUSES USED FOR CONTROL OF TREE SIZE

High density plantings of dwarf trees have been very successful in improving efficiency and economy in plantings of deciduous fruit with increased productivity per unit area (Tukey 1964; Mendel 1969; Castle 1979; Childers 1979). Smaller trees may be obtained by using virus-free dwarfing rootstocks or may also result from inoculation of trees with viruses (or viroids) which reduce tree size but not fruit quality or tree longevity (Phillips 1980). In the United States, Australia, and South America there has been considerable interest in the use of selected strains of exocortis viroid to control tree size of citrus trees on certain rootstocks. Cohen (1968), describing a citrus rootstock experiment in Florida, cited data which demonstrated that some exocortis-stunted 'Ruby Red' grapefruit trees on 'Rangpur' lime rootstock would, if solidly planted, be more productive than conventional plantings of non-stunted trees.

Long *et al.* (1972) described work with exocortis-dwarfed 'Valencia' and 'Washington' navel orange trees on trifoliate orange rootstocks. They found no deficiencies of fruit size, fruit quality, or foliage density and demonstrated that such trees were highly productive when planted at high density. They also showed that size reduction in trees was related to the time when exocortis inoculation took place. The later the time of inoculation, the larger the eventual size of the trees. Thus, tree size could be controlled by varying the time of inoculation. Additional information on this experiment was given by Stannard *et al.* (1975).

More recent dwarf citrus plantings at Dareton and Yanco in Australia were described in Bevington and Bacon (1978). Effective size limitation of 'Bellamy' navel orange trees on 'Troyer' citrange, 'Rangpur' lime, and trifoliate orange rootstocks has been obtained in experimental plantings.

A large exploratory study involving duplicate experiments at Lake Alfred and Fort Pierce, Florida was initiated in 1973 by Cohen *et al.* (1980). Five scion cultivars—'Valencia' and 'Pineapple' oranges, 'Marsh' and 'Ruby' grapefruit, and 'Dancy' tangerine—on the exocortis-sensitive rootstocks trifoliate orange, 'Carrizo' citrange, and 'Rangpur' lime and two "non-sensitive" stocks, sour orange and rough lemon, were included. These trees were inoculated, during the first year after planting, with eight different strains of exocortis viroid, although not in all combinations. In general the same strains gave the greatest size reduction for all stock-scion combinations and for both locations.

In Brazil also (Pompeu *et al.* 1976), the presence of exocortis in trees on a sensitive rootstock resulted in tree size reduction without impairment of fruit quality, tree size, or loss of yield on an area basis.

Extensive commercial plantings of dwarfed citrus trees have not yet been set out due, in part, to an understandable reluctance on the part of both growers and experimentalists to promote the use of an agent which causes a "disease." In part, it is also due to fear of the possibility of mutations in the viroid which might be harmful to the trees.

Exocortis viroid is widely distributed, however, in older citrus plantings around the world, usually symptomlessly in trees on non-sensitive rootstocks. In some cases dwarf exocortis-infected trees have been inadvertently successfully planted on exocortis-sensitive rootstocks. The advent of new understanding of citrus viruses and viroids has made it possible for citrus budwood schemes in many countries to provide propagative material free of exocortis (Bridges 1976; Nauer *et al.* 1978). The same budwood agencies could distribute budwood containing useful strains of exocortis. In New South Wales, Australia, guidelines have been set up for release of dwarfing citrus budwood for the establishment of demonstration commercial plantings (Duncan *et al.* 1980). Citrus growers elsewhere are also likely to begin utilizing appropriate strains of exocortis viroid to produce productive trees of reduced size as demonstrations of dwarfing become more familiar.

VII. CONCLUSIONS

Viruses and virus diseases are usually detrimental to plants, but examples have been given of a number of ways in which viruses can be utilized for man's benefit in his work with horticultural plants. Most of these used have been the product of the ingenuity and imagination of virologists and, as knowledge grows in the young science of virology, newer and more imaginative ideas can be expected.

New techniques are developing which may make possible improvements not even dreamed of in the past. One method in genetic engineering uses bacterial viruses—bacteriophages—to introduce desired segments of chromosomes or plasmids into bacteria to enable them to synthesize new compounds or to achieve other chemical effects (Murray 1978; Cohen 1975). Other genetic engineering methods will produce improvements in plant quality and productivity.

Despite the great advances in virology it seems clear that we are only beginning to comprehend the principles which operate in this important branch of science. Progress must be made cautiously because virologists deal with powerful forces which could do tremendous damage if unleashed in the wrong direction. With patience and restraint, advances in understanding will be safely coupled with increased ability to find ways of using viruses beneficially.

VIII. LITERATURE CITED

ALLEN, G.E., C.M. IGNOFFO, and P. JAQUES (eds.) 1978. Microbial control of insect pests: future strategies in pest management systems. Selected Papers from NSF-USDA-Univ. of Florida Workshop, Jan. 10–12, 1978, Gainesville, Fla. Univ. of Florida, Gainesville.

ANAGNOSTAKIS, S.L. 1978. The American chestnut: new hope for a fallen giant. *Conn. Agr. Expt. Sta. Bul.* 777.

ANAGNOSTAKIS, S.L. and R.A. JAYNES. 1973. Chestnut blight control: use of hypovirulent cultures. *Plant Dis. Rptr.* 57:225–226.

ANON. 1971. Tulipomania: marketing's maddest moment. *Marketing/Communications* 299:48–53.

ANON. 1979. New Elcar plant. *Agrichem. Age* 23 (7):8.

BAKER, K.F. and R.J. COOK. 1974. Biological control of plant pathogens. W.H. Freeman and Company Publishers, San Francisco.

BALARAMAN, K. and K. RAMAKRISHNAN. 1980. Strainal variation and cross protection in citrus tristeza virus on acid lime *Citrus aurantifolia* (Christm.) Swing. p. 60–68. *In* E.C. Calavan, S.M. Garnsey, and L.W. Timmer (eds.) Proc. 8th Conf. Intern. Organ. Citrus Virol., May 23–25, 1979, Mildura, Victoria, Australia. Intern. Organ. Citrus Virol., Riverside, Calif.

BALCH, R.E. and F.T. BIRD. 1944. A disease of the European spruce saw fly *Gilpinia hercyniae* (Htg.), and its place in natural control. *Sci. Agr.* 25:65–80.

BAWDEN, F.C. 1964. Plant viruses and virus diseases. 4th ed. Ronald Press, New York.

BEDFORD, G.O. 1976. Use of a virus against the coconut palm rhinoceros beetle in Fiji. *PANS* 22:11–25.

BELL, M.R. and R.F. KANAVEL. 1978. Tobacco budworm: development of a spray adjuvant to increase effectiveness of a nuclear polyhedrosis virus. *J. Econ. Entomol.* 71:350–352.

BENNETT, C.W. 1953. Interactions between virus strains. *Adv. Virus Res.* 1:39–89.

BERGER, T. 1977. 'Tulipomania' was no Dutch treat to gambling burghers. *Smithsonian* 8:70–76.

BEVINGTON, K.B. and P.E. BACON. 1978. Effect of rootstocks on the response of navel orange trees to dwarfing inoculations. p. 567–570. *In* W. Grierson (ed.) 1977 Proc. Intern. Soc. Citriculture, Vol. II. May 1–8, 1977, Orlando, Fla. Intern. Soc. Citriculture, Lake Alfred, Fla.

BIRAGHI, A. 1953. Possible active resistance to *Endothia parasitica* in *Castanea sativa*. Rpt. 11th Congr. Intern. Union of Forest Res. Organisation-Rome. p. 643–645.

BIRD, F.T., J.C. CUNNINGHAM, and G.M. HOWSE. 1973. The possible use of viruses in the control of spruce budworm. *Proc. Entomol. Soc. Ont.* 103:69–75.

BLUNT, W. 1950. Tulipomania. Penguin Books, New York.

BRIDGES, G.D. 1976. Florida's budwood registration program. *Citrograph* 61:413–415.

BRIDGES, G.D. and C.D. YOUTSEY. 1972. Natural tristeza infection of citrus species, relatives and hybrids at one Florida location from 1961–1971. *Proc. Fla. State Hort. Soc.* 85:44–47.

CASTANHO, B. and E.E. BUTLER. 1978a. Rhizoctonia decline: a degenerative disease of *Rhizoctonia solani. Phytopathology* 68:1505–1510.

CASTANHO, B. and E.E. BUTLER. 1978b. Rhizoctonia decline: studies on hypovirulence and potential use in biological control. *Phytopathology* 68: 1511–1514.

CASTANHO, B., E.E. BUTLER, and R.J. SHEPARD. 1978. The association of double-stranded RNA with Rhizoctonia decline. *Phytopathology* 68:1515–1519.

CASTLE, W.S. 1979. Controlling citrus tree size with rootstocks and viruses (exocortis) for higher density plantings. *Proc. Fla. State Hort. Soc.* 91:46–50.

CAYLEY, D.M. 1928. 'Breaking' in tulips. *Ann. Appl. Biol.* 15:529–539.

CHARUDATTAN, R., F.W. ZETTLER, H.A. CORDO, and R.G. CHRISTIE. 1980. Partial characterization of a potyvirus infecting the milkweed vine *(Morrenia odorata). Phytopathology* 70:909–913.

CHILDERS, N.F. 1979. Trend toward high-density plantings with dwarfed trees in deciduous orchards. *Proc. Fla. State Hort. Soc.* 91:34–36.

COHEN, M. 1968. Exocortis virus as a possible factor in producing dwarf citrus trees. *Proc. Fla. State Hort. Soc.* 81:115–119.

COHEN, M. 1976. A comparison of some tristeza isolates and a cross protection trial in Florida. p. 50–54. *In* E.C. Calavan (ed.) Proc. 7th Conf. Intern. Organ. Citrus Virol., Sept. 9–Oct. 4, 1975, Athens, Greece. Intern. Organ. Citrus Virol. Univ. of California, Plant Pathology Dept., Riverside.

COHEN, M., W.S. CASTLE, R.L. PHILLIPS, and D. GONSALVES. 1980. Effect of exocortis viroid on citrus tree size and yield in Florida. p. 195–200. *In* E.C. Calavan, S.M. Garnsey and L.W. Timmer (eds.) Proc. 8th Conf. Intern. Organ. Citrus Virol., May 23–25, 1979, Mildura, Victoria, Australia. Intern. Organ. Citrus Virol., Riverside, Calif.

COHEN, S.N. 1975. The manipulation of genes. *Sci. Amer.* 233:24–33.

COSTA, A.S. and G.W. MULLER. 1980. Tristeza control by cross protection: a U.S.-Brazil cooperative success. *Plant Dis.* 64:538–541.

COX, J.E., L.R. FRASER, and P. BROADBENT. 1976. Stem pitting of grapefruit: field protection by the use of mild strains, an evaluation of trials in two climatic districts. p. 68–74. *In* E.C. Calavan (ed.) Proc. 7th Conf. Intern. Organ. Citrus Virol., Sept. 9–Oct. 4, 1975, Athens, Greece. Intern. Organ. Citrus Virol. Univ. of California, Plant Pathology Dept., Riverside.

DAVID, W.A.L. 1975. The status of viruses pathogenic for insects and mites. *Annu. Rev. Entomol.* 20:97–117.

DAY, P.R. and S.L. ANAGNOSTAKIS. 1973. The killer system in *Ustilago maydis*: heterokaryon transfer and loss of determinants. *Phytopathology* 63: 1017–1018.

DAY, P.R., J.A. DODDS, J.E. ELLISTON, R.A. JAYNES, and S.L. ANAGNOSTAKIS. 1977. Double-stranded RNA in *Endothia parasitica. Phytopathology* 67:1393–1396.

DODDS, J.A. 1978. Double-stranded RNA and virus-like particles in *Endothia parasitica.* p. 108–110. *In* W. MacDonald (ed.) American Chestnut Symp. Proc. Jan. 4–5, 1978, Morgantown, West Va. West Virginia Univ. Agr. Expt. Sta. and USDA, Morgantown.

DUNCAN, J.H., R.S. SPROULE, and K.B. BEVINGTON. 1980. Commercial application of virus induced dwarfing. p. 317–319. *In* P.R. Cary (ed.) 1978 Proc. Intern. Soc. Citriculture, Aug. 15–23, 1978, Sydney. Intern. Soc. Citriculture, Griffith, New South Wales, Australia.

ENTWISTLE, P.F. 1977. The development of an epizootic of a nuclear polyhedrosis virus disease in European spruce saw fly, *Gilpinia hercyniae.* p. 184–188. *In* Intern. Colloq. on Invertebrate Pathology. IXth Annu. Mtg., Soc. for Invertebrate Pathology. Kingston, Ontario.

FALCON, L.A. 1976. Problems associated with the use of anthropod viruses in pest control. *Annu. Rev. Entomol.* 21:305–324.

FENNER, F.J. 1959. Myxomatosis. p. 6–16. *In* The rabbit problem in Australia. Conference convened by CSIRO, Sept. 11–12, 1958, Melbourne.

FENNER, F.J. and K. MEYERS. 1978. Myxoma virus and myxomatosis in retrospect: the first quarter century of a new disease. p. 539–623. *In* E.E. Kurstack and K. Maramorosch (eds.) Viruses and environment. Academic Press, New York.

FERNOW, K.H. 1967. Tomato as a test plant for detecting mild strains of potato spindle tuber virus. *Phytopathology* 57:1347–1352.

FRASER, L.R., K. LONG, and J. COX. 1968. Stem pitting of grapefruit—field protection by the use of mild virus strains. p. 27–31. *In* J.F.L. Childs (ed.) Proc. 4th Conf. Intern. Organ. Citrus Virol., Oct. 10–12, 1966, Catania, Sicily. Univ. of Florida Press, Gainesville.

GARNSEY, S.M. and J.L. JACKSON, JR. 1976. A destructive outbreak of tristeza in central Florida. *Proc. Fla. State Hort. Soc.* 88:65–69.

GIBBS, A. 1980. A plant virus that partially protects its wild legume host against herbivores. *Intervirology* 13:42–47.

GRENTE, J. and S. BERTHELAY-SAURET. 1978. Biological control of chestnut blight in France. p. 30–34. *In* W. MacDonald (ed.) American Chestnut Symp. Proc. Jan. 4–5, 1978, Morgantown, West Va. West Virginia Univ. Agr. Expt. Sta. and USDA, Morgantown.

GRENTE, J. and S. SAURET. 1969. L'hypovirulence exclusive est-elle controlée par des determinants cytoplasmiques? *C. R. Acad. Sci. Paris, Ser. D* 268:3173–3176.

HANKIN, L. and J.E. PUHALLA. 1971. Nature of a factor causing interstrain lethality in *Ustilago maydis. Phytopathology* 61:50–53.

HOLLINGS, M. 1978. Mycoviruses: viruses that infect fungi. *Adv. Virus Res.* 22:2–54.

IGNOFFO, C.M. 1978. Guest editorial: Strategies to increase the use of entomopathogens. *J. Invertebrate Pathol.* 31:1–3.

IGNOFFO, C.M., D.L. HOSTETTER, K.D. BIEVER, C. GARCIA, G.D. THOMAS, W.A. DICKERSON, and R. PINNELL. 1978. Evaluation of an entomopathogenic bacterium, fungus and virus for control of *Heliothis zea* on soybeans. *J. Econ. Entomol.* 71:165–168.

IZHEVSKY, S.S. 1980. Application of pathogenic microorganisms in control of weeds in the USSR. Proc. Joint American-Soviet Conf. on the Use of Beneficial Organisms in the Control of Crop Pests. Aug. 13–14, 1979, Washington, D.C. (in press)

KAPUR, S.P., L.G. WEATHERS, and E.C. CALAVAN. 1974. Studies on strains of exocortis virus in citron and *Gynura aurantiaca*. p. 105–109. *In* L.G. Weathers and M. Cohen (eds.) Proc. 6th Conf. Intern. Organ. Citrus Virol., Aug. 21–28, 1972, Mbabane, Swaziland. Univ. of California, Div. Agr. Sci., Richmond.

KOIZUMI, M. 1969. Ecological studies on citrus canker caused by *Xanthomonas citri*. III. Seasonal changes in number of causal bacteria and its bacteriophages CP_1 in rain water flowing down from the diseased trees. (Japanese with English summary). Bul. Hort. Res. Sta., Okitsu, Japan. Series B. no. 9 (December) p. 129–144.

KOLTIN, Y. and P.R. DAY. 1975. Specificity of *Ustilago maydis* killer proteins. *Appl. Microbiol.* 30:694–696.

LEMKE, P.A. 1977. Fungal viruses and agriculture. p. 159–175. *In* J. A. Romberger (ed.) Virology in agriculture. Allanheld Osmun, Montclair, N.J.

LEMKE, P.A. and C.H. NASH. 1974. Fungal viruses. *Bacteriol. Rev.* 38:29–56.

LONG, J.K., L.R. FRASER, and J.E. COX. 1972. Possible value of close-planted virus-dwarfed orange trees. p. 262–267. *In* W.C. Price (ed.) Proc. 5th Conf. Intern. Organ. Citrus Virol., Oct. 30–Nov. 8, 1969, Japan. Univ. of Florida Press, Gainesville.

LURIA, S.E., J.E. DARNELL, JR., D. BALTIMORE, and A. CAMPBELL. 1978. General virology. 3rd ed. John Wiley and Sons, New York.

MAGNOLER, A. 1974. Field dissemination of a nucleopolyhedrosis virus against the gypsy moth, *Lymantria dispar* L. *Z. Pflanzenkrankheiten und Pflanzenschutz* 81:511.

MARAMOROSCH, K. 1960. Friendly viruses. *Sci. Amer.* 203:138–144.

MARTIGNONI, M.E. and J.E. MILSTEAD. 1962. Trans-ovum transmission of the nuclear polyhedrosis virus of *Colias eurytheme* Boisduval through contamination of the female genitalia. *J. Insect Pathol.* 4:113–121.

MATTHEWS, R.E.F. 1970. Plant virology. Academic Press, New York.

MAZZONE, H.M. and G.H. TIGNOR. 1976. Insect viruses: serological relationships. *Adv. Virus Res.* 20:237–270.

MCKINNEY, H.H. 1929. Mosaic diseases in the Canary Islands, West Africa, and Gibralter. *J. Agr. Res.* 39:557–578.

MENDEL, K. 1969. New concepts in stionic relations of citrus. p. 387–390. Proc. First Intern. Citrus Symp., Mar. 16–26, 1968, Univ. of California, Riverside.

MITTEMPERGHER, L. 1978. The *Endothia parasitica* epidemic in Italy. p. 34–37. *In* W. MacDonald (ed.) American Chestnut Symp. Proc. Jan. 4–5, 1978, Morgantown, West Va. West Virginia Univ. Agr. Expt. Sta. and USDA, Morgantown.

MORRIS, O.N. 1963. The natural and artificial control of the Douglas fir tussock moth, *Orgyia pseudotsuga* McDunnough, by a nuclear polyhedrosis virus. *J. Insect Pathol.* 5:410–414.

MULLER, G.W. and A.S. COSTA. 1968. Further evidence on protective interference in citrus tristeza. p. 71–82. *In* J.F.L. Childs (ed.) Proc. 4th Conf. Intern. Organ. Citrus Virol., Oct. 2–4, 1966, Rome. Univ. of Florida Press, Gainesville.

MULLER, G.W. and A.S. COSTA. 1972. Reduction in yield of Galego lime avoided by preimmunization with mild strains of tristeza virus. p. 171–175. *In* W.C. Price (ed.) Proc. 5th Conf. Intern. Organ. Citrus Virol., Oct. 30–Nov. 8, 1969, Japan. Univ. of Florida Press, Gainesville.

MURRAY, N.E. 1978. Bacteriophage lambda as a vector in recombinant DNA research—advantages and limitations. p. 31–52. *In* A.M. Chakrabarty (ed.) Genetic engineering. CRC Press, West Palm Beach, Fla.

NAUER, E.M., E.C. CALAVAN, and C.N. ROISTACHER. 1978. Update on the CCPP budwood program. *Citrograph* 63:163–165.

NIBLETT, C.L., E. DICKSON, K.H. FERNOW, R.K. HORST, and M. ZAITLIN. 1978. Cross protection among four viroids. *Virology* 91:198–203.

PHILLIPS, R.L. 1980. Citrus tree spacing and size control. p. 319–324. *In* P.R. Cary (ed.) 1978 Proc. Intern. Soc. Citriculture, Aug. 15–23, 1978, Sydney. Intern. Soc. Citriculture, Griffith, New South Wales, Australia.

POMPEU, J., JR., O. RODRIGUEZ, J. TEOFILO SOB, J.P. NEVES JORGE, and A.A. SALIBE. 1976. Behavior of nucellar and old clones of Hamlin sweet orange on Rangpur lime rootstock. p. 96–97. *In* E.C. Calavan (ed.) Proc. 7th Intern. Organ. Citrus Virol., Sept. 9–Oct. 4, 1975, Athens. Intern. Soc. Citrus Virol., Univ. of California, Riverside.

RAST, A.T.B. 1972. M II-16, an artificial symptomless mutant of tobacco mosaic virus for seedling inoculation of tomato crops. *Neth. J. Plant Pathol.* 78:110–112.

RAST, A.T.B. 1979. Infection of tomato seed by different strains of tobacco mosaic virus with particular reference to the symptomless mutant M II-16. *Neth. J. Plant Pathol.* 85:223–233.

SHAW, J.G. and J.B. BEAVERS. 1970. Introduced infections of viral disease in populations of citrus red mites. *J. Econ. Entomol.* 63:850–855.

SMITH, D.B., D.L. HOSTETTER, and C.M. IGNOFFO. 1977. Laboratory performance specifications for a bacterial *(Bacillus thuringensis)* and a viral *(Baculovirus heliothis)* insecticide. *J. Econ. Entomol.* 70:437–441.

STANNARD, M.C., J.C. EVANS, and J.K. LONG. 1975. Effect of transmission of exocortis dwarfing factors into Washington Navel orange trees. *Austral. J. Expt. Agr. Animal Husb.* 15:136–141.

STOLTZ, D.C. and S.B. VINSON. 1979. Viruses and parasitism in insects. *Adv. Virus Res.* 24:125–171.

THORNTON, I.R., R.W. EMMETT, and L.L. STUBBS. 1980. A further report on the grapefruit tristeza preimmunization trial at Mildura, Victoria. p. 51–53. *In* E.C. Calavan, S.M. Garnsey, and L.W. Timmer (eds.) Proc. 8th Conf. Intern. Organ. Citrus Virol., May 23–25, 1979, Mildura, Victoria, Australia. Intern. Organ. Citrus Virol., Riverside, Calif.

TINSLEY, T.W. 1979. The potential of insect pathogenic viruses as pesticidal agents. *Annu. Rev. Entomol.* 24:63–87.

TUKEY, H.B., SR. 1964. Dwarfed fruit trees. Macmillan, New York. (Reissued 1978 Cornell University Press, Ithaca, N.Y.).

TURCHETTI, T. 1978. Some observations on the "hypovirulence" of chestnut blight in Italy. p. 92–94. *In* W. MacDonald (ed.) American Chestnut Symp. Proc., Jan. 4–5, 1978, Morgantown, West Va. West Virginia Univ. Agr. Expt. Sta. and USDA, Morgantown.

WALLACE, J.M. and R.J. DRAKE. 1974. Field performance of tristeza-susceptible citrus trees carrying virus derived from plants that recovered from seedling yellows. p. 67–74. *In* L.G. Weathers and M. Cohen (eds.) Proc. 6th Conf. Intern. Organ. Citrus Virol., Aug. 21–28, 1972, Mbabane, Swaziland. Univ. of California, Div. Agr. Sci., Richmond.

WILSON, F. 1963. The effectiveness of a granulosis virus applied to field population of *Pieris rapae* (Lepidoptera). *Austral. J. Agr. Res.* 11:485–487.

WORLD HEALTH ORGANIZATION (WHO). 1973. The use of viruses for the control of insect pests and disease vectors. WHO Tech. Rpt. Ser. 531, Geneva.

YOUNG, E.C. 1974. The epizootiology of two pathogens of the coconut palm rhinoceros beetle. *J. Invertebrate Pathol.* 24:82–92.

10

Effects of Modified Atmospheres on Postharvest Pathogens of Fruits and Vegetables

M. A. El-Goorani
Department of Plant Pathology, Faculty of Agriculture, University of Alexandria, Alexandria, Egypt
N. F. Sommer
Department of Pomology, University of California, Davis, California, 95616

I. INTRODUCTION

Creating a controlled atmosphere (CA) or a modified atmosphere (MA) involves reducing oxygen and/or increasing carbon dioxide concentration. This modification can slow respiration and ethylene production, and delay ripening of fruits. When that is done, the physiological life of a commodity is lengthened (Dewey 1977, 1979; Dewey *et al.* 1969; Grierson 1969). But conflicting results cloud assessment of CA's value in controlling decay in fruits and vegetables during and after storage.

Published reviews have mentioned the effects of CA on postharvest pathogens and the postharvest diseases they cause (Smith 1963, 1976; Harvey 1978; Tabak and Cooke 1968; Eckert and Sommer 1967; Tomkins 1966b; Lougheed *et al.* 1978). Isenberg (1979) and Smock (1979) have provided reviews of the CA storage of vegetables and fruits, respectively. Burton (1974) has outlined biophysical principles relevant to CA storage of fruits and vegetables. Indexed reference lists on modified atmospheres contain valuable references on the subject (Murr *et al.* 1974; Kader *et al.* 1976; Kader and Morris 1977; Morris *et al.* 1971). This review covers the extensive literature on the effects of controlled atmospheres on postharvest pathogens and diseases.

II. METHODOLOGY

A. Atmosphere Maintenance

The creation and maintenance of test atmospheres of precise compositions are crucial to critical studies. In early studies desired atmospheres for test organisms often were created in containers that were sealed until the end of the test. For most studies, such techniques have been found to be inadequate, yielding, as often as not, results subject to misinterpretation. In those closed systems, respiration of test commodities usually caused a sharp increase in CO_2 concentration in the atmosphere, while O_2 was depleted. In experiments with fruit, the physiologically active gas ethylene (C_2H_4) may accumulate also along with certain volatile esters and aldehydes evolved by the fruit.

In modern test systems, gases are usually provided as a continuous flow at a rate sufficient to avoid appreciable accumulation of CO_2, C_2H_4, or other respiratory gases, or to prevent depletion of O_2. Mixed gases are sometimes obtained in pressure cylinders, from which desired flows can be metered using reducing valves and flow meters. Atmospheres may also be prepared by using cylinders of O_2, CO_2, and N_2 bought commercially without mixing before tests. When separate cylinders are used, each gas is metered in the proportions calculated to achieve the desired synthetic atmospheres (Claypool and Keefer 1942), and mixed before entered into test containers. Atmospheres low in O_2 are often obtained by diluting air with N_2. Although O_2 obtained by air dilution is satisfactory for many studies, it should be noted that CO_2 and other air contaminants are present in dilute quantities. Furthermore, gases from commercial sources may contain contaminants.

B. Disease Suppression

Disease suppression attributable to atmosphere modification has two effects: on the resistance of the host commodity and on the altered growth of the pathogen.

1. Commodity Maturity or Condition.—Fruits that go through pronounced ripening (apples, bananas, mangos, peaches, pears, papayas, etc.) usually are highly resistant to postharvest pathogens during most of their lives. Susceptibility increases as they approach maturity and proceed through the ripening process to senescence. Care must be taken to ensure that fruits are of about the same physiological age at harvest. Comparisons of surface or flesh color, penetrometer readings, or soluble solids measurements are generally satisfactory ways to compare uniformity of maturity. With climacteric-type fruits, it may be desirable to determine whether the fruits are truly pre-climacteric at the start of extended storage tests. Similarly, nonclimacteric fruits, leafy vegetables, stems, and roots or tubers undergo changes leading to senescence with a likely decrease in resistance to diseases. Lesions frequently develop much more rapidly in senescent fruit than in recently harvested fruits of proper maturity.

2. Inoculations.—Occasionally, tests can be conducted satisfactorily with uninoculated fruit. An experienced researcher can sometimes be reasonably certain that an acceptably uniform natural infection has occurred before harvest and is undeveloped, quiescent, or latent in the fruit. For example, pathogens causing Gloeosporium rots of pome fruits, Botrytis rots of strawberries and grapes, and anthracnose of various subtropical and tropical fruits may heavily infect fruit before harvest but disease development occurs after harvest.

Fruits usually must be inoculated before being placed in chambers supplied with test gases. Inoculation methods vary and no one method is appropriate in all circumstances. Inoculation is often done by piercing the epidermis of the commodity with a needle contaminated with spores or cells of the fungal or bacterial pathogen. Uniformity of the inoculum may be improved by standardizing spore or cell concentrations using a hemocytometer. The inoculated commodity may be placed immediately in test chambers. To achieve greater uniformity of lesion size within treatments, however, fruits may be held in air at 20° to 25°C in high humidity for 12 to 24 hours to permit infection to occur before they are placed in chambers.

Once an atmosphere has been established, it should not be interrupted, and fruit should not be removed from a chamber until the end of the test. Tests at a particular atmosphere normally are ended before a lesion occupies a major portion of a fruit's surface (generally less than 6 cm in diameter for apples, pears, or peaches). Lesion diameters can be measured and average radial growth rates are expressed as millimeters per day. Such data, however, suggest that growth has been linear when, in fact, it has really been sigmoidal with a lag phase before establishment of a rapid and steady rate of lesion development. Researchers occasionally have removed and weighed rotted tissue. Such fresh or dry weights would reflect, in a large measure, host tissue rather than mycelium. Similarly, measurement of the volume of the lesion does not indicate the density of fungal mycelium within the lesion.

Alternative methods have been suggested as possible indicators of the extent of pathogenic growth. Some have suggested respiration measurement (Okazaki and Sugama 1979), but most methods have involved chemical analyses of mycelial components in rotted host tissue. Chitin, a polymer of acetyl glucosamine, is a major component of mycelial walls of many fungi and is found in animals, but is generally not found in higher plant tissues. Chitin content of lesions can be used as a measure of mycelium within host tissue. In analyses, chitin is hydrolyzed to glucosamine and measured colorimetrically or chromatographically (Ride and Drysdale 1972; Wu and Stahmann 1975; Donald and Mirocha 1977).

Seitz *et al.* (1977, 1979) have suggested that the amount of ergosterol could be a measure of fungus growth in higher plant tissues. They have pointed out that ergosterol is the predominant sterol of most fungi and is either absent or sparse in most higher plants. Studies with *Alternaria alternata* (Fr.) Keissler, *Aspergillus flavus* Link, and *A. amstelodami* (Mangin) Thom & Church in cereal seeds led to the development of an analytical procedure for high-pressure liquid chromatography. Extraction procedures for ergosterol are said to be simpler and less time-consuming than for chitin.

C. *In vitro* Studies

Studies of postharvest pathogens in culture provide a direct assessment of the effects of an atmosphere on the physiology of an organism. Eliminated are the indirect suppressive effects of host resistance, which may vary with maturity or condition of the host commodity (which, in turn, may be affected by the atmosphere).

Most studies reviewed herein employed common complex media such as potato-dextrose or malt-extract agar. Chemically defined media are required for many biochemical or physiological studies. Fungi have sometimes been observed to be more dramatically affected by atmospheric modification when grown on defined media instead of complex media (Brown 1922).

Solid media generally have been employed, but some researchers have used liquid media. Fungal growth curves resulting from various O_2 concentrations are often quite different in liquid than in solid media (Mitchell and Zentmyer 1970, 1971a,b; Wells and Uota 1970). Consequently, it is appropriate to review briefly some advantages and problems of solid versus liquid media.

1. Solid Media.—Growth of surface colonies of fungi on solid media has been the subject of many studies, of which only a few of the most pertinent have been included (Trinci 1971a,b; Pirt 1967; Righelato *et al.* 1968). Bull and Trinci (1977) covered the subject of filamentous fungi growth in solid versus liquid media in a recent review.

Trinci (1971a) and Carter and Bull (1971) studied in detail the growth on solid media of *Penicillium chrysogenum* Thom, *Geotrichum lactis* (= *G. candidum* Lk. ex Pers.), *Rhizopus stolonifer* (Ehrenb. ex Fr.) Lind., *Mucor racemosus* Fres., *Absidia glauca* Hagem, *Aspergillus niger* Van Tiegh., *A. wentii* Wehmer, and *A. nidulans* Eidam. Elongation of the hyphae occurred only at the apices. However, a much larger zone, including hyphal segments that have septa with yet unplugged pores, influenced growth by contributing protoplasm by mass transfer to the apex. Thus, the rate of extension was believed to be a function of the length of the terminal portion and the rate of duplication of protoplasm within the peripheral growth zone surrounding the colony behind the apices. Growth behind the peripheral area resulted in increased hyphal density, formation of fruiting structures, and bodies such as sclerotia but did not contribute to the colony's radial expansion.

Bull and Trinci (1977) pointed out that peripheral growth zones are much wider for colonies of fungi (8.5 mm for *Rhizopus stolonifer*) than for bacterial colonies (90 μm for *Escherichia coli* [Migula] Castellani and Chalmers). The wide growth zones result in radial growth rates that usually exceed the rate of diffusion of a chemical within the medium. Thus, secondary metabolites and other products formed at the center of the colony diffuse through the medium more slowly than the colony expands. The peripheral growth zone is thus advancing constantly into fresh, unexploited areas of the medium. Hence, exhaustion of nutrients and accumulation of toxic metabolites affect the central portion of the colony but not the periphery. Bacterial colonies, on the other hand, expand very slowly and secondary metabolites formed at the center of the colony diffuse through the medium and inhibit growth at the periphery.

In experiments involving effects of test atmospheres, hyphae of the peripheral zone are exposed to the atmosphere on or just below the agar surface. Hyphae submerged beneath the central portion of the colony, however, may be subject to anoxia and staling of the medium by secondary metabolites.

In tests of various atmospheres, measurements of colony diameters or radii presumably reflect peripheral growth accurately. However, measurements do not reflect differences in colony height, density, or of those growth processes involved in sporulation or formation of sclerotia. Presumably, methods proposed for estimating mycelia in host tissue could be adapted for growth measurements in solid media.

2. Liquid Media.—Problems associated with use of liquid media are exacerbated when tests involve atmosphere modifications. Difficulties with liquid media include (1) the low solubility of gases, (2) the growth habit of filamentous fungi, and (3) accumulation of secondary metabolites.

a. Solubility of Gases.—The solubility of gases in liquids follows definite patterns (Golding 1945; Brancato and Golding 1953), as follows:

(a) As indicated by Henry's law (Weast *et al.* 1964), the solubility of a gas that will dissolve in a liquid is proportional to the partial pressure of that gas over the liquid, and is independent of the partial pressures of other gases or of total pressure.

(b) Gases are more soluble at low than at high temperatures.

(c) In salt or sugar solutions, gases change in solubility in inverse proportion to the concentration of the solute, provided that the gases are not also soluble in the solute.

Finn (1954), Meynell and Meynell (1965), Carter and Bull (1969), Corman *et al.* (1957), Darby and Goddard (1950), Lockhart and Squires (1963), Roels *et al.* (1974), and Starks and Koffler (1949) have reviewed problems of O_2 supply and have pointed out that O_2 is relatively insoluble in water. Distilled water saturated with air at one atmosphere at 30°C contains only 240 μM/liter of dissolved O_2 (Harrison 1976). Atmospheres of 2% O_2 (16 mm Hg) or 1% O_2 (8 mm Hg) contain 24 μM/liter and 12 μM/liter, respectively. However, fungal respiration and growth are usually about equal between O_2 tensions of 160 mm and 20 mm Hg. The critical tension at which growth rates start to drop rapidly was found to be about 7 mm Hg (10.5 μM/liter) for *Aspergillus nidulans* (Carter and Bull 1969, 1971), although the critical tensions appeared to vary with the previous growth history of the fungus.

Commonly, 2% O_2 is used in MA tests with fruits and vegetables. Culture broths saturated with such an atmosphere may approach the critical dissolved O_2 tension for many fungi. If the rate of O_2 diffusion into the medium cannot meet respiratory demands of an organism, levels of dissolved O_2 in the medium may fall drastically. Inadequate diffusion causes the O_2 concentration of liquid media to be far below equilibrium with the partial pressure of the O_2 of the atmosphere. A lower-than-expected growth rate is a consequence. Thus, problems of O_2 diffusion in liquid media could result in apparent growth suppression at relatively high O_2 atmospheric pressures.

CO_2 dissolves in liquid media to form carbonic acid, which disassociates to form bicarbonate and carbonate ions as follows:

$$CO_2 + H_2O \rightleftharpoons H_2CO_3$$
$$H_2CO_3 \rightleftharpoons H^+ + HCO_3^-$$
$$HCO_3^- \rightleftharpoons H^+ + CO_3^{2-}$$

The concentration of bicarbonate ions, which affect growth, increases about 10-fold for every increase of one pH unit. Therefore, it is essential that cultures be buffered to avoid important pH shifts.

b. Fungal Growth Habit.—In liquid media, filamentous fungi clump under the usual conditions of a shake-flask culture. Spores added to such liquid media may be well dispersed initially but under conditions favorable to germination, clumping becomes noticeable after germ-tube protrusion. Germ tubes become elongated and, after additional growth, the resulting mycelium becomes clumped into masses of various sizes and shapes (Trinci 1971b). Researchers have long suspected that conditions of low O_2, or of anoxia, existed within such clumps because O_2 uptake exceeds penetration of dissolved O_2. Bungay *et al.* (1969) obtained direct supporting evidence for that supposition by using an O_2 electrode 1.5 μ in diameter. The very low O_2 level within a 200 μ-thick bacterial film was characteristic of respiration limited by O_2 concentration. Therefore, mycelial clumping results in less growth than would otherwise be expected from the dissolved oxygen concentration of the fluid or from the O_2 partial pressure of the atmosphere.

c. *Metabolite Accumulation.*—Fungi produce secondary metabolites that accumulate in the media. Some of these either kill or otherwise affect the fungus. Bu'Lock (1975) reviewed the effects of secondary metabolites on growth and development. Secondary metabolites are dispersed thoroughly in a liquid when the medium is agitated for aeration. To prevent an excessive accumulation of these growth-suppressing products, fresh medium must be added constantly to replace an equal volume of old medium with its suspended cells. Single-celled organisms such as bacteria and yeasts are adapted to such a system. Unfortunately, the growth habit of filamentous fungi seriously complicates the establishment of a steady state. Chemostats for continuous culture of several filamentous fungi have been developed in the fermentation industry for the commercial production of antibiotics and other chemicals. Considerable effort has been devoted to studies of fungal growth kinetics and the problems of use of continuous cultures. These have been reviewed by Righelato (1975), Righelato and Pirt (1967), and Bull and Trinci (1977).

3. Anaerobic Conditions.—Tests designed to determine if an organism can grow in the absence of molecular O_2 may lead to erroneous results if O_2 contaminates the culture from unsuspected sources. Although serious problems are posed by the low solubility of O_2 in aqueous solutions and inadequate diffusion throughout the media and fungus structures, the removal of all molecular O_2 poses equally serious difficulties. Small amounts of O_2, unnoticed and unimportant in comparisons of O_2 levels in atmospheres for fruit or vegetable storage, may prevent the achievement of anaerobic conditions. Sources of contamination are varied. Media contain dissolved O_2 unless thoroughly purged. Air is introduced when flasks are opened to permit inoculation of cultures. Commercially supplied N_2 contains O_2 that can be removed by passing it through hot copper filings, bubbling it through alkaline pyrogallol, or both. Small amounts of O_2 may permeate rubber tubing. Furthermore, Claypool and Keefer's (1942) flow meter system, widely used in postharvest laboratories, provides considerable contact with water that may contain dissolved O_2 (Bussel *et al.* 1968, 1969a,b, 1971).

III. ATMOSPHERIC EFFECTS *IN VITRO*

A. Oxygen Effects

1. General.—Respiration accounts for the greatest involvement of O_2 in aerobic organisms, where it serves as an electron acceptor during substrate oxidation. In addition to a tricarboxylic acid cycle, most fungi have an electron-transport system that closely resembles that of animals or higher plants. Fungal electron transport involves a flow of electrons from $NADH_2$ and substrate through a system that includes cytochromes b, c, and aa_3, and contains up to three energy-conserving sites (Bull and Trinci 1977). The cytochromes are hemoproteins found in mitochondrial membranes. Their principal biological function is to transport electrons, which they do by means of a reversible valency change of their prosthetic group, tetrapyrollic chelate of iron ($Fe^{++} \rightleftharpoons Fe^{+++}$) (Bull and Trinci 1977; Harrison 1976; Lindenmayer 1965; Meynell and Meynell 1965).

In contrast with eukaryotic organisms such as fungi, bacteria have a great variety of cytochrome systems. Aside from the variety of cytochromes found in different species, it is not unusual for bacteria to have up to three different terminal cytochrome oxidases (Harrison 1976; Lindenmayer 1965). Other oxidative enzyme systems exist in cells and are probably affected by low-oxygen tensions, but their impact on growth suppression is less well known and normally involves much less O_2 than does the cytochrome system (Harrison 1976).

Bussel *et al.* (1969a) studied the effects of absence of O_2 on the ultrastructure of *Rhizopus stolonifer* sporangiospores. Spores were incubated for 3 hours with or without subsequent periods of anoxia. Of particular note was the condition of mitochondria, which were distributed randomly throughout the cell and exhibited well developed parallel cristae. After 72 hours of anoxia, mitochondria were few, small with indistinct cristae, and they were located at the periphery of the cytoplasm. The cytoplasm had become very mottled and had many vacuoles and vesicles. Lipid bodies were abundant. Upon returning the spores to air, mitochondria were again seen in the central portion of the cell. They had enlarged, developed prominent cristae, and otherwise resembled mitochondria of spores that had not been subjected to anoxia. In further studies, Bussel (1969b) showed that spores of *R. stolonifer* do not germinate if O_2 is rigorously excluded. Spores incubated 1, 2, or 3 hours, and spores given no such incubation, were insensitive to 72 hours of anoxia in liquid medium at 23° to 25°C. However, spores incubated for 4, 5, and 6 hours were very sensitive to 72 hours of anoxia; only 8, 0.2, and 0.09% were viable, respectively.

Research on the effects of O_2 at partial pressures above the 21% O_2 content in air has been reported by ZoBell and Hittle (1967) and reviewed by Haugaard (1968) and Harrison (1976). Toxicity to high O_2 tensions is widespread among animals, higher plants, and microorganisms. Among the latter, the susceptibility to high O_2 tensions varies, but probably no microorganism is completely insensitive to the toxic effects of O_2. Among anaerobic bacteria, which cannot grow except under strictly reduced conditions, any O_2 is toxic. Some aerobic bacteria and fungi can grow in atmospheres of near 100% O_2.

Haugaard (1968) attributed O_2 toxicity to such damaging events as the oxidation of sulfhydryl groups in various compounds, such as gluthion, or of enzymes. He also suggested the possibility of peroxide formation, which can cause extensive destruction (particularly of lipids). Harrison (1976) mentioned work suggesting the possibility of free radical accumulation. Chance *et al.* (1965) reported on the toxic effects of high O_2 on intracellular oxidation states of reduced pyridine nucleotide in general, and on the energy-linked pathway for pyridine nucleotide reduction in particular. The belief that O_2 exerts its toxic effect through the formation of superoxide radicals (O_2^-), which destroy some aspects of cell metabolism (Harrison 1976; Gregory and Fridovich 1973), has recently gained support.

Robb (1966) subjected 103 species of fungi to 10 atmospheres pressure of O_2 at 25°C for 7 or 14 days. Among postharvest disease fungi studied, *Aspergillus flavus* and *A. niger* grew after cultures were exposed to high O_2 for 14 days. *Cladosporium herbarum* Link ex Fr., *Colletotrichum dematium* (Fr.) Grove, *Trichoderma viride* Pers. ex Fr., and *Mucor racemosus* survived 7 days. The following fungi were not able to withstand 7 days of high O_2 pressure: *Phytophthora cactorum* (Leb. & Cohn) Schroet., *Rhizopus arrhizus* Fischer, *R. oryzae* Went & Prin., *R. stolonifer, Glomerella cingulata* (Stonem.) Spauld. & Schr., *Sclerotinia fructigena* (Pers.) Schr., *Alternaria tenuis* (=*A. alternata*), *Botrytis cinerea* Pers. ex Fr., and *Trichothecium roseum* Link. A lag before resumption of normal growth was frequently observed in fungi surviving the high O_2 treatments.

2. Low Oxygen at Normal Atmospheric Pressure.—The quantitative relations of postharvest fungi and O_2 supply vary considerably among species. The reduction of O_2 required for growth and/or spore-germination inhibition varies with the species. Table 10.1 includes a summary of the literature on the effects of various O_2 concentrations, and from them the following may be concluded:

Oxygen concentrations of about 1% or less are required to obtain appreciable reduction of growth, spore formation, and germination in many postharvest fungi. In most fungi, no growth occurs without molecular O_2 (Ashworth *et al.* 1969; Brown 1922; Brown and Kennedy

TABLE 10.1. THE EFFECTS OF LOW O_2 CONCENTRATIONS IN THE PRESENCE OR ABSENCE OF ABOUT 0.03% CO_2 ON COMMON POSTHARVEST PATHOGENS (LM=liquid medium, SM=solid medium, FS=flowing atmosphere system and SS=static atmosphere system)

Species	O_2 Concentration (%)	Remarks	Reference
Alternaria alternata (Fr.) Keissler. Syn., *A. tenuis* Nees.	0.5	Growth reduced by 45% SM-FS at 15°C	Follstad 1966
	0	Growth absent SM-FS at 15°C	Follstad 1966
	4	Growth reduced by 69% and decreased linearly with decreasing O_2 below 4%. Sporulation inhibited at low O_2 LM-FS at 19°C	Wells and Uota 1970
	0	No growth LM-FS at 19°C	Wells and Uota 1970
	2.3	Growth not affected SM-FS at 5.5° or 12.5°C	El-Goorani and Sommer 1979
Ascochyta caricae-papayae Pat. Probable sexual state, *Mycosphorella caricae* Syd.	2.3	Growth reduced by 20% SM-FS at 5.5° or 12.5°C	El-Goorani and Sommer 1979
Aspergillus niger Van Tiegh.	Low	Good growth and spore germination SM-SS	Brancato and Golding 1953
Aspergillus flavus Link	2	Growth and aflatoxin production low in cotton seed	Ashworth *et al.* 1969
	2	Good growth and aflatoxin production SM-FS at 20°C	Buchanan *et al.* (unpublished data)
Botryodiplodia theobromae Pat.	2.3	Growth reduced by 15% SM-FS at 12.5°C	El-Goorani and Sommer 1979

Botrytis cinerea Pers. ex Fr. Sexual state, *Botryotinia fuckeliana* (DeBary) Whetz.	0.5	Growth reduced by 50%. Sporulation absent in 1% or less SM-FS at 15°C	Follstad 1966
	4	Growth reduced by 55%. Sporulation inhibited at low O_2 LS-FS at 19°C	Wells and Uota 1970
	2.3	Growth not affected SM-FS at 5.5° and 12.5°C	El-Goorani and Sommer 1979
Cladosporium herbarum Link ex Fr.	0.5	Growth reduced by 65%. Sporulation at low O_2 SM-FS at 15°C	Follstad 1966
	4	Growth reduced by 50% LS-FS at 19°C	Wells and Uota 1970
Colletotrichum gloeosporioides (Penz.) Van Arx Sexual state, *Glomerella cingulata* (Stonem.) Spauld. & Schr.	2.3	Growth reduced *ca.* 10% SM-FS at 12.5°C	El-Goorani and Sommer 1979
Fusarium roseum (Link) emend. Snyd. & Hans.	4	Growth reduced by 62% LM-FS at 19°C	Wells and Uota 1970
Geotrichum candidum Link ex Pers. Syn., *Oospora lactis* (Fres.) Sacc.	1 and 3	Growth as great as or greater than in air SM-FS at 21°C	Wells and Spalding 1975
	2.3	Growth stimulated by 10% and 20% SM-FS at 5.5° and 12.5°C	El-Goorani and Sommer 1979
Gloeosporium album Osterw. Syn., *Phlyctaena vagabunda* Desm. Sexual state, *Pezicula alba* Guth.	2.5	Growth retarded LM	Lockhart 1969

TABLE 10.1. *(Continued)*

Species	O_2 Concentration (%)	Remarks	Reference
Monilinia fructicola (Wint.) Honey Syn., *Sclerotinia fructicola* (Wint.) Rehm.	Anoxia	Viability not lost after 24 or 48 hours, but after 72 hours only 50% grew when placed in air LM-FS at 23° to 25°C	Bussel *et al.* 1971
	2.3	Growth reduced by 40% SM-FS at 5.5° and 12.5°C	El-Goorani and Sommer 1979
Penicillium digitatum Sacc. and *P. italicum* Wehmr.	2.3	Growth slightly reduced SM-FS at 5.5° and 12.5°C	El-Goorani and Sommer 1979
Penicillium expansum Link ex Thom.	Low	Good germination and growth SM-SS	Brancato and Golding 1953
	2.3	Growth unaffected SM-FS at 5.5° and 12.5°C	El-Goorani and Sommer 1979
Phomopsis citri Faw. Sexual state, *Diaporthe citri* (Faw.) Wolf	2.3	Growth not affected SM-FS at 5.5° and 12.5°C	El-Goorani and Sommer 1979
Phytophthora cactorum (Leb. & Cohn) Schroet.	21	Maximum growth LM-FS at 22° to 24°C	Covey 1970
	2.3	Growth stimulated by 16% and 20% at 5.5° and 12.5°C SM-FS	El-Goorani and Sommer 1979
Phytophthora citrophthora (R.E. Sm. & Br.) Leonian	1.6	Growth occurred LM-FS at 25.5° to 27.5°C	Klotz *et al.* 1963

	0.04	Zoospores did not germinate	Klotz *et al.* 1963
	5	Growth more rapid than in air SM-FS at 25°C	Mitchell and Zentmyer 1971a
Phytophthora parasitica Dast. Syn., *P. nicotianae* var. *parasitica* (Dast.) Waterh.	0.0	No germination LM-FS	Klotz *et al.* 1963
	1.6	Good growth occurred LM-FS at 25.5° to 27.6°C	Klotz *et al.* 1963
	6	Growth greater than in air SM-SS at 28°C	Dukes and Apple 1965
	0.1	Growth very little SM-SS at 28°C	Dukes and Apple 1965
	21	Growth maximum LM-FS at 25°C	Mitchell and Zentmyer 1971a
Phytophthora spp.	Low O_2	Zoospores formed SM-SS	Uppal 1924, 1926
	1	Growth occurred SM and LM-FS at 25°C	Mitchell and Zentmyer 1971a
Pythium sp.	1.3	Growth reduced LM-FS at 20° to 22°C	Brown and Kennedy 1966
	4 and above	No growth reduction LM-FS at 20° to 22°C	Brown and Kennedy 1966
Rhizopus stolonifer (Ehrenb. ex Fr.) Lind. Syn., *R. nigricans* Ehren.	0	No germination LM-FS at 20°C	Wood-Baker 1955
	2	Germination reduced 18% from air control LM-FS at 20°C	Wood-Baker 1955
	0.5	Growth reduced by 50%. No mature sporangia in 5 days SM-FS at 15°C	Follstad 1966
	0	Growth occurred SM-FS at 15°C	Follstad 1966
	2	Growth reduced 60 to 70% LM-FS at 18°C	Wells 1967, 1968
	0.25	No growth LM-FS at 18°C	Wells 1967, 1968
	0	No growth LM-FS at 18°C	Wells 1967, 1968
	Anoxia	Germinating sporangiospores inactivated but tolerated by non-germinating spores	Buckley *et al.* 1967

TABLE 10.1. *(Continued)*

Species	O_2 Concentration (%)	Remarks	Reference
	4	Growth reduced by 15% LM-FS at 19°C	Wells and Uota 1970
	0	Growth observed LM-FS at 19°C	Wells and Uota 1970
	2.3	Growth not affected SM-FS at 5.5° and 12.5°C	El-Goorani and Sommer 1979
Thielaviopsis paradoxa (de Seynes) Höhnel Sexual state, *Ceratocystis paradoxa* (Dade) Moreau.	2.3	Growth not affected SM-FS at 12.5°C	El-Goorani and Sommer 1979
Whetzelinia sclerotiorum (Lib.) Korf & Dumont Syn., *Sclerotinia sclerotiorum* (Lib.) Mass.	2.3	Growth reduced by 50% SM-FS at 5.5° and 12.5°C	El-Goorani and Sommer 1979
Erwinia carotovora (Jones) Holland, *Erwinia atroseptica* (Van Hall) Jennison, and *Pseudomonas fluorescens* Migula	3	Growth *ca.* 50 to 60% of air decreased linearly to 0.25% LM-FS at 21°C	Wells 1974

1966; Follstad 1966; Wells and Uota 1970). However, some fungi were found to grow as rapidly in low O_2 atmospheres as in air. According to Hawker *et al.* (1960), a pressure as low as 0.01 atmosphere " . . . will suffice for many species of Penicillium growing on a solid medium." Follstad (1966) and Wells and Uota (1970) reported that *Rhizopus stolonifer* grew significantly at 0% O_2. Follstad (1966) added that gas chromatography detected a trace of O_2 in highly purified N_2, and that this amount possibly accounted for the measurable growth at 0% O_2.

These studies effectively demonstrated the ability of *R. stolonifer* sporangiospores to germinate at very low O_2 levels. Buckley *et al.* (1967) reported that holding *R. stolonifer* sporangiospores in anoxia not only prevented germination and growth but inactivated those that were germinating. Bussel *et al.* (1968, 1969b) showed that the viability of *R. stolonifer* spores was related to the stage of germination and length of exposure to anaerobiosis. In his studies, he purified the commercial N_2 (99.998% pure) of contaminating air by passing it twice through red-hot copper turnings before passing it through flow meters. Subsequently, the N_2 was bubbled twice through reduced cysteine hydrochloride and resazurin before being passed through the headspace of shaken flasks. The system was purged of O_2 by passage of O_2-free N_2 gas through the flow meters and culture flask. Glass or Tygon tubing was used throughout to minimize absorption of O_2, which can occur through rubber. Bussel *et al.* (1969a) showed that *R. stolonifer* underwent prominent ultrastructural changes under anoxia. For example, mitochondria were found scattered randomly throughout the cytoplasm in spores incubated aerobically for 3 hours, but they moved to the periphery of the cytoplasm when spores were exposed to anoxia.

It is clear from Table 10.1 that *Phytophthora cactorum* (El-Goorani and Sommer 1979), *P. citrophthora* (R.E. Sm. & Br.) Leonian (Mitchell and Zentmyer 1971a), and *P. parasitica* Dast. (Dukes and Apple 1965) have optimum growth rates at low O_2 concentrations when grown on agar, but have optimum growth rates at 21% O_2 in liquid medium (Covey 1970; Klotz *et al.* 1963). Mitchell and Zentmyer (1971a) reported that the reasons for this difference are unknown. Mycelium is more exposed to the atmosphere when grown on agar than in the submerged colony in the liquid medium. However, because only 15 ml of liquid medium per 250 ml flask was used in their study, the upper surface of the mat was always exposed to the atmosphere. Therefore, they believed that O_2 was probably not limiting. They suspected that inhibitory compounds accumulate at low O_2 levels and that these compounds cannot diffuse as rapidly through agar as through liquid.

Atmospheres low in O_2 permitted *Geotrichum candidum* to grow faster than in air when grown in liquid (Wells and Spalding 1975) or on solid media (El-Goorani and Sommer 1979).

3. Low Oxygen at Low Pressure (Hypobaric) Conditions.—Wu and Salunkhe (1972a) compared mycelial extension in potato-dextrose agar at 278 mm and 102 mm Hg with the normal atmospheric pressure of 646 mm Hg at their location. When air was withdrawn to produce the low pressure, the amount of available O_2 was effectively reduced. An atmosphere of 278 mm and 102 mm Hg would contain O_2 equivalent to 7.7% and 2.7% O_2, respectively, at normal atmospheric pressure. Generally, as pressures decreased, there was a slight suppression of colony size, amount of sporulation, or an increase in time for growth to be detected. When compared with 2.7% O_2 at their normal atmospheric pressure (646 mm Hg), growth and sporulation at 102 mm Hg were less and the time required for growth to appear was greater.

Borecka and Olak (1978) compared the growth and sporulation of *Penicillium expansum* Link ex Thom, *P. spinulosum* Thom, *P. diversum* Raper and Fennell, *Botrytis cinerea, Trichothecium roseum,* and *Rhizopus nigricans* (=*R. stolonifer*) at normal atmospheric pressure to that at 0.1 and 0.05 atmospheres. They reported that growth of *Penicillium expansum* at 0.1 atmosphere was reduced to about 85% of that in air, while at 0.05 atmosphere it was about 40% of the growth in air. Sporulation of the Penicillia was not affected by low pressure. *Rhizopus stolonifer* developed normally at 0.1 atmosphere, but it did not sporulate at 0.05 atmosphere. With *Botrytis cinerea* and *Trichothecium roseum*, no suppression was noted in growth at 0.1 atmosphere, but growth was reduced by one-half or more at 0.05 atmosphere. Although mycelial growth of *Botrytis cinerea* was normal at 0.1, no sclerotia were produced.

It has been possible to compare (Fig. 10.1) the growth of *Botrytis cinerea* at low pressures (Borecka and Olak 1978) to growth at normal barometric pressure in atmospheres of low O_2 composition (Couey *et al.* 1966; Follstad 1966; Sommer *et al.*, in press). Growth was remarkably similar at comparable O_2 levels regardless of barometric pressure. Therefore, the evidence suggests that suppression of *Botrytis cinerea* was due to a reduction of the partial pressure of O_2, and not to low total pressure. Definitive conclusions must await further comparisons with other postharvest pathogens.

B. Carbon Dioxide Effects

Although CO_2 is low in air (*ca.* 0.03%), Hartman *et al.* (1972) showed that growth of *Verticillium albo-atrum* Reinke & Berth. was severely curtailed in a CO_2-free atmosphere in which glucose or glycerol was the sole carbon source. With increasing concentration, the role of CO_2 changes from stimulatory to inhibitory. While fungi can fix CO_2, it cannot be used as an exclusive source of carbon for metabolism (Burnett

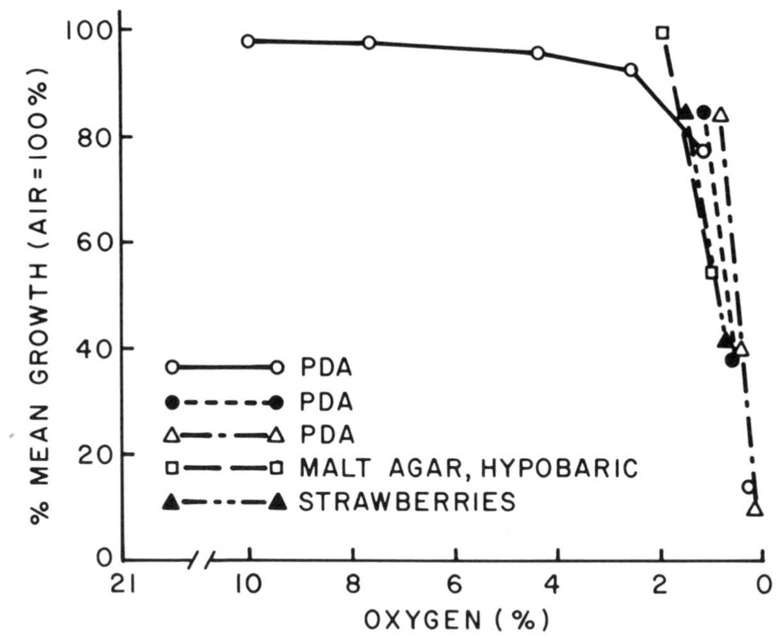

From Borecka and Olak 1978; Couey et al. 1966; Follstad 1966; Sommer et al. (in press)

FIG. 10.1. EFFECT OF LOW OXYGEN ON GROWTH OF *BOTRYTIS CINEREA* IN CULTURE (POTATO-DEXTROSE AGAR [PDA]) OR ON STRAWBERRY FRUITS

All tests except hypobaric were at normal barometric pressure in an atmosphere of low oxygen composition. Hypobaric was in air at 0.1 atmospheres and 0.05 atmospheres.

1968). Several enzymes have been implicated in CO_2 fixation in fungi (Bull and Trinci 1977). Similarly, elevated CO_2 evidently suppresses metabolic functions at multiple locations. Acidification of media by CO_2 was once considered to be an important cause for fungal suppression (Thornton 1934), but more recent evidence suggests that pH changes do not necessarily play a dominant role (Bull and Trinci 1977; Lwoff and Monod 1947).

The amount of CO_2 required to inhibit growth and/or spore-germination varies with the species (Cochrane 1958; Cochrane *et al.* 1963). Table 10.2 summarizes the reported effects of various concentrations; from it, the information suggests that high levels of CO_2 are required to slow growth and/or germination of fruit-rotting fungi. Some effects of CO_2 on bacterial growth are reported by Coyne (1933), and Lwoff and Monod (1947) provided an early review of the role of CO_2 in microbial growth.

TABLE 10.2. EFFECTS OF HIGH CO_2 CONCENTRATION IN AIR ON COMMON POSTHARVEST PATHOGENS (LM=liquid medium, SM=solid medium, FS=flow system and SS=static atmosphere system)

Species	CO_2 Concentration (%)	Remarks	Reference
Alternaria alternata (Fr.) Keissler. Syn., *A. tenuis* Nees	20	Growth reduced by about 50% LM-FS at 19°C	Wells and Uota 1970
	32	Spore germination inhibited LM-FS at 19°C	Wells and Uota 1970
	5	Growth not reduced SM-FS at 19°C	El-Goorani and Sommer 1979
	18	Growth reduced by 65% and 15% at 5.5° and 12.5°C, respectively SM-FS at 5.5° and 12.5°C	El-Goorani and Sommer 1979
Alternaria solani (Ell. and G. Martin) Sor.	>0.5	Growth inhibition increased with increasing concentration	Klaus 1941
Ascochyta caricae-papayae Pat. Probable sexual state, *Mycosphorella caricae* Syd.	5	Growth not reduced SM-FS at 5.5° and 12.5°C	El-Goorani and Sommer 1979
	18	Growth reduced by about 70% SM-FS at 5.5° and 12.5°C	El-Goorani and Sommer 1979
Aspergillus niger Van Tiegh.	0	Spores germinated very poorly	Rippel and Bortels 1927
	0.5	Spore germination stimulated LM-SS	Vakil *et al.* 1961
	3	Spore germination inhibited SM-SS	Vakil *et al.* 1961
		Growth inhibited SM-SS	Brancato and Golding 1953
	20	Growth inhibited SM-SS (gases changed daily)	Golding 1945
Aspergillus flavus Link	<80	Growth and sporulation not visibly affected. Peanuts—FS at 15°C	Landers *et al.* 1967

	20 to 80	Aflatoxin production suppressed. Peanuts—FS at 15°C	Landers *et al.* 1967
Botryodiplodia theobromae Pat.	5	Growth unaffected SM-FS at 12.5°C	El-Goorani and Sommer 1979
	18	Growth reduced by 25% SM-FS at 12.5°C	El-Goorani and Sommer 1979
Botrytis cinerea Pers. ex Fr. Sexual state, *Botryotinia fuckeliana* (DeBary) Whetzl.	10 and 20	Germination of conidia suppressed LM-SS at 15° to 18°C	Brown 1922
	30	Germination prevented LM-SS at 15° to 18°C	Brown 1922
	10	Growth suppressed LM-SS at 15° to 18°C	Brown 1922
	23	Growth suppressed almost completely	Brooks *et al.* 1932
	20	Growth suppressed by about 50% LM-FS at 19°C	Wells and Uota 1970
	5	Growth unaffected SM-FS at 5.5° and 12.5°C	El-Goorani and Sommer 1979
	18	Growth suppressed 86% at 5.5°C and 69% at 12.5°C SM-FS	El-Goorani and Sommer 1979
Cladosporium herbarum Link ex Fr.	20	Growth reduced by 50% LM-FS at 19°C	Wells and Uota 1970
	16	Spore germination inhibited by about 90% LM-FS at 19°C	Wells and Uota 1970
Colletotrichum gloeosporioides (Penz.) Van Arx Sexual state, *Glomerella cingulata* (Stonem.) Spauld. & Schr.	5	Growth reduced by 10% SM-FS at 12.5°C	El-Goorani and Sommer 1979
	18	Growth reduced by 77% SM-FS at 12.5°C	El-Goorani and Sommer 1979

TABLE 10.2. *(Continued)*

Species	CO_2 Concentration (%)	Remarks	Reference
Colletotrichum lindemuthianum (Sacc. & Magn.) Boriosi and Cav.			
		Little response to CO_2 treatment Beans and SM-SS	Brooks *et al.* 1936
Dothiorella gregaria Sacc.			
	18	Growth reduced 85% and 28% at 5.5° and 12.5°C, respectively SM-FS	El-Goorani and Sommer 1979
Fusarium sp.			
	20	Growth reduced 15 to 21% SM-FS at 25°C	Durbin 1959
	60	Spores germinated strongly SM-SS	Brown 1922
Fusarium roseum (Link) emend. Snyd. & Hans.			
	10	Growth stimulated LM-FS at 19°C	Wells and Uota 1970
	45	Growth inhibited about 50% LM-FS at 19°C	Wells and Uota 1970
	16	Spore germination stimulated LM-FS at 19°C	Wells and Uota 1970
Geotrichum candidum Link ex Pers. Syn., *Oospora lactis* (Fres.) Sacc.			
	5	Growth unaffected SM-FS at 5.5° and 12.5°C	El-Goorani and Sommer 1979
	18	Growth suppressed by about 30% SM-FS at 5.5° and 12.5°C	El-Goorani and Sommer 1979
Monilinia fructicola (Wint.) Honey Syn., *Sclerotinia fructicola* (Wint.) Rehm.			
		Spore germination suppressed by high CO_2 LM-SS at 15° to 18°C	Brown 1922
	10 to 20	Growth inhibited LM and SM	Thornton 1934

	5	Growth stimulated SM-FS at 5.5° and 12.5°C	El-Goorani and Sommer 1979
	18	Growth reduced by 65% SM-FS at 5.5° and 12.5°C	El-Goorani and Sommer 1979
Mucor sp.		Spore germination suppressed by high CO_2 LM-SS at 15° to 18°C	Brown 1922
Mucor mucedo Fr.	10	Spore germination inhibited	Lopriore 1895
	100	Spore germination totally inhibited. Spores not killed by 3-month exposure	Lopriore 1895
Nectria galligena Bers.	5	Growth stimulated SM-FS at 22°C	Swinburne 1974
	10	Growth suppressed SM-FS at 22°C	Swinburne 1974
Penicillium digitatum Sacc. and *P. italicum* Wehmer	5	Growth unaffected SM-FS at 5.5° and 12.5°C	El-Goorani and Sommer 1979
	18	Growth suppressed by 70% and 37% at 5.5° and 12.5°C, respectively SM-FS	El-Goorani and Sommer 1979
Penicillium expansum Link ex Thom	5	Growth unaffected SM-FS at 5.5° and 12.5°C	El-Goorani and Sommer 1979
	18	Growth suppressed 55 to 65% SM-FS at 5.5° and 12.5°C	El-Goorani and Sommer 1979
Penicillium glaucum Link	10	Growth suppressed SM-SS at 15°C	Brown 1922
Phomopsis citri Faw. Sexual state, *Diaporthe citri* (Faw.) Wolf	5	Growth unaffected SM-FS at 5.5° and 12.5°C	El-Goorani and Sommer 1979

TABLE 10.2. *(Continued)*

Species	CO_2 Concentration (%)	Remarks	Reference
	18	Growth suppressed 65% (5.5°C) and 15% (12.5°C) SM-FS at 5.5° and 12.5°C	El-Goorani and Sommer 1979
Phomopsis vexans (Sacc. and Syd.) Harter		Growth completely suppressed at high CO_2	Brooks *et al.* 1936
Phytophthora cactorum (Leb. & Cohn) Schroet.	20	Growth inhibited 25% SM-FS at 25°C	Durbin 1959
	5	Growth unaffected SM-FS at 5.5° and 12.5°C	El-Goorani and Sommer 1979
	18	Growth suppressed 70 to 80% SM-FS at 5.5° and 12.5°C	El-Goorani and Sommer 1979
Phytophthora citrophthora (R.E. Sm. & Br.) Leonian	15	Growth reduced about 30% LM and SM-FS at 25°C	Mitchell and Zentmyer 1971a
Phytophthora parasitica Dast. Syn., *P. nicotianae* var. *parasitica* (Dast.) Waterh.	15	Growth unaffected SM-SS at 28°C	Dukes and Apple 1965
	99	Growth completely suppressed. Mycelia not killed after 7 days SM-SS at 28°C	Dukes and Apple 1965
	15	Growth suppressed by 1/3 LM and SM-FS at 25°C	Mitchell and Zentmyer 1971a
	18	Growth suppressed by 40% SM-FS at 5.5° and 12.5°C	El-Goorani and Sommer 1979

Phytophthora spp.		Sporangia and oospore formation reduced with increasing CO_2 when O_2 was 1, 5 or 20% LM and SM-FS at 22° to 25°C	Mitchell and Zentmyer 1971b
Rhizoctonia solani Kuhn	20	Growth suppressed 31 to 80% SM-FS at 25°C	Durbin 1959
		Growth suppressed by high CO_2	Brooks *et al.* 1936
	20	Growth moderately suppressed in soil FS	Papavizas and Davey 1962a,b
	30	Growth suppressed drastically in soil FS	Papavizas and Davey 1962a,b
Rhizopus stolonifer (Ehrenb. ex Fr.) Lind. Syn., *R. nigricans* Ehren.		Germination suppressed by high CO_2 LM-SS at 15° to 18°C	Brown 1922
	23	Growth nearly completely suppressed	Brooks *et al.* 1932
	20	Growth reduced by about 50% LM-FS at 19°C	Wells and Uota 1970
	16	Spore germination suppressed *ca.* 90% LM-FS at 19°C	Wells and Uota 1970
	18	Growth suppressed by 93% (5.5°C) and 75% (12.5°C) SM-FS	El-Goorani and Sommer 1979
Sclerotinia minor Jagger	23.5	Growth suppressed 85% SM-FS at 21°C	Louvet and Bulit 1964
Sclerotium rolfsii Sacc.	20	Growth suppressed 55% SM-FS at 25°C	Durbin 1959
		Growth suppressed by high CO_2	Brooks *et al.* 1936

TABLE 10.2. *(Continued)*

Species	CO_2 Concentration (%)	Remarks	Reference
Thielaviopsis paradoxa (de Seynes) Höhnel Sexual state, *Ceratocystis paradoxa* (Dade) Moreau.	18	Growth suppressed by 55% SM-FS at 12.5°C	El-Goorani and Sommer 1979
Whetzelinia sclerotiorum (Lib.) Korf & Dumont Syn., *Sclerotinia sclerotiorum* (Lib.) Mass.	18	Growth reduced by 80% SM-FS at 12.5°C	El-Goorani and Sommer 1979
Erwinia carotovora (Jones) Holland, *Erwinia atroseptica* (Van Hall) Jennison, *Pseudomonas fluorescens* Migula	30	Growth reduced by 15 to 30% LM-FS at 21°C	Wells 1974
	0	*Erwinia* spp. did not grow at any concentration of O_2 within 24 hours LM-FS at 21°C	Wells 1974

C. Low Oxygen with High Carbon Dioxide

Several investigators studied the effect of low O_2 in the presence of
CO_2 concentrations on suppression of growth of postharvest pathogen
Organisms studied included the following: *Botrytis alii* Munn, *Rhizopus stolonifer, Penicillium expansum* (Littlefield *et al.* 1966), *Gloeosporium album* Osterw. (Lockhart 1967), *Alternaria tenuis, Botrytis cinerea, Sclerotinia sclerotiorum* (=*Whetzelinia sclerotiorum* (Lib.) Korf & Dumont) (Adair 1971), *Erwinia* spp., and *Pseudomonas fluorescens* Migula (Wells 1974). On the other hand, others have observed that atmospheres of low O_2 and high CO_2 have little effect in slowing the mycelial growth of *Rhizoctonia solani* Kuhn, *Fusarium roseum* (Link) emend. Snyd. & Hans. (Adair 1971), and *Pezicula alba* Guth. (Bompeix 1978a). In fact, Wells and Spalding (1975) found that in an atmosphere of 3% O_2 with 3% CO_2, the average growth of *Geotrichum candidum* in 24 hours was about twice that in air. They added that at concentrations of 3% or more, CO_2 repressed growth of *G. candidum* in the presence of 21% O_2. In a low O_2 atmosphere, however, 30% CO_2 was necessary to repress growth.

Yackel *et al.* (1971) showed that the effect of CA (10.5% CO_2 + 2% O_2) on growth and development of 7 spoilage molds (*Mucor hiemalis* Wehmer, *Rhizopus oryzae, Aspergillus niger, Cladosporium herbarum, Penicillium expansum, Alternaria* sp., and *Fusarium bulbigenum* Cooke & Massee) is quite variable and, for the most part, is temperature related.

When O_2 concentrations were 1% or 2%, and thereby limiting to growth, CO_2 at low levels stimulated growth and/or germination of *Gloeosporium album* (Lockhart 1967), *Alternaria tenuis, Botrytis cinerea, Cladosporium herbarum, Rhizopus stolonifer,* and *Fusarium roseum* (Wells and Uota 1970). Mitchell and Zentmyer (1970, 1971a) found that growth of *Phytophthora* spp. on solid media was stimulated by the addition of 5% CO_2 when the O_2 level was 1%. However, CO_2 concentrations above 5% reduced growth of most isolates in liquid or solid media.

IV. ATMOSPHERIC EFFECTS ON POSTHARVEST DISEASES

Pathogenic growth of bacteria and fungi may be affected by MA through altered physiology of the organism. An important influence on disease development is the physiological condition of the host commodity. Commonly, fruits are resistant to most postharvest pathogens until the initiation of ripening. Similarly, tissues other than fruits may be very resistant until they become senescent. Consequently, MA-induced delay of ripening or senescence to thereby maintain host resistance may be highly important in disease suppression.

438 stigators have studied the effect of low O_2 concentrations vest decay during storage of fruits and vegetables. Decay of awberries caused by *Botrytis cinerea* was decreased by lowering concentration to 0.5% or less. Berries were evaluated after 5 days C in the low O_2 atmosphere and after an additional 2 days at 15°C in (Couey *et al.* 1966). Wells (1967, 1968) reported that decayed areas on strawberries inoculated with *Rhizopus stolonifer*, and held at 15°C in 1% O_2, averaged half the size of those on strawberries held in air. The least decay occurred under 0% O_2. However, strawberries developed off-flavors at O_2 concentrations that reduced decay (Couey *et al.* 1966; Tomalin and Robinson 1971a,b).

Decay of cranberries caused by *Botrytis cinerea, Pullularia pullulans* (de Bary) Berk., and *Rhizopus* sp. was decreased by storage in an atmosphere of N_2 rather than in air (Lockhart *et al.* 1971). A marked increase in the amount of physiological breakdown of cranberries caused by storage in N_2, however, precluded that type of storage for fresh fruit.

Lockhart and Eaves (1967) found that tomatoes rotted more when held in atmospheres of 100% N_2 at 10°C than in those containing 2.5% O_2. *Rhizopus* spp., *Fusarium* spp., *Alternaria tenuis*, and bacteria were the major rot organisms. Lockhart and Eaves added that tomatoes from air and 2.5% O_2 had excellent flavor, but those from 100% N_2 had a strong off-flavor. Parsons *et al.* (1970) found that decay of tomatoes by *Rhizopus* spp. and *Alternaria* spp. was significantly less after storage in a 3% O_2 atmosphere than in air.

Aharoni and Lattar (1972) found that atmospheres containing low concentrations of O_2 (2.5%, 5%) greatly reduced rotting of 'Shamouti' oranges caused chiefly by *Alternaria citri* Ell. & Pierce, *Penicillium digitatum* Sacc., *P. italicum* Wehmr., and *Diplodia natalensis* P. Evans. They added that under these conditions the peel of developing fruits harvested before October became injured; this did not occur on fruits harvested in October or later. Mature fruits harvested from October to March showed very little rot under low O_2 atmospheres, and rot that was present was caused by *Penicillium digitatum, P. italicum*, and *Diplodia natalensis*. The developing fruit harvested earlier were largely rotted by *Alternaria citri.*

Lipton (1972) reported that atmospheres of 2% O_2 at 10°C inhibited the growth of aerial mycelium by 50% more than air on radishes with lesions of downy mildew (*Peronospora parasitica* Pers. ex Fr.).

Several investigators have studied the effect of low O_2 concentrations on the bacterial soft rots of vegetables. Lipton (1965) observed that O_2 atmospheres in the range of 1 to 21% had little effect on soft rot (*Erwinia carotovora* [Jones] Holland) of asparagus spears stored for 17 days at 6°C. Atmospheres lacking O_2 caused severe injury. It was shown that 'White Rose' potatoes did not tolerate low O_2 concentrations (1%, 0.5%) when held at temperatures used in transcontinental shipments (Lipton 1967). These atmospheres prevented suberization and periderm formation below skinned areas or small wounds that permitted unchecked bacterial action at 15°C and 20°C. He found that even 5% O_2 may be undesirable in practice because the thinner periderm at this O_2 concentration favors mechanical injury and subsequent decay. Lipton also reported (1972) serious bacterial soft rot in radishes stored 8 days or 15 days in 0.25% O_2 at 10°C and no reduction of bacterial soft rot of lettuce by low O_2 (Lipton 1968).

B. High Carbon Dioxide

Brown (1922) suggested that, within CO_2 concentrations permissible in practice, the gas was not as important as the lowering of temperature in reducing the amount of fungal growth. Brooks *et al.* (1932, 1936) and Brooks and McColloch (1937) studied effects on some transit and storage diseases of exposing fruits and vegetables to high CO_2 levels (20 to 30%) for short periods as a possible substitute for precooling. Diseases in most cases were noticeably suppressed. Host tolerance to CO_2 varied considerably and the suppression of diseases also varied greatly, presumably as a consequence of differences among pathogens. Ryall (1935) obtained similar results in studies with prunes.

With strawberry fruits, Brooks *et al.* (1932) examined the effect of CO_2 after inoculation with *Botrytis cinerea* and *Rhizopus stolonifer*. They found that 10 to 13% CO_2 had little effect upon growth, but that 23% almost stopped growth of both organisms, and 37% stopped growth entirely. Similar results were obtained by Smith (1957) and Couey and Wells (1970). Sommer *et al.* (1973) pointed out that the effectiveness of high CO_2 (5 to 15%) is striking in suppressing gray mold rot at 5°C or above. At lower temperatures (2.2°, 0°C), the benefits of CO_2 were essentially undetectable during the 7-day simulated transit and marketing period, because little fungus developed. Brooks *et al.* (1932) concluded that to inhibit strawberry rots satisfactorily, concentration of CO_2 should be approximately 25%. Couey and Wells (1970) detected off-flavors caused by CO_2 only at 30% CO_2 when fruits were held at 15°C. No off-flavors were detected at either 10% or 20% CO_2 at 10°C. Botrytis rot was reduced to about one-third and Rhizopus rot to about half of that in air.

Brooks *et al.* (1932) and Luvisi and Sommer (1960) found that major reductions in peach decay caused by *Monilinia fructicola* (Wint.) Honey and *Rhizopus stolonifer* could be achieved only at high CO_2 levels, which have been shown to harm fruit quality.

significant effect on decay development caused by *Monilinia fructicola, Botrytis cinerea, Penicillium expansum*, and *Alternaria* spp.

3. Citrus Fruits.—Harding (1969) reported that storage in CA (3% or 5% O_2—about 2% CO_2) was deleterious to citrus fruits (lemons, grapefruit, and Valencia oranges). Decay by *Penicillium digitatum* and *P. italicum* was increased in all three fruits by such storage. Moreover, sporulation in these molds was not controlled. Besides the tendency to increase fungal decay and rind breakdown, the low O_2 atmosphere (3.0 to 3.5%) caused a fermented odor and flavor. McGlasson and Eaks (1972) found that *Penicillium digitatum* rot was hardly affected by storage in 3% O_2-5% CO_2. Stem-end decay (mainly *Fusarium* sp.) was significantly lower in air treatment than in the CA treatment.

4. Avocados.—CA storage of fruit of cold-sensitive avocados in 2% O_2-10% CO_2 prevented development of both anthracnose (*Colletotrichum gloeosporioides* Penz.) and chilling injury at 7.2°C. The suppression of decay development appears to be on delay of fruit ripening and maintenance of resistance rather than on the metabolism of the fungus. It was concluded that the action of the CA in slowing the rate of softening would serve to keep the fungus dormant (Reeder and Hatton 1970; Spalding and Reeder 1972, 1975).

5. Other Fruits.—Anderson *et al.* (1963) showed no benefit to cranberries from MA in relation to unspecified decays or breakdown. Ceponis and Cappellini (1979) reported a reduction of blueberry fruit decays by MA. Diplodia rot of mangos was not controlled by MA. Hatton and Reeder (1969) found that papayas stored in 1% O_2 + 5% CO_2 at 12.5°C for 21 days were less infected with anthracnose *(Colletotrichum gloeosporioides)* and *Diplodia natalensis* than were papayas stored in air.

6. Vegetables.—Rots of tomatoes caused by *Botrytis cinerea, Rhizopus* spp., *Penicillium* spp., *Sclerotinia sclerotiorum,* and *Colletotrichum coccodes* (Wallroth) Hughes were reduced by CA storage (3% O_2 and 3% CO_2). Although *Alternaria tenuis* was retarded slightly in CA, it was still a problem since it developed rapidly on CA-stored tomatoes when they were transferred to the ripening room (Lockhart 1969; Eaves and Lockhart 1961). Tomatoes inoculated with *Erwinia carotovora* (Jones) Holland and held 6 days at 12.5°C kept better in a CA with 3% O_2 and 5% CO_2 than in air (Parsons and Spalding 1972). On the other hand, Fusarium rot (Lockhart 1969; Eaves and Lockhart 1961) and *Geotrichum candidum* rot of tomato fruits were stimulated by 3% O_2 + 3% CO_2 and 3% O_2 with or without 5% CO_2 atmospheres, respectively (Wells and Spalding 1975).

Adair (1971) found that the fungi important in postharvest decay of cabbage (*Botrytis cinerea* and *Sclerotinia sclerotiorum*) were apparently controlled in 1.4% O_2 + 4.7% CO_2, but *Fusarium roseum* was an active pathogen under these conditions and invaded cabbage leaf tissue in atmospheres of 0.8% O_2.

Nielsen (1968) found that in potato tubers aerated with humidified air, little decay developed during a 3-day test period. By contrast, when potatoes were placed in sealed containers, O_2 was depleted, CO_2 exceeded 30%, and soft rot developed in nearly all tubers. Lund and Wyatt (1972) and Lund and Nicholls (1970) found that the most extensive potato tuber rots caused by *Erwinia carotovora* occurred in anaerobic conditions. Accumulation of CO_2 due to tuber respiration did not affect the production of rots significantly. In addition to *Erwinia carotovora*, pectolytic *Clostridia* spp. could be recovered from the spreading rots.

D. Low Pressure (Hypobaric) Storage

Recently, much research interest has been devoted to the possibilities of storage or transport of fruit and vegetables under hypobaric conditions (Burg 1967; Dilley 1977; Salunkhe and Wu 1973; Spalding and Reeder 1977). Lougheed *et al.* (1978) prefer the term low-pressure storage (LPS), because the term is less medicinal in implication and is readily abbreviated. This storage method consists of placing the commodity in a slowly moving stream of humidified air maintained at low pressure. Under these conditions, the commodity is partially degassed at a controlled-atmosphere environment low in O_2. As air is bled continually into the low-pressure chamber from the exterior and then expelled, a slow-moving flowing system is created. As a consequence, respiratory gases which would otherwise accumulate are constantly removed. The storage life is extended because the ripening hormone, ethylene, escapes from the tissue, while its production and action, like other metabolic processes, are greatly retarded (Apelbaum and Barkai-Golan 1977; Lougheed *et al.* 1978). Several investigators suggested that LPS may suppress postharvest pathogens (Burg 1973; Wu and Salunkhe 1972a,b; Adams *et al.* 1976; Spalding and Reeder 1976; Tolle 1969; Borecka 1976; Apelbaum and Barkai-Golan 1977; Borecka and Olak 1978). Wu and Salunkhe (1972a) found that the growth and sporulation of several postharvest storage fungi, such as *Penicillium expansum, Rhizopus stolonifer, Aspergillus niger, Botrytis allii,* and *Alternaria* spp., were retarded when grown at a pressure of 278 mm Hg; suppression was more pronounced at 102 mm Hg. In these tests, potato-dextrose agar at pH 5.6 was inoculated at the center of petri dishes with suspended spores before the dishes were placed in storage at several pressures. Growth at 102 mm Hg, the lowest used, was compared with growth under normal atmospheric pressure in which the O_2 level was maintained at 2.7%. Thus, similar O_2 supplies were provided in the two atmospheres. Relative to 2.7% O_2 at normal atmospheric pressure, all fungi tested grew sooner, sporulated earlier, and showed greater growth at 102 mm Hg. As a consequence of these results, the authors concluded that "... the inhibition of growth by sub-atmospheric pressure is due not only to lower oxygen concentration but also lower pressure exerted on fungi."

Apelbaum and Barkai-Golan (1977) found that spore germination, mycelial growth, and sporulation of tested fungi were inhibited under LPS at 23°C. Inhibition increased with the decrease in pressure below 150 mm Hg. Mycelial growth of *Penicillium digitatum, Alternaria alternata, Botrytis cinerea*, and *Diplodia natalensis* was inhibited 5 to 25%, 45 to 80%, and 100% after 5 days at 100, 50, and 25 mm Hg, respectively. Delay in fungal sporulation was recorded under 50 mm and 25 mm Hg. Inhibition of *Geotrichum candidum* was less pronounced under these conditions. Borecka and Olak (1978), in a similar study, inoculated 2% malt extract agar in petri dishes with the following fungi: *Penicillium expansum, P. spinulosum, P. diversum, Botrytis cinerea, Trichothecium roseum,* and *Rhizopus nigricans* (=*R. stolonifer*). Incubation was at 15°C for 10 days in air at 0.1 and 0.05 atmospheres pressure, which have the same partial pressures of O_2 as do 2% and 1% O_2 at normal atmospheric pressure, respectively. A comparison was made with CA at normal atmospheric pressure. *Penicillium* spp. were significantly suppressed at both LPS atmospheres, with suppression at 0.05 greater than at 0.1 atmospheres. On the other hand, *Botrytis cinerea* grew equally well at 0.1 atmosphere as in air, but sclerotia were absent. At 0.05 atmosphere growth was reduced to about 55% that of growth in air (*ca.* 21% O_2).

Couey *et al.* (1966) determined growth of *Botrytis cinerea* on potato-dextrose agar at 15°C at normal barometric pressure but at a low partial pressure of O_2. They reported that in atmospheres containing 1, 0.5, and 0.25% O_2, growth was 83, 38, and 1.4% of that in air, respectively. These results, at normal atmosphere at 15°C, can be compared to results of Borecka and Olak (1978) who grew *Botrytis cinerea* at the same temperature and similar partial pressures of O_2 achieved by LPS conditions. The results between the two are remarkably similar (Fig. 10.1).

The retardation of fungal development *in vitro* under LPS can be attributed to the reduction in partial O_2 tension prevailing at the reduced pressure. However, it was found that mycelial growth rates under normal atmospheric pressure were significantly higher than at subatmospheric pressures (Wu and Salunkhe 1972b; Apelbaum and Barkai-Golan 1977). Thus, it has been suggested that growth inhibition under low-pressure conditions may be partially due to effects other than the reduction of O_2.

E. Carbon Monoxide

Recently, CO has been used as an added component to CA or MA, primarily as a discoloration inhibitor, especially during transit of lettuce (Kader *et al.* 1973, 1977; Stewart 1978; Stewart and Uota 1976; Stewart *et al.* 1970). Boarini and Buonocore (1973) reported that 1% CO added to 2 to 3% O_2-2 to 3% CO_2 during storage of endive at 0°C for periods of up to 39 days reduced bacterial activity, wilting, and growth of the flower stem. Kader *et al.* (1977) observed less decay (diseases not specified) in CO-treated tomato and pepper lots than in those held in air. Woodruff (1977) reported that CO (5 to 20%) was effective in controlling unspecified decays of several fruits and vegetables, but presented no data.

Kader *et al.* (1978) found that 5 to 10% CO retarded growth of *Botrytis cinerea in vitro* and rot incidence and severity in tomatoes. Suppression was greater when CO was added to an atmosphere containing 4% O_2 instead of air.

El-Goorani and Sommer (1979) studied *in vitro* and *in vivo* effects of atmospheres enriched with 9% CO on 18 postharvest pathogens held at 5.5° or 12.5°C. Test fungi differed greatly in response to CO. The mean percentage of growth in air plus CO ranged from 20 to 100% of that in air alone. The effect of CO was generally much greater if the atmosphere was low in O_2. The mean percentage of growth in 9% CO plus 2.3% O_2 was 4.8 to 89.5% that of air. Suppression was sometimes increased when CO_2 (5% or 18%) was added. The test fungi most sensitive to CO were *Monilinia fructicola, Penicillium expansum, P. italicum, P. digitatum,* and *Whetzelinia sclerotiorum*. Disease development was similarly suppressed. In comparison with air, CO added to 2.3% O_2 + 5% CO_2 reduced rot development by 80 to 90% in strawberries *(B. cinerea)*, apples *(P. expansum)*, lemons *(W. sclerotiorum)*, and oranges (*P. italicum* and *P. digitatum*) that were inoculated and then incubated for 11 to 23 days at 5.5° or 12.5°C. No phytotoxicity was observed. Occasional off-flavors appeared to be associated with O_2 and CO_2 modification rather than CO addition. CO combined with low O_2 effectively reduced decay in sweet cherries (Ogawa *et al.* 1978) and peaches inoculated with *Monilinia fructicola*. Freshly harvested pistachio nuts inoculated with *Aspergillus flavus* and having high moisture (67% on a dry-weight basis) remained bright and free of fungus growth after 18 days at 20°C when exposed to 3% O_2 plus 10% CO. Colonization did occur in air or in MA without CO. The fungus also grew in 10% CO + 3% O_2, duration and temperature as above.

Burg and Burg (1969) and Kader *et al.* (1977) have shown CO to mimic the biological effects of C_2H_4. Solomos and Laties (1973) showed that CO hastened the climacteric and ripening of avocados and bananas. Similar results were observed with tomato fruits (Morris and Kader 1976).

Commercial use of MA containing CO may, therefore, be restricted to commodities relatively insensitive to C_2H_4 at storage or transport temperatures.

V. EFFECTS OF MODIFIED ATMOSPHERES ON MYCOTOXIN ACCUMULATION

Mycotoxins are toxic secondary metabolites of fungi. They are of special concern if they are also carcinogenic and, therefore, any amount is objectionable. In fresh fruits and vegetables the greatest concern has been for the mycotoxin patulin, a suspected carcinogen, which is produced by several species of *Aspergillus* and *Penicillium*, including the blue mold organism of deciduous fruit, *P. expansum*.

Because *P. expansum* is sometimes prevalent in apples for processing, the presence of patulin in juice or other apple products is of concern. Unlike with fresh fruits, consumers of apple products are unable to visually avoid the blue mold disease lesions. Accumulation of patulin in deciduous fruits has been studied by Buchanan *et al.* 1974, Sommer *et al.* 1974, Sommer and Buchanan 1978, and Wilson and Nuovo 1973.

The effect of MA on fungus growth and patulin production was tested by Lovett *et al.* (1975), Orth (1976), and Sommer *et al.* (1977). With atmospheres commonly used for apples (2 to 3% O_2 - 5% CO_2), growth of *P. expansum* was only modestly suppressed. Patulin accumulated in apples stored in MA, but in air at the same temperature (0°C) levels were from 5- to 60-fold greater.

Aflatoxin, produced by *Aspergillus flavus* and *A. parasiticus* Speare, is a potent carcinogen of test animals. Although the elaborating fungi are not common postharvest pathogens of fresh fruit or vegetables, they pose a more serious threat to commodities to be processed by drying or to various tree nuts (almonds, Brazil nuts, cashews, filberts, macadamias, pecans, pistachios, and walnuts).

Studies with figs for drying showed that fruits on the tree were resistant to *A. flavus* until they became completely ripe. As fruits dried on the tree or ground considerable aflatoxin was produced before fruits were sufficiently dry to prevent further fungal growth (Buchanan *et al.* 1975). Pistachios supported growth of *A. flavus* when it was inoculated into the mesocarp of developing fruits (Sommer *et al.* 1976). In California, the incidence of aflatoxin in almonds and walnuts is generally associated with insect activity in the mature nuts, particularly that of the navel orangeworm, *Myelois venipars* Dyar. (Sommer and Buchanan 1978).

MA's effects on *A. flavus* and aflatoxin accumulation were studied by Landers *et al.* (1967), Wilson and Jay (1976), El-Goorani and Sommer

(1979), Shih and Marth (1973), and Buchanan *et al.* (unpublished data). In general, the effects of MA on storage of freshly harvested or only partially-dried nuts have been as follows: levels of 2% O_2 or higher are only modestly suppressive of *A. flavus* growth and aflatoxin production; CO_2 concentrations of 20% or higher suppressed fungal growth and aflatoxin accumulation considerably, but much less suppression was observed in 10% CO_2 or less. The possible adverse effects of MA on high moisture nuts are not known. Presumably taste, odor, or onset of rancidity might be affected.

Possible utilization of CO in atmospheres to inhibit aflatoxin accumulation was suggested by El-Goorani and Sommer (1979) and Buchanan *et al.* (unpublished data). When accompanied by low O_2 (2 to 3%), CO (*ca.* 10%) suppressed fungal growth beyond the effects of low O_2 and greatly suppressed aflatoxin accumulation in high moisture pistachios. However, atmospheres approaching those commonly used for fresh fruits and vegetables slowed but did not stop growth of *A. flavus*, even with the addition of CO. Aflatoxin generally could be detected given sufficient time.

VI. SUMMARY AND CONCLUSIONS

The rate of postharvest disease development in fruits and vegetables depends upon the pathogenic powers of the bacterium or fungus when arrayed against the resistance of the host commodity. The environment may influence the antagonists differently. Low temperatures, for example, may drastically suppress the pathogen's activity. At the same time the physiological activity of the commodity is suppressed. Often the result is an extension of the period before the normal resistance of the fruit or vegetable host is lost. Similarly, elevated CO_2 in atmospheres may suppress the fungus pathogen and simultaneously delay ripening of fruit to thereby retain a greater resistance to postharvest diseases. It is essential to delay the onset of senescence in the host as long as possible.

Lowering the O_2 composition of the atmosphere from about the 21% of air to the 2 to 3% common in modified atmospheres evidently suppresses pathogen activity very little. Postharvest pathogens tested have usually been strikingly suppressed only after the O_2 had been lowered to less than 1%, a level at which commodities risk injury.

Levels of CO_2 commonly are restricted to little more than about 5% in extended storage for fear that higher concentrations would lead to injury of the commodity. The CO_2's effect is suppression of the fruit or vegetable's respiration. The growth of pathogens *in vitro* is generally suppressed moderately in an atmosphere of 5% CO_2. Fruits or vegetables may tolerate high CO_2 levels for short periods. CO_2 at 10 to 20% has been used successfully with strawberries to suppress *Botrytis cinerea* and with sweet cherries to suppress *Monilinia fructicola* and *Botrytis cinerea* during transit periods of up to 8 to 10 days.

MA may increase disease incidence. Examples where MA worsens the disease problem can be found in potato tubers or various root crops which have highly active wound-healing processes involving periderm formation. Atmospheres low in O_2 slow wound healing to thereby give pathogens increased time to establish infections and colonize the wound while it is still highly susceptible.

The effect of MA on phytoalexin production evidently has been less well studied. However, it is likely that phytoalexin formation would be slowed by MA as a consequence of lowered metabolic activity of the commodity.

It is unrealistic, where serious postharvest diseases are encountered, to expect MA to substitute for other disease control measures. Fungistatic gases, such as CO, suppress but do not completely arrest fungal growth. However, MA can be an important adjunct to chemical controls by reducing the disease pressure in the fruit or vegetable during storage or transport. The delay in ripening of fruits and senescence of vegetables by MA may be especially effective against those weak pathogens which attack mostly ripe or senescing fruits and vegetables to sometimes cause serious losses.

VII. LITERATURE CITED

ADAIR, C.N. 1971. Influence of controlled-atmosphere storage conditions on cabbage postharvest decay fungi. *Plant Dis. Rptr.* 55:864–868.

ADAMS, K.B., M.T. WU, and D.K. SALUNKHE. 1976. Effects of subatmospheric pressures on the growth and patulin production of *Penicillium expansum* and *Penicillium patulum. Lebensm.-Wiss. u.-Technol.* 9:153–155.

AHARONI, Y. and F.S. LATTAR (LITTAUER). 1972. The effect of various storage atmospheres on the occurrence of rots and blemishes on Shamouti oranges. *Phytopathol. Z.* 73:371–374.

ANDERSON, R.E., R.E. HARDENBURG, and H.C. VAUGHT. 1963. Controlled-atmosphere storage studies with cranberries. *Proc. Amer. Soc. Hort. Sci.* 83:416–422.

ANDROSOVA, O.G. 1962. A study of control measures for gray rot of cabbage in storage. *Tr. Khar'Kovsk Sel'skokhoz. Inst.* 38. 75. 194–205. (*Biol. Abstr.* 45:14424.)

ANON. 1940. Fruit cold-storage research. New Zealand Dept. Sci. Ind. Res. Annu. Rpt. 14:52–54. (*Chem. Abstr.* 35:218.)

APELBAUM, A. and R. BARKAI-GOLAN. 1977. Spore germination and mycelial growth of postharvest pathogens under hypobaric pressure. *Phytopathology* 67:400–403.

ASHWORTH, L.J., JR., J.L. MCMEANS, and C.M. BROWN. 1969. Infection of cotton by *Aspergillus flavus*: the influences of temperature and aeration. *Phytopathology* 59:669–673.

BIALE, J.B. 1953. Storage of lemons in controlled atmosphere. *Calif. Citrograph* 38:429, 436–438.

BOARINI, F. and C. BUONOCORE. 1973. La conservazione della scarola in atmosfera controllata con l'impiego dell'ossido di carbonio. *Notiziario del CRIOF* 4(2):1–6. (*Hort. Abstr.* 44:9485.)

BOMPEIX, G. 1978a. The comparative development of *Pezicula alba* and *P. malicorticis* on apples in vitro (air and controlled atmosphere). *Phytopathol. Z.* 91:97–109.

BOMPEIX, G. 1978b. Quelques aspects physiologigues des relations hôte-parasite durant la conservation des pommes. *Fruits* 33(1):22–26.

BORECKA, H.W. 1976. Effect of BCM fungicides on the incidence of fungi rots of apple, strawberry and raspberry fruit, and degradation of BCM residues in fruits in CA storage. Research Institute of Pomology, Skierniewice, Poland. PL-ARS-29, FG-Po-308.

BORECKA, H. and J. OLAK. 1978. The effect of hypobaric storage conditions on the growth and sporulation of some pathogenic fungi. *Fruit Sci. Rpt. (Poland)* 5:39–41.

BRAMLAGE, W.J., P.H. BAREFORD, G.D. BLANPIED, D.H. DEWEY, S. TAYLOR, S.W. PORRITT, E.C. LOUGHEED, W.H. SMITH, and F.S. MCNICHOLAS. 1977. Carbon dioxide treatments for 'McIntosh' apples before CA storage. *J. Amer. Soc. Hort. Sci.* 102:658–662.

BRANCATO, F.P. and N.S. GOLDING. 1953. Gas requirements of molds: the importance of dissolved oxygen in the medium for germination and growth of several molds. *Northw. Sci.* 27(1):33–38.

BROOKS, C., C.O. BRATLEY, and L.P. MCCOLLOCH. 1936. Transit and storage diseases of fruits and vegetables as affected by initial carbon dioxide treatments. *USDA Tech. Bul.* 519.

BROOKS, C. and L.P. MCCOLLOCH. 1937. Some effects of storage conditions on certain diseases of lemons. *J. Agr. Res.* 55:795–809.

BROOKS, C., E.V. MILLER, C.O. BRATLEY, J.S. COOLEY, P.V. MOOK, and H.B. JOHNSON. 1932. Effect of solid and gaseous carbon dioxide upon transit diseases of certain fruits and vegetables. *USDA Tech. Bul.* 318.

BROWN, G.E. and B.W. KENNEDY. 1966. Effect of oxygen concentration on Pythium seed rot of soybean. *Phytopathology* 56:407–411.

BROWN, W. 1922. On the germination and growth of fungi at various temperatures and in various concentrations of oxygen and of carbon dioxide. *Ann. Bot.* 36:257–283.

BUCHANAN, J.R., N.F. SOMMER, and R.J. FORTLAGE. 1975. *Aspergillus flavus* infection and aflatoxin production in fig fruits. *Appl. Microbiol.* 30: 238–241.

BUCHANAN, J.R., N.F. SOMMER, R.J. FORTLAGE, E.C. MAXIE, F.G. MITCHELL, and D.P.H. HSIEH. 1974. Patulin from *Penicillium expansum* in stone fruits and pears. *J. Amer. Soc. Hort. Sci.* 99:262–265.

BUCKLEY, P.M., N.F. SOMMER, T.T. MATSUMOTO, and M. DALLY. 1967. Responses of germinating *Rhizopus stolonifer* sporangiospores to heating, chilling, gamma irradiation, and anoxia. *Phytopathology* 57:806. (Abstr.)

BULL, A.T. and A.P.J. TRINCI. 1977. The physiology and metabolic control of fungal growth. *Adv. Microbiol. Physiol.* 15:1–84.

BU'LOCK, J.D. 1975. Secondary metabolism in fungi and its relationship to growth and development. p. 33–58. *In* J.E. Smith and D.R. Berry (eds.) The filamentous fungi, Vol. I. Edward Arnold (Publishers) Ltd, London.

BUNGAY, H.R., W.J. WHALEN, and W.M. SANDERS. 1969. Microprobe techniques for determining diffusivities and respiration rates in microbial slime systems. *Biotechnol. Bioeng.* 11:765–772.

BURG, S.P. 1967. Method for storing fruit. U.S. Patent Office. No. 3,333,967. Filed Sept. 26, 1963.

BURG, S.P. 1973. Hypobaric storage of cut flowers. *HortScience* 8:202–205.

BURG, S.P. and E.A. BURG. 1969. Internation of ethylene, oxygen and carbon dioxide in the control of fruit ripening. *Qual. Plant. Mater. Veg.* 19:185–200.

BURNETT, J.H. 1968. Fundamentals of mycology. St. Martin's Press, New York.

BURTON, W.G. 1974. Some biophysical principles underlying the controlled atmosphere storage of plant material. *Ann. Appl. Biol.* 78:149–168.

BUSSEL, J., P.M. BUCKLEY, and N.F. SOMMER. 1968. Effect of anaerobiosis on germination and viability of *Rhizopus stolonifer* sporangiospores. *Phytopathology* 58:1046. (Abstr.)

BUSSEL, J., P.M. BUCKLEY, N.F. SOMMER, and T. KOSUGE. 1969a. Ultrastructural changes in *Rhizopus stolonifer* sporangiospores in response to anaerobiosis. *J. Bacteriol.* 98:774–783.

BUSSEL, J., M. MIRANDA, and N.F. SOMMER. 1971. Response of *Monilinia fructicola* conidia to individual and combined treatments of anoxia and heat. *Phytopathology* 61:61–64.

BUSSELL, J., N.F. SOMMER, and T. KOSUGE. 1969b. Effect of anaerobiosis upon germination and survival of *Rhizopus stolonifer* sporangiospores. *Phytopathology* 59:946–952.

CAMERON, A.C. 1978. Carbon dioxide and potato wound metabolism. M.S. Thesis, Colorado State University, Fort Collins.

CARTER, B.L.A. and A.T. BULL. 1969. Studies of fungal growth and intermediary carbon metabolism under steady and non-steady state conditions. *Biotechnol. Bioeng.* 11:785–804.

CARTER, B.L.A. and A.T. BULL. 1971. The effect of oxygen tension in the medium on the morphology and growth kinetics of *Aspergillus nidulans*. *J. Gen. Microbiol.* 65:265–273.

CEPONIS, M.J. and R.A. CAPPELLINI. 1979. Control of postharvest decays of blueberry fruits by precooling, fungicide, and modified atmospheres. *Plant Dis. Rptr.* 63(12):1049–1053.

CHANCE, B., D. JAMIESON, and H. COLES. 1965. Energy-linked pyridine nucleotide reduction: inhibitory effects of hyperbaric oxygen *in vitro* and *in vivo*. *Nature* 206:257–263.

CLAYPOOL, L.L. and R.M. KEEFER. 1942. A colorimetric method for CO_2 determination in respiration studies. *Proc. Amer. Soc. Hort. Sci.* 40:177–186.

COCHRANE, J.C., V.W. COCHRANE, F.G. SIMON, and J. SPAETH. 1963. Spore germination and carbon metabolism in *Fusarium solani*. I. Requirements for spore germination. *Phytopathology* 53:1155–1160.

COCHRANE, V.W. 1958. Physiology of fungi. John Wiley & Sons, New York.

CORMAN, J., H.M. TSUCHIYA, H.J. KOEPSELL, R.G. BENEDICT, S.E. KELLEY, V.H. FEGER, R.G. DWORSCHACK, and R.W. JACKSON. 1957. Oxygen absorption rates in laboratory and pilot plant equipment. *Appl. Microbiol.* 5:313–318.

COUEY, H.M., M.N. FOLLSTAD, and M. UOTA. 1966. Low-oxygen atmospheres for control of postharvest decay of fresh strawberries. *Phytopathology* 56:1339–1341.

COUEY, H.M. and K.L. OLSEN. 1975. Storage response of 'Golden Delicious' apples after high carbon dioxide treatment. *J. Amer. Soc. Hort. Sci.* 100:148–150.

COUEY, H.M. and J.M. WELLS. 1970. Low-oxygen or high-carbon dioxide atmospheres to control postharvest decay of strawberries. *Phytopathology* 60: 47–49.

COUEY, H.M. and T.R. WRIGHT. 1977. Effect of a prestorage CO_2 treatment on the quality of 'd'Anjou' pears after regular or controlled atmosphere storage. *HortScience* 12:244–245.

COVEY, R.P., JR. 1970. Effect of oxygen tension on the growth of *Phytophthora cactorum*. *Phytopathology* 60:358–359.

COYNE, F.P. 1933. The effect of carbon dioxide on bacterial growth. *Proc. Roy. Soc. London Ser. B*: 113:196–217.

DARBY, R.T. and D.R. GODDARD. 1950. Studies of the respiration of the mycelium of the fungus *Myrothecium verrucaria*. *Amer. J. Bot.* 37:379–387.

DEWEY, D.H. 1977. Controlled atmospheres for the storage and transport of perishable agricultural commodities. Mich. State Univ. Hort. Rpt. 28. East Lansing.

DEWEY, D.H. 1979. Three remarkable generations of postharvest horticulture. *HortScience* 14(3-Sect. 2):342–344.

DEWEY, D.H., R.C. HERNER, and D.C. DILLEY. 1969. Controlled atmospheres for the storage and transport of horticultural crops. Mich. State Univ. Hort. Rpt. 9. East Lansing.

DILLEY, D.R. 1977. Hypobaric storage of perishable commodities—fruits, vegetables, flowers and seedlings. *Acta Hort.* 62:61–70.

DONALD, W.W. and C.J. MIROCHA. 1977. Chitin as a measure of fungal growth in stored corn and soybean seed. *Cereal Chem.* 54:466–474.

DUKES, P.D. and J.L. APPLE. 1965. Effect of oxygen and carbon dioxide tensions on growth and inoculum potential of *Phytophthora parasitica* var. *nicotianae. Phytopathology* 55:666–669.

DULLUM, V.N. and P.M. RASMUSSEN. 1951. Apple storing experiments 1940–48. *Tidsskr. Planteavl* 54:249–317.

DURBIN, R.D. 1959. Factors affecting the vertical distribution of *Rhizoctonia solani* with special reference to CO_2 concentration. *Amer. J. Bot.* 46:22–25.

EAVES, C.A. 1964. Nitrogen fertilization in relation to the keeping quality of McIntosh apples in controlled atmosphere storage. A.R. Nova Scotia Fruit Growers Assoc., p. 35–36. (*Hort. Abstr.* 36:6180.)

EAVES, C.A., F.R. FORSYTH, J.S. LEEFE, and C.L. LOCKHART. 1964. Effect of varying concentrations of oxygen with and without CO_2 on senescent changes in stored McIntosh apples grown under two levels of nitrogen fertilization. *Can. J. Plant Sci.* 44:458–465.

EAVES, C.A. and C.L. LOCKHART. 1961. Storage of tomatoes in artificial atmospheres using calcium hydroxide absorption method. *J. Hort. Sci.* 36:85–92.

ECKERT, J.W. and N.F. SOMMER. 1967. Control of diseases of fruits and vegetables by postharvest treatment. *Annu. Rev. Phytopathol.* 5:391–432.

EDNEY, K.L. 1956. The rotting of apples by *Gloeosporium perennans* Zeller and Childs. *Ann. Appl. Biol.* 44:113–128.

EDNEY, K.L. 1964. The effect of the composition of the storage atmosphere on the development of rotting of Cox's Orange Pippin apples and the production of pectolytic enzymes by *Gloeosporium* spp. *Ann. Appl. Biol.* 54:327–334.

EDNEY, K.L. 1973. Postharvest deterioration of fruit. *Chem. and Indust.* 22:1054–1056.

EL-GOORANI, M.A. and N.F. SOMMER. 1979. Suppression of postharvest plant pathogenic fungi by carbon monoxide. *Phytopathology* 69:834–838.

FINN, R.K. 1954. Agitation-aeration in the laboratory and in industry. *Bacteriol. Rev.* 18:254–274.

FOLLSTAD, M.N. 1966. Mycelial growth rate and sporulation of *Alternaria tenuis, Botrytis cinerea, Cladosporium herbarum,* and *Rhizopus stolonifer* in low-oxygen atmospheres. *Phytopathology* 56:1098–1099.

FURLONG, C.R. 1946. The storage and ripening of green tomatoes with special reference to open-air fruit and end-of-season fruit from glasshouses. *J. Hort. Sci.* 22:197–208.

GOLDING, N.S. 1945. The gas requirements of molds. IV. A preliminary interpretation of the growth rates of four common mold cultures on the basis of absorbed gases. *J. Dairy Sci.* 28:737–750.

GREGORY, E.M. and I. FRIDOVICH. 1973. Oxygen toxicity and the superoxide dismutase. *J. Bacteriol.* 114(3):1193–1197.

GRIERSON, W. 1969. Some random thoughts on research. Mich. State Univ. Hort. Rpt. 9. p. 77–79. East Lansing.

GRIERSON, W., H.M. VINES, M.F. OBERBACHER, S.V. TING, and G.J. EDWARDS. 1966. Controlled atmosphere storage of Florida and California lemons. *Proc. Amer. Soc. Hort. Sci.* 88:311–318.

HARDING, P.R., JR. 1969. Effect of low oxygen and low carbon dioxide combination in controlled atmosphere storage of lemons, grapefruit and oranges. *Plant Dis. Rptr.* 53:585–588.

HARRISON, D.E.F. 1976. The oxygen metabolism of microorganisms. Meadowfield Press Ltd., Durham, England.

HARTMAN, R.E., N.T. KEEN, and M. LONG. 1972. Carbon dioxide fixation by *Verticillium albo-atrum*. *J. Gen. Microbiol.* 73:29–34.

HARVEY, J.M. 1978. Reduction of losses in fresh market fruits and vegetables. *Annu. Rev. Phytopathol.* 16:321–341.

HASSAN, F.M.E. 1966. The effect of controlled atmospheres on the keeping quality of sweetpotatoes. Ph.D. Thesis, Virginia Polytechnic Institute. (*Diss. Abstr.* 27B:650–651.)

HATTON, T.T., JR. and W.F. REEDER. 1969. Controlled atmosphere storage of papayas 1968. *Proc. Amer. Soc. Hort. Sci. (Tropical Region)* 13:251–256.

HATTON, T.T., JR., J.J. SMOOT, and R.H. CUBBEDGE. 1972. Influence of carbon dioxide exposure on stored mid- and late-season 'Marsh' grapefruit. *Proc. Amer. Soc. Hort. Sci. (Tropical Region)* 16:49–58.

HAUGAARD, N. 1968. Cellular mechanisms of oxygen toxicity. *Physiol. Rev.* 48(2):311–373.

HAWKER, L.E., A.H. LINTON, B.F. FOLKES, and M.J. CARLILE. 1960. An introduction to the biology of microorganisms. St. Martin's Press, New York.

HUELIN, F.E. and C.G. TINDALE. 1947. The gas storage of Victorian apples. *J. Dept. Agr. Victoria* 45:74–80.

ISENBERG, F.M.R. 1979. Controlled atmosphere storage of vegetables. p. 337–394. *In* J. Janick (ed.) Horticultural reviews, Vol. 1. AVI Publishing, Westport, Ct.

JULIEN, J.B. and W.R. PHILLIPS. 1963. Note on the effect of CO_2 and O_2 mixtures on the growth of apple scab cultures. *Can. J. Plant Sci.* 43:227.

KADER, A.A., P.E. BRECHT, R. WOODRUFF, and L.L. MORRIS. 1973. Influence of carbon monoxide, carbon dioxide, and oxygen levels on brown stain, respiration rate, and visual quality of lettuce. *J. Amer. Soc. Hort. Sci.* 98:485–488.

KADER, A.A., G.A. CHASTAGNER, L.L. MORRIS, and J.M. OGAWA. 1978. Effects of carbon monoxide on decay, physiological responses, ripening and composition of tomato fruits. *J. Amer. Soc. Hort. Sci.* 103:665–670.

KADER, A.A., J.A. KLAUSTERMEYER, L.L. MORRIS, and R.F. KASMIRE. 1975. Extending storage life of mature-green chili peppers by modified atmosphere. *HortScience* 10:335.

KADER, A.A. and L.L. MORRIS. 1977. Modified atmospheres: an indexed reference list with emphasis on horticultural commodities. Suppl. 2. Veg. Crops Series, Univ. of Calif. 187. Davis.

KADER, A.A., L.L. MORRIS, and J.A. KLAUSTERMEYER. 1976. Postharvest handling and physiology of horticultural crops. A list of selected references. Veg. Crops Series, Univ. of Calif. 169. Davis.

KADER, A.A., L.L. MORRIS, and J.A. KLAUSTERMEYER. 1977. Physiological responses of some vegetables to carbon monoxide. Mich. State Univ. Hort. Rpt. 28. p. 197–202. East Lansing.

KLAUS, H. 1941. Untersuchungen über *Alternaria solani* (Ell. & G. Martin) Sor. Insbesondere über seine Pathogenilät an Kartoffelknollen in Abhangigkeit von den Aussenfaktoren. *Phytopathol. Z.* 13:126–195.

KLOTZ, L.J., L.H. STOLZY, and T.A. DEWOLFE. 1963. Oxygen requirements of three root-rotting fungi in a liquid medium. *Phytopathology* 53:302–305.

LANDERS, K.E., N.D. DAVIS, and U.L. DIENER. 1967. Influence of atmospheric gases on aflatoxin production by *Aspergillus flavus* in peanuts. *Phytopathology* 57:1086–1090.

LAU, O.L. and N.E. LOONEY. 1978. Effects of a pre-storage high CO_2 treatment on British Columbia and Washington State 'Golden Delicious' apples. *J. Amer. Soc. Hort. Sci.* 103:341–344.

LIN, K.-H. 1948. The effect of modified air on the rotting of apples in storage. *Lingnan Sci. J.* 22:133–138.

LINDENMAYER, A. 1965. Carbohydrate metabolism 3 terminal oxidation and electron transport. p. 301–348. *In* G.C. Ainsworth and A.S. Sussman (eds.) The fungi, Vol. 1. Academic Press, New York.

LIPTON, W.J. 1965. Postharvest responses of asparagus spears to high carbon dioxide and low oxygen atmospheres. *Proc. Amer. Soc. Hort. Sci.* 86:347–356.

LIPTON, W.J. 1967. Some effects of low-oxygen atmospheres on potato tubers. *Amer. Potato J.* 44:292–299.

LIPTON, W.J. 1968. Low O_2 atmospheres benefits and dangers. Yearb. United Fresh Fruit and Veg. Assoc. 1968. p. 99–100, 103. Washington, D.C.

LIPTON, W.J. 1971. Controlled atmosphere effects on lettuce quality in simulated export shipments. U.S. Dept. Agr., Agr. Res. Serv. (Rpt.) 51–45.

LIPTON, W.J. 1972. Market quality of radishes stored in low O_2-atmospheres. *J. Amer. Soc. Hort. Sci.* 97:164–167.

LITTLEFIELD, N.A., B.A. WANKIER, D.K. SALUNKHE, and J.N. MCGILL. 1966. Fungistatic effects of controlled atmospheres. *Appl. Microbiol.* 14: 579–581.

LOCKHART, C.L. 1967. Influence of controlled atmospheres on the growth of *Gloeosporium album in vitro. Can. J. Plant Sci.* 47:649–651.

LOCKHART, C.L. 1969. Effect of CA storage on storage rot pathogens. Mich. State Univ. Hort. Rpt. 9. p. 113–121. East Lansing.

LOCKHART, C.L. and C.A. EAVES. 1967. The influence of low oxygen levels and relative humidity on storage of green tomatoes. *J. Hort. Sci.* 42:289–294.

LOCKHART, C.L., C.A. EAVES, and E.W. SHIPMAN. 1969. Suppression of rots on four varieties of mature green tomatoes in controlled atmosphere storage. *Can. J. Plant Sci.* 49:265–269.

LOCKHART, C.L., F.R. FORSYTH, R. STARK, and I.V. HALL. 1971. Nitrogen gas suppresses microorganisms on cranberries in short term storage. *Phytopathology* 61:335–336.

LOCKHART, W.R. and R.W. SQUIRES. 1963. Aeration in the laboratory. *Adv. Appl. Microbiol.* 5:157–187.

LOPRIORE, G. 1895. Ueber die Einwirkung der Kohlensäure auf das Protoplasma der lebenden Pflanzenzelle. *Jahrb. Wiss. Bot.* 28:531–626.

LOUGHEED, E.C., D.P. MURR, and L. BERARD. 1978. Low pressure storage for horticultural crops. *HortScience* 13:21–27.

LOUVET, J. and J. BULIT. 1964. Recherches sur l'écologie des champignons parasites dans le sol. I. action du gaz carbonique sur la croissance et l'activité parasitaire de *Sclerotinia minor* et de *Fusarium oxysporum* f. *melonis*. *Ann. Epiphyties* 15(1):21–44.

LOVETT, J., R.G. THOMPSON, JR., and B.K. BOUTIN. 1975. Patulin production in apples stored in a controlled atmosphere. *J. Assoc. Official Anal. Chem.* 58:912–914.

LUND, B.M. and J.C. NICHOLLS. 1970. Factors influencing the soft-rotting of potato tubers by bacteria. *Potato Res.* 13:210–214.

LUND, B.M. and G.M. WYATT. 1972. The effect of oxygen and carbon dioxide concentrations on bacterial soft rot of potatoes. I. King Edward potatoes inoculated with *Erwinia carotovora* var. *atroseptica*. *Potato Res.* 15:174–179.

LUVISI, D.A. and N.F. SOMMER. 1960. Polyethylene liners and fungicides for peaches and nectarines. *Proc. Amer. Soc. Hort. Sci.* 76:146–155.

LWOFF, A. and J. MONOD. 1947. Essai d'analyse du rôle de l'anhydride carbonique dans la croissance microbienne. *Ann. Inst. Pasteur* 73(4):323–347.

MAGIE, R.O. 1971. Carbon dioxide treatment of gladiolus corms reveals latent *Fusarium* infections. *Plant Dis. Rptr.* 55:340–341.

MARTIN, D. and J. CERNY. 1956. Low oxygen gas storage trials of apples in Tasmania. Tech. Pap. Div. Pl. Ind., C.S.I.R.O. Melbourne 6. p. 19. (*Hort. Abstr.* 26:2498.)

MCGLASSON, W.B. and I.L. EAKS. 1972. A role for ethylene in the development of wastage and off-flavors in stored 'Valencia' oranges. *HortScience* 7: 80–81.

MELLENTHIN, W.M. and C.Y. WANG. 1977. Storage response of d'Anjou pears following short-term high carbon dioxide treatment. *Acta Hort.* 69: 323–326.

MEYNELL, G.G. and E. MEYNELL. 1965. Theory and practice in experimental bacteriology. Cambridge University Press, New York.

MITCHELL, D.J. and G.A. ZENTMYER. 1970. Effects of CO_2 and O_2 on growth and sporulation of several species of *Phytophthora*. *Phytopathology* 60:1304–1305.

MITCHELL, D.J. and G.A. ZENTMYER. 1971a. Effects of oxygen and carbon dioxide tensions on growth of several species of *Phytophthora*. *Phytopathology* 61:787–791.

MITCHELL, D.J. and G.A. ZENTMYER. 1971b. Effects of oxygen and carbon dioxide tensions on sporangium and oospore formation by *Phytophthora* spp. *Phytopathology* 61:807–811.

MONTGOMERY, H.B.S. 1958. Effect of storage conditions on the incidence of *Gloeosporium* rots of apple fruits. *Nature* 182:737–738.

MORRIS, L.L., L.L. CLAYPOOL, and D.P. MURR. 1971. Modified atmospheres: an indexed reference list through 1969 with emphasis on horticultural commodities. Univ. of Calif. Div. Agr. Sci. Davis.

MORRIS, L.L. and A.A. KADER. 1976. Postharvest physiology of tomato fruits. p. 69–84. *In* Fresh Market Tomato Research—1975. Veg. Crop Series, Univ. of Calif. 176. Davis.

MURR, D.P., A.A. KADER, and L.L. MORRIS. 1974. Modified atmospheres: an indexed reference list with emphasis on horticultural commodities. Veg. Crops Series, Univ. of Calif. 168. Davis.

NIELSEN, L.W. 1964. Pathogenesis of three *Erwinia* species to potato tuber tissue in CO_2-N atmosphere. *Phytopathology* 54:902.

NIELSEN, L.W. 1968. Accumulation of respiratory CO_2 around potato tubers in relation to bacterial soft rots. *Amer. Potato J.* 45:174–181.

NILSSON, F. *et al.* 1956. Carbon dioxide storage of apples and pears, 1951–54. *J. Roy. Swedish Acad. Agr.* 95:319–347. (*Hort. Abstr.* 27:1174.)

NYHLÉN, Å. and J. JOHANSSON. 1964. Controlled atmosphere storage of apples, 1957–1960. *K. Skogs-Lantbr. Akad. Tidskr.* 103:307–337. (*Hort. Abstr.* 35:2889.)

NYHLÉN, Å. and T. NILSON. 1960. Gas storage of apples, 1954–1957. *K. Skogs-Lantbr. Akad. Tidskr.* 99:171–211. (*Hort. Abstr.* 31:2059.)

OGAWA, J.M., B.T. MANJI, and D.C. JANECKE. 1978. Effects of modified oxygen and carbon monoxide environments on Monilinia decay of sweet cherries. The American Phytopathological Society, 70th Annu. Mtg., Tucson, Ariz. p. 173. The American Phytopathological Society, St. Paul, Minn. (Abstr.)

OKAZAKI, N. and S. SUGAMA. 1979. A new apparatus for automatic growth estimation of mold cultured on solid media. *J. Ferment. Technol.* 57(5):413–417.

OLSSON, K. 1965. The occurrence of various species of *Gloeosporium* on apples in different types of store. *Växtskyddsnotiser* 29:59–61. (*Hort. Abstr.* 36: 4234.)

ORTH, R. 1976. Wachstum un Toxinbildung von Patulin- und Sterigmatocystein-bildenden Schimmelpilzen unter Kontrollierter Atmosphere. *Z. Lebens. Unter. Forsch.* 160:359–366.

OSTROWSKI, W., I. RZEPECKA, and J. GDOWSKI. 1958. Transport of sweet cherries in freight cars cooled with natural and dry ice. *Prace Inst. Sadown. Skierniewice* 3:101–129. (*Hort. Abstr.* 29:2181.)

PAPAVIZAS, G.C. and C.B. DAVEY. 1962a. A mycostatic role for CO_2 in the suppression of *Rhizoctonia* in soil. *Phytopathology* 52:165. (Abstr.)

PAPAVIZAS, G.C. and C.B. DAVEY. 1962b. Activity of *Rhizoctonia* in soil as affected by carbon dioxide. *Phytopathology* 52:759–766.

PARSONS, C.S., R.E. ANDERSON, and R.W. PENNEY. 1970. Storage of mature-green tomatoes in controlled atmospheres. *J. Amer. Soc. Hort. Sci.* 95:791–794.

PARSONS, C.S. and D.H. SPALDING. 1972. Influence of a controlled atmosphere, temperature, and ripeness on bacterial soft rot of tomatoes. *J. Amer. Soc. Hort. Sci.* 97:297–299.

PETT, B., D. KLEINHAMPEL, and J. GOTZ. 1977. Fusarium dry and mixed rot in potatoes are influenced by environmental conditions. *Nachrichtenblatt fur den Pflanzenschutz in der DDR* 31:4–7. (*Rev. Plant Pathol.* 56:622–623.)

PHILLIPS, W.R. and J.B. JULIEN. 1966. Note on the effect of CO_2 and O_2 mixtures on the growth of apple scab lesions on McIntosh apple. *Phytoprotection* 47:116–117. (*Hort. Abstr.* 37:2409.)

PHILLIPS, W.R., W.M. RUTHERFORD, and J.B. JULIEN. 1959. The development of scab in CA storage. Rpt. Can. Committee Fruit Veg. Pres. for 1958, 1959. p. 3–4. (*Hort. Abstr.* 30:1785.)

PIRT, S.J. 1967. A kinetic study of the mode of growth of surface colonies of bacteria and fungi. *J. Gen. Microbiol.* 47:181–197.

REEDER, W.F. and T.T. HATTON, JR. 1970. Storage of Lula avocados in controlled atmosphere—1970 test. *Proc. Fla. State Hort. Soc.* 83:403–405.

RIDE, J.P. and R.B. DRYSDALE. 1972. A rapid method for the chemical estimation of filamentous fungi in plant tissue. *Physiol. Plant Pathol.* 2:7–15.

RIGHELATO, R.C. 1975. Growth kinetics of mycelial fungi. p. 79–103. *In* J.E. Smith and D.R. Berry (eds.) The filamentous fungi, Vol. 1. Edward Arnold (Publishers) Ltd., London.

RIGHELATO, R.C. and S.J. PIRT. 1967. Improved control of organism concentration in continuous cultures of filamentous micro-organisms. *J. Appl. Bacteriol.* 30(1):246–250.

RIGHELATO, R.C., A.P.J. TRINCI, S.J. PIRT, and A. PEAT. 1968. The influence of maintenance energy and growth rate on the metabolic activity, morphology and conidiation of *Penicillium chrysogenum*. *J. Gen. Microbiol.* 50:399–412.

RIPPEL, A. and H. BORTELS. 1927. Vorläufige versuche über die allgemeine Bedeutung der Kohlensäure für die Pflanzenzelle (Versuche an *Aspergillus niger*). *Biochem. Ztschr.* 184:237–244.

ROBB, S. 1966. Reactions of fungi to exposure to 10 atmospheres pressure of oxygen. *J. Gen. Microbiol.* 45:17–29.

ROELS, J.A., J. VAN DEN BERG, and R.M. VONCKEN. 1974. The rheology of mycelial broths. *Biotechnol. Bioeng.* 16:181–208.

RYALL, A.L. 1935. Certain physiological effects of carbon dioxide treatments of plums. *Proc. Amer. Soc. Hort. Sci.* 32:164–169.

RYGG, G.L. and A.W. WELLS. 1962. Experimental storage of California lemons in controlled atmospheres. USDA, Agr. Mktg. Ser. 475.

SALUNKHE, D.K. and M.T. WU. 1973. Effects of subatmospheric pressure storage on ripening and associated chemical changes of certain deciduous fruits. *J. Amer. Soc. Hort. Sci.* 98(1):113–116.

SCHOLZ, E.W., H.B. JOHNSON, and W.R. BUFORD. 1960. Storage of Texas red grapefruit in modified atmospheres. USDA, Agr. Mktg. Ser. 414.

SCHULZ, F.A. 1974. Uber das Auftreten von Apfellagerfäulen unter kontrollierten Bedingungen. *Z. Pflanzen. Pflanzen.* 81:550–558.

SEITZ, L.M., H.E. MOHR, R. BURROUGHS, and D.B. SAUER. 1977. Ergosterol as an indicator of fungal invasion in grains. *Cereal Chem.* 54:1207–1217.

SEITZ, L.M., D.B. SAUER, R. BURROUGHS, H.E. MOHR, and J.D. HUBBARD. 1979. Ergosterol as a measure of fungal growth. *Phytopathology* 69(11):1202–1203.

SHIH, C.N. and E.H. MARTH. 1973. Aflatoxin produced by *Aspergillus parasiticus* when incubated in the presence of different gases. *J. Milk Food Technol.* 38:421–425.

SMITH, W.H. 1957. Accumulation of ethyl alcohol and acetaldehyde in blackcurrants kept in high concentrations of carbon dioxide. *Nature* 179:876–877.

SMITH, W.H. 1963. The use of carbon dioxide in the transport and storage of fruits and vegetables. *Adv. Food Res.* 12:95–146.

SMITH, W.L., JR. 1976. Non-chemical control of deterioration of fresh produce. p. 577–587. *In* J.M. Sharpley and A.M. Kapland (eds.) Proc. 3rd Intern. Biodegradation Symp. Applied Science Publishers, London.

SMITH, W.L., JR. and R.E. ANDERSON. 1975. Decay control of peaches and nectarines during and after controlled atmosphere and air storage. *J. Amer. Soc. Hort. Sci.* 100:84–86.

SMOCK, R.M. 1979. Controlled atmosphere storage of fruits. p. 301–336. *In* J. Janick (ed.) Horticultural reviews, Vol. 1. AVI Publishing, Westport, Ct.

SOLOMOS, T. and G.G. LATIES. 1973. Cellular organization and fruit ripening. *Nature* 245:390–392.

SOMMER, N.F. and J.R. BUCHANAN. 1978. Mycotoxin production by postharvest pathogens of fruits and vegetables. p. 819–828. *In* P. Rosenberg (ed.) Toxins: animal, plant and microbial. Proc. 5th Intern. Symp. Intern. Soc. Toxicology. Pergamon Press, New York.

SOMMER, N.F., J.R. BUCHANAN, and R.J. FORTLAGE. 1976. Aflatoxin and sterigmatocystin contamination of pistachio nuts in orchards. *Appl. Environ. Microbiol.* 32(1):64–67.

SOMMER, N.F., J.R. BUCHANAN, and R.J. FORTLAGE. 1977. Patulin in CA-stored apples. Mich. State Univ. Hort. Rpt. 28. East Lansing.

SOMMER, N.F., J.R. BUCHANAN, R.J. FORTLAGE, and D.P.H. HSIEH. 1974. Patulin, a mycotoxin, in fruit products. Proc. IV Intern. Congr. Food Sci. and Technol., Madrid, 1974. 3:266–271. Instituto Nacional de Ciencia y Tecnologia de Alimentos, Valencia, Spain.

SOMMER, N.F., R.J. FORTLAGE, J.R. BUCHANAN, and A.A. KADER. Carbon monoxide suppression of postharvest pathogens of fruits as affected by oxygen. *Plant Dis.* (in press)

SOMMER, N.F., R.J. FORTLAGE, F.G. MITCHELL, and E.C. MAXIE. 1973. Reduction of postharvest losses of strawberry fruits from gray mold. *J. Amer. Soc. Hort. Sci.* 98:285–288.

SPALDING, D.H. and W.F. REEDER. 1972. Quality of 'Booth 8' and 'Lula' avocados stored in a controlled atmosphere. *Proc. Fla. State Hort. Soc.* 85: 337–341.

SPALDING, D.H. and W.F. REEDER. 1975. Low-oxygen high-carbon dioxide controlled atmosphere storage for control of anthracnose and chilling injury of avocados. *Phytopathology* 65:458–460.

SPALDING, D.H. and W.F. REEDER. 1976. Low pressure (hypobaric) storage of limes. *J. Amer. Soc. Hort. Sci.* 101:367–370.

SPALDING, D.H. and W.F. REEDER. 1977. Low pressure (hypobaric) storage of mangos. *J. Amer. Soc. Hort. Sci.* 102:367–369.

STARKS, O.B. and H. KOFFLER. 1949. Aerating liquids by agitating on a mechanical shaker. *Science* 109:495–496.

STEVENSON, C.D. 1957. Apple cool storage investigations in 1955. *Queens. J. Agr. Sci.* 14:167–181. (*Hort. Abstr.* 28:1161.)

STEWART, J.K. 1978. Influence of oxygen, carbon dioxide and carbon monoxide levels on decay of head lettuce after harvest. *Sci. Hort.* 9:207–213.

STEWART, J.K., M.J. CEPONIS, and L. BERAHA. 1970. Modified-atmosphere effects on the market quality of lettuce shipped by rail. USDA Mktg. Res. Rpt. 863.

STEWART, J.K. and M. UOTA. 1976. Postharvest effect of modified levels of carbon monoxide, carbon dioxide, and oxygen on disorders and appearance of head lettuce. *J. Amer. Soc. Hort. Sci.* 101:382–384.

SWINBURNE, T.R. 1970a. Fungal rotting of apples. I. A survey of the extent and cause of current fruit losses in Northern Ireland. Min. Agr. N. Ireland. *Record Agr. Res.* 18:15–19.

SWINBURNE, T.R. 1970b. Fungal rotting of apples. II. A preliminary survey of the effect of storage conditions on the development of rots. Min. Agr. N. Ireland. *Record Agr. Res.* 18:89–94.

SWINBURNE, T.R. 1974. The effect of store conditions on the rotting of apples, cv. Bramley's seedling, by *Nectria galligena*. *Ann. Appl. Biol.* 78:39–48.

TABAK, H.H. and W.B. COOKE. 1968. The effects of gaseous environments on the growth and metabolism of fungi. *Bot. Rev.* 34:126–252.

TALVIA, P. 1960. Various species of *Gloeosporium* in stored apples in Finland. *Maataloust Aikakausk.* 32:239–246. (*Hort. Abstr.* 31:4041.)

THORNTON, N.C. 1934. Carbon dioxide storage. VI. Lowering the acidity of fungal hyphae by treatment with carbonic acid. *Contrib. Boyce Thomp. Inst.* 6:395–405.

TOLLE, W.E. 1969. Hypobaric storage of mature-green tomatoes. USDA Mktg. Res. Rpt. 842.

TOMALIN, A.W. and J.E. ROBINSON. 1971a. Refrigeration to store soft fruit. *The Grower* 76:634.

TOMALIN, A.W. and J.E. ROBINSON. 1971b. Cool storage only suitable for top-class fruit. *The Grower* 76(14):674–675.

TOMKINS, R.G. 1963. The effects of temperature, extent of evaporation, and restriction of ventilation on the storage life of tomatoes. *J. Hort. Sci.* 38:335–347.

TOMKINS, R.G. 1966a. The storage of fruits and vegetables. Small scale gas storage experiments. Agr. Res. Counc. Ditton Lab. Annu. Rpt. p. 8. East Malling, England.

TOMKINS, R.G. 1966b. The choice of conditions for the storage of fruits and vegetables. Amos Memorial Lect., Annu. Rpt. East Malling Res. Sta., 1965. p. 60–76. East Malling, England.

TRINCI, A.P.J. 1971a. Influence of the width of the peripheral growth zone on the radial growth rate of fungal colonies on solid media. *J. Gen. Microbiol.* 67:325–344.

TRINCI, A.P.J. 1971b. Exponential growth of the germ tubes of fungal spores. *J. Gen. Microbiol.* 67:345–348.

UPPAL, B.N. 1924. Spore germination of *Phytophthora infestans. Phytopathology* 14:32–33.

UPPAL, B.N. 1926. Relation of oxygen to spore germination in some species of Peronosporales. *Phytopathology* 16:285–292.

VAKIL, J.R., M.R.R. RAO, and P.K. BHATTACHARYYA. 1961. Effect of carbon dioxide on the germination of conidiospores of *Aspergillus niger. Arch. Mikrobiol.* 39:53–57.

VAN DEN BERG, L. and C.P. LENTZ. 1973. High humidity storage of carrots, parsnips, rutabagas, and cabbage. *J. Amer. Soc. Hort. Sci.* 98:129–132.

WEAST, R.C., S.M. SELBY, and C.D. HODGMAN. 1964. Handbook of chemistry and physics. 45th ed. The Chemical Rubber Co., Cleveland.

WEICHMANN, J. 1973. Die qualität von möhren nach lagerung in einseitig kontrollierter atmosphäre. *Gartenbauwissenschaft* 38:75–84.

WELLS, J.M. 1967. Growth and production of pectic and cellulolytic enzymes by *Rhizopus stolonifer* in low-oxygen atmospheres. *Phytopathology* 57:1010. (Abstr.)

WELLS, J.M. 1968. Growth of *Rhizopus stolonifer* in low-oxygen atmospheres and production of pectic and cellulolytic enzymes. *Phytopathology* 58:1598–1602.

WELLS, J.M. 1974. Growth of *Erwinia carotovora, E. atroseptica* and *Pseudomonas fluorescens* in low oxygen and high carbon dioxide atmospheres. *Phytopathology* 64:1012–1015.

WELLS, J.M. and D.H. SPALDING. 1975. Stimulation of *Geotrichum candidum* by low oxygen and high carbon dioxide atmospheres. *Phytopathology* 65:1299–1302.

WELLS, J.M. and M. UOTA. 1970. Germination and growth of five fungi in low-oxygen and high-carbon dioxide atmospheres. *Phytopathology* 60:50–53.

WILSON, D.M. and E. JAY. 1976. Effect of controlled atmosphere storage on aflatoxin production in high moisture peanuts (ground nuts). *J. Stored Prod. Res.* 12:97–100.

WILSON, D.M. and G.J. NUOVO. 1973. Patulin production in apples decayed by *Penicillium expansum. Appl. Microbiol.* 26:124–125.

WOOD-BAKER, A. 1955. Effects of oxygen-nitrogen mixtures on the spore germination of mucoraceous moulds. *Trans. Brit. Mycol. Soc.* 38:291–297.

WOODRUFF, R.E. 1977. Use of carbon monoxide in modified atmospheres for fruits and vegetables in transit. Mich. State Univ. Hort. Rpt. 28. East Lansing.

WORKMAN, M., E. KERSCHNER, and M. HARRISON. 1976. The effect of storage factors on membrane permeability and sugar content of potatoes and decay by *Erwinia carotovora* var. *atroseptica* and *Fusarium roseum* var. *sambucinum. Amer. Potato J.* 53:191–204.

WORKMAN, M. and J. TWOMEY. 1969. The influence of storage atmosphere and temperature on the physiology and performance of 'Russet Burbank' seed potatoes. *J. Amer. Soc. Hort. Sci.* 94:260–263.

WORKMAN, M. and J. TWOMEY. 1970. The influence of storage on the physiology and productivity of Kennebec seed potatoes. *Amer. Potato J.* 47: 372–378.

WU, L. and M.A. STAHMANN. 1975. Chromatographic estimation of fungal mass in plant materials. *Phytopathology* 65:1032–1034.

WU, M.T. and D.K. SALUNKHE. 1972a. Fungistatic effects of subatmospheric pressures. *Experientia* 28:866–867.

WU, M.T. and D.K. SALUNKHE. 1972b. Subatmospheric pressure storage of fruits and vegetables. *Utah Sci.* 33:29–31.

YACKEL, W.C., A.I. NELSON, L.S. WEI, and M.P. STEINBERG. 1971. Effect of controlled atmosphere on growth of mold on synthetic media and fruit. *Appl. Microbiol.* 22:513–516.

ZAGORYANSHII, V.S. 1933. Influence of carbon dioxide on the keeping quality of fruits. *Schriften Zentral. Biochem.* 3:141. *Forschungsinst. Nahr. Genussmittelind. (U.S.S.R.).* (*Chem. Abstr.* 28:1417.)

ZOBELL, C.E. and L.L. HITTLE. 1967. Some effects of hyperbaric oxygenation on bacteria at increased hydrostatic pressures. *Can. J. Microbiol.* 13: 1311–1319.

Index (Volume 3)

Cumulative Index (Volumes 1–3 Inclusive)

Other AVI Books

BEES, BEEKEEPING, HONEY AND POLLINATION
Gojmerac

BREEDING FIELD CROPS
2nd Edition *Poehlman*

FIELD CROP DISEASES HANDBOOK
Nyvall

FOODBORNE AND WATERBORNE DISEASES: THEIR EPIDEMIOLOGIC CHARACTERISTICS
Tartakow and Vorperian

FUNDAMENTALS OF ENTOMOLOGY AND PLANT PATHOLOGY
2nd Edition *Pyenson*

HEALTHY PLANT HANDBOOK
Pyenson

LEAFY SALAD VEGETABLES
Ryder

PLANT DISEASE CONTROL
Sharvelle

PLANT PHYSIOLOGY IN RELATION TO HORTICULTURE
American Edition *Bleasdale*

PLANT PROPAGATION AND CULTIVATION
Hutchinson

RICE: PRODUCTION AND UTILIZATION
Luh

SMALL FRUIT CULTURE
5th Edition *Shoemaker*

TREE NUTS: PRODUCTION, PROCESSING, PRODUCTS
2nd Edition *Woodroof*

TROPICAL AND SUBTROPICAL FRUITS
Nagy and Shaw

VEGETABLE GROWING HANDBOOK
Splittstoesser